The Study of Fast Processes and Transient Species by Electron Pulse Radiolysis

NATO ADVANCED STUDY INSTITUTES SERIES

Proceedings of the Advanced Study Institute Programme, which aims
at the dissemination of advanced knowledge and
the formation of contacts among scientists from different countries

The series is published by an international board of publishers in conjunction
with NATO Scientific Affairs Division

A	Life Sciences	Plenum Publishing Corporation
B	Physics	London and New York
C	Mathematical and Physical Sciences	D. Reidel Publishing Company Dordrecht, Boston and London
D	Behavioural and Social Sciences	
E	Engineering and Materials Sciences	Martinus Nijhoff Publishers The Hague, London and Boston
F	Computer and Systems Sciences	Springer Verlag Heidelberg
G	Ecological Sciences	

Series C – Mathematical and Physical Sciences

Volume 86 – The Study of Fast Processes and Transient Species
by Electron Pulse Radiolysis

The Study of Fast Processes and Transient Species by Electron Pulse Radiolysis

Proceedings of the NATO Advanced Study Institute held at Capri, Italy, 7-18 September, 1981

edited by

JOHN H. BAXENDALE
Chemistry Department, The University, Manchester, U.K.

and

FABIO BUSI
Instituto F.R.A.E., C.N.R., Bologna, Italy

D. Reidel Publishing Company

Dordrecht : Holland / Boston : U.S.A. / London : England

Published in cooperation with NATO Scientific Affairs Division

Library of Congress Cataloging in Publication Data

NATO Advanced Study Institute (1981 : Capri, Italy)
    The study of fast processes and transient species by electron pulse radiolysis.

    (NATO advanced study institutes series. Series C, Mathematical and physical sciences ; v. 86)
    Includes bibliographical references and index.
    "Published in cooperation with NATO Scientific Affairs Division."
    1. Radiation chemistry—Congresses. I. Baxendale, John H., 1917–   . II. Busi, Fabio. III. Title. IV. Series.
QD625.N37   1980   541.3'8   82–9067
                                            AACR2

ISBN-13: 978-94-009-7854-6   e-ISBN-13: 978-94-009-7852-2
DOI: 10.1007/978-94-009-7852-2

Published by D. Reidel Publishing Company
P.O. Box 17, 3300 AA Dordrecht, Holland

Sold and distributed in the U.S.A. and Canada
by Kluwer Boston Inc.,
190 Old Derby Street, Hingham, MA 02043, U.S.A.

In all other countries, sold and distributed
by Kluwer Academic Publishers Group,
P.O. Box 322, 3300 AH Dordrecht, Holland

D. Reidel Publishing Company is a member of the Kluwer Group

CONTENTS

Preface

This volume contains the lectures given at the NATO Advanced
Study Institute "The Study of Fast Processes and Labile Species
in Chemistry and Molecular Biology Using Ionising Radiation" held
in Capri, Italy, September 7-18th 1981.

The aim of the Institute was to summarise the present
position of the use of pulsed ionising radiation in chemical and
biological chemical research.   For background an outline of the
basic radiation chemistry and physics involved and descriptions
of techniques and equipment in current use was presented.   It
was followed by comprehensive coverage of the state of this
research to date in various areas of chemistry and biological
chemistry.

It was hoped to demonstrate to researchers not directly
involved with ionising radiation how this technique is now at a
stage in its development where it can have wider applications in
various branches of chemistry and biology. The fifty participants
did indeed form a wide spectrum of scientific interest covering
inorganic, physical and organic chemistry, molecular physics,
molecular biology, radiobiology and bacteriology.   They also
represented a wide variety of countries viz. Belgium, China,
Denmark, France, Germany, Greece, Holland, Hungary, India, Italy,
Poland, Turkey, U.S.A., U.K. and Yugoslavia.

The 13 lecturers came from Canada, France, Germany, Italy,
The Netherlands, U.S.A. and the U.K. In addition to presenting
the formal material required of the courses they were also
actively engaged in discussion seminars on their material and
contributed papers to a research colloquium which completed the
school.   It was considered preferable to publish the latter in
the normal journals.

We are grateful for the support, financial and otherwise,
received from the Consiglio Nationale delle Ricerche and the
Comitato Nazionale per l'Energia Nucleare, Italy, the regional
government of Compania in Naples, the municipalities of Capri and
Anacapri, and the Banco di Napoli.   The generous activity of the
members of the Capri Organisation Committee, A. Scognamiglio,

J. H. Baxendale and F. Busi (eds.),
The Study of Fast Processes and Transient Species by Electron Pulse Radiolysis, xv–xvi.
Copyright © 1982 by D. Reidel Publishing Company.

F. Serao, C. Federico, F. Arcucci and T. Boniello, contributed
enormously to the warm reception of the participants to Capri
and to the very friendly atmosphere which prevailed throughout
the meeting. We were honoured to be welcomed to Italy by On.
Francesco Compagna representing the Italian government and also
by the President of C.N.R. Italy, Prof. Ernesto Quagliariello
who agreed to act as President of our Institute.

J.H. Baxendale   Manchester University,
                        Chemistry Department,
                        MANCHESTER,
                        U.K.

F. Busi   Instituto FRAE, CNR,
                        BOLOGNA
                        Italy

                        Editors and Directors of the Institute.

December 1981

ADSORPTION OF ENERGY FROM IONIZING RADIATION

Gordon R. Freeman

Chemistry Department, University of Alberta Edmonton, Canada T6G 2G2

ABSTRACT

The radiation in radiation chemistry is simply the energy source that initiates the reactions. Activation of molecules by high energy radiation is briefly compared with activation by heat and ultraviolet light. The transfer of energy from various types of high energy radiation to matter is discussed. X and γ rays: photoelectric effect, Compton scattering, $e^- e^+$ pair production. High energy electrons and ions: electronic excitation and ionization of matter, emission of electromagnetic radiation, elastic scattering. Rates of energy loss. Penetration ranges. Charged particles below 1000 eV. Neutrons.

J. H. Baxendale and F. Busi (eds.),
The Study of Fast Processes and Transient Species by Electron Pulse Radiolysis, 1–17.
Copyright © 1982 by D. Reidel Publishing Company.

INTRODUCTION

Radiation chemistry is the subject that deals with the chemical changes induced by high energy radiation.

The radiations usually used in radiation chemical studies are high energy photons (γ rays, X rays) and charged particles (such as 2 MeV electrons from a Van de Graaff generator, or 20 MeV protons from a cyclotron). High energy neutrons can also be used. These radiations cause ionization of the medium in which they are absorbed, so they are often called "ionizing radiation".

Although research in radiation chemistry has been particularly active in the past three decades, the field is much older. The effects of electric discharges on gases were studied in the middle of the nineteenth century. Natural radiolysis reactions have been occurring since early in the history of the universe. One of the theories of the origin of life suggests that the action of cosmic rays on seas and lakes, which are aqueous solutions, caused reactions to occur that produced larger and larger molecules. Eventually some of the large molecules had special properties that allowed them to propagate themselves. Thus life began, the theory says.

The radiation in radiation chemistry is simply the energy source that initiates the reactions. In general, energy can be given to molecules in many different ways. For example, energy can be added to a system by heating it (furnace, shock waves), by irradiating it with electromagnetic radiation (microwaves, light, X rays), or by irradiating it with high energy particles (electrons helium ions, and so on). Irradiation of liquids with ultrasonic waves also leads to molecular decomposition; energization of the molecules might result from charge separation in the cavitation process (triboelectricity).

Consider the decomposition of a molecule. The molecule must possess a certain amount of internal energy to decompose. When energy is fed to a molecule thermally, the molecule accumulates vibrational and rotational energy in fairly small increments during collisions. For example, at 800K the average energy of a molecule containing 10 atoms is roughly $3 \times 10\ kT/2 \approx 1.0$ eV.

$$A \quad + \quad A \quad \longrightarrow \quad A' \quad + \quad A$$
$$1.0\ eV \quad 0.8\ eV \qquad 1.5\ eV \quad 0.3\ eV$$
$$A' \quad + \quad A \quad \longrightarrow \quad A'' \quad + \quad A$$
$$1.5\ eV \quad 1.2 \qquad 2.3 \qquad 0.4$$
$$A'' \quad + \quad A \quad \longrightarrow \quad A''' \quad + \quad A$$
$$2.3 \qquad 1.0 \qquad 2.8 \qquad 0.5$$
$$A''' \quad + \quad A \quad \longrightarrow \quad A'^v \quad + \quad A$$
$$2.8 \qquad 1.2 \qquad 3.3 \qquad 0.7$$

When an A* has enough energy, say 4-5 eV, it can decompose.
        Thus, a molecule begins in one of its lower energy levels
and is gradually raised, through favorable collisions, to an ener
gy level where it can react (Fig. 1).

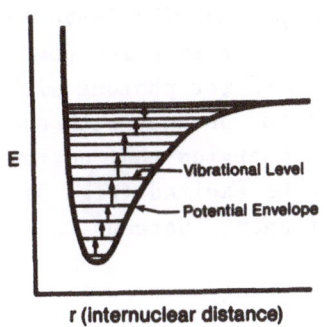

Fig. 1 - Thermo-excitation of a molecule.

        When energy is fed to a molecule photolytically, the mo-
lecule absorbs a photon and is raised to an excited electronic
state. If the equilibrium internuclear distances in the excited
state are different from those in the ground state molecule, then
the excited molecule might also have vibrational excitation energy (Fig. 2).

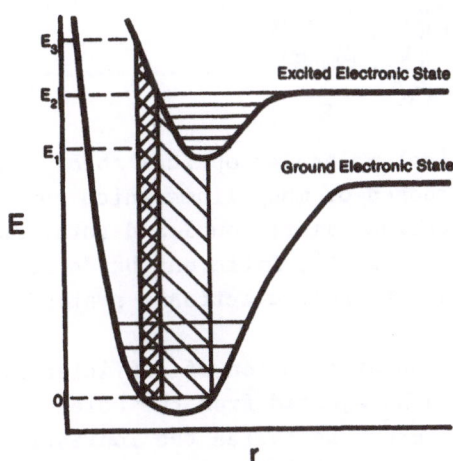

Fig. 2 - Photo-excitation of a molecule. The lowest energy level
        of the molecule is arbitrarily chosen as the zero of the
        energy axis. Molecules decompose within one vibration
        time after being excited above the vibrational dissocia-
        tion limit of the corresponding electronic state.

   At room temperature, most of the ground electronic state
molecules are in their lowest vibrational level. Electronic motion
is much more rapid than is nuclear motion, so the electronic exci
tation process occurs with nearly no alteration in the internuclear
distances of the atoms in the molecule. The excitation process
only occurs if absorption of the photon will leave the molecule
in a well-defined excited state. The absorption coefficient of
the molecule will therefore be large for photons with energy between E_1 and E_3 and will be small for photons with energy less than
E_1 or greater than E_3, except at energies that correspond to other
states to which the molecule can be excited (Fig. 3). If the mole
cule is excited with a photon of energy between E_2 and E_3, it will
decompose.

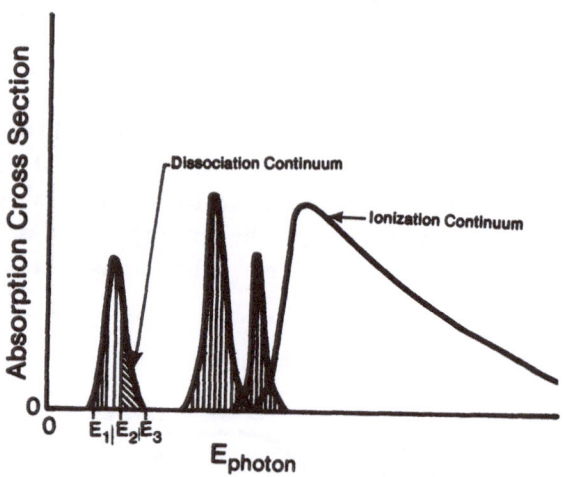

Fig. 3 – A hypothetical molecular optical absorption spectrum. A
         band is composed of many lines which correspond to exci-
         tation to various vibrational and rotational levels of
         the electronic state. Different bands correspond to ex-
         citation to different electronic states.

   The absorption of a photon of sufficiently high energy
causes an electron to be ejected from the molecule. This portion
of the absorption spectrum is called the ionization continuum,
because the possible energy levels of the ejected electrons are
essentially continuous (Fig. 3).

   High energy charged particles tend to excite molecules
through processes similar to photo-excitation.

INTERACTION OF RADIATION WITH MATTER

The energy of each particle in a high energy beam is much greater than the ionization potential of a molecule. Thus ionization occurs in a substance when high energy radiation passes through it. The ionization is sometimes used to measure the amount of energy absorbed by the system.

Beams of high energy radiations are composed of discrete particles. The discrete γ and X ray photons can be considered as particles in this context. The distance between individual particles in the radiation beams used in radiolysis experiments is large compared with the distance over which each particle can exercise its influence during its passage through matter. For example, if the particle velocity were 10^{10} cm/s and the beam intensity were 10^{19} particles/cm^2 s ($\sim$ 1 amp of electrons/cm^2), the average distance between particles would be about 10^{-3} cm. The excitation interaction distance of a charged particle in a liquid is of the order of 10^{-7} cm.

It is therefore possible to consider a beam of high energy radiation as a series of independent particles each interacting separately with the atoms and molecules of the medium through which it passes. After their passage, the molecules affected by the particles may react, whence arises radiation chemistry.

1. X and γ Rays

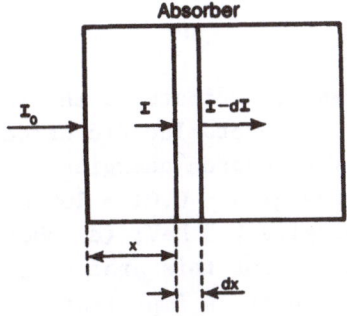

Fig. 4 – Absorption of photons in matter. I_0 is the number of photons/cm^2 incident upon the absorber.

The number of photons absorbed in a slice of absorber of thickness dx is proportional to the intensity of the impinging beam, I, and to the thickness of the slice dx (Fig. 4).

$$dI = -\mu \, Idx \qquad\qquad (1)$$

where dI = number of photons absorbed from the beam in the slice of absorber of thickness dx, μ = linear absorption coefficient, and I = number of photons/cm^2 incident upon the slice of absorber at depth x. The minus sign in equation (1) indicates that I decreases as x increases. Integrating equation (1) between the proper limits we get

$$\int_{I_o}^{I} \frac{dI}{I} = \int_{o}^{x} - \mu dx$$

$$I = I_o e^{-\mu x} \tag{2}$$

where I_o = intensity of the beam incident upon the absorber. The units of μ are the inverse of those of x (Table 1).

TABLE 1

Units and Symbols of Absorption Coefficients

x units	abs'n. coef.	symbol	coef. units
cm	linear	μ	cm^{-1}
g/cm^2	mass	μ/ρ	cm^2/g, ρ = density, g/cm^3
electrons/cm^2	electronic	$_e\mu$	cm^2/electron
atoms/cm^2	atomic	$_a\mu$	cm^2/atom

Several processes contribute to the absorption of energy from photons by matter: (a) photoelectric effect – in light atoms this is important mostly at photon energies < 0.1 MeV; (b) Compton effect – important for energies $\sim 0.01 - 100$ MeV; (c) pair production – important for energies > 3 MeV; (d) photo-disintegration of the nucleus – for many atoms this process is important only at energies > 10 MeV, although it is important even at 2 MeV for such nuclides as ^{2}D or ^{9}Be. The last process is usually unimportant under the conditions of radiolysis studies so we may consider that

$$\mu = \tau + \sigma + \pi \tag{3}$$

where τ = photoelectric absorption coefficient, σ = Compton absorption coefficient, and π = pair production absorption coefficient.

(a) Photoelectric effect

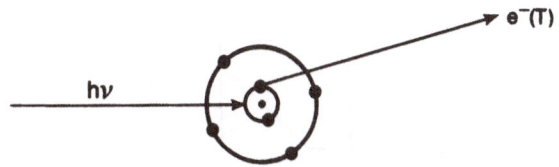

Fig. 5 – Photoelectric absorption of photon $h\nu$ by an inner electron shell of an atom.

The photon of energy $h\nu$ is completely absorbed by an atom
or molecule and an extranuclear electron is ejected (Fig. 5). The
kinetic energy T of the ejected electron is equal to the photon
energy minus the binding energy ω of the electron in the atom or
molecule.

$$T = h\nu - \omega \tag{4}$$

The electron ejected is usually that with a binding ener
gy nearest to but below the energy of the photon. Therefore when
$h\nu > \omega_K$ a K electron is usually ejected. Examples of ω_K values
are given in Table 2.

TABLE 2

Ionization potentials of K electrons in atoms

atom	Z	ω_K(eV)	atom	Z	ω_K(eV)
H	1	13.6	S	16	2,500
C	6	284	Cu	29	9,000
N	7	400	Pb	82	88,000
O	8	531			

For photon energies well above the highest ω_K in a series
of atoms the atomic photoelectric absorption coefficient, $_a\tau$, is
approximately proportional to Z^5, where Z is the atomic number of
the atom. Furthermore τ decreases rapidly with increasing photon
energy.

$$_a\tau \propto Z^5 (h\omega)^{-n} \tag{5}$$

where n = 3.5 at low energies and decreases to 1 at high energies
[1].

(b) Compton effect

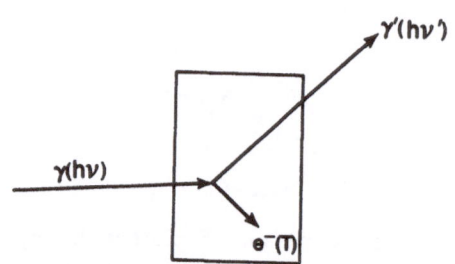

Fig. 6 – Compton scattering of a photon by an electron.

The photon "collides" with an electron and gives up part of its energy to the electron (Fig. 6).

$$T = h\nu - h\nu' - \omega$$

$$= h(\Delta\nu) - \omega \tag{6}$$

where $\Delta\nu$ = change in frequency of the photon.

Whereas the photoelectric process can occur only with a bound electron ("resonance" process), the Compton process can occur with either a bound or a free electron. The atomic Compton absorption coefficient $_a\sigma$ depends only on Z, the number of electrons in the atom. $_a\sigma \propto Z$; $_e\sigma$ is independent of Z.

The intensity, I, of a beam is the number of photons per second that pass through a 1 cm^2 window (photons/cm^2s). The beam energy flux density, E, (eV/cm^2s) is given by E = Ihν, where hν is the energy of each photon.

When a beam of γ rays passes through a slice of absorber of thickness dx, the change in beam intensity dI due to the Compton effect is given by dI = -I σ dx. If τ and π are negligible, then $I = I_0 e^{-\sigma x}$. The change in beam intensity caused by passage through thickness x of absorber is:

$$\Delta I = I_0 - I = I_0 (1 - e^{-\sigma x}). \tag{7}$$

The corresponding change in energy flux density is:

$$\Delta E = \Delta I h\nu = E_0 (1 - e^{-\sigma x}) \tag{8}$$

where $E_0 = I_0 h\nu$.

One must distinguish between the change in energy flux density of the beam and the actual rate of energy absorption in the absorber. Consider an absorber that is large enough to completely absorb the energy from the Compton electron but small enough not to absorb the scattered photon. When a photon of energy $h\nu$ interacts by the Compton process, the total amount of energy lost from a narrow beam is $h\nu$. The amount of energy actually absorbed by the absorber is only $(h\nu-h\nu')$. The degraded γ, with energy $h\nu'$, is scattered out of the beam but, if the absorber is small, is not absorbed. Thus the actual rate of energy absorption by the absorber is

$$\Delta E_{abs} = \Delta I(h\nu-h\nu') = \Delta I\ A\ h\nu \tag{9}$$

where $A = (h\nu-h\nu')/h\nu$ = fraction of incident photon energy imparted to the Compton electron. Similarly, the energy carried off by the scattered photons is

$$\Delta E_s = \Delta I\ h\nu' = \Delta I\ S\ h\nu \tag{10}$$

where $S = h\nu'/h\nu$ = fraction of incident photon energy remaining in the scattered photon. It follows that

$$A + S = 1 \tag{11}$$

It is customary to arbitrarily divide the Compton coefficient into two parts [1] :

$$\sigma_t = \sigma_a + \sigma_s \tag{12}$$

where σ_t = total Compton coefficient

$\quad\quad = \sigma$, as used above,

$\quad \sigma_a$ = Compton absorption coefficient

$$\quad\quad = A\ \sigma_t, \tag{13}$$

and σ_s = Compton scatter coefficient

$$\quad\quad = S\ \sigma_t. \tag{14}$$

At low gamma energies ($\lesssim 0.01$ MeV) $\sigma_a \approx 0$ and $\sigma_t \approx \sigma_s$ (Fig. 7). This means that gammas are scattered from the beam but little energy is transferred to the electrons. The process is also known as Rayleigh scattering and is what gives rise to X ray diffraction patterns from crystals.

At high gamma energies ($\gtrsim 20$ MeV), $\sigma_a \approx 3\sigma_s$. In the vicinity of 1 MeV (^{60}Co γ region), $\sigma_a \approx \sigma_s$, in which case the average energy of the ejected electron is about 0.5 $h\nu$.

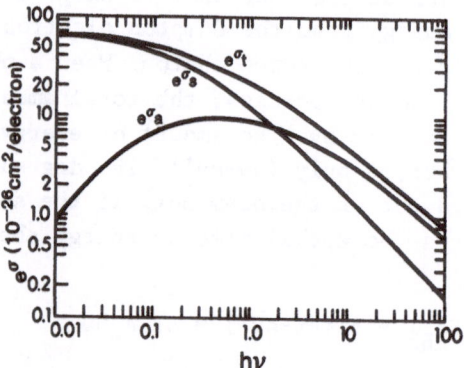

Fig. 7 – Compton coefficients per electron of absorber. Total, $_e\sigma_t$; scatter $_e\sigma_s$; absorption $_e\sigma_a$. Ref. 1.

(c) Pair production

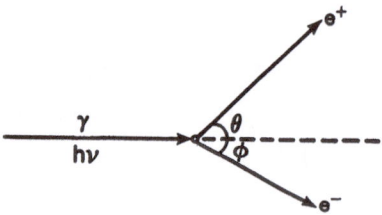

Fig. 8 – Pair production.

$T(e^+ + e^-) = h\nu - 1.02$ MeV – nuclear recoil energy (small)

The photon interacts with the field of a nucleus and pro‐
duces an electron-positron pair (Fig. 8). The energy equivalent
to the rest mass of an electron is 0.51 MeV ($E = m_o c^2$). Since two
electrons, with a total rest mass of 1.02 MeV, are created, the
energy of the gamma must be greater than 1.02 MeV. The electron
and positron carry off essentially all of the energy in excess of
the 1.02 MeV required to create them. The nucleus that suffered
the photon interaction recoils somewhat, but the recoil energy is
small because of the large mass of the nucleus compared to that
of the electron-positron pair (conservation of momentum).

The value of the pair production cross section increases
with increasing photon energy, and $_a\pi \propto Z^2$ [1] .

When the positron has been slowed to thermal energy, it
annihilates with an electron, thereby producing two gammas of ener‐
gy 0.51 MeV each. Many systems are of such a size that they absorb

all the excess energy from most of the electrons and positrons,
but not much from the annihilation gammas.

If more than one type of photon absorption process occurs
at the same time in a given system, ΔE_{abs} can be calculated from

$$\Delta E_{abs} = I_o A' \, h\nu \, (1 - e^{-\mu x}) \tag{15}$$

where $\mu = \tau + \sigma_a + \sigma_s + \pi'$, $A' = \dfrac{\tau + \sigma_a + \pi'}{\mu}$, and $\pi' =$

the pair production absorption coefficient slightly reduced to
account for the loss from the system of the annihiliation radiation
of the thermalized positron and an electron.

The only process that contributes appreciably to energy
absorption from ^{60}Co γ rays (E $\approx$ 1.25 MeV) is the Compton process.
The various cross sections for water are shown in Figure 9.

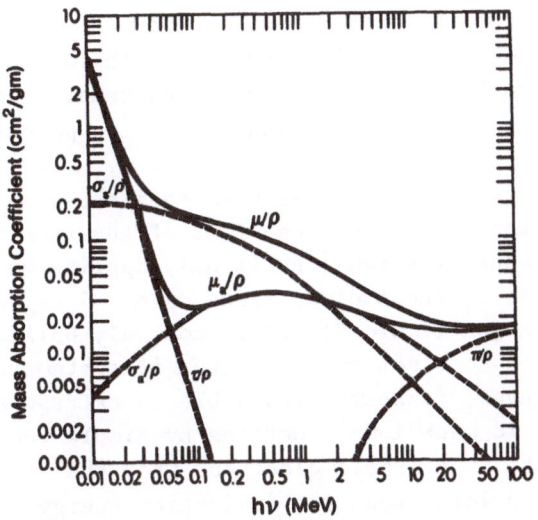

Fig. 9 – Mass absorption coefficients for water as a function of
photon energy.
$\mu/\rho = (\tau + \sigma_a + \sigma_s + \pi)/\rho$. Ref. 1.

2. Electrons

High energy electrons lose energy by a variety of proces‐
ses. The main one is excitation of the medium through which they
pass.

(a) Excitation and ionization of molecules

Bethe derived an equation for the rate of energy loss of a high energy electron per unit path length in a medium [2]:

$$-\left(\frac{dT}{dx}\right)_{coll} = \frac{2\pi e^4}{m_o v^2} ZN \left[\ln \frac{m_o v^2 T}{2I^2(1-\beta^2)} - (2\sqrt{1-\beta^2} - 1 + \beta^2) \ln 2 + (1-\beta^2) + \frac{1}{8}(1 - \sqrt{1-\beta^2})^2 \right]$$

(16)

where $-dT/dx$ = rate of loss of kinetic energy (erg/cm), e = electronic charge, N = number of atoms/cm^3, Z = number of electrons/atom, T = relativistic kinetic energy of electron, m_o = rest mass of electron, v = velocity of electron, β = v/c, c = velocity of light in a vacuum, and I = average excitation potential of the atom. This expression assumes that the most energetic electron after a collision is the incident electron.

The rate of energy loss to the medium is proportional to the electron density ZN, it decreases with increasing electron velocity up to β = 0.9 (T = 1 MeV) then remains nearly constant, and it decreases with increasing excitation potential I of the material.

The Bethe equation assumes that the velocity of the incident electron greatly exceeds that of the atomic orbital electrons (Born approximation), so it only applies when T > ω_K of the heaviest atoms in the medium. Since ω_K = 284 eV for carbon and 531 eV for oxygen (Table 2), it is commonly believed that the Bethe equation does not apply in organic systems for T below about 1000 eV. However, it seems reasonable to extrapolate the use of the equation to much lower energies by simply removing the contributions of the inner shell electrons from the values of N and I when T becomes lower than the ionization energy ω of electrons in the given shell. Thus the value of I for all the electrons (Z = 48) in cyclohexane is 51 eV, but at T < 300 eV one may use Z = 36 and I $\approx$ 15 eV to extend the use of the Bethe equation down to T $\approx$ 30 eV (Fig. 10 and ref. 3).

(b) Emission of radiation

X rays are emitted when a high energy electron is rapidly decelerated in the field of a nucleus. The probability of such an event increases as the square of the charge on the nucleus, Z^2. The generation of X rays increases the protective shielding requi

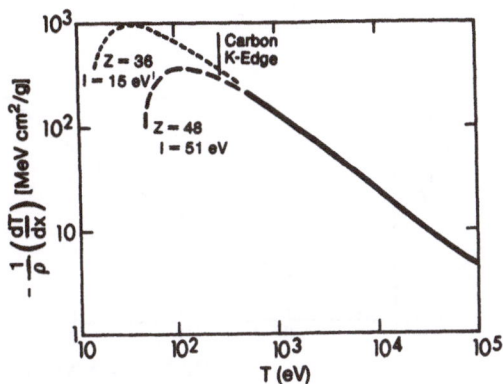

Fig. 10 – Rate of energy loss of electrons in cyclohexane, for 15
eV < T < 10⁵ eV. ρ = liquid density. The curves were
calculated from the Bethe equation, using the values of
Z and I indicated. Ref. 3.

rements of the electron beam, because the X rays are usually more
penetrating than the parent electrons. However, the fraction of
T lost in this manner is small in radiolysis systems and we will
not consider it further.

A still smaller fraction of T is lost by the emission of
Čerenkov radiation, which is seen as bluish-white light. It is
emitted when the electron speed is greater than that of light in
the medium. The speed of light is c divided by the refractive index of the medium. Water at 20°C has a refractive index of 1.33,
so electrons with T > 0.26 MeV emit Čerenkov radiation.

(c) Electron range

The depth of penetration of high energy electrons into a
material increases more than linearly with increasing energy up
to 1 MeV. The reason is that the rate of energy loss –dT/dx decreases with increasing T in this range. At T > 1 MeV the range increases approximately linearly with T. As an example of the range
of electrons in a material of low atomic number, that of 1 MeV
electrons in aluminum is 0.41 g Al/cm² (Fig. 11 and ref. 4), which
corresponds to a thickness of 1.52 mm of the metal (ρ = 2.70 g/cm³).

(d) Low energy electrons, T ≤ 10 eV

At T ≤ 30 eV in water or an organic material the rate of
energy loss decreases (Fig. 10), but the rate remains relatively
large as long as the electron has enough energy to electronically

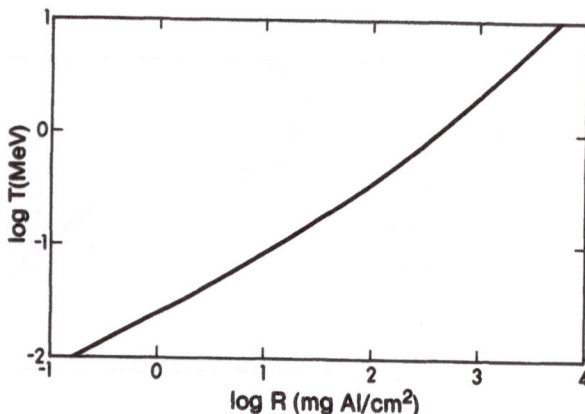

Fig. 11 - Penetration ranges of high energy electrons in aluminum.
       Ref. 4.

excite the medium. When T drops below the lowest electronic excitation energy of the medium, which is ~ 5 eV in many organic systems, the rate of energy loss decreases drastically [5]. Energy
is then transferred only to vibrational and rotational modes of
the molecules, which is much less efficient than transfer to elec
tronic mode. When T becomes less than the vibrational excitation
energies, $\leq$ 0.2 eV, energy can only be lost to rotational modes,
and to intermolecular modes in condensed media; -dT/dx drops shar
ply again.
        Most of the electrons in radiolysis systems are generated
with T $\approx$ 10 eV. As these electrons travel away from the parent
ions they lose energy by collision with molecules of the medium,
and eventually become thermalized. Most of the distance travelled
away from their parent ions during thermalization is attained in
the sub-electronic excitation energy region. In liquid hydrocarbon:
most of the electrons are reduced to the sub-electronic excitation
region within one or two collisions, during which the electron
would have moved < 1 nm from its parent ion. The most probable
thermalization ranges in liquid hydrocarbons are 4-50 nm [6,7],
the largest value occurring for the most spherelike molecules,
methane. In the noble liquids argon and xenon, which have spherical molecules, the most probable ranges are ~ 250 and ~ 160 nm,
respectively [8]. The greatest part of the distance is travelled
while the electron is skittering along in a conduction band a
few tenths of an eV above thermal energy [9]. When the molecules
are more spherelike the potential boundaries of the conduction
band are smoother and the electrons are scattered less, so the

thermalization distance is greater.

3. Heavy Positive Particles

(a) High energy

Protons and helium ions are a few thousand times heavier
than electrons. The primary mode of energy loss by heavy particles
at energies of interest, up to 10^8 eV, is through inelastic collisions with the bound electrons of the medium. Loss by emission
of radiation is completely negligible, even from the point of view
of radiation hazard.

For relativistic energies, Bethe developed the following
equation [2]:

$$-\left(\frac{dT}{dx}\right) = \frac{4\pi\, e^4 z^2}{m_o v^2}\; NZ \left[\ln \frac{2m_o v^2}{I} - \ln\,(1 - \beta^2) - \beta^2 \right] \tag{17}$$

where ze = charge on the incident heavy particle, v = heavy particle velocity. Thus the rate of energy loss varies as z^2. The
dependences on NZ, β and I are similar to those for high energy
electrons.

The maximum amount of energy transferable from a heavy
particle to an electron per collision is $2m_o v^2$, or 0.002 T for a
proton.

(b) Low energy, $T \leq 10^5$ eV

When a positive particle slows down enough it can pick
up an electron from the medium.

$$H^+ + \text{medium} \longrightarrow H\ast + \text{medium}^+ \tag{18}$$

$$He^{2+} + \text{medium} \longrightarrow He^{+\ast} + \text{medium}^+ \tag{19}$$

$$He^+ + \text{medium} \longrightarrow He\ast + \text{medium}^+ \tag{20}$$

where * represents kinetic energy. This occurs when the particle
velocity is reduced to near or below the equivalent velocity of
an electron in an orbital of H or He^+ or He, respectively. Thus
equation (17) is not valid unless $v > 2.2 \times 10^8 z$ cm/s. For photons
this means $T > 24$ keV, and for α particles $T > 400$ keV.

The rapidly moving atom can lose the electron in a subsequent collision. For example,

$$H\ast + \text{medium} \longrightarrow H^+ + e^- \text{ (medium)} \tag{21}$$

The net result is that the moving particle is alternately a proton and a hydrogen atom until it has been slowed down enough that reaction (21) can no longer occur, that is, when $v \ll 2.2 \times 10^8$ cm/s. The probabilities of reactions (18) and (21), σ_{pu} and σ_ℓ respectively, are equal when the H^+ and H^* velocities are 2.2×10^8 cm/s, and the value of σ_{pu}/σ_ℓ increases with decreasing velocity. The rate of energy loss of a neutral hydrogen atom is much smaller than that of a proton of the same energy. The theoretical calculation of the average rate of energy loss of a proton in the energy region below 0.1 MeV is therefore complex. For an α particle it is complex below about 2 MeV.

The diagnostic velocity $v = 2.2 \times 10^8 z$ cm/s strictly applies only in the gas phase. The corresponding velocity is probably somewhat lower in the liquid phase, due to the larger interaction energy of the electron with the dense medium. Furthermore, the (Pauli principle) volume requirement of H^* may cause the species to be perturbed as it passes between the close packed molecules of the liquid.

Penetration ranges of positive particles are therefore slightly greater than one might expect from the equation (17) alone. To indicate the magnitude of positive particle ranges, that of a 10 MeV proton in carbon is 0.13 g/cm^2, or 0.6 mm of graphite ($\rho = 2.25$ g/cm^3) [1].

4. Neutrons

Neutrons have no electric charge, so they do not interact with the electrons of a system. High energy neutrons lose their kinetic energy through collisions with nuclei. Enough energy is usually transferred to a nucleus during such a collision that the atom minus some of its electrons is knocked out of the molecule, thereby becoming an energetic positive ion. The positive ion then behaves in the manner described in the preceding section.

REFERENCES

1. H.E. Johns and J.S. Laughlin, in "Radiation Dosimetry", G.J. Hine and G.L. Brownell (eds.), Academic Press, New York, 1956, p. 83.

2. H.A. Bethe and J. Ashkin, in "Experimental Physics", ed. E. Segré, Vol. 1, J. Wiley & Sons, New York, 1953, pp. 166-357.

3. G.R. Freeman, Quaderni dell'Area di Ricerca dell'Emilia-Romagna, 2, 55 (1972).

4. G. Friedlander and J.W. Kennedy, "Nuclear and Radiochemistry", Wiley, New York, 1955.

5. A. Mozumder and J.L. Magee, J. Chem. Phys. 47, 939 (1967).

6. M.G. Robinson, P.G. Fuochi and G.R. Freeman, Can. J. Chem. 49, 3657 (1971).

7. K. Shinsaka and G.R. Freeman, Can. J. Chem. 52, 3495 (1974).

8. S.S.-S. Huang and G.R. Freeman, Can. J. Chem. 55, 1838 (1977).

9. T.G. Ryan and G.R. Freeman, J. Chem. Phys. 68, 5144 (1978).

BASICS OF RADIATION CHEMISTRY

Gordon R. Freeman

Chemistry Department, University of Alberta Edmonton,
Canada T6G 2G2

ABSTRACT

The reactive species generated by high energy radiation
are not distributed randomly in space, but are strung out along
the tracks of high energy electrons or positive ions. This requires the use of concepts somewhat different from those of conventional reaction kinetics, and which are classified under the general title "nonohomogeneous kinetics". The timescale of events
in the radiolysis of a liquid are developed: energy absorption,
spur formation, spur reactions and dissipation by diffusion, reac
tions in the bulk fluid. Electron localization and solvation is
also briefly described.

J. H. Baxendale and F. Busi (eds.),
The Study of Fast Processes and Transient Species by Electron Pulse Radiolysis, 19 34.

INTRODUCTION

The preceding article was an overiew of processes by
which energy is transferred from radiation to a system. We will
now examine how the energy is distributed spatially in the system
and what effects it causes.

The interaction of a high energy photon with matter gene
rates a high energy electron and a low energy positive ion (conservation of momentum). For example,

$$\gamma\ (1.0\ \text{MeV}) + \text{medium}\ \longrightarrow\ \gamma'\ (0.5\ \text{MeV}) + e^-\ (0.5\ \text{MeV}) + \text{medium}^+$$

The 0.5 MeV electron then ionizes many more molecules, producing
lower energy electrons which then ionize and excite more molecules,
and so on, until there are about 2×10^4 low energy electrons and
positive ions. All of the ions and electrons can cause chemical
change, so stastically the single ion formed by the photon interaction is negligible by comparison with the 2×10^4 ions generated by the subsequent energetic electrons. Radiolysis by X and γ
rays is therefore similar to radiolysis by high energy electrons.

Similarly, high energy neutrons generate energetic positive ions by colliding with nuclei. Radiolysis by neutrons therefore produces effects similar to those of high energy, heavy, positively charged particles.

The present examination will therefore be confined to ef
fects of high energy electrons and heavy, positive particles.

SPACIAL DISTRIBUTIONS AND REACTION KINETICS

1. Charged Particle Tracks and Track Densities

High energy charged particles do not transfer energy con
tinuously to the matter through which they pass. They transfer it
in packets of various sizes, with an average size of the order of
10^2 eV. A packet of energy creates a small zone of excited and
ionized species in the medium. The species are capable of chemical reaction, so radiation chemical processes may be considered
to begin with the creation of these zones. A series of zones mark
the path that a particle traversed, so they are called the particle track.

The Bethe equations in the previous article describe the
mean rate of energy loss by the particle per unit path length. The
jargon term for the mean rate of energy absorbed by the medium

along a particle track is "linear energy transfer", or LET. Average values of LET can be estimated by dividing the particle's initial energy by its total path length [1,2]. Examples are given in Table 1. The value of LET increases rapidly in the sequence electron, proton (H^+) and alpha particle (He^{2+}), for the same T_0. This is because the value of LET increases with decreasing particle velocity (when $\beta < 0.9$), which decreases with increasing particle mass for the same T_0, and because LET increases as the square of the particle charge. The value of LET has important consequences for radiation chemistry and biology, because the reaction kinetics are affected by the track density.

TABLE 1

Average LET values in water [1,2]

Particle	T_0(MeV)	LET(eV/nm)
e^-	10	0.19
	1	0.24
H^+	10	8.3
	1	43.
He^{2+}	10	92.
	1	190.

The average distance between the small reactive zones along a track is roughly 10^2 eV/LET. Considering particles with T_0 = 10 MeV, one obtains ~ 500 nm between zones along an electron track, ~ 12 nm along a proton track, and ~ 1 nm along an α particle track. The corresponding distances for T_0 = 1 MeV particles are ~ 400 nm, 2 nm and 0.5 nm. The average radius of the reactive zone created by the deposition of ~ 10^2 eV is ~ 2 nm in liquid water, and is larger in liquid hydrocarbons. There is therefore little overlap of the zones along a high energy electron track, but nearly complete overlap along an alpha particle track. Due to the overlap, the concept of ~ 10^2 eV zones is not useful for α particles in liquids and solids; the entire track may be considered to be a long, cylindrical reactive zone. A high energy proton track can be treated as a mixture of roughly spherical small zones and long cylindrical zones with radii similar to those of the spheres.

A zone of reactivity is commonly called a "spur". Spurs are not all of the same size or composition. They initially con-

tain one or more ion-electron pairs and neutral excited molecules.
Most of the spurs along a high energy electron track contain only
one ion-electron pair, with progressively fewer containing two,
three, and so on [3-5]. Roughly spherical spacial distribution
may be assumed up to about ten pairs. The long cylindrical spur
produced by an α particle contains tens of thousands of ion-electron pairs and excited molecules. The length of the spur is several orders of magnitude greater than its radius, the latter being
similar to that of a small spherical spur. For each type of geometry one considers the density of reactive species as a function
of spur radius.

2. Spur Reactions

The molecules affected by a given small packet of energy
are in close proximity to one another. The ions, electrons and
radicals that result from the packet are also relatively close to
each other. As these species diffuse about, the random tendency
is for the distance between them to increase. However, there is
an appreciable probability that some of their paths will cross
and that they will react together in the zone where they were
formed. An example of such events is illustrated in Figure 1.

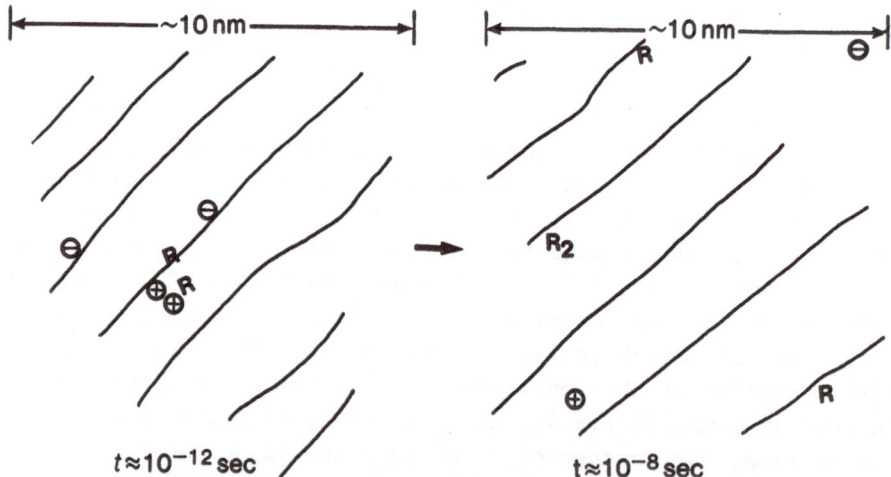

Fig. 1 - An example of the dissipation of a spur over a period of
several ns in a liquid. Time is measured from the instant
of deposition of the packet of energy. $\oplus$, ion; $\ominus$, electron; R, free radical; R_2, product molecule. The spur
that existed at 10^{-12} s no longer exists at 10^{-8} s because
the volume occupied by the remaining reactive species is
too large to be called a spur. The oblique lines represent the thousand of other molecules in the $\sim(10 \text{ nm})^3$
volume of the depicted zone.

To minimize the proliferation of jargon terms I use a general definition for spur. *A spur is a grouping of reactive intermediates in which there is a significant probability that some of the intermediates will react with each other before they can diffuse into the bulk medium.* This definition includes everything from an isolated ion-electron pair in an electron track to an entire α particle track. Some people use different terms for spurs of different sizes, but it seems preferable to speak of the distribution of spur sizes and of spur population densities.

3. Nonhomogeneous Kinetics

Conventional reaction kinetics deals with homogeneous systems, such as liquid or gas phase solutions. It also deals with heterogeneous systems, such as a mixture of a gas and a solid, with the reaction occurring at the interface between the phases.

In a homogeneous system the reactants are distributed randomly throughout a single phase. In the radiolysis of a condensed medium the reactive species are not generated at random throughout the medium, but are initially concentrated in spurs which are strung out along the tracks of the high energy particles. The reactive species are initially not homogeneously distributed in space, so the interpretation of some of their reactions requires the use of nonhomogeneous kinetics.

We thus distinguish the following categories of kinetics.

Homogeneous kinetics   – *single phase system; reactants distributed randomly.*

Nonhomogeneous kinetics – *single phase system; some of the reactants not randomly distributed.*

Heterogeneous kinetics   – *two (or more) phase system; reaction occurs at interface between phases.*

It is perhaps worth noting that the prefix *non-* in nonhomogeneous was picked in preference to *un-*, *in-* or *a-* after examining many words that conventionally use the various prefixes. The simple negation *non-* has the appropriate flavor for the variety of applications of nonhomogeneous kinetics concepts.

Nonhomogeneous kinetics apply mainly to condensed phase systems. In low density gases the molecules are so far apart that the concept of a spur does not apply. The average free distance between molecules is greater than a molecular diameter, so the ion and electron or the two radicals formed from an energized mo

lecule fly apart with negligible probability of being reflected
back together.

Several models of the nonhomogeneous kinetics of radiolysis reactions have been devised. Some are based on a macroscopic
diffusion equation (Fick's Second Law of Diffusion) and homogeneous kinetics reaction rate constants [6-8] . Another is based
upon a stochastic (probability), molecular treatment [9,10] . In
water, which has a large dielectric constant, the coulombic interaction between the ions and electrons can be neglected without
too severe consequences [6,7] . In hydrocarbons and other low dielectric constant liquids coulombic interactions tend to dominate
the diffusion processes [9,10] . The coulombic interaction, stocha
stic model has also been applied to high dielectric constant liquids, such as alcohols [11] and water [12] .

The nonhomogeneous kinetics models that are based upon
equations for macroscopic behavior [6-8,13] are the easiest for
chemists to deal with. I will use such equations in the development of the time scale of radiolysis processes. A deeper understanding of the processes can be obtained from stochastic models
[9,10,14] , but they involve more parameters and are less suitable
for a general introduction to the subject.

TIMESCALE OF EVENTS IN A LIQUID

1. Development

The processes that occur during radiolysis can be grouped
according to the time interval during which they occur.

Consider the events that occur during and after the pas
sage of a fast charged particle through a small group of molecules in a liquid.

(a) Velocities of electrons and α particles are illustra
ted in Table 2. The neutral He* is included as a reminder of the
electron pickup that complicates the energy loss processes of
heavy, positive particles near the ends of their tracks; pickup
is appreciable for α particles at energies below 10^6 eV and is
dominant below about 10^5 eV.

The diameters of many simple molecules are in the region
0.5 ± 0.3 nm. The fast charged particles therefore interact with
a given molecule for only 10^{-18} - 10^{-16} s. Thus the electronic
excitation and ionization of molecules, along a segment of track
equal to a few molecular diameters, occur in a time $\lesssim 10^{-15}$ s.

$$RS \quad \longrightarrow\!\!\!\sim\!\!\!\sim\!\!\!\sim\!\!\!\longrightarrow \quad RS\cdot^{+} + e^{-} \tag{1}$$

$$RS \quad \longrightarrow\!\!\!\sim\!\!\!\sim\!\!\!\sim\!\!\!\longrightarrow \quad RS^{*} \tag{2}$$

where the dot on $RS\cdot^{+}$ indicates the unpaired electron.

TABLE   2

Particle   velocities

Particle	Energy (eV)	Velocity (m/s)
e^{-}	10^{6}	2.8×10^{8}
e^{-}	10^{6}	1.9×10^{6}
α	10^{6}	6.9×10^{6}
He*	10^{4}	6.9×10^{5}

(b) Bond vibration frequencies are $10^{13} - 10^{14}$ s^{-1}, so molecules and ions that were excited to levels above their dissociation limit dissociate in $\leq 10^{-13}$ s.

$$RS^{*} \quad \longrightarrow \quad R\cdot + S\cdot \tag{3}$$

where $R\cdot$ and $S\cdot$ are free radicals.

(c) The electrons generated in reaction (1) transfer their excess energy to a highly polar liquid such as water in $10^{-14} - 10^{-13}$ s [15-17]. In a nonpolar liquid comprised of molecules that have anisotropic polarizability, such as n-hexane, the electrons become thermalized in ~ 10^{-12} s [18,19]. Judging from the relative thermalization ranges of secondary electrons in n-hexane, neopentane and liquid argon, which are 7 nm [20], 21 nm [21], and ~ 250 nm [22], respectively, the electron thermalization time would be ~ 10^{-11} s in liquid neopentane and ~ 10^{-9} s in liquid argon [19]. The thermalization time increases roughly as the square of the thermalization range, due to the large changes of flight direction that occur during the scattering events.

Electrons form stable localized states in liquids that are made up of polar or anisotropically polarizable molecules, such as water or n-hexane. In these liquids the electrons generated by reaction (1) become trapped (form a localized state) as soon as, or slightly before, they reach thermal energy. Electron localization therefore requires ~ 10^{-13} s in water and alcohols, and ~ 10^{-12} s in most hydrocarbons. In spherelike molecular systems such as neopentane, methane, argon and xenon the thermalized electrons remain in the quasifree state, because localized states are not stable in these liquids.

      (d) The initially formed localized state of the electron relaxes to a lower energy level as the medium around it shifts in reaction to the charge and zero point energy requirement of the electron. Just before the electron becomes localized, its kinetic energy has probably been reduced to < 1 eV. Its velocity would therefore be $< 6 \times 10^5$ m/s and it would take $\sim 10^{-15}$ s to traverse a localization site. The site would have a size similar to that of a molecule. Electronic polarization of molecules requires only $\sim 10^{-16}$ s, so when an electron becomes trapped it has already electronically polarized the medium adjacent to it. Thus the trapping site can be considered to be initially electronically polarized. The potential energy of the localized electron then decreases as the site undergoes other kinds of polarization (Fig. 2).

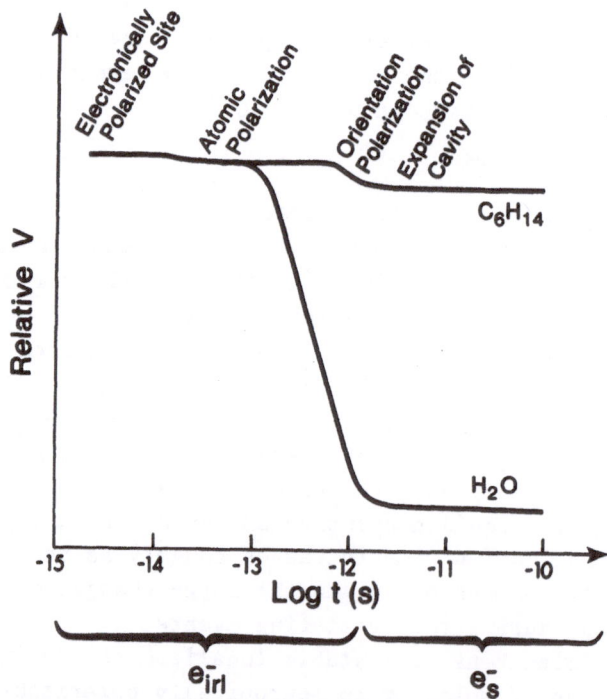

Fig. 2 - Variation of the relative potential energy V of an electron as a function of time after its localization in a nonpolar liquid such as hexane, or in a polar liquid such as water. The subscripts irl and s mean "incompletely relaxed localized" and "solvated", respectively.

Atomic polarization occurs in a few vibration times, or $\sim 10^{-13}$ s. The resultant reduction of V is expected to be relatively

small. If the electron is localized in an interstitial cavity in
the liquid, there would be a tendency for the cavity to enlarge.
This is due to the zero point energy requirement of the electron
($T_0 \approx h^2/8mr^2$, where r is the effective radius of the cavity) and
the short range electron-molecule repulsions. There is a balance
between the tendency of the internal pressure of the liquid to
localize the electron in a small volume and the zero-point energy
requirement which tends to enlarge the volume occupied by the
electron. The time needed for the molecular shifts would be similar
to a diffusive jump time, or ~ 10^{-12} s in water and hexane. The
largest reduction of V in polar liquids is due to the tendency
for the electric dipoles to align themselves with the field around
the electron. The time τ' required is similar to the relaxation
time of the medium in the field of a fixed charge, which is shorter than the usually reported relaxation time τ in a fixed field;
$\tau' \approx 10^{-12}$ s in water. In nonpolar liquids comprised of molecules
that have anisotropic polarizabilities there would be a tendency
for the axis of largest polarizability to align with the field.
However, the alignment effect on V would be much smaller for
hexane than for water.

The final electron state is in thermal equilibrium with
the medium and is called the solvated electron, e^-_s (Fig. 2). The
earlier, higher energy states have several jargon names, including
"trapped electrons", but could be referred to un-ambiguously as
incompletely relaxed localized states, e^-_{irl}.

(e) Reaction can sometimes occur between a newly formed
intermediate species and one of its nearest neighbors.

$$RS\cdot^+ + RS \longrightarrow RSR^+ + S\cdot \tag{4}$$

$$R\cdot + S\cdot \longrightarrow RS \tag{5}$$

$$R\cdot + S\cdot \longrightarrow P_1 + P_2 \tag{6}$$

where P_1 and P_2 are product molecules. Reaction (4) is an ion-molecule reaction, such as $H_2O^+ + H_2O \rightarrow H_3O^+ + \cdot OH$; (5) and (6) are
free radical combination and disproportionation reactions, respec-
tively, such as $\cdot CH_3 + \cdot C_2H_5 \rightarrow C_3H_8$ and $\cdot CH_3 + \cdot C_2H_5 \rightarrow CH_4 + C_2H_4$.
In some cases (5) is possible but not (6), for example, $H + \cdot OH \rightarrow H_2O$.
When (5) and (6) involve radicals that were formed from the same
molecule, as in (3), it occurs because the radicals were prevented
by the surrounding closely packed molecules from flying apart
(Fig. 3). This type of confinement in solids and liquids is called the "cage effect", in the jargon of reaction kinetics.

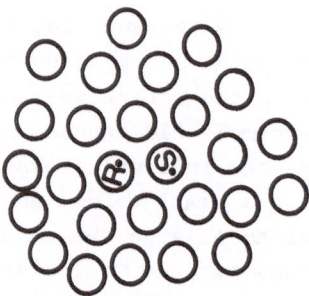

Fig. 3 – An R· and S· formed by reaction 3 in a liquid cannot im
        mediately fly apart, because translational motion is re
        stricted by the availability of space between the sur-
        rounding molecules.

        The collision frequency between nearest neighbors in a
liquid is ~ 10^{13} s^{-1}. A given configuration, such as that in Fig.
3, is terminated when one of the species undergoes a diffusive
jump. In liquids that have viscosities similar to those of water
or n-hexane the jump frequencies are 10^{11} – 10^{12} s^{-1}, so if (4) –
(6) are to occur between initial nearest neighbors they must occur
in $\leq 10^{-11}$ s.
        (f) The positive and negative charges in reaction (1)
eventually undergo neutralization. Their lifetimes have a wide
range of values in each system. The lifetime spectrum can differ
greatly from one system to another, depending on such things as
the distance that the electron travels from its parent ion during
thermalization, the mobilities of the thermalized electron and ion,
and the dielectric constant and temperature of the liquid.
        In a given system, some of the electrons become therma-
lized before they have moved beyond the coulombic attraction of
their parent ions.

$$M \ \longrightarrow\!\!\!\sim\!\!\!\!\sim\!\!\!\!\rightarrow \ [M^+ + e^-] \tag{7}$$

where the square brackets represent the coulombic attraction. The
attractive interaction tends to draw the charges back together.

$$[M^+ + e^-] \rightarrow M^*, \text{ geminate neutralization,} \tag{8}$$

Random, thermal energy tends to cause the charges to diffuse apart.

$$[M^+ + e^-] \rightarrow M^+ + e^-, \text{ free ions,} \tag{9}$$

The free ions from diverse pairs would ultimately neutralize each
other at random.

$$M^+ + e^- \longrightarrow M^*, \text{ random neutralization,} \tag{10}$$

The lifetimes t_{gn} of the species that undergo geminate neutralization can be estimated as follows.

$$t_{gn} = \int_y^o dr/v_d = \int_y^o \frac{\varepsilon r^2 dr}{\xi(\mu_+ + \mu_-)}$$

$$= 2.3 \times 10^6 \; \varepsilon y^3/(\mu_+ + \mu_-) \tag{11}$$

where $v_d = E(\mu_+ + \mu_-)$ is the drift velocity of the species towards each other in their mutual field E, $(\mu_+ + \mu_-)$ is the sum of their mobilities in units of cm^2/Vs, $E = \xi/\varepsilon r^2 = (4.8 \times 10^{-10}/\varepsilon r^2)$ stat volts/cm $= (1.44 \times 10^{-7}/\varepsilon r^2)$ V/cm, ε is the dielectric constant of the medium, r is the ion-electron separation distance and y is the initial separation distance of the thermalized electron and ion. In n-hexane at 25°C y $\approx 7 \times 10^{-7}$ cm, $\varepsilon = 1.9$, $(\mu_+ + \mu_-) = 0.08$ cm^2/Vs, which give $t_{gn} \approx 2 \times 10^{-11}$ s. The distribution of y values in hexane would spread t_{gn} values an order of magnitude on either side of this most probable value.

(g) The radii of the spurs in high energy particle tracks in water are usually considered to be 1-2 nm. The concentration of radicals and ions in a spur decreases with increasing time because of the occurrence of two processes: reaction of the intermediates in the spur; expansion of the effective volume of the spur by random diffusion of the intermediates (Fig. 1). In pure water, the intermediates can only react with each other. Radicals in the same liquid cage can react together in about 10^{-12} s. If radicals escape from the cage or are formed in adjacent cages, they have a finite probability of meeting and reacting during their random diffusion in the spur. A spur has been effectively dissipated when its volume has increased by more than an order of magnitude, that is, when the radius of a spherical spur has roughly tripled, or when that of a cylindrical spur has increased roughly four-fold. The effective lifetimes of spurs can therefore be calculated.

Consider a spherical spur that has an initial radius of 1.5 nm. For three dimensional random molecular motion, the root-mean-square displacement r_a of a molecule in time t is:

$$r_a = \sqrt{6Dt} \tag{12}$$

where D is the diffusion coefficient of the molecule in the medium. The value of D for many small molecules in water is about 1×10^{-5} cm^2/s. Therefore, the average time required for a radical to diffuse about $3 \times 1.5 = 4.5$ nm in water is

$$t = r_a^2/6D \approx (4.5 \times 10^{-7})^2/6 \times 10^{-5} = 3 \times 10^{-9} \text{ s.}$$

At the end of about 3 ns there is a negligible chance that two radicals from the same spherical spur will meet again, so the spur is said to be dissipated in ~ 3 ns.

Now consider a cylindrical spur that has an initial radius of 1.5 nm. For two dimensional random walk, the root-mean-square displacement of a molecule in time t is:

$$r_a = \sqrt{4Dt} \tag{13}$$

The average time required for a radical to diffuse about $4 \times 1.5 = 6.0$ nm in water is 9×10^{-9} s, so a cylindrical spur is dissipated in about 10^{-8} s in water.

The same is true for neutral radicals in other liquids that have viscosities in the vicinity of 1 centipoise.

Reactions between charged intermediates in spurs have shorter lifetimes, due to the coulombic interactions between them.

(h) The spurs have dissipated after 10^{-8} s and any inter mediates that have not reacted in that time are available for reaction in the "bulk" solution. If six reactive intermediates are generated in a spur with a radius of 1.5 nm, the local concen tration of intermediates in the spur is

$$\frac{6 \times 10^3/(6 \times 10^{23})}{(4/3)\ 3.1 \times (1.5 \times 10^{-7})^3} \approx 0.7 \text{ M}$$

After 10^{-8} s the volume of the spur has expanded by an order of magnitude, so the local concentration of intermediates has decrea sed by an order of magnitude. Reactions of the intermediates with each other in the spur might have reduced their concentration still further, so the local concentration after 10^{-8} s is about 10^{-2} M and still decreasing with increasing time. Concentration $\propto r_a^{-3}$, but $r_a \propto t^{\frac{1}{2}}$, so concentration $\propto t^{-3/2}$. Diffusion causes the local concentration of intermediates to decrease by 1½ orders of magnitude for each order of magnitude of increase in time, either until reaction with a solute terminates the life of the intermediates, or until they become homogeneously distributed throughout the solution.

The steady state concentration of reactive intermediates R· in the bulk medium, if they reacted only with each other, would be dependent on the dose rate. For the mechanism

$$A \xrightarrow{\quad I \quad} 2R\cdot + \text{products} \qquad (14)$$

$$2R\cdot \longrightarrow R_2 \qquad (15)$$

at steady state,

$$d[R\cdot]/dt = 2I - 2k_{15}[R\cdot]^2 = 0$$

Therefore,

$$[R\cdot] = (I/k_{15})^{\frac{1}{2}} \qquad (16)$$

At a dose rate of 10^{16} eV/cm^3 s (the approximate dose rate in water from a thousand curie ^{60}Co Gammacell) and with $G(R\cdot) \approx 10$ radicals/100 eV, one has $I = 10^{15}$ R$\cdot$/cm^3 s. Assuming $k_{15} \approx 10^{-11}$ cm^3/molecule s, the steady state concentration of R$\cdot$ is 10^{13}/cm^3, or about 10^{-8} molar. The average lifetime of these species is

$$t_a = 1/2k_{15}[R\cdot] \approx 5 \times 10^{-3} \text{ s.}$$

The steady state concentrations of intermediates vary as the square-root of the dose rate if they react only with each other. If they can also react with a solute in the medium, or with the medium itself, their steady state concentrations and average lifetimes will be smaller than those mentioned above. Consider the reaction

$$R\cdot + X \longrightarrow RX\cdot \qquad (17)$$

where X is a solute or the solvent. Since the concentration of radicals in the bulk medium is several orders of magnitude less than that in the spurs, an added solute is able to compete much more efficiently with reaction (15) in the bulk medium than it is in the spurs.

Quantitatively, the value of $k_{17}[X]$ must be greater than $\sim 10^2$ s^{-1} for reaction (17) to compete effectively with (15) in the bulk solution at a dose rate of 10^{16} eV/cm^3 s. Since a spur is dissipated in about 10^{-8} s, $k_{17}[X]$ must be greater than $\sim 10^8$ s^{-1} for reaction (17) to compete effectively with (15) in the spurs.

The extents of reactions (15) and (17) at different concentrations of X are shown in Figure 4. The same plots should be obtained for several scavengers X if the concentration of each scavenger is multiplied by the value of its rate constant.

Point a in Figure 4 is moved to higher concentrations by higher dose rates, and by the presence of a reactive impurity.

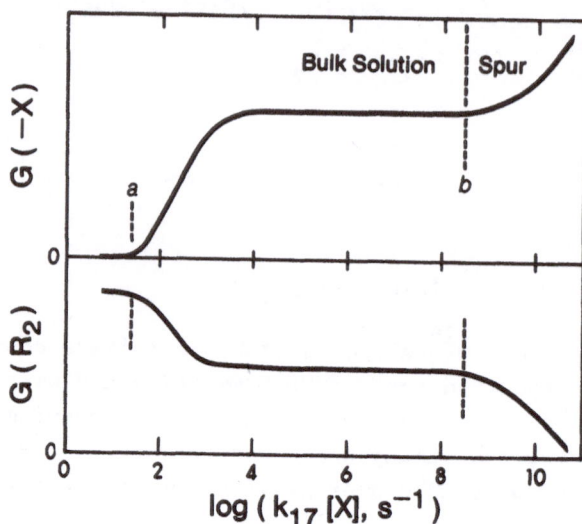

Fig. 4 – Schematic variations of G(-X) and $G(R_2)$ with concentration of scavenger X. G = no. of molecules formed or de-
stroyed/100 eV absorbed by the system. Steady state radiolysis dose rate $\approx 10^{16}$ eV/cm^3. See text. Reaction(17)
becomes significant in the bulk solution at
$\log(k_{17}[X]) \approx \underline{a}$, and in the spurs at $\log(k_{17}[X]) \approx \underline{b}$.

The left hand end of the $G(R_2)$ curve is sometimes greatly lowered
due to impurities or to R_2 itself being a scavenger of R (as H_2O_2
is in water). Point $\underline{b}$ in Figure 4 is independent of dose rate,
except at extremely high dose rates where track overlap occurs.
Point $\underline{b}$ is also insensitive to traces of impurities because [X]
is relatively large.

2. <u>Summary of time scale</u>
        From the preceding discussion the following approximate
time scale can be set up for liquids that have viscosities ~ 1 cP.

log t (s)

	-15	-12	-11		-8
Particle-system interactions	(a) Radical formation	(a) Expansion of spurs by diffusion		Reactions in bulk medium	
	(b) Cage reactions				
Spur formation	(c) Ion-molecule	(b) Reaction inside spurs			
	(d) Solvation of electrons and ions				

REFERENCES

1. H.E. Johns and J.S. Laughlin, in "Radiation Dosimetry", G.J. Hine and G.L. Brownell (eds.), Academic Press, New York, 1956, chapter 2.

2. J.W.T. Spinks and R.J. Woods, "An Introduction to Radiation Chemistry", 2nd edn., Wiley-Interscience, Toronto, 1976, p.29.

3. D.E. Lea, "Actions of Radiations on Living Cells", 2nd edn., Cambridge University Press, 1955, chapter 1.

4. G.R. Freeman and J.M. Faydh, J. Chem. Phys. $\underline{43}$, 86 (1965).

5. G.R. Freeman, "Radiation Chemistry", Quaderni de la Ricerca Scientifica, no. 67, Rome, 1970, p. 45, and assume that $E \approx 20$ eV is needed to produce an ion-electron pair.

6. A.H. Samuel and J.L. Magee, J. Chem. Phys. $\underline{21}$, 1080 (1953).

7. H.A. Schwarz, J. Phys. Chem. $\underline{73}$, 1928 (1969).

8. J.L. Magee and A. Chatterjee, J. Phys. Chem. $\underline{84}$, 3529 (1980).

9. G.R. Freeman, J. Chem. Phys. $\underline{46}$, 2822 (1967).

10. J.-P. Dodelet and G.R. Freeman, Can. J. Chem. $\underline{49}$, 2643 (1971); $\underline{50}$, 2729 (1972).

11. K.N. Jha and G.R. Freeman, J. Chem. Phys. $\underline{51}$, 2839, 2846 (1969).

12. J.C. Russell and G.R. Freeman, J. Chem. Phys. $\underline{48}$, 90 (1968).

13. A. Hummel, J. Chem. Phys. $\underline{48}$, 3268 (1968), and later modifications.

14. W.P. Helman and K. Funabashi, J. Chem. Phys. $\underline{66}$, 5790 (1977).

15. R.L. Platzman, in "Physical and Chemical Aspects of Basic Mechanisms in Radiobiology", NRC publication 305, Washington, D.C., 1953, p. 22.

16. H. Fröhlich and R.L. Platzman, Phys. Rev. $\underline{92}$, 1152 (1953).

17. J.L. Magee, in "Physical and Chemical Aspects of Basic Mechanisms in Radiobiology", NRC publication 305, Washington, D.C. 1953, p. 51.

18. A. Mozumder and J.L. Magee, J. Chem. Phys. $\underline{47}$, 939 (1967).

19. G.R. Freeman, Quaderni dell'Area di Ricerca dell'Emilia-Roma-

gna $\underline{2}$, 55 (1972).

20. J.-P. Dodelet, K. Shinsaka, U. Kortsch and G.R. Freeman, J. Chem. Phys. $\underline{59}$, 2376 (1973).

21. J.-P. Dodelet and G.R. Freeman, Can. J. Chem. $\underline{50}$, 2667 (1972).

22. S.S.-S. Huang and G.R. Freeman, Can. J. Chem. $\underline{55}$, 1838 (1977).

SOURCES OF PULSED RADIATION[1]

Myran C. Sauer, Jr.

Chemistry Division
Argonne National Laboratory
Argonne, Illinois 60439, USA

ABSTRACT

   Characteristics of various sources of pulsed radiation are
examined from the viewpoint of their importance to the radiation
chemist, and some examples of uses of such sources are mentioned.
A summary is given of the application of methods of physical dosimetry to pulsed sources, and the calibration of convenient chemical
dosimeters by physical dosimetry is outlined.

1. INTRODUCTION

   An attempt will be made here to describe the ways in which a
variety of pulsed sources have been used. The characteristics of
typical sources will be discussed from the viewpoint of parameters
important to the experimentalist. As will be seen, several different sources have been used with success in studying radiation
induced reactions despite orders of magnitude differences in parameters such as beam current, pulse length, and dose per pulse. A
low dose per pulse is sufficient if detection methods of high sensitivity can be used.

   The time scale of the process being studied has an important
bearing on the requirements for the pulsed source. It is clear
that the advantages of the pulsed method would be diminished if
measurements were possible only during the pulse because of short
transient lifetime relative to the pulse length. Sometimes the
experimental conditions can be arranged such that the lifetimes of
species are lengthened, in which case the use of a source which is
limited to long pulse lengths is no disadvantage. However, for

J. H. Baxendale and F. Busi (eds.),
The Study of Fast Processes and Transient Species by Electron Pulse Radiolysis, 35–47.

second-order transient decay, obtaining longer lifetimes requires
that lower transient concentrations be used, and the limit of detection sensitivity therefore becomes important. Problems of
reaction of transient species with impurities or with the solvent
can arise if low transient concentrations are used to slow down
second-order reactions. In general, the shorter the radiation
pulse available, the more flexible are the experimental requirements.

Types of pulsed sources which have been used will be discussed in the following with examples of experiments being given
to point out areas where a particular pulsed source may be especially useful. Published material used in preparing this summary includes references 1-5.

2. TYPES OF PULSED RADIATION SOURCES

In Table I a list is presented of pulsed sources which have
been used, and approximate values of relevant parameters of these
sources. A parameter of prime importance is the dose per pulse,
which is seen to vary over many orders of magnitude. Values are
given for typical conditions such as distance from the beam exit
window and degree of beam focusing.

Table I.   Examples of Types of Pulsed Radiation Sources

Type	Particle	Energy (MeV)	Pulse Length (ns)	Peak Amps	Dose per Pulse (krad)	Repetition Rate (s^{-1})
Linac (L-band)	e^-	20	0.03	200	2	800
			10	20	60	100
			3000	2	2000	30
Febetron 705	e^-	1.8	50	5000	2000	0.01
Febetron 706	e^-	0.5	4	7000	800	0.01
Van de Graaff	e^-	3	1	5	1	10^5
			100	5	100	10^5
Van de Graaff	H^+	2	0.2	10^{-5}	10^{-4}	10^6
Tandem Van de Graaff	H^+	9	1	10^{-4}	10^{-3}	10^6
Cyclotron	H^+	10	10^4	10^{-6}	0.05	10^3
X-ray unit	-	0.12	10^6	-	10^{-4}	10

The energy of an electron beam is important mainly in two respects. If the energy is much below 0.5 MeV, difficulties arise with respect to absorption and scattering in windows and in the inconveniently small penetration depths in the case of condensed phase samples. From the standpoint of radiation chemistry, there is no use in beams of more than about 20 MeV, and at energies above this induced radioactivity can become a problem.

The repetition rate is an important parameter because many experiments are possible only with the use of signal averaging techniques. Such techniques are now widely used in current pulse-radiolysis work due to developments in signal averaging devices and especially to the general computerization of experiments. Extensive signal averaging is practical with most of the sources given in Table I. The Febetrons can only be pulsed about once a minute, so only a limited amount of signal averaging is practical; however, the large pulse from a Febetron often obviates the necessity of signal averaging.

2.1. Microwave Linear Accelerators

Details concerning operating versions of "L-band" and "S-band" accelerators are available in the literature (6-8) and will not be given here. A characteristic of these machines is that without "prebunching", i.e., a technique by which several fine-structure pulses are combined to form one pulse (7), the pulses produced (duration ~5 ns to ~10 μs) actually consist of a series of fine-structure pulses of width on the order of 20-30 picoseconds. Figure 1 shows part of such a series for an "S-band" machine. In the case of the "S-band" machine the spacing between pulses is 350 ps, while the "L-band" machine gives 770 ps spacing. The "L-band"

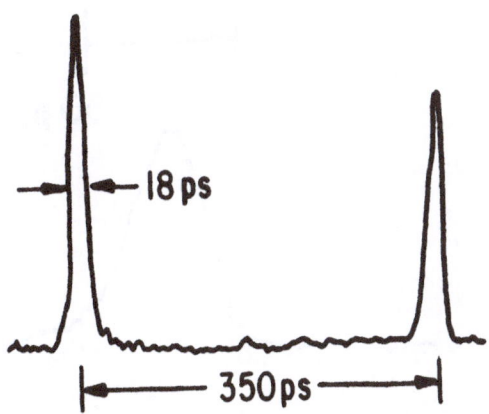

Figure 1. Streak camera record of S-band linac fine structure pulses (8), using Cerenkov light.

machines have the advantage of considerably higher currents, while the "S-band" machines have somewhat narrower fine-structure pulses.

In using such chains of fine-structure pulses in pulse-radiolysis experiments, one must bear in mind the possibility of intensity effects due to the pulse structure. The approximation that the pulse is "uniform" can only be made if the lifetime of transient species of importance in the chemical reactions being considered is long compared with the fine-structure spacing.

To gain a better understanding of radiation induced transient species and their reactions, it is often desirable to examine proc-esses occurring at as early a time as possible. The 20-30 ps fine-structure pulses afford the opportunity to attain time resolution on the order of 20 ps, and experiments of this type were first done (9) with water where the hydrated electron was measured in the time between fine structure pulses. The development of the "subharmonic prebunching" technique (7) to produce single 30 ps pulses (singly or separated by milliseconds) as shown in Figure 2, removed the limitation of having a second pulse occur after 770 ps.

Detection methods are a determining factor in the utilization of the 30 ps pulse. In 1970, rise-times of optical detection (photomultipliers, photodiodes) were not better than about 100 ps. Therefore, methods were devised (9,11,12), which will be discussed elsewhere in this volume, for producing 30 ps pulses of Cerenkov light (using part of the electron pulse), and measuring the absorb-ance in liquid samples using this Cerenkov light, which was delayed a time t (variable) with respect to the electron pulse. The time resolution is therefore essentially determined by the electron

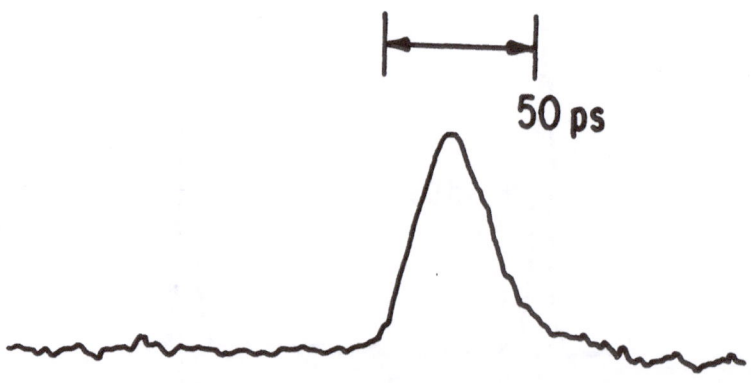

50 ps

Figure 2.   Single pulse after subharmonic prebunching; measured by streak camera recording of Cerenkov light (10).

pulse width, because the Cerenkov light has the same width as the electron pulse.

More recently, pulses of 30 picosecond duration have been used to investigate fluorescence from gas and liquid samples, using a photodiode or crossed-field photomultiplier to obtain time resolution of about 100 ps (13). Improvement on this time resolution has been attained by the use of streak camera techniques (14) for recording fluorescence from hydrocarbon solutions of organic scintillators, where time resolution approaching the pulse width has been obtained. The use of streak camera techniques in the picosecond time regime has been primarily in studies of light emission because of problems caused by the high light intensities needed for good signal-to-noise ratio in optical absorption measurements made in the picosecond regime.

Linear accelerators with or without the capability of single psec pulses can be used under conditions where the pulse is many microseconds long. This capability allows doses per pulse to be obtained comparable to doses obtained from Febetrons (see Table I). For gas phase work, this is convenient in the following way. Most gas phase studies of atomic and free radical reactions do not require time resolution better than several microseconds, because of the fact that radical-radical recombination times are in the range of hundreds of microseconds for transient concentrations which are typically attainable. Therefore, the linear accelerator can be used, for example, for studies where it is desired to put the electron beam through the wall of a high-temperature (ca. 1000°C) oven (15). The 15-20 MeV energy of the electrons is convenient in this respect, when compared with the (maximum) 2 MeV electrons from a Febetron, which create an undesirable situation when a high temperature cell is used. That is, if the windows are made thin enough to allow the 2 MeV beam to pass through, too much heat loss to the Febetron occurs; therefore, a temperature limit of about 200°C exists for Febetron work. The lower concentration of species produced by the "weaker" linac pulse, and the smaller useable irradiated volume can be compensated for by making use of the greater pulse repetition rate available and using signal averaging techniques.

2.2. Febetrons

Febetrons, the trade-name for pulsed sources produced originally by the Field Emission Corporation, operate on field emission principle, using a Marx surge circuit to supply a large DC voltage pulse which causes electrons to be emitted from the cathode and accelerated to the anode. These machines produce single pulses of electrons at a maximum rate of about one per minute. The pulse produced is quite different from a typical linac pulse; it does not consist of a chain of fine structure pulses and is orders of

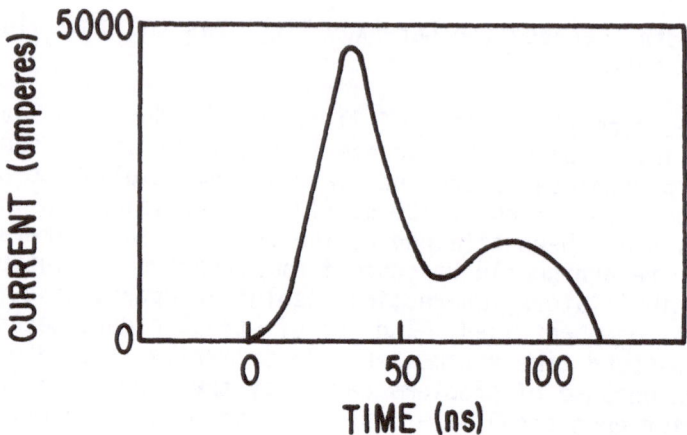

Figure 3. Current waveform of electron pulse from
Febetron 705 measured with a Faraday cup (16).

magnitude higher in current and lower in energy (see Table I).
The lower energy and high current make the Febetron more suitable
for gas phase work, although there have been significant applications to condensed phases. The small penetration depth in condensed phases leads to concentrations of transients which are
usually unnecessarily high.

The pulse shape from a Febetron 705 is shown in Figure 3.
Because of the fact that the use of shorter pulses from the Febetron 706 or linear accelerator allows time resolution in the nsec
region, studies involving the 705 have been mainly of processes
occurring on time scales of many microseconds.

Figure 4 shows an application to the reaction of HO_2 with NO,
producing NO_2 which is observed at 404.7 nm (17). The absorbance,
$A_{404.7}$, is rather weak and requires a blank correction to be made
in order to correct for absorbance in the cell window and a reproducible "noise" pattern on the analyzing light (see Figure 4).

A problem which often arises, particularly with the Febetrons,
is the creation of "noise" due to the electromagnetic fields produced coincident with the very high current, short duration electron pulse. The fields cause disturbances in the electronics used
to measure changes in absorbed or emitted light. The use of double
copper screened rooms, in which the fast detection and amplification equipment is placed, has been found to reduce the interference
to tolerable levels.

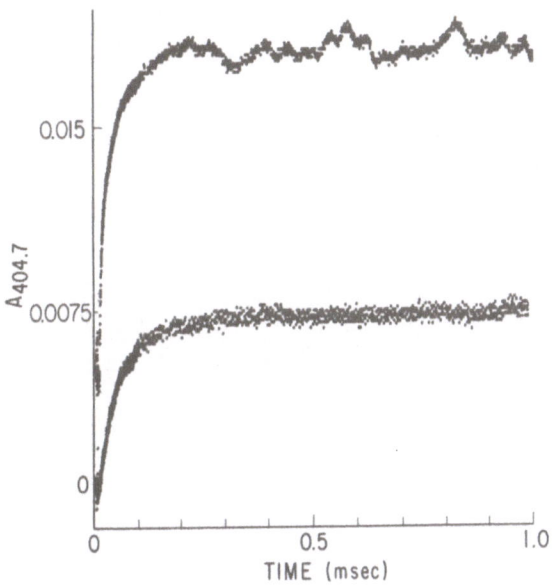

Figure 4. Example of use of Febetron 705 to observe NO_2 production in the reaction of HO_2 with NO (1190 torr H_2, 2.5 torr O_2, 0.13 torr NO). The lower curve results from subtracting the "blank" (not shown) on 1200 torr H_2 alone from the upper curve (which has been shifted vertically for convenience in presentation).

2.3. Other Sources of Pulsed Radiation

Van de Graaff generators have been used successfully to produce nanosecond (and shorter) pulses of electrons with peak current of about 5 amps (18). The Van de Graaff accelerator normally produces a continuous beam (with no fine-structure pulses) but is modified to produce pulses by controlling the voltage on the grid of the electron gun. Pulses of length between 1 and 100 ns can readily be obtained, maintaining up to about 5 amp current during the pulse. Thus, the dose per pulse is comparable to that from a linac for an equivalent pulse length. A disadvantage is that the beam divergence is greater due to the lower electron energy (3 MeV) typical of Van de Graaff sources.

A pulsed Van de Graaff source has also been used recently (19) to study the yield of the hydrated electron in water at nanosecond times using a 1 ns beam of 3 MeV protons. Because the penetration depth for 3 MeV protons is only about 0.15 mm, an ingenious method of putting a laser beam along the axis of a narrow jet of water was used to measure the optical absorbance due to the hydrated electron.

CYCLOTRON PULSE-RADIOLYSIS

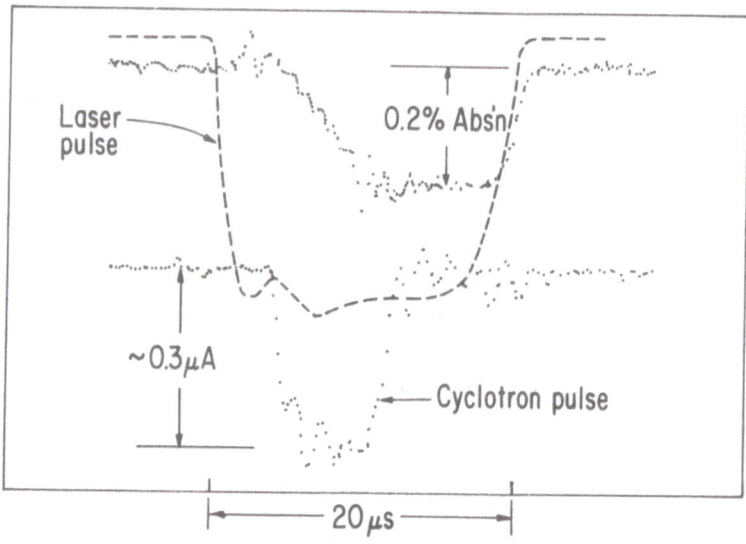

Figure 5. An example of the use of an 8 μs beam of 20 MeV deuterons from a cyclotron (21). The absorption of $(CNS)_2^-$ at 514.5 nm is measured in an aqueous KCNS solution with an argon ion laser.

        In comparison with pulsed-electron radiolysis, few studies have been made in condensed phases using pulses of protons or heavier ions. This is partially due to the problem just mentioned of small penetration depth. High energy particles with greater penetration depth can be produced by other means, but in general currents are low and/or pulse lengths less than microseconds are not available. In the past several years, G-values of primary species in water have been studied at microsecond times using 10 microsecond pulses of deuterons and helium ions from a cyclotron (20) (see Figure 5). Information is obtained on the effect of LET (linear energy transer, i.e., the amount of energy transferred per unit distance along the track of the particle) and particle type on G-values, which is used to improve the theories of diffusion kinetics and track structure which are used to model such results.

        Proton and helium ion beams of nanosecond length from Van de Graaff accelerators have also been used in recent years to investigate LET effects by observing the time dependence of fluorescence from benzene in cyclohexane (22). However, there is a need for more work at nanosecond times in this area of high LET radiation chemistry, to determine whether time dependences of transient G-values follow the predictions of models being used. The results in Figure 5 indicate how a rather weak absorption can be measured by appropriate signal averaging; hopefully, the effects of a variety of high LET particles will be investigated using such techniques with nanosecond pulses in the future.

Pulses of 2 MeV protons of 15 nsec duration from a Van de Graaff have been used to study processes occurring in rare gases by means of examining the time dependences of optical emissions from the vacuum uv through the visible region (23). The small penetration depth is not a problem in the gas phase; in fact, the higher LET is an advantage.

Pulses of low intensity x-rays (see Table I) have also been used in a series of experiments on ion-recombination in the gas phase (24). The reason that a dose per pulse at least five orders of magnitude lower than is typical of electron pulses can be used is that very sensitive techniques exist for measuring charge.

3. PHYSICAL DOSIMETRY FOR PULSED SOURCES

For most practical purposes, in the case of an experimental set-up where optical absorptions are being measured, there is no need for physical dosimetry, i.e., calorimetric methods, or ionization chamber methods. The reason for this is that considerable effort has been devoted to the problem of accurate physical dosimetry, and the G-values of products in various chemical dosimeters have been determined on the basis of this physical dosimetry (3-5). Thus, for both gas and condensed phase irradiations, convenient chemical systems can be routinely used, by optically monitoring product formation after a pulse, to accurately establish the dose under the same geometrical conditions to be used for other samples. Even in the case where optical absorption is not used as a measuring technique, the dose can be established by measuring chemical products (e.g., N_2 from N_2O) from established chemical dosimeters.

Because of the localized and directional nature of a pulsed source, consideration has to be given to the effect of beam scattering on the dose deposited. For an electron beam entering a liquid sample, the axial dose will decrease with depth if the initial beam diameter is much smaller than the range, due to scattering. But, if broad beam geometry is used, the axial dose actually increases significantly with depth before decreasing (1). In the case of gas samples, the scattering of the beam is more significant; a 13 MeV beam passing through 1.5 mm quartz windows was found to scatter enough that the dose decreased a factor of two in about 3 cm (25). The point to be made is that beam scattering must be kept in mind in order that the dosimetry is appropriate for a particular experimental situation.

For high dose pulsed sources, calorimetry has been the primary method used for absolute dosimetry (2). Ionization chamber methods can also be applied to pulsed sources (26); the charge produced by a pulse is measured and can be related to the dose if the energy required to produce an ion pair is known for the gas in the ionization chamber. However, for high intensity pulses the use of ion chambers requires corrections to relate the observed currents to dose, and has not been a frequently applied method (26).

In the case of water, a pulse of 10^5 rads will produce enough of a temperature rise (2.4°C) to be readily measured. The general procedure is to use a thermocouple to measure the temperature rise. Correction must be made for the small amount ($\sim$5%) of energy which produces chemical change rather than heat. If the pulse is short and thermocouple wires of diameter less than about 0.07 mm (4) are used in water, the system is essentially adiabatic, i.e., heat loss corrections are not appreciable during the fraction of a second needed for the thermocouple to reach equilibrium with the water. An example of a calorimetric measurement of dose used in calibrating the Fricke dosimeter at high dose rates from a linear accelerator is given by the work of Anderson (27). A recent publication by Panta (28) describes a calorimetric method for measurement of single linac pulses of 100 rads to 130 krads, similar in response characteristics to the method described below; and gives a thorough discussion of the theoretical principles involved in such measurements.

The dose from a pulse can be measured by directly irradiating a thermocouple (or several thermocouples in series). This method has been used by Willis and co-workers (16,29,30) for conditions applicable to both gas and liquid phase irradiations using the high intensity pulses from Febetrons. For the liquid phase, using a Febetron 705, a thin disc (graphite, aluminum, or nickel) was used, with thickness being varied such that all or part of the beam was absorbed. The response from a thermocouple attached to an aluminum disc appeared qualitatively as shown in Figure 6. The response time of the recorder is responsible for the initial rise, and the sharp peak at the beginning is due partly to the lower specific heat of the thermocouple compared with that of aluminum, but primarily to the fact that the thermocouple was at the center of the disc, where the radiation intensity was higher, and hence, the temperature change was initially greater. The negative peak is an artifact due to the switching off of the focussing magnet on the Febetron 705 about 2 sec after the pulse. The subsequent cooling curve is extrapolated back to time zero, and that reading of the thermocouple is converted to a temperature rise. The dose is then determined from the known specific heat of aluminum.

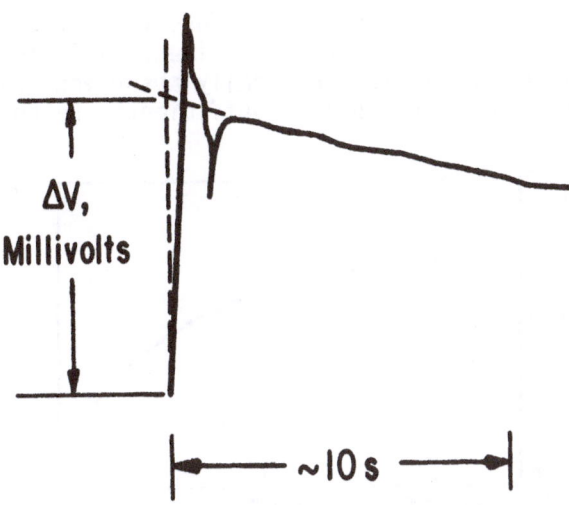

Figure 6. Thermocouple output from an aluminum calorimeter (16) following a 100 ns electron pulse from a Febetron 705.

From data obtained as a function of the thickness of the calorimeter material, the dose to a liquid sample put in the same position was calculated.

For gas samples, irradiated with either the Febetron 705 or 706, similar measurements were made using aluminum calorimeters, yielding curves similar to that shown in Figure 6. In the case of the Febetron 705, measurements were made at several positions near the Febetron window over a range of about 3 cm. The mean dose measured over this range as a function of calorimeter thickness is shown in Figure 7. Extrapolation to zero thickness allows the mean dose to a sample gas in the cell to be obtained.

Using the dose determined in this way, the $G(N_2)$ from N_2O was determined to be 12.4 ± 0.2 molecules per 100 eV for the Febetron 705 (16) ($\sim$100 ns pulse, 2.3 Mrad). In the case of the Febetron 706, the same calorimetric method was applied (30), and $G(N_2)$ from N_2O was determined to be 12.3 ± 0.3 molecules per 100 eV ($\sim$5 ns pulse, 0.9 Mrad). This chemical dosimeter can, of course, be used by the experimenter instead of calorimetric procedures. However, it should be noted that when an optical absorption experiment is being carried out and the dose in the path traversed by the optical beam is desired, it is much more convenient to measure the concentration of O_3 formed in an atmosphere of O_2 containing $\sim$0.5% SF_6. This system has been determined to have $G(O_3) = 6.2$ (31,32). The known absorption coefficient of ozone can be used to determine its concentration after its formation is complete (a few msec after a pulse), and hence the dose to that part of the sample in the optical path can be determined. This is convenient for measuring G-values of other optically absorbing transient species, assuming their absorption coefficients are known.

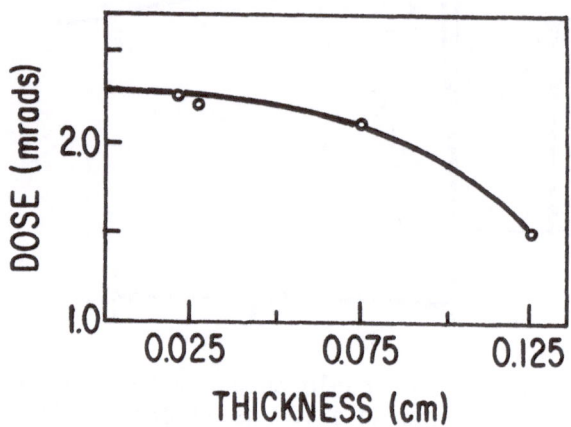

Figure 7. Mean dose for a gas cell, using a 100 ns pulse from a Febetron 705, as a function of calorimeter thickness (16).

NOTES AND REFERENCES

[1]Work supported by the Office of Basic Energy Sciences, Division of Chemical Sciences, U. S. Department of Energy, under Contract W-31-109-Eng-38.

(1) Matheson, M. S. and Dorfman, L. M.: 1969, *Pulse Radiolysis*, M.I.T. Press, Cambridge, MA.
(2) Firestone, R. F. and Dorfman, L. M.: 1971, in *Actions Chimiques et Biologiques des Radiations*, ed. M. Haissinsky, Vol. 15, Masson, Paris, pp. 7-46.
(3) Sauer, M. C.,Jr.: 1976, Adv. Radiat. Chem. 5, pp. 97-184.
(4) Holm, N. W. and Berry, R. J., Editors: 1970, *Manual on Radiation Dosimetry*, Dekker, New York.
(5) Spinks, J. W. T. and Woods, R. J.: 1976, *An Introduction to Radiation Chemistry*, 2nd Edition, Wiley, New York.
(6) Gallagher, W., Johnson, K., Mavrogenes, G., and Ramler, W.: 1971, IEEE Trans. Nucl. Sci. NS-18 (3), pp. 584-588.
(7) Ramler, W., Mavrogenes, G., and Johnson, K.: 1975, *Fast Processes in Radiation Chemistry and Biology*, ed. G. E. Adams *et al.*, Inst. Phys., Wiley, London, pp. 25-32.
(8) Katsumura, Y., Tagawa, S., and Tabata, Y.: 1980, J. Phys. Chem. 84, pp. 833-839.

(9)   Hunt, J. W., Wolff, R. K., Bronskill, M. J., Jonah, C. D.,
      Hart, E. J., and Matheson, M. S.: 1973, J. Phys. Chem. 77,
      pp. 425-426.
(10) Mavrogenes, G. S., Jonah, C., Schmidt, K. H., Gordon, S.,
      Tripp, G. R., and Coleman, L. W.: 1976, Rev. Sci. Instrum.
      47, pp. 187-189.
(11) Bronskill, M. J., Taylor, W. B., Wolff, R. K., and Hunt,
      J. W.: 1970, Rev. Sci. Instrum. 41, pp. 333-340.
(12) Jonah, C. D.: 1975, Rev. Sci. Instrum. 46, pp. 62-66.
(13) Jonah, C. D., Sauer, M. C.,Jr., Cooper, R., and Trifunac,
      A. D.: 1979, Chem. Phys. Lett. 63, pp. 535-538.
(14) Katsumura, Y., Tagawa, S., and Tabata, Y.: 1980, J. Phys.
      Chem. 84, pp. 833-839.
(15) Jonah, C. D., private communication, work in progress.
(16) Willis, C., Miller, O. A., Rothwell, A. E., and Boyd, A. W.:
      1968, Advan. Chem. Series 81, pp. 539-549.
(17) Lii, R., Sauer, M. C.,Jr., and Gordon, S., unpublished re-
      sults.
(18) Hunt, J. W. and Thomas, J. K.: 1967, Radiat. Res. 32,
      pp. 149-163.
(19, Burns, W. G., May, R., Buxton, G. V., and Tough, G. S.:
      1977, Faraday Discuss. Chem. Soc. 63, pp. 47-54.
(20) Sauer, M. C.,Jr., Schmidt, K. H., Hart, E. J., Naleway,
      C. A., and Jonah, C. D.: 1977, Radiat. Res. 70, pp. 91-106.
(21) Sauer, M. C.,Jr., Schmidt, K. H., and Jonah, C. D., unpub-
      lished results.
(22) Miller, J. H. and West, M. L.: 1977, J. Chem. Phys. 67,
      pp. 2793-2797.
(23) Nayfeh, M. H., Chen, C. H., and Payne, M. G.: 1976, Phys.
      Rev. A14, pp. 1739-1744.
(24) Sennhauser, E. S. and Armstrong, D. A.: 1978, Radiat. Phys.
      Chem. 11, pp. 17-28.
(25) Sauer, M. C.,Jr., and Mulac, W. A.: 1972, J. Chem. Phys. 56,
      pp. 4995-5004.
(26) Dinter, H. and Tesch, K.: 1974, Nucl. Instr. Meth. 120,
      pp. 113-119.
(27) Anderson, A. R.: 1962, J. Phys. Chem. 66, pp. 180-182.
(28) Pañta, P. P.: 1979, Nucleonika 24, pp. 927-939.
(29) Willis, C., Miller, O. A., Rothwell, A. E., and Boyd, A. W.:
      1968, Radiat. Res. 35, pp. 428-436.
(30) Willis, C., Boyd, A. W., and Miller, O. A.: 1971, Radiat.
      Res. 46, pp. 428-443.
(31) Willis, C., Boyd, A. W., Young, M. J., and Armstrong, D. A.:
      1970, Can. J. Chem. 48, pp. 1505-1514.
(32) Johnson, G. R. A. and Wilkey, D. D.: 1969, J. Chem. Soc. D,
      pp. 1455-1456.

CHEMICAL DOSIMETRY OF PULSED ELECTRON AND X-RAY SOURCES IN THE 1-20 MeV RANGE

E. Martin Fielden

Radiobiology Unit, Physics Department, Institute of
Cancer Research, Clifton Avenue, Sutton, Surrey.

1.1. Units of Absorbed Dose

Since only that fraction of the incident radiation flux
which is absorbed by a system can induce chemical change, the
dose absorbed by the system is the important quantity.

The preferred SI unit of absorbed dose is the GRAY,
abbreviation Gy, which is defined as:

$$1 \text{ Gray} = 1 \text{ Joule/kg}$$

Prior to the adoption of the Gray, absorbed dose was variously
reported in terms of eV/g or erg/g.   The most commonly used
unit, which is still met in publications was the rad, where
$1 \text{ rad} = 100 \text{ erg/g}$ or 6.24×10^{13} eV/g.   Thus $1 \text{ Gray} = 10^2$ rad.

N.B. The centigray ($cGy = 10^{-2}$ Gy) is equivalent to the rad,
but although permissible under the SI conventions its use is to
be discouraged.

1.2. Units of Radiation Chemical Yield

At the time of writing (1981) the unit has been redefined
in SI compatible units but both new and old units appear in the
literature.   Unfortunately, the same symbol, G, is used for
both new and old units although since they differ in magnitude
by approximately a factor of 10^7 it should be possible to avoid
confusion.

The recently adopted system defines the radiation chemical

49

J. H. Baxendale and F. Busi (eds.),
The Study of Fast Processes and Transient Species by Electron Pulse Radiolysis, 49–62.
Copyright © 1982 by D. Reidel Publishing Company.

yield, $G(x)$, as $G(x) = \dfrac{n(x)}{\varepsilon}$ where $n(x)$ is the amount of a specific substance, x, which is produced, destroyed or changed by the mean energy imparted, ε, to a specified material.

$G(x)$ is measured in mol J^{-1}.

The value of $G(x)$ can be calculated from the G value as originally defined (number of molecules changed per 100 eV energy absorbed) by multiplying the G value by 1.036×10^{-7},

i.e. 1 mol $J^{-1} \equiv 9.65 \times 10^6$ molecules $(100 \text{ eV})^{-1}$.

2.1. Principles of Chemical Dosimetry

Chemical dosimetry is based on the quantitative measurement of chemical change, generally at the completion of a series of radiation-induced reactions involving radicals. The accuracy of chemical dosimetry depends upon the invariance of the yield of primary radicals in a given system per unit of absorbed dose. A chemical dosimeter typically consists of a bulk component (usually a liquid or gas) in which effectively all of the absorbed energy is deposited, and a minority component (solute) which reacts with radiation-induced species (including primary radicals) formed in the bulk component to produce the observed chemical change.

In addition to reactions of the primary radicals, with either solute or solvent, yielding the desired product there is the possibility that a fraction of the primary radicals may react with themselves resulting in a net loss of product. This effect is negligible under conditions of low dose per pulse but ultimately limits the useful range of the dosimeter. The maximum dose per pulse that can be accommodated within the linear response range of a chemical dosimeter is usually considerably less than the maximum measurable dose for the same dosimeter when used for low dose rate dosimetry. Further descriptions of the limitations of chemical dosimeters for pulsed radiation will be found in refs. (1) and (2).

Pulses from a linear accelerator are peculiar in that they possess a fine structure since they are composed of a train of pulses each of a few picoseconds' duration. The spacing of these fine structure pulses is set by the frequency of the accelerating microwave field. Thus a nominal 1 μs pulse from an S band linear accelerator (3000 MHz) consists of 3000 short pulses spaced 3.3×10^{-10} sec. apart. Fortunately, the question: What is the dose rate of such a pulse? is irrelevant provided that the measuring system is not capable of resolving the fine structure. This is true of the dosimeters described in this section and linear accelerator pulses are indistinguishable from the D.C. pulses of a Van de Graaff generator.

2.2. The Ferrous Sulphate or Fricke Dosimeter

The radiation-induced oxidation of ferrous ions to the ferric state has long formed the basis of a standard chemical dosimeter (3). In the presence of oxygen and at low pH, the reactions occurring in an irradiated solution of ferrous sulphate can be summarised thus:

$$H_2O \xrightarrow[\text{radiation}]{\text{ionizing}} e^-_{aq}, \; H, \; OH, \; H_2O_2 \, (H_2)$$

$$e^-_{aq}/H + O_2 \longrightarrow HO_2$$

$$OH + Fe^{2+} \longrightarrow Fe^{3+} + OH^-$$

$$Fe^{2+} + HO_2 \xrightarrow{H+} Fe^{3+} + H_2O_2$$

$$2Fe^{2+} + H_2O_2 \xrightarrow{2H+} 2Fe^{3+} + 2H_2O$$

resulting in $G(Fe^{3+}) = 3G(e^-_{aq}) + 3G(H) + G(OH) + 2G(H_2O_2)$

The presence of organic impurities can lead to a chain reaction involving organic-peroxy radicals resulting in too high a yield of Fe^{3+}. For dosimetry at low dose rates it is usual to add sodium chloride to the system which desensitizes it towards the adverse effects of organic impurities. However, for pulse dosimetry chloride must not be added as it introduces such a strong dose rate dependence as to render the system unusable for pulsed irradiation. Thus careful attention must be made to reducing organic impurities in the Fricke solution for pulse dosimetry (4).

The standard Fricke solution consisting of 1 mol m^{-3} Fe^{2+} (SO_4), 400 mol m^{-3} H_2SO_4 in air saturated water maintains its value of $G(Fe^{3+}) = 1.61 \times 10^{-6}$ mol J^{-1} up to 10 Gy/pulse (2,5). Above 10 Gy/pulse the yield of Fe^{3+} falls quite sharply reaching 1.4×10^{-6} mol J^{-1} at 100 Gy/pulse.

Various means of measuring the quantity of ferric ion produced have been used (4) with spectrophotometric estimation being the most commonly used. The importance of a careful calibration of the spectrophotometer used, has been emphasised by Fricke and Hart (4) who list standard extinction values and their temperature coefficient for the Fe^{3+}/Fe^{2+} system.

2.3.   The "Super Fricke Dosimeter"

By increasing the ferrous ion concentration to 10 mol m^{-3} and saturating the solution with oxygen, the pulse-dose response of the system can be significantly improved.   This dosimeter, known as the "Super Fricke" has a slightly greater yield of ferric ion than the standard version (2,5):

Super Fricke G(Fe^{3+}) = 1.67 x 10^{-6} mol J^{-1}.

The dosimeter is usable up to 100 Gy/pulse and the yield is only 10% down at 700 Gy/pulse.   Because of its tendency to aut-oxidise it is advisable to make the solution on the day it is required.   The same care with respect to the avoidance of organic impurities and chloride apply as with the normal Fricke solution.   If this dosimeter is used to measure the total dose accumulated from a number of separate pulses, care must be taken to ensure that significant depletion of the ferrous or oxygen concentrations does not occur.   In practice, provided the pulses are not too closely spaced a total dose of 1750 Gy can be accumulated before loss of linearity is observed.

2.4.   Other Dosimeters Suitable for High Intensity Pulsed Sources

The dosimetry of pulses larger than 1000 Gy is probably best approached by calorimetric methods.   However, some chemical systems have been described and calibrated by calorimetry that maintain a linear response up to this dose level.   To avoid problems due to solute depletion all these dosimeters consist of a single pure component (see section 2.1).   Only a brief description of each dosimeter is given and the reader is referred to original articles for fuller details.

Liquid systems

(i)   When neutral or acidic water which has been degassed is irradiated, hydrogen is formed by the recombination of the reducing radicals in addition to the molecular yield of H$_2$.

G(H$_2$) = 7.3 x 10^{-8} mol J^{-1} is reported to be constant (6) from 1 Gy/pulse up to 10^4 Gy/pulse.

(ii)   The radiolysis of pure, degassed, benzene also results in the production of hydrogen.

G(H$_2$) = 4.1 x 10^{-9} mol J^{-1} is reported to be constant (7) up to 5 x 10^4 Gy/pulse.

Gaseous systems

For the pure gases listed below, the yield of a specified product has been shown to be pulse-dose independent up to 1000 Gy/pulse for $G(O_3)$ and up to 1.6×10^4 Gy/pulse for the remainder.

Gas	Product and yield ($\times 10^{-6}$ mol J^{-1})	Reference
O_2	$G(O_3) = 1.43 \pm 0.01$	8
N_2O	$G(N_2) = 1.27 \pm 0.3$	9
CO_2	$G(CO) = 0.8 \pm 0.2$	9
H_2S	$G(H_2) = 1.36 \pm 0.04$	9

3.1. Dosimetry by the Measurement of Fugitive Species

Instead of allowing the radiation-induced free radicals and other highly reactive species to be converted by reactions to persistent and stable species before dosimetric assay, it is possible to measure directly the concentration of fugitive, or transient, species on the time scale of the pulse itself. In principle, such a method should improve dosimetric accuracy since the possible loss of primary radicals in side reactions which do not contribute to the measured yield becomes irrelevant. In practice, the techniques to make time resolved measurements of short-lived species are highly specialised, although inherent in the pulse radiolysis technique (see other sections for descriptions of these techniques).

Thus the direct measurement of the concentration of radicals produced by a short pulse of radiation is a special form of chemical dosimetry which is particularly appropriate in radiation chemistry research. As well as being a rapid and convenient method of dosimetry for pulse radiolysis with optical detection, it has the advantage of measuring the linear average dose along the path of the analysing light beam (see Fig. 1). Due to focussing considerations, the analysing light does not usually sample the irradiation vessel uniformly and the spatial average dose throughout the whole cell may be different from the linear average dose in the light path. This is particularly likely for pulsed electron irradiation where a uniform dose distribution is difficult to achieve. The spatial average dose can be measured by the usual type of chemical dosimeter (e.g. the Fricke) whereas the linear average is given by one of the radical dosimeters described below.

A typical oscillogram is shown in Fig. 2. The short horizontal line at the lower left represents 100% light transmission before the radiation pulse. A rapid decrease in trans-

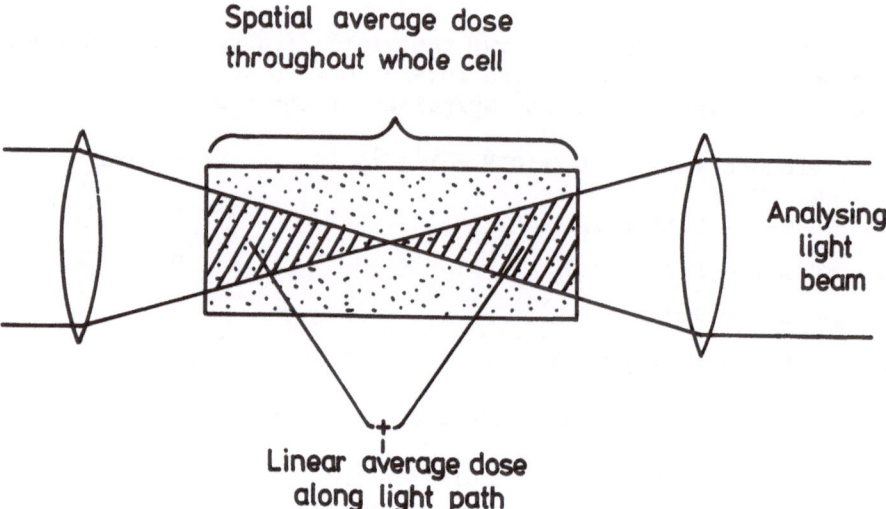

Fig. 1. Schematic diagram of a typical irradiation cell used in pulse radiolysis and the path of the analysing light beam.

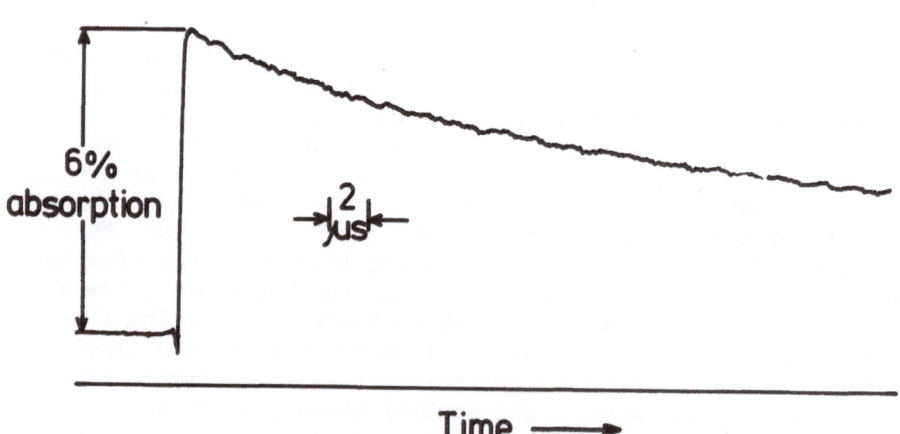

Fig. 2. Oscillogram of $(CNS)_2^-$ production and decay following a 0.6 Gy pulse of 0.2 μs duration on a 10 mol m^{-3} KCNS solution. Optical path length 2 cm.   Radiation: 4 MeV electrons.

mission occurs at the instant of the pulse due to the formation of the absorbing species, in this case the $(CNS)_2^-$ radical anion, which is proportional to the absorbed dose.   As the radicals disappear by subsequent reactions the transmission returns towards the 100% level.

In calculating the dose absorbed by the system corrections may have to be applied for the duration of the pulse and the finite response time of the measuring system.   In an ideal situation the pulse would be of zero width and the recording system infinitely fast.   In practice, one of these ideal conditions can be approached and the effect of not satisfying the remaining condition can be corrected for if necessary.

3.2. Sources of Error

1.   Correction due to pulse duration

As a pulse of finite duration is absorbed so the radical concentration builds up.   If the pulse is large an appreciable fraction of the radicals may disappear by reacting during the pulse.   If this happens the peak absorption recorded at the end of the pulse does not represent all the radicals that were produced.   In a practical dosimeter, where competing solutes are avoided, the loss of radicals is due to a second order reaction of the type:- $R + R \xrightarrow{\ k\ }$ Products ---- (1) where R represents the species of radical observed and k is the bimolecular rate constant for the reaction.

Provided that the intensity-time profile of the radiation pulse is of a simple geometrical form and the response time of the recording system is short compared with the pulse duration, a correction for the fraction of radicals lost during the pulse can be calculated.   Pulse profiles are commonly rectangular, at least for pulses of 1 μsec or longer duration where corrections are most frequently needed.   For the duration of such a pulse radicals are produced at a constant rate, H.   Simultaneously, radicals are disappearing at a rate kc^2 by reaction(1) where c is the concentration of R at time t after the start of the pulse. The rate of increase of the radical concentration during the pulse is then given by:-

$$\frac{dc}{dt} = H - kc^2$$

Integration of this expression for a pulse of length 'a' gives

$$c(a) = \sqrt{\frac{H}{k}} \ \tanh\left(a\sqrt{Hk}\right)$$

where c(a) is the concentration of radicals actually observed at the end of the pulse. Since the total yield of radicals produced by the pulse is Ha, the observed concentration c(a) has to be multiplied by a correction factor f_1 given by

$$f_1 = a\sqrt{Hk} \cot h \quad (a\sqrt{Hk}) \qquad \text{(see Fig. 3)}$$

The rate of formation of radicals, H, is obtained from the initial rate of rise of the optical density during the pulse, and the quantity k, if not already known, is obtained from the decay kinetics after the pulse. It should be remembered that data taken from the oscilloscope has to be converted to optical density before applying the correcting factor. Correction factors for first order decay kinetics have also been calculated (1,2).

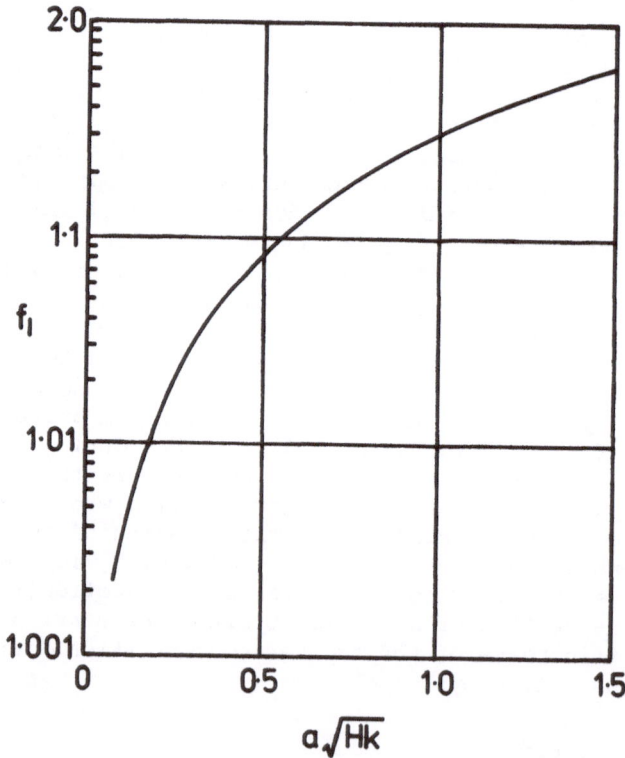

Fig. 3. Pulse duration correction factor, f_1, for the radical concentration measured at the end of a pulse of duration a, where H is the rate of production of radicals during the pulse and k the bimolecular rate constant for the radical-radical reaction.

2. Correction for response time of system

The response time of the recording system (photometer, oscilloscope, etc.) is set by the rise time of the electronic circuits and when this becomes appreciable in comparison with the decay time of the radicals following the pulse, a correction is required. In such a case the pulse duration can be neglected and the pulse regarded as being instantaneous.

The decrease of peak height has been calculated exactly (10) for the case of an instantaneous pulse, a resistive-capacitative (RC) circuit time constant τ and a first order decay of the absorption signal. The observed peak height has to be multiplied by a correction factor f' to obtain the undistorted peak height, where:-

$$f' = (k\tau)^{-k\tau/(1-k\tau)}$$ (see Fig. 4)

k is the first order rate constant (sec^{-1}) for the radical disappearance reaction and τ is the time constant of the recording system.

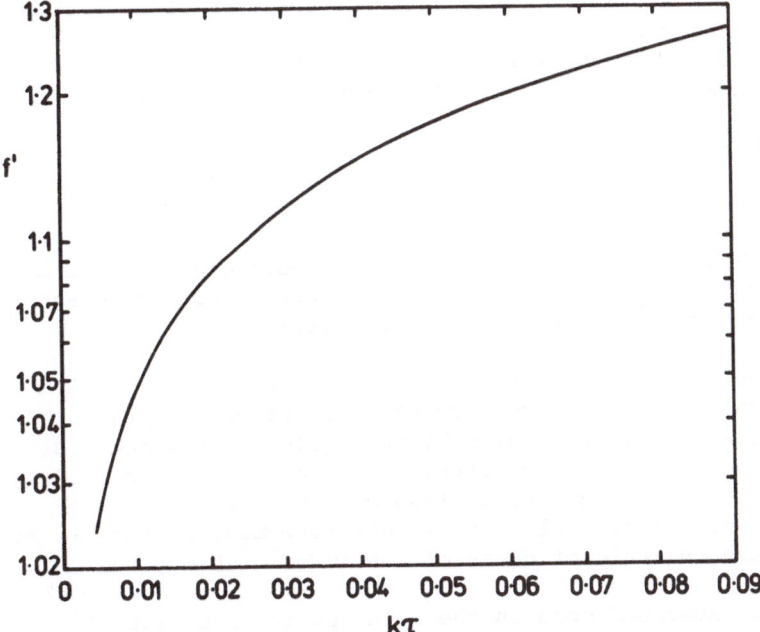

Fig. 4. Correction factor, f', for the peak radical absorption at the end of an instantaneous pulse. The electronic recording system has an RC time constant τ and the radicals disappear with a first order rate constant k.

Response time corrections are frequently important: e.g. even if τ is 1/70 of the half life of the radical decay a 5% correction for the peak height is required.

An exact solution for the correcting factor when the absorption decays by a second order process is much more complicated but a satisfactory approximation is obtained by substituting for k the reciprocal of the first half life of the second order decay curve $k = 1/t_{\frac{1}{2}}$, in the formula for f'.

When applying rise time corrections caution should be exercised if large optical absorptions are encountered, since even a relatively small correction to the oscillogram peak height given by f' represents a much larger correction to the radical concentration.

The formula for f' given above was calculated for an RC rise time but it should be noted that some electronic equipment, e.g. fast rise time digitizers, do not have a simple RC equivalent rise time.

3.3.   The Thiocyanate Dosimeter

Pulse irradiation of an aqueous solution of potassium thiocyanate produces a transient optical absorption with a peak at 480 nm.   The absorption is due to the $(CNS)_2^-$ ion formed in a two step sequence from the OH radical:-

$$OH + CNS^- \longrightarrow OH^- + CNS$$

$$CNS + CNS^- \longrightarrow (CNS)_2^-$$

The $(CNS)_2^-$ species has an extinction coefficient of 710 m^2mol^{-1} and decays by a second order process with a rate constant of 3×10^6 m^3mol^{-1} in neutral solution (11).

The solution used for dosimetry should be a 10 mol m^3 aerated solution of KCNS.   Such a solution is stable and the presence of dissolved oxygen removes hydrated electrons which would otherwise present an interfering optical absorption.   Oxygen depletion should thus be avoided but the use of a low pH to remove the hydrated electron is not recommended since it also increases the rate of decay of the $(CNS)_2^-$ absorption.

The absorbed dose in the solution is calculated from the expression

$$\text{Absorbed dose} = \frac{A}{G(CNS)_2^-.\epsilon.l.p} \quad Gy$$

where A = change in optical density of the solution as a result
        of the pulse (corrected for artefacts where necessary -
        see 3.2),
     $G(CNS)_2^-.\epsilon$ = product of $G(CNS)_2^-$ and the molar linear
        absorption coefficient, ϵ, in appropriate units, for
        the absorbing species,
     l = optical path length in the solution,
     p = density of the solution.

The recommended value (2) for the quantity $G(CNS)_2^-.\epsilon$ at a
wavelength of 500 nm (bandwidth 5 nm) is:-

$$G(CNS)_2^-.\epsilon = 2.23 \times 10^4 \ m^2 \ J^{-1}$$

At 480 nm the value is 11% higher.

     The thiocyanate radical-dosimeter is the one most commonly
used in pulse radiolysis due to its simplicity and wide dose/
pulse range.   Doses of 0.1 Gy are readily measured and corrections for radicals reacting during the pulse (section 3.2.1) are
modest for quite large pulses, e.g. A 1.6 μs pulse delivering
170 Gy requires only an 8% correction for radical loss during
the pulse.   The precision obtainable with the thiocyanate dosimeter is about ± 2 percent.

3.4.   The Hydrated Electron Dosimeter

     The hydrated electron, e_{aq}^-, is a primary radical species
produced directly on the radiolysis of water.   It has an absorption band, with a high peak extinction at 720 nm which extends
through the visible and UV spectrum.   It reacts rapidly with
many solutes and to obtain the maximum lifetime for the hydrated
electron, oxygen and other impurities must be carefully excluded
and the hydrogen ion concentration kept low.

     The e_{aq}^- dosimeter has been calibrated down to 0.05 Gy/pulse
(12) and the measured value for the product $G(e_{aq}^-).\epsilon$ at a
wavelength of 578 nm was:-

$$G(e_{aq}^-).\epsilon = 3.05 \times 10^{-4} \ m^2 \ J^{-1}$$

For pulses of less than 1 Gy the yield of e_{aq}^- can be approximately
doubled by converting the OH radicals to hydrogen atoms via a
reaction with dissolved hydrogen:-

$$OH + H_2 \longrightarrow H + H_2O$$

The poor solubility of H_2 in water sets the practical upper dose
limit for this system.   A unique feature of the H_2/OH^- hydrated
electron dosimeter is that no overall chemical change has taken

place after the decay of e_{aq}^-. The solution can therefore be
permanently sealed into an optical cell and used indefinitely
(12). The product $G(e_{aq}^-).\varepsilon$ for the H_2/OH^- system at a wavelength of 578 nm is

$$G(e_{aq}^-).\varepsilon = 7.38 \times 10^{-4} \, m^2 \, J^{-1}$$

The bimolecular decay of the hydrated electron:

$$e_{aq}^- + e_{aq}^- \xrightarrow{\quad 2H_2O \quad} H_2 + 2OH^-$$

has a rate constant of $1.2 \times 10^7 \, m^3 \, mol^{-1} \, s^{-1}$. The two versions
of the e_{aq}^- dosimeter together cover a useful range from 100 Gy/
pulse down to 0.01 Gy/pulse.

3.5. The Ferrocyanide Dosimeter

In aqueous solution the ferrocyanide ion reacts with radiation induced OH radicals to give the ferricyanide ion:-

$$Fe(CN)_6^{4-} + OH \longrightarrow Fe(CN)_6^{3-} + OH^-$$

The oxidation of ferro- to ferricyanide results in an increased
optical absorption at 420 nm where the magnitude of the molar
absorption coefficient change is $100 \, m^2 mol^{-1}$ (13). In contrast
to the other radical dosimeters described here, the immediate
optical absorption change in this system is due to the formation
of a stable transition metal radical ion. However, subsequent
reactions involving HO_2 (or O_2^-) and H_2O_2 increase the yield of
$Fe(CN)_6^{3-}$ in the system. These reactions are very much slower
than the initial OH oxidation step so that provided the
concentration of ferricyanide is measured between 10 µs and
100 µs after the pulse, this concentration is due only to the
oxidation of $Fe(CN)_6^{4-}$ by OH radicals (13). This period of
stability should obviate the need for pulse-duration and risetime corrections.

The dosimeter solution consists of 5 mol m^{-3} of $K_4Fe(CN)_6$
saturated with oxygen (1,2). The optical absorption change at
420 nm is measured 10 µs after the pulse. At this wavelength
the product $G(Fe(CN)_6^{3-}).\varepsilon = 3.3 \times 10^{-5} \, m^2J^{-1}$ in a system with a
bandwidth of 10 nm. The absorbed dose is calculated by
substituting the appropriate values in the expression given for
the thiocyanate dosimeter (section 3.3).

Since the G.ε value is lower than for the CNS^- and e_{aq}^-
dosimeters the system is not suitable for absorbed doses below
1 Gy but the upper limit should be comparable.

The magnitude of the absorption change may be approximately doubled (as with the CNS^- dosimeter also) by saturating the solution with nitrous oxide gas. This has the effect of converting the yield of hydrated electrons into OH radicals.

$$N_2O + e^-_{aq} \xrightarrow{H_2O} N_2 + OH^- + OH$$

However, the enhancement factor is not precisely 2 so that recalibration of the dosimeters is required if nitrous oxide is employed.

References

1. Fielden, E.M., and Hohn, N.W.: 1970, Chapter X in Manual on Radiation Dosimetry. Eds N.W. Hohn and R.J. Berry (Marcel Dekker, New York).

2. ICRU: 1982, International Commission on Radiation Units and Measurements,"Dosimetry of Pulsed Radiation"(report in press). ICRU Washington.

3. Fricke, H., and Morse, S.: 1929, "The actions of x rays on ferrous sulphate solutions", Philos. Mag. 7, pp. 129.

4. Fricke, H., and Hart, E.J.: 1966, "Chemical dosimetry", Chapter 12 in Radiation Dosimetry, Vol.II, Eds F.H. Attix and W.C. Roesch (Academic Press, New York).

5. Sehested, K., Bjergbakke, E., Hohn, N.W., and Fricke, H.: 1973, "The reaction mechanism of the ferrous sulphate dosimeter at high dose rates", in Dosimetry in Agriculture, Industry, Biology and Medicine (International Atomic Energy Agency, Vienna).

6. Hart, E.J.: 1963, "Chemical dosimetry at high dose rates" in International Conference on Radiation Research (U.S. Army Natick Laboratories, Natick, Massachusetts).

7. Willis, C., Miller, O.A., Rothwell, A.E. and Boyd, A.W.: 1968, "The dosimetry of very high intensity pulsed electron sources used for radiation chemistry: I. Dosimetry for liquid samples", Radiat. Res. 35, pp. 428.

8. Ghormley, J.A., Hochanadel, C.J., and Boyle, J.W.: 1969, "Yield of ozone in the pulse radiolysis of gaseous oxygen at very high dose rates. Use of this system as a dosimeter", J. Chem. Phys. 50, pp. 419.

9. Willis, C., Boyd, A.W., and Miller, O.A.: 1971, "The absolute dosimetry of high intensity, 600 kV pulsed electron accelerator used for radiation chemistry studies of gaseous samples", Radiat. Res. 46, pp. 428.

10. Michael, B.D.: 1968, PhD Thesis, London University.

11. Adams, G.E., Boag, J.W., and Michael, B.D.: 1965, "Reactions of the hydroxyl radical, Part I", Trans. Faraday Soc. 61, pp. 1674.

12. Fielden, E.M., and Hart, E.J.: 1968, "Hydrated electron
     and thermoluminescent dosimetry of pulsed x-ray beams",
     Advances in Chemistry Vol.81, part I, American Chemical
     Society.
13. Adams, G.E., Boag, J.W., and Michael, B.D.: 1965, "Spectro-
     scopic studies of reactions of the OH radical in aqueous
     solutions", Trans. Faraday Soc. 61, pp. 492.

OPTICAL MONITORING TECHNIQUES

G. Roffi

C.N.R. F.R.A.E. Bologna Italy

The investigation of the properties of transient chemical species
produced by means of exciting sources such as laser or electron-
beam pulses, can be carried out by a variety of methods. Kinetic
spectrophotometry is one very powerful method and is probably the
most commonly used in this field. It is used to make quantitative
measurements of optical density as a function of time and wavelength and enables absorption or emission spectra of short lived
species e.g. radicals, ions, excited states to be obtained.
The technique is very simple in principle but requires a carefully assembled set-up in order to provide accurate measurements.
Here we briefly discuss the most important aspects of the factors
required to operate such an experimental set-up.

1. GENERAL CONSIDERATIONS

The light absorption characteristics of a substance are expressed
in terms of optical density and spectrophotometers are used to
measure optical density as a function of wavelength. Optical density is defined as:

$$O.D. = \log_{10} \frac{P_o}{P_t} = \log_{10} \frac{1}{T}$$

where:
P_o = incident light
P_t = transmitted light
T = fractional transmission

63

J. H. Baxendale and F. Busi (eds.),
The Study of Fast Processes and Transient Species by Electron Pulse Radiolysis, 63–89.
Copyright © 1982 by D. Reidel Publishing Company.

A typical spectrophotometric set-up for observing the optical density of a transient species produced by an electron pulse from an
accelerator may be schematically represented as in fig. 1.

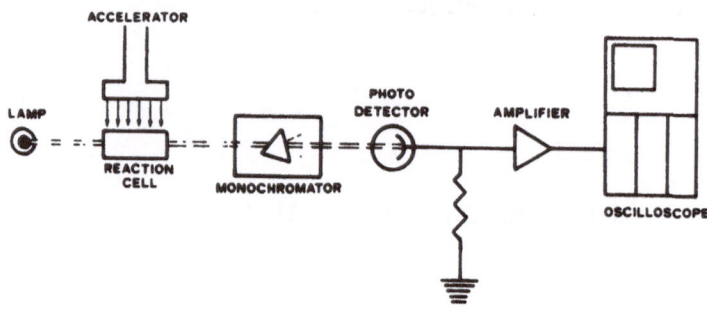

fig. 1. Schematic arrangement for kinetic spectrophotometry.

In this system, change in light intensity due to the contents of
the cell during and after an excitation pulse, are monitored by
the photodetector and recorded as an oscilloscope trace.
Typical shape of oscilloscope traces are shown in fig. 2.

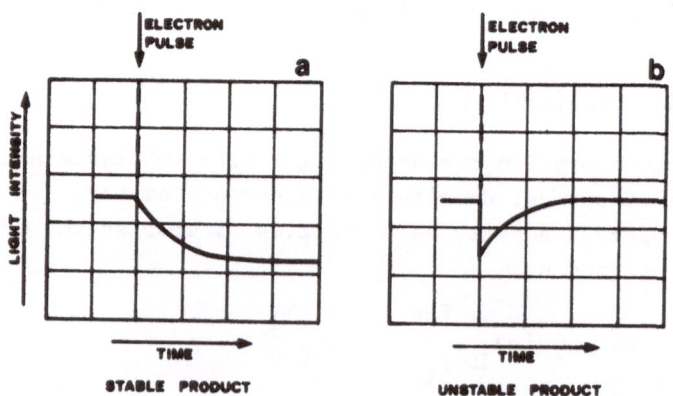

fig. 2. a) formation of an absorbing species after the
pulse
b) formation of an absorbing species within the
time of the pulse and its subsequent decay.

The system works like a conventional spectrophotometer without
the need of a reference cell. The cell acts as reference cell be
fore the excitation pulse giving P_o, and as sample cell after the
excitation pulse giving P_t as a function of time. This arrangement
with appropriate electronics can give a system of high sensitivity.
Variations of the order of a fraction of 1% may be accurately mea
sured.

The smallest signal that can be detected and analyzed is essential
ly limited by the fluctuations of the analysing light generally
referred to as noise (N).

Fig. 3 shows noisy traces, with and without superimposed signal.

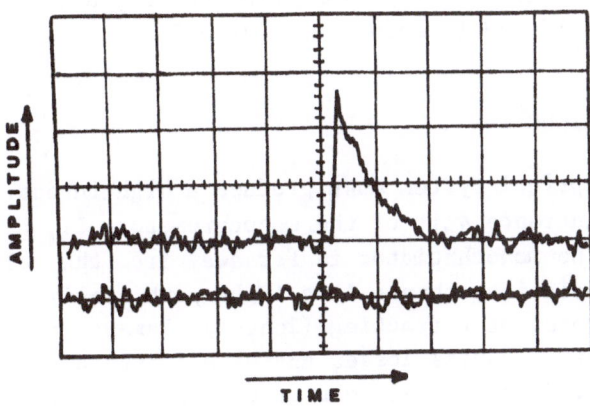

fig. 3. Noisy traces.

The parameter describing the efficiency of the system is the signal
to noise ratio S/N.

At very short times e.g. nanoseconds the noise is essentially due
to statistical fluctuation in the number of photoelectrons produced
in the detector by the relatively small number of photons, which
are incident during the short time.

The fluctuations are inversely proportional to the square root of
the number of photoelectrons produced by the light and hence:

$$S/N \propto i^{\frac{1}{2}}$$

where i is the photocurrent.

For a given situation, S/N may therefore be maximized by increasing
the light intensity passing through the cell and by using high

quantum efficiency photodetectors. It will be seen however that,
in order to get an increase of 10 in the value of S/N, the light
intensity must increase by a factor of 100.
As will be seen in section 2.3, high light intensity can be produced in some lamps by overrunning for a short time.
To measure such very fast signals the response of the monitoring
system must also be very fast but when response time is not a critical parameter e.g. for changes over µsec, suppression of the
noise may be brought about by "smoothing" the signal electronically which reduces the response to rapid fluctuations. This is usually an R.C. (resistor-capacitor) electronic circuit which has a response time τ. As a general rule τ must be kept 0.1% of the total duration of the oscilloscope sweep to ensure that the signal
is not distorted by the R.C. circuit
As we will see later, at very long time scales, other factors
have to be considered to ensure good signals.

2. OPTICAL SYSTEMS

An efficient optical system must produce a high intensity cell
image at the entrance slit of the monochromator. If, as is usually the case, the monochromator is far away from the cell to avoid
exposure to radiation, then a long light path requiring many reflecting (mirrors) or refracting (lenses) elements is required.
Any lens or mirror causes losses and with lenses aberration causes
further losses. When many mirrors have to be used, the overall
amount of transmitted light may be drastically reduced as shown
in fig. 4. In this figure R represents the reflectivity of the
mirror surface and the transmission in a logarithmic scale is
given as a function of the number of mirrors. The use of lenses
has the disadvantage of requiring continuous focussing changes
for each wavelength to get maximum light. Transmission losses in
mirrors and lenses may be partially reduced by using appropriate
coatings and problems arising from lens aberration may be minimized by choosing "best form" composite lens systems.
The irradiation cell may be considered as a part of the optical
system, and the optical path may be either at right angles to or
in the same direction as the electron beam. High purity fused
silica is an almost ideal material for sample cells.

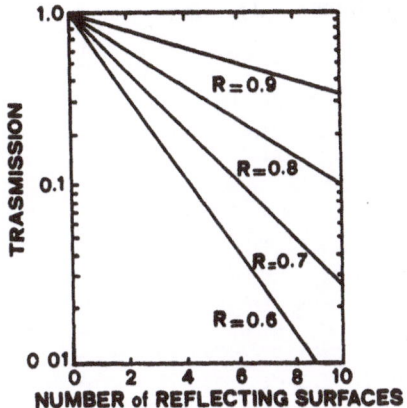

fig. 4. Transmission of an optical system as a function
        of the number of mirrors and their reflectivity
        R. (1)

.1. Lenses

s a background aid in the design of an optical system it is con-
enient to remember some imaging properties of the simple lens.

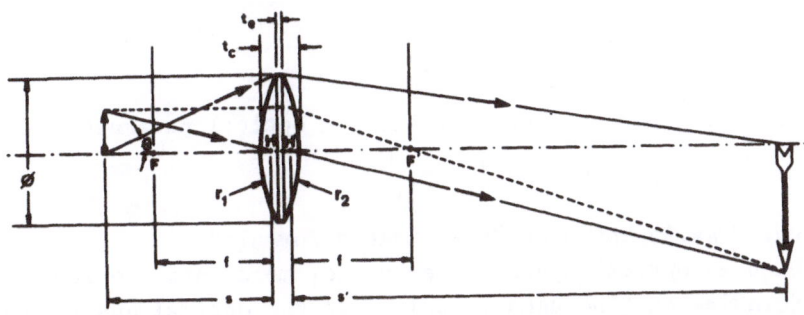

r_1 = radius of curvature r_2 = radius of curvature
       of 1st surface   of 2nd surface
s = object distance   s' = image distance
t_c = center thickness t_e = edge thickness
f = focal length $\emptyset$ = lens diameter

θ = arctan $\dfrac{\emptyset}{2s}$

fig. 5. Imaging properties and parameters of a simple
        lens.

A complete discussion of the optical properties and formulas of the lenses is given in optics books (2). Here we present the most important formulas:

Conjugate law:

$$\frac{1}{f} = \frac{1}{s} + \frac{1}{s'} \qquad (2.1)$$

Magnification:

$$m = \frac{s'}{s} \qquad (2.2)$$

Object distance:

$$s = \frac{s'}{m} = f(\frac{1}{m} + 1) = \frac{s'f}{s'-f} \qquad (2.3)$$

Image distance:

$$s' = s \cdot m = f(m + 1) = \frac{sf}{s-f} \qquad (2.4)$$

Sum of conjugate distances:

$$s + s' = f(m + \frac{1}{m} + 2) \qquad (2.5)$$

Another important relation between optical parameters is:

$$\frac{1}{f} = (n - 1) \cdot (\frac{1}{r_1} - \frac{1}{r_2}) \qquad (2.6)$$

universally known as "Lens makers formula".

When an optical system makes use of more than one lens, combination formulas must be used to calculate the optical quantity of interest. Combination conjugate distances are expressed by the same simple lens formula 2.1 but in this case f will be:

$$f = \frac{f_1 \cdot f_2}{f_1 + f_2 - d} \qquad (2.7)$$

where:

d = distance of the lens
f_1 = focal length of lens 1
f_2 = focal length of lens 2

Finally we define f-number or relative aperture:

$$\text{f-number} = \frac{f}{\emptyset} \tag{2.8}$$

as the ratio between focal length and diameter of the lens, and numerical aperture N.A. :

$$\text{N.A.} = n \sin \theta \tag{2.9}$$

where n is the refractive index of the object space and θ is arctg $\emptyset/2s$. For small angles i.e. s ——— ∞ :

$$\text{f-number} = \frac{1}{2N.A} \tag{2.10}$$

The last defined parameters play an important role in the design of an optical system because they apply to all parts of the optical equipment i.e. mirrors, lenses, monochromator and detector should have the same aperture.

Lens shape and their best use
A large choice of lens shape is commercially available. Generally the correct choice of lens shape minimize the aberration always present in any optical system. It is possible to define the best form of a lens according to the conjugate ratio $\frac{s}{s'}$.
Convex lenses.
The plano-convex lens is the best form for parallel light entering or leaving the lens. The biconvex lens is the best form for conjugate ratio approaching unity i.e. s = s' .
Concave lenses.
Plano-concave and biconcave lenses are divergent lenses, they produce only virtual images and are generally used to extend focal length in existing systems.
Meniscus lenses.
Meniscus lenses are especially convenient in rapid assembly of a prototype optical system. Combination focal length may be continuously adjusted in a stepwise manner by substituting different positive or negative meniscus.
Cylindrical lenses.
These are used when magnification along one axis only is to be achieved, e.g. changing a point to a line or changing the width but not the height of an image. The standard shape is plano-convex. This lens is often used to adjust a light beam profile exactly to the rectangular slit of a monochromator.

Lens materials
The choice of materials for the lenses in an optical system depends
on the application. If the interest is restricted to the visible
and the near infrared, soft or hard silicate glass is adequate.
However, all silicate glasses are permanently radiation darkened
and they cannot be used where they receive electron beam or x-ray
irradiation. They also give absorption transients.
Fused silica is the best material for U.V. transmission, but for
use with high energy radiation it must be the purest quality since
impurities can lead to transient absorbing species in the silica.
For measurements in the far infrared, sapphire is a commonly used
material but here again it must be checked for transient absorbing
species in the condition of the experiment.

Lens coating
A lens loses light due to reflection on the surface and this can
amount to 4 or 5%.
The reflected light may be reduced by the use of a surface coating
of a material of intermediate refractive index between glass and
air. Magnesium floride, Mg F_2, is the commonly used coating material.

2.2. Mirrors

Mirrors may substitute lenses in many cases.
Flat and concave mirrors are generally used. For mirrors, coating
is more important than for lenses. Two coatings are usually applied
a reflection coating and a protection coating. Transmission and
reflectance can be tightly controlled through control of film
thickness of the coating. The percent of reflectance of a mirror
at a given wavelength is strongly dependent on the material used
as a reflection coating. The latter is frequently subject to de-
terioration by corrosive materials in the atmosphere and to maintai:
good reflectance a protection coating is frequently applied.

Materials for coating
Metals commonly used as reflecting or partially reflecting coatings
include aluminum, ultraviolet enhanced aluminum, silver, gold,
protected gold, rhodium. These materials are applied by vacuum de
position usually on a glass former.
The characteristic properties of these coatings are as follows:

Aluminum:
Aluminum offers good reflectance from the U.V.to the near I.R.
Oxidation of the coating readily occurs and a loss in the U.V. region as well as slight scattering throughout the spectrum occurs
with time.

Ultraviolet-enhanced-aluminum:
This is obtained by an additional U.V. transparent dielectric
coating, Mg F_2, on aluminum.

Silver:
Silver has higher reflectance than aluminum at least when new but
because the advantage is temporary it is rarely used in external
reflection. An overcoating of inconel or copper prevents fast degradation.

Gold:
Gold is very well suitable for the I.R. and its use is preferred in
application where thermal radiation wavelengths are involved. There
is poor reflectance in the U.V. and visible.

Rhodium:
Rhodium is by far the most resistant coating.It is generally elec-
troplated rather than deposited and can withstand cleaning without degradation. However it shows a reflectance lower than other
materials over the whole spectrum.

Any reflecting coating may be protected by means of additional
layers of protecting material usually silicon monoxide or MgF_2,
magnesium floride. Reduction in reflection is invariably produced.

2.3. Light sources

High intensity and high stability are primary requirements for
any analyzing light source. Wide spectral emission is desirable and
for use in specific spectral bands or lines, special choices are
available.
Commercially available high pressure Xenon lamps emitting in the
range I.R. to U.V., and with power input 75-1000 watts are well
suited for short time scales experiments. When long time scales
have to be used in the visible spectrum, tungsten lamps are chosen
because of the high long term stability. In this case the lower

intensity may be acceptable because the unfavorable signal to
noise ratio can be offset by electronic integration of the photodetector signal.
Mercury lamps, deuterium lamps and most low pressure gas-filled
lamps are very stable low intensity sources and give a choice of
spectral emission from visible to U.V. Low power gas-lasers or
dye-lasers may be also be used as analysing sources when particular wavelength are required.
The high pressure Xenon lamp is generally used because of its
wide spectrum (quasi-solar), very good U.V. emission as can be
seen in fig. 6, and adaptability to pulsing.

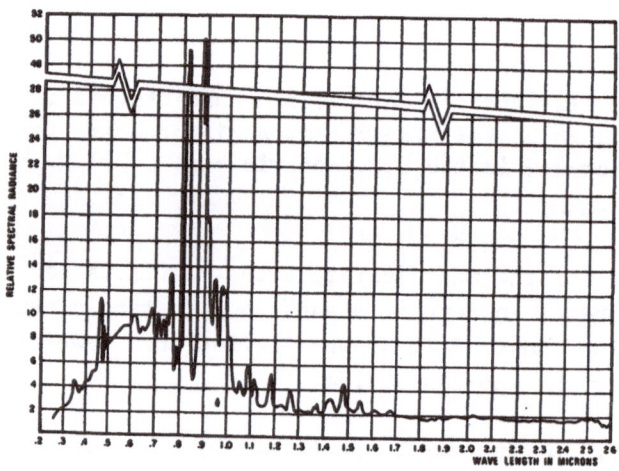

fig. 6. Xenon emission spectrum (3)

Recently developed Xenon lamps with sapphire windows make them
useful for far infrared investigation. The possibility of producing short very intense light flashes by means of suitable electronic circuitry increases its value for the observation of very
fast small signals. Fig. 7 shows a typical increase in emission
produced in a Xenon lamp by a short high intensity current pulse.
When used in pulsed mode the Xenon lamp shows poor long term stability of light due to physical displacement of the arc inside
the bulb. This sets an upper limit to the time scale which can be
used with the pulsed lamp (usually up to ca. 5 µsec). A schematic

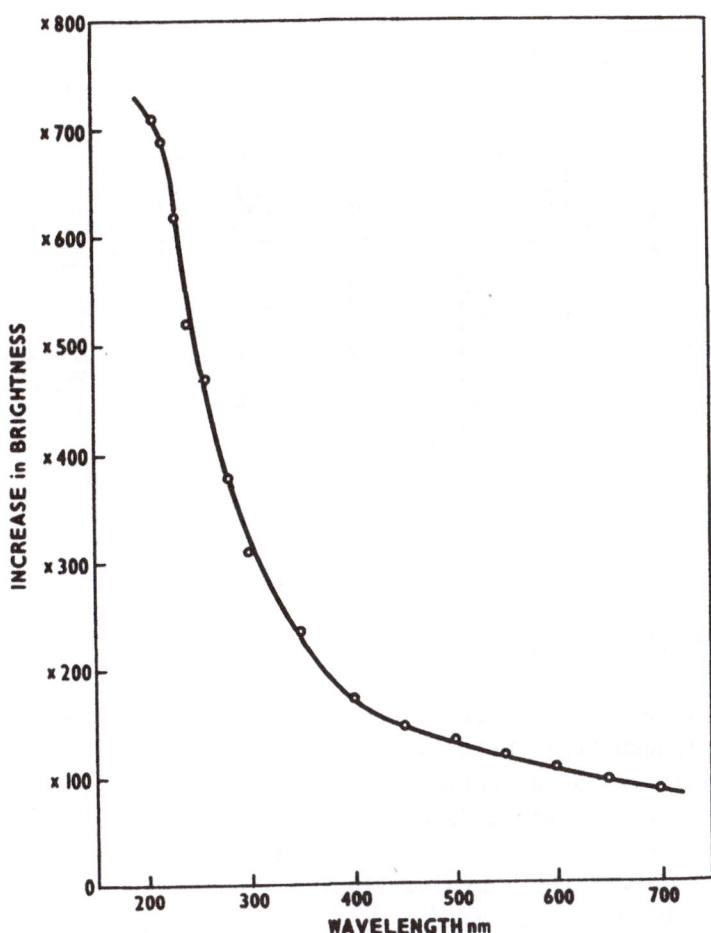

fig. 7. Increase in Xenon lamp intensity produced
by a 500 Amp. pulse. (1)

diagram of a circuit producing a 500 Amp. pulse for a 450 Xenon
lamp is shown in fig. 8.
The lamp, normally running at 20 Amp, is forced to take 500 Amp
from the capacitor bank C_1 by the switching-on of the SCR_1. A pulse
length of the order of a few msec is achieved by means of SCR_2
which when triggered, inverts the polarity of SCR_1 through capacitor C_2 and switches it off.
The fall in intensity of the light output during the flat part of
the pulse may be in the order of 1% per millisecond providing the
bank capacitor has acapacity of ca. 0.4 Farad.

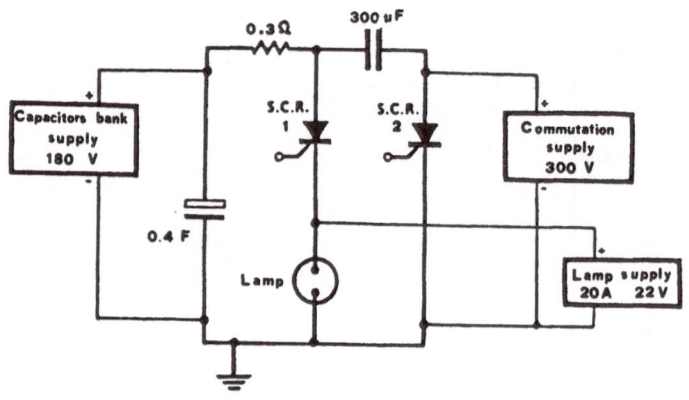

Fig. 8. Pulsing circuit for a Xenon lamp.

2.4. Monochromators

Modern monochromators use ruled gratings to select a narrow band
of wavelength from white light.
Light incident on a grating surface is diffracted from the grating
grooves because each groove becomes a source of reflected or transmitted light. For a given groove spacing there is a unique series
of angles where the emerging light is in phase for one wavelength
This is shown in fig. 9.
The process is governed by the equation:

$$m\lambda = a \, (\sin \alpha + \sin \beta) \qquad (2.11)$$

where:

m = diffracting order
λ = diffracted wavelength
α = angle of incident light
β = angle of diffraction
a = groove spacing
For   m = 0 the grating act as a mirror and for
$\alpha = \beta$, 2.11 becomes:

$$m\lambda = 2 \, a \, \sin \beta \qquad (2.12)$$

Thus a given incident angle determines the wavelength of the diffracted light.

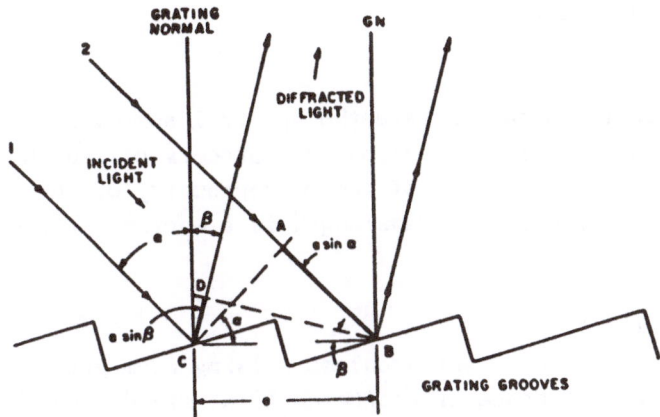

fig. 9. Phase relation between rays diffracted from
adjacent grating grooves. (5).

Dispersion
Angular dispersion in a grating may be defined as the ratio betwe-
en the angular separation dβ and the corresponding dλ.
A characteristic parameter generally associated with a monochromator is the linear dispersion defined as the product of angular
dispersion by the monochromator focal length.
Linear dispersion D, is expressed in nm/mm.

Resolution
The resolution of a monochromator or generally of a grating is a
measure of its capability to separate small differences in wavelength and is defined as:

$$R = m.N \qquad (2.13)$$

where:

m = order
N = number of grooves

substituting m we get:

$$R = 2.a.N \frac{\sin\beta}{\lambda} \qquad (2.14)$$

where   a.N = W is the width of the grating.

Equation 2.14 shows that R is independent of the order m_o and the groove number. Monochromators makers give resolution in Ångstroms Å, or in nanometers.

Blazing
It is possible to concentrate the spectral energy into one angular and therefore wavelength region by a process called "blazing". It is obtained by adjustment of the microgeometry of the grooves. The choice of the blazed wavelength is a matter for the user of an optical system.

Scattered light
One of the most important problems arising from grating monochromators is the presence of scattered light mixed with that of the chosen wavelength essentially because of mechanical imperfections. It is responsible for a loss of efficiency and for impurity of the transmitted light. It can lead to large errors in the determinatio of optical density since the intensity of the incident and transmi ted light levels at a particular wavelength are less than those measured. The measured values contain a contribution from the scat tered light of other wavelengths. It is a particular problem in the U.V. region using photodetectors whose sensitivity is much greater at the longer scattered wavelengths.
The effect can be minimized using a double monochromator or narrow band filters.

Operation of monochromators
The diagram of a monochromator in fig. 10 shows an image of the

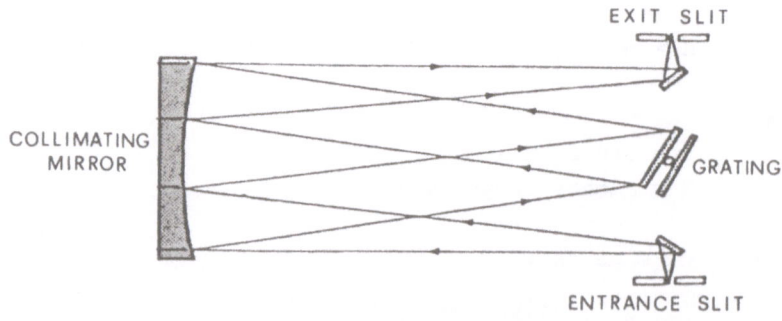

fig. 10. Monochromator diagram.

entrance slit produced at the exit slit.
The light rays drawn correspond to one wavelength only.
The grating gives an angular dispersion of the light into different wavelengths at the exit slit.
By rotating the grating we can select a given output wavelength
at the exit. This will in fact consist of a band of wavelengths
and the dispersion, expressed in nm per mm, and the width of the
slits determine the purity of the output or, as it is usually called the bandwidth of the output light:

$$BW = D \cdot d \qquad\qquad\qquad\qquad (2.15)$$

where:

   BW = bandwidth (nm)
   D = linear dispersion (nm/mm)
   d = exit slit width

Reduction of the slit width improves the resolution but at the same time it reduces the amount of monitoring light passing through
the monochromator to the detector and hence the signal to noise
ratio.
Generally it is necessary to make a compromise between these factors according to the specific aim of the experiment in hand.
To summarize the choice of a monochromator should match the optical parameters of the optical system, in particular the f-number.
Gratings and their "blazing" should be carefully selected to suit
the spectral region of interest.
Suitable filters must be chosen to minimize the effects of stray
light and the presence of harmonics from the grating e.g. a setting of 500 nm on a monochromator will allow some 250 nm light to
pass the detector as well. Some types of filters now available
are: neutral density filters which simply reduce the light inten-

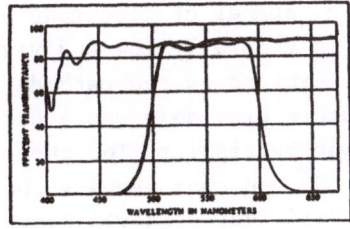

fig. 11. Band pass filter (2)

sity and prevent overload of the detectors at any wavelength, long
wave pass L.W.P, short wave pass S.W.P and interference filters.
A combination of L.W.P and S.W.P is generally used in order to get
a band-pass or a band-rejection filter as shown in fig. 11.

3 MONITORING TECHNIQUES

The measurement of light intensities is a fundamental part of kinetic spectrophotometry and to be effective, these measurements
must be made on intensities which are rapidly changing as chemical reaction proceeds.
With present equipment, changes occurring in the nanosecond and
picosecond time ranges may be detected.

3.1. Photo detectors

In any kind of spectrophotometry, photodetectors are used to produce an electrical signal, usually a current, which is proportional to the light intensity. Two devices are commonly used viz,
photomultipliers and photodiodes.

3.2. Photomultipliers

A photomultiplier consists essentially of a photocathode or primary emitting electode, a series of plates called dynodes which
on impact by electrons emit more electrons and multiply these up
considerably, and an anode or collecting electrode. Thus light
produces primary cathode photoemission of electrons which is followed by secondary emission from the dynodes as the electrons are
bounced from plate to plate as shown in fig. 12.
Amplification of the initial photoelectic signal is thus accompli·
shed internally and can be 10^6 times or more to suit the application. The quantum efficiency i.e. number of electrons produced
per photon, of a photocathode, is generally low and varies with
wavelength of light in a way determined by the cathode material.
Thus photomultiplier manufacturers provide a characteristic spectral response curve for each photomultiplier. The photocathode types are described as S_1, S_2 ... S_n and each shows a peak at different wavelength as well as a different spectral bandwidth.
Fig. 13 shows a typical spectral response of different cathodes
as a curve of sensitivity against wavelength.

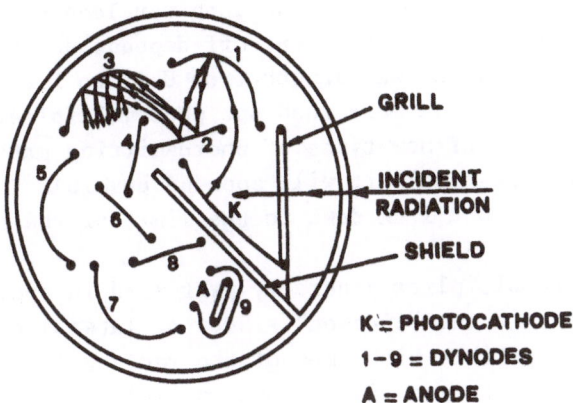

fig. 12. Typical structure of a photomultiplier (6).

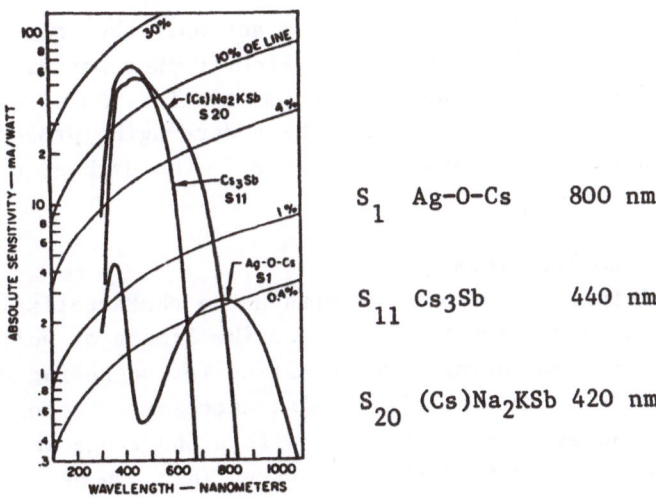

S_1 Ag-O-Cs   800 nm

S_{11} Cs3Sb   440 nm

S_{20} (Cs)Na$_2$KSb 420 nm

fig. 13. Typical spectral response of some photocathodes,
their composition and wavelength of maximum
response (6).

The spectral response of a photomultiplier is also determined by
the window material. essentially the long wavelength cut-off depend
on the photocathode and the U.V. cut-off depends on the window.
In order to cover all the wavelength from U.V. to I.R., it is
usually neccessary to use more than one photomultiplier. However,
with the development of new types of photoemitting materials such
as GaAs (Gallium arsenide), it will soon be possible to cover all
the spectrum from about 1 μm down to the limit of the photomultiplier window.
The type of photomultiplier generally preferred in spectrophotometric applications is the "Side-on" window or lateral cathode. Its
small size is compatible with that of the rectangular monochromator slit. "Head-on" or end window photomultipliers, the other common type, have photocathode deposited on the flat end of the tube.
They are generally used more in photon counting applications although they are also used in spectrophotometry with steady light
at relatively low level. With high intensity pulsed light sources
they readily saturate due to the thin cathode layer being unable
to provide sufficient electrons. The "Side-on" window solid photocathodes are very superior in this respect. Dark current in photomultipliers arises from the existence of stray electrons liberated
by heat or cosmic rays. It is important only when measuring very
low light levels which is rare in kinetic photometry. Dark current
may be reduced by cooling the photomultiplier. Particularly noisy
photomultipliers may be improved by a dark-aging process accomplished by operating the photomultiplier in the dark at the maximum
rating voltage.

Use of photomultipliers
A photomultiplier must be used within the characteristics laid
down by the manufacturers which describe limits of voltage, current etc. Maximum ratings are specified for operating voltage and
average anode current but it is also neccessary to control the cathode current at a level which is within the range where the output
current is linear with light intensity. This can be established
by experiments over a range of light intensities.
The usual source of voltage for the dynode chain is a resistor
voltage circuit as shown in fig. 14.
The current flowing into the resistors should be at least 10 times
greater than the average anode current. This prevents lack of sensitivity and loss of linearity due to variation in the voltage of
the later stages.

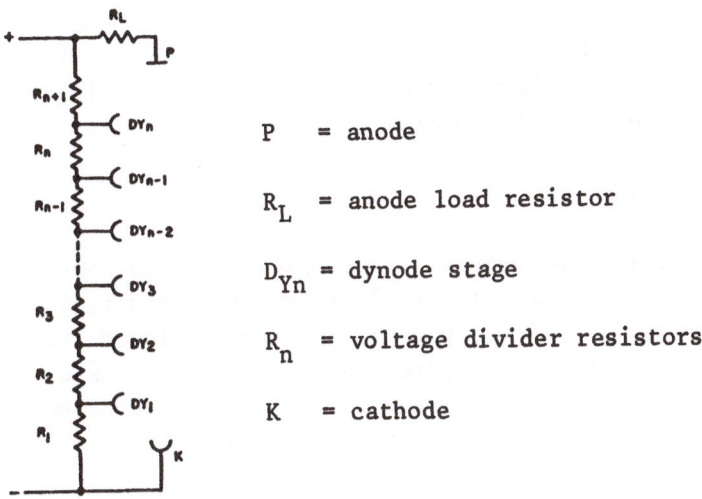

P   = anode

R_L = anode load resistor

D_{Yn} = dynode stage

R_n = voltage divider resistors

K   = cathode

fig. 14. Voltage dynode chaine (6).

The overall gain of a photomultiplier is controlled by the overall
voltage applied. Generally a zener diode in place of R_L provides
a very constant voltage for the 1st stage which is a critical amplification stage.
In pulsed operation the important parameter is the anode current.
The voltage supply network must be designed to provide the maximum
to be used, which can be as much as 10 mA while keeping dynode
potential constant for the short pulse time. This is achieved by
means of storage capacitors across the dynode chain mounted closed
to the photomultiplier socket and of a value calculated to limit
the voltage change on the dynode to about 1%. For example for the
last dynode to anode capacitor:

$$C = 100 \frac{q}{V} \qquad\qquad (3.1)$$

where:

C = value of capacitor (Farad)
q = total charge of the anode pulse (Coulombs)
V = anode to last dynode voltage (Volts)

For very fast kinetics an important parameter of the photomultiplier is the frequency response which is determined by the time
taken by the anode to respond to a fast signal and is generally

referred as the rise time.
Rise time is dependent on the physical configuration of the multi
plier electrodes and is also a function of the overall voltage ap
plied, decreasing with increased voltage. Some configurations usir
fewer dynodes at higher than the recommended voltages can give ri
se times of less than 1 nsec.

3.3. Photodiodes

Semiconductor photodiodes are complementary to photomultipliers i
that their peak sensitivity is in the infrared region. Compared
with photomultipliers, photodiodes require low voltage supplies,
have higher quantum efficiencies, 80% in some cases, lower noise
and the ability to withstand very high peak currents. The main di
sadvantage is the absence of multiplication so that external amplification is neccessary.
Photodiodes commercially available cover all the visible spectrum
and the I.R. or from near to very far I.R.
For example:

        Silicon photodiodes   200 - 1150 nm,   30% of Q.E.
        Germanium photodiodes   500 - 1800 nm,
        Indium antimonide   1 -   5 μm
Typical spectral responses are shown in fig. 15.

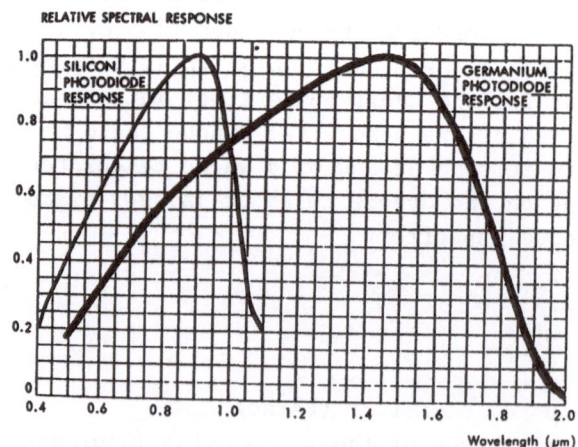

fig. 15. Spectral response of Si and Ge photodiodes (7)

For specific applications in the far infrared region special devices such as Hg Cd Te photodiodes have been developed. All these

semiconductor devices may be used in the photovoltaic and photoconductive modes, i.e. with or without external biassing. In the
spectrophotometric application the photoconductive mode is more
frequently used. Fig. 16 shows the circuitry for photovoltaic and
photoconductive operation.

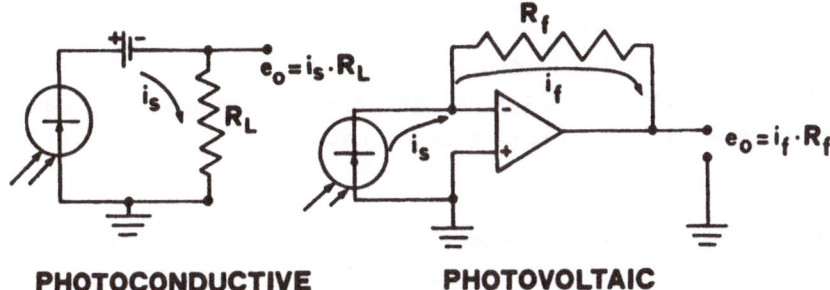

PHOTOCONDUCTIVE   PHOTOVOLTAIC

fig. 16. Photodiode application (7).

In both photovoltaic and photoconductive mode the current flowing
to the reversed polarized junction is composed of a photocurrent
and a dark current (reverse leakage current).
The dark current is constant independent of the external bias for
a given temperature while the photocurrent varies linearly with
the intensity of the incident light.
The time response of a photodiode has a quite unpredictable behaviour and in some case, under certain biassing condition, varies
with wavelength. Two distinct factors determine the time response
of a photodiode, viz, the charge collecting time and the R.C. time
constant of the circuit. The first term depends on the biassing
voltage, the second on the circuit time constant which takes into
account the diode internal series resistance and any stray capacitance. (9)
From an operational point of view the fastest response is always
achieved by a bias voltage near the inverse breakdown voltage.
This voltage is device-dependent and for example is ca. 10 Volt
for the Ge and usually ca. 100 Volt for the Si diodes.
The characteristic performance of a photodiode is usually described in terms of the following parameters:

Noise equivalent power (N.E.P.)
N.E.P. is the minimum power that may be detected. It is expressed
by:

$$N.E.P = \frac{E \cdot A \cdot N}{S \cdot (f)^{\frac{1}{2}}} \quad Watts/Hz^{\frac{1}{2}} \tag{3.2}$$

where:

E = radiation energy incident on the detector surface $(Watts/cm^2)$
A = sensitive surface area (cm^2)
S = signal voltage (Volts)
N = noise voltage (Volts)
f = noise bandwidth (Hz)

When the noise bandwidth is 1 Hz equ. 3.2 gives the power required to produce a signal to noise ratio of unity.

Detectivity

Since N.E.P depends on the sensitive area of the detector, a second parameter is introduced called detectivity, expressed by:

$$D = \frac{A^{\frac{1}{2}}}{N.E.P} \quad (Watts^{-1} \cdot cm) \tag{3.3}$$

For many detectors the data sheet defines D as "factor of merit".

Responsivity

Another important parameter is the responsivity R expressed in V/W or A/W. Responsivity is the measure of the voltage or current generated for a given incident light intensity. It is generally defined for the wavelength of maximum response. At other wavelength responsivity can be calculated using of the spectral response curve

Dark current

For a given temperature the dark current is specified in µA. It usually doubles for every 10°C increase. Because it contributes to noise N it should be kept as low as possible but noise is seldom a problem in kinetic spectrophotometry except in special photodiodes when liquid nitrogen cooling is used.

3.4. Detector circuits

The simplest circuits are shown in fig. 17 for photomultipliers and in fig. 16 for photodiodes.

Both circuits show the output of the device going to earth connected through a load resistor R or R_L, the value of which may range from 50 Ω to many K Ω. Changes in the light intensity caused by a transient absorbing species results in a change in the photocurrent and hence in the voltage across R. This voltage signal is fed into a voltage monitoring device, usually an oscilloscope. In

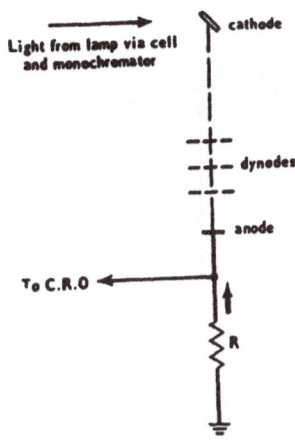

Light from lamp via cell
and monochromator

cathode

dynodes

anode

To C.R.O

R

fig. 17. Basic photomultiplier circuit (1).

absorption measurement the variation of the transmitted light due
to absorption by a transient may be of the order of a fraction of
1% of the total steady monitoring light. This means that the signal
to be observed is superimposed onto a continuous light level which
is greater by more than a hundred times, e.g. a few mV change in
a 1 V level . It is impossible to analyse such a small change con‾
veniently using normal oscilloscope operation.
A method of overcoming this problem is to provide a steady current
equal and opposite to the steady anode current and thus leave only
the small difference from the change in light intensity due to
absorption which can now be examined at high amplification. This
arrangement is described as a back-off circuit and is shown in
fig. 18.
In this circuit a current generated by an external source (e.g.
a battery) is injected into the anode via a high value resistor.
This current is adjusted to be equal and opposite to that generated
by the initial light intensity I_o thus giving zero voltage across
the resistor R. The value of this back-off current may be read from
a meter in the circuit and is a measure of the initial light intensity I_o.
This technique requires careful adjustment when using very high
sensitivity i.e., for small signals and the position of the trace
on the oscilloscope screen may wander due to small changes in the

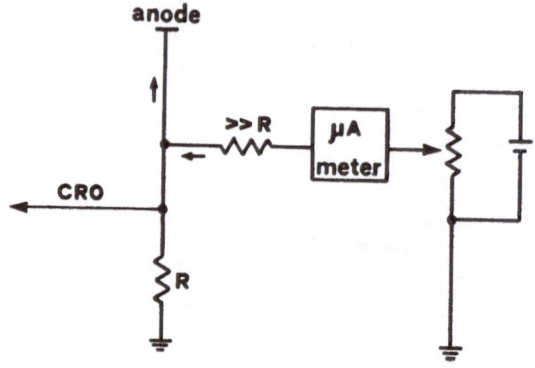

fig. 18. Back-off circuit (1).

monitoring light intensity.

When a pulsed light source is used, manual back-off is very inconvenient due to light variation and an automatic back-off circuit is used. This in principle is an electronic circuit that provides a feed-back which generates the exact amount of the current required to back-off the initial anode current. Fig. 19 shows such a circuit:

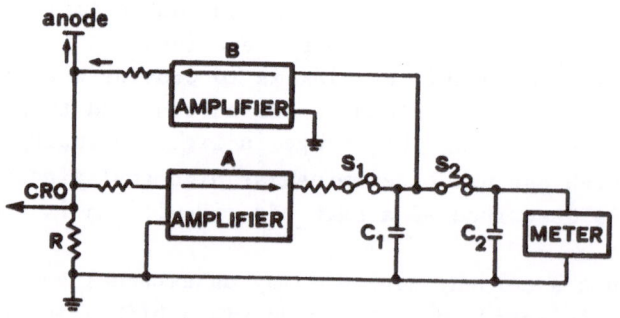

fig. 19. Automatic back-off circuit (1).

The photodetector current corresponding to the initial light intensity I_o, produces a signal at the input of the amplifier A which charges capacitor C_1 if S_1 is closed. This will drive amplifier B which generates a feed-back current of the right size and

polarity to reduce the input signal to zero. The operation is performed in a time of the order of a few μseconds or less depending
on the time constant of the circuit. The potential of the conden-
ser C_1 and of C_2 in parallel with C_1 is a measure of the anode
current. A meter connected across C_2, separated by a second switch
S_2 may be directly calibrated in anode current units.
The circuit shown is a schematic arrangement. A real circuit uses
sophisticated high impedence operational amplifiers and solid state ultra fast switches.
A more detailed explanation of the circuit operation may be helpful. When switch 1 and 2 are closed the circuit provides the feed-
back loop that keeps the anode potential at zero. Any variation occurring at the input is automatically removed in a few μsec. The
meter reads the steady anode current I_o. If S_1 is opened just before the input signal varies due to light absorption, the loop
becomes inactive, but amplifier B still provides the previous feed-
back current which is determined by the potential of capacitor C_1
and zero anode potential will be maintained. Any subsequent variation due to the absorption signal will be displayed on the oscilloscope. In this condition the oscilloscope may be used at maximum
sensitivity without any offset problem due to the steady light.
The time of opening and closing S_1 is controlled by the sweep gate of the oscilloscope itself or by an independent signal gate syn-
cronized with the event of interest. The switch S_2 provides the
possibility of maintaining the reading on the meter as long as we
need to note its value. In the case of pulsed lamp operation, S_1
and S_2 will be opened on the flat part of the light pulse just before the transient generated by an accelerator pulse. S_1 may be
closed again after the signal has disappeared while S_2 stays open
until the meter is read.
Fig. 20 shows the response of an automatic back-off circuit for
an input signal of 5 Volt produced by a pulsed lamp.
Fig. A is a profile of the pulsed lamp input on a sensitivity of
1 V.cm^{-1} while B shows the effect of the back-off on the much
higher sensitivity of 0.05 V.cm^{-1}. Any signal will appear in the
noisier part of the trace.
Finally in fig. 21 there is a typical transient signal produced
by an electron pulse absorbed in aerated water which is an absorption of light by the hydrated electron at 600 nm. The signal has
been produced using an optical system designed as in section 2
using a photomultiplier and a pulsed Xenon lamp.
Such signal may be displayed on an oscilloscope and recorded by
photographing. The initial linear portion is the level of the

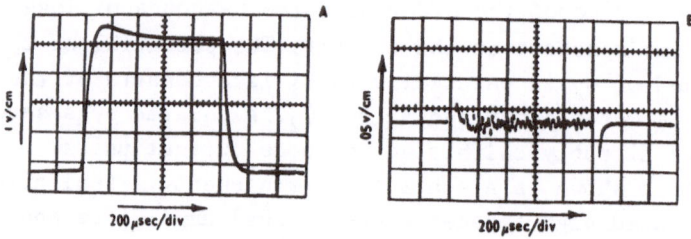

fig. 20. Response of an automatic back-off circuit to
         a pulse: A servo loop open, B servo loop clo-
         sed.

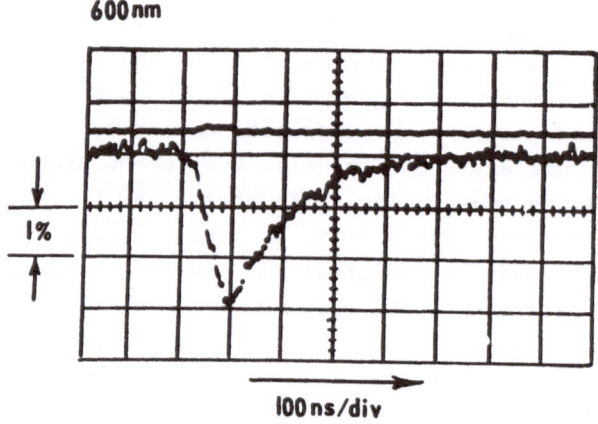

fig. 21. Absorption of hydrated electron in aerated
         water. Upper trace, signal with light off
         lower trace, signal with light on (1).

backed-off monitoring light intensity I_0 which has been displayed
on a meter. The following part of the trace records the decrease
in light intensity over the duration of the electron pulse when
the absorbing species is forming and at the end of the pulse we
see the increase in light intensity as the species decays. At
any time the difference from the initial baseline level is a mea-

sure of I_{abs}, so that from I_o the value of I_{trans} can be calculated and hence the optical density obtained at various times. Such data are obtained better from enlarged photographs or by direct digiti sation and computer analysis as will be described later.

Acknowledgements

I am greatly indebted for the help and advice of prof. J.H. Baxen dale in the preparation of these notes.

References

1) J.P. Keene "Some practical aspects of kinetic spectrophotometry in pulse radiolysis" Quaderni dell'area di ricerca dell'Emilia e Romagna n° 1 C.N.R. Bologna 1972 Italy.

2) Melles-Griot "Optics Guide" 1975 Arnhem, Holland.

3) VARIAN Associates "product bulletin".

4) Hutton, Roffi, Martelli "Experimental apparatus at the Bologna linac" Quaderni dell'area della ricerca dell'Emilia e Romagna n° 5 C.N.R. Bologna 1975 Italy.

5) Bausch & Lomb "Diffraction grating handbook" Rochester, New York.

6) RCA "Photomultiplier manual" PT-61. Harrison N.J.

7) Rofin Laser & Instruments "Germanium photodiode" data sheet Egham, Surrey.

8) EG & G "Silicon photodiodes application notes" D 3000 C-1 SALEM, MASS.

9) J. H. Baxendale, C. Bell and J. Mayer "Some solid state photo-detectors used in pulse radiolysis studies" Int. J. Radiat. Phys. Chem. 1974, 6, 117, Pergamon Press G.B.

CONDUCTIVITY MONITORING TECHNIQUES

K.-D.Asmus and E.Janata

Hahn-Meitner-Institut für Kernforschung Berlin GmbH,
Bereich Strahlenchemie, Glienicker Strasse 100
D-1000 Berlin 39, Fed.Republic of Germany

ABSTRACT

A survey is given on fundamental principles, experimental
set-ups and limitations of pulse radiolysis conductivity techniques.
In addition, some chemical examples are presented and discussed.

INTRODUCTION

Pulse radiolysis is known to be an excellent technique for
the direct observation of fast chemical processes and short lived
intermediates. Most commonly the species involved are detected
through their colour, i.e. their optical absorption. Often, however, optical absorptions may be uncharacteristic,or may be too
small in the experimentally accessible wavelength range, or simultaneous presence of various species with similar absorptions may
complicate unambiguous interpretation of optical data.

A number of physical parameters other than optical absorption
have therefore been used to obtain additional information on fast
processes and intermediates. Thus, the creation or destruction of
charged species (ions) can conveniently be monitored by follow -
ing the time dependence of the conductance of a solution.

Starting with rather simple, straight-forward devices in the
sixties (1-3), a number of more complex and highly sophisticated
conductivity methods have now been developed in various laboratories (4-9). A survey of these techniques will be the subject of
this lecture. It includes the presentation and discussion of some
fundamental principles, experimental set-ups, the limitations of

J. H. Baxendale and F. Busi (eds.),
The Study of Fast Processes and Transient Species by Electron Pulse Radiolysis, 91–113.
Copyright © 1982 by D. Reidel Publishing Company.

conductivity techniques, and some chemical examples.

ELECTRICAL PRINCIPLES

    The basic idea is to apply a voltage to a pair of electrodes
in solution - the measuring cell - and to measure the radiation induced change in conductance (resistance) by monitoring the
change in current flowing through the cell. The source of radiation usually is a pulsed electron accelerator or laser.

    Fig.1a shows the principle arrangement. The voltage V_B is
applied to a voltage divider consisting of the cell and the load
resistor $R_L = 1/G_L$. The conductance of the cell due to the intrinsic concentration of ions in the solution is G_c. A change in
ion concentration due to the radiation results in an additional conductance $\Delta G_c(t)$ which may be time dependent. The voltage across
the load resistor is

$$V_L(t) = V_{L_o} + \Delta V_L(t) = \frac{G_c \times V_B}{G_L + G_c + \Delta G_c(t)} + \frac{\Delta G_c(t) \times V_B}{G_L + G_c + \Delta G_c(t)} \quad (1)$$

    The first term of equ.1 is constant under normal experiment-
al conditions ($\Delta G_c(t) \ll G_L$, $G_c \ll G_L$), while the change in voltage $\Delta V_L(t)$ is directly proportional to the change in conductance
$\Delta G_c(t)$ ($\Delta G_c(t) \ll G_L$).

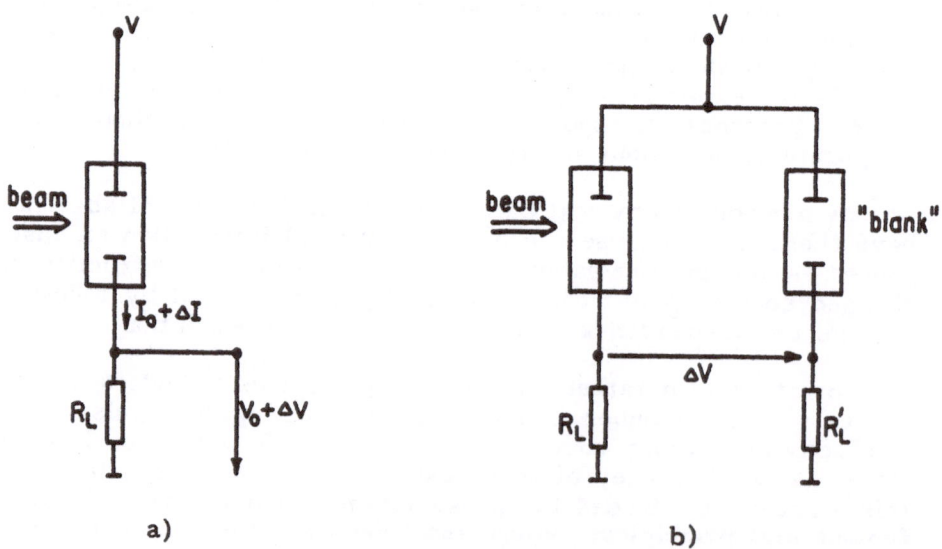

Figure 1. General arrangement for conductivity measurements.
     a) voltage divider string; b) bridge configuration

A bridge configuration is preferred in those cases where it is difficult to compensate the initial voltage V_{L_0}, for example in AC-conductivity set-ups or for long time DC-methods without back-off techniques. The bridge configuration, shown in Fig. 1b, is built by two divider strings each consisting of a cell and a load resistor but only one cell is irradiated while the other acts as a blank. The measured voltage V is, if both strings are identical $(G_{L_1} = G_{c_2} = G_c, R_{L_1} = R_{L_2} = R_2)$ and $\Delta G_c(t) \ll G_L$

$$V = \Delta V_L(t) = \frac{\Delta G_c(t) \times V_B}{G_L + G_c + \Delta G_c(t)} \qquad (2)$$

It could be easily seen that the change in voltage $\Delta V_L(t)$ due to a change in conductance is the same as it is for the voltage divider method. The blank cell may in principle be replaced by its electrical equivalents, i.e. by a resistor (and a capacitor in parallel), but any, even small changes in temperature or concentration of the solution will result in a change in conductance G_c, and a new balancing of the bridge would therefore be necessary. These effects are compensated, of course, if two identical cells both filled with solution are used.

The voltage applied in both methods may either be an AC- or a DC-voltage, which yields four basic configurations:

AC-voltage divider method
AC-bridge configuration
DC-voltage divider method
DC-bridge configuration.

The AC-voltage divider method has so far not been reported, probably due to its practical lacks. Among the DC-voltage divider methods a particular one should be mentioned here: the constant base current through the cell could be compensated by an additional feedback loop similar to that used for the compensation of the photomultiplier steady state current I_0 in optical detection systems. Thus, the "unwanted" voltage V_{L_0} of eqn. 1 is always close to zero and a change in sensitivity at high ion concentrations - as discussed below - is not observable as it follows from equ. 3

$$\Delta V_L(t) = \frac{\Delta G_c(t) \times V_B}{G_L + \Delta G_c(t)} \qquad (3)$$

The change in conductance, which reflects the change in concentration of the observed species, is

$$\Delta G_c(t) = \frac{F}{k_z \times 10^3} \sum_i \Delta c_i |z_i| \mu_i \tag{4}$$

where F is the Faraday constant in As, k_z the cell constant in cm^{-1} representing the geometric dimensions of the cell, Δc_i the change in concentration in M of the ith charged species, $|z_i|$ its net charge and μ_i its mobility in $cm^2 \ V^{-1}s^{-1}$. Eqn.4 could also be written as

$$\Delta G_c(t) = \frac{1}{k_z \times 10^3} \sum_i \Delta c_i |z_i| \ l_i \tag{5}$$

with l_i being the specific conductance in $\Omega^{-1}cm^2$ of the involved species. (Equ.4 and 5 are also valid for G_c if Δc_i is replaced by c_i.) For non-aqueous solutions it is common to give the mobility in terms of μ $(cm^2V^{-1}s^{-1})$ which is related with l through the Faraday constant: $\mu = l/F$. Under normal experimental conditions $G_L \gg G_C$ and $G_L \gg \Delta G_C(t)$. This allows to simplify equation 1 and to combine it with equ.5 to

$$\Delta V_L(t) = \frac{V_B \times R_L}{k_z \times 10^3} \sum_i \Delta c_i |z_i| l_i \tag{6}$$

The change in voltage measured across the load resistor R_L under these conditions depends linearily on the change in concentration of the charged species.

An important parameter is the specific conductance of the various ions involved. Some representative numbers for l(at 18°C) of ions in aqueous solutions listed below show that

$$H^+_{aq} \ : \ 314 \ \Omega^{-1}cm^2 \qquad F^- \ : \ 46.5 \ \Omega^{-1}cm^2$$

$$OH^-_{aq} \ : \ 172 \quad " \qquad\qquad NO_3^- : 61.7 \quad "$$

$$e^-_{aq} \ : \ 190 \quad "$$

$$Na^+ \ : \quad 43.5 \ "$$

$$NH_4^+ \ : \quad 64.5 \ "$$

major contributions to an overall conductivity signal result whenever protons, hydroxide ions, and - particularly relevant for radiation chemical processes - hydrated electrons are involved.

LIMITATIONS

The following discussion will be concerned with the limitations of the conductivity techniques. They include the smallest detectable change in ion concentration, the time resolution, and the
range of intrinsic ion concentration over which the sensitivity
will not change considerably. These parameters are not independent from each other and the variation of one of them may well effect the others. Some specific examples which are given with absolute numbers refer to aqueous solutions, the most commonly used system in pulse radiolysis conductivity studies. The considerations apply in principle, however, also to any other solvent.

The smallest detectable change in ion concentration could be
calculated from equ.6. A change in voltage of 0.5mV, for example,
could be conveniently observed if an oscilloscope with a sensitivity of 1 mV/div is used. If this change in voltage is caused by a
proton and anion ($\Sigma l_i \approx 360 \Omega^{-1} cm^2$), and $V_B = 200V$, $R_L =$
50 Ω, $k_z = 1$ and $|z| = 1$, this refers to a change in concentrat-
ion of $\Delta c = 1.4 \times 10^{-7}M$.

This example shows that the sensitivity of conductometric experiments is better than or at least the same as for optical measurements. The sensitivity may even still be increased by appropriate measures since the noise created in the cell by the current
flowing through it is some orders of magnitude lower and may
correspond to a change in concentration of about 10^{-9} to $10^{-10}M$,
depending on the band width of the detection system.

The time resolution of DC-conductivity experiments is limited
by 1) the time constant of the detection system and 2) the recovery time of the detection system from overload by spurious signals due to electron beam stopping in the cell (the pulse width of
the irradiating pulse is considered to be shorter or of the same
order than the time constant of the detection system). The time
constant of the whole detection system will not be discussed here
since it depends on the individual experimental set-up. But in
many cases, the overall time constant is essentially controlled by
the time constant of just the cell. This time constant can be calculated from the capacitance of the cell C_c, i.e. the capacitance
between the electrodes in solution and additional stray capacitances, the conductance of the cell and the load resistor, by

$$\tau = R \times C_c = \frac{C_c \times R_L}{R_L \times G_c + 1} \qquad (7)$$

or

$$\tau \simeq R_L \times C_c \quad \text{if } G_c \times R_L << 1 \qquad (7a)$$

The time constant thus is the product of cell capacitance times the parallel combination of load resistor and cell resistance (or if the load resistor is much smaller than the cell resistance, only of the load resistor).

The cell capacitance itself is, at first approximation,

$$C_c \approx \varepsilon \times \varepsilon_o \times \frac{A}{d} \qquad (8)$$

with A the area of the electrodes, d their distance, ε_o the absolute $(8.854 \times 10^{-14} \text{ F cm}^{-1})$ and ε the relative dielectric constant of the solution. The cell constant is, also at first approximation,

$$k_z \approx \frac{d}{A} \qquad (9)$$

Both C_c and k_z are seen to depend on the geometry of the cell. Combining equ. 8 and 9 finally yields

$$C_c \times k_z = \varepsilon_o \times \varepsilon \qquad (10)$$

which means that for a given dielectric constant, the product of cell capacitance times cell constant is constant. An improvement of the time constant can be achieved either by reducing the load resistor or the cell capacitance, but this goes at the expense of sensitivity, as follows from equs.6 and 10.(A reduction in cell capacitance increases the cell constant which in turn leads to a lower $\Delta V(t)$ signal).

The time resolution of AC-conductivity measurements is also governed by the time constant of the detection circuity including the cell. Of even more importance in this case is the frequency of the applied voltage since rectification of the bridge signal requires some filtering of the remaining unwanted AC-components. The time constant of that low pass filter corresponds - in practice - to at least ten periods of the applied voltage, hence

$$\tau_{AC} \geq \frac{10}{2 \pi f} \qquad (11)$$

The useful range of total ion concentration in the unirrad-
iated solution is defined here as the range where the sensitivity
of the set-up does not decrease considerably. The denominator
of the term for $\Delta V_L(t)$ in equ.1 contains the sum of G_L, G_c, and
$\Delta G_c(t)$. At low and medium ion concentrations G_L is much larger
than G_c and $\Delta G_c(t)$, at higher ion concentrations G_c cannot, however, anymore be neglected against G_L, and the sensitivity which
is constant at low and medium concentration starts to decrease.
The range of useful ion concentrations depends largely on the
value of the load resistor, the higher the load resistance, the
smaller the accessible range of ion concentrations. It should be
pointed out that the above mentioned effects are not only dependent on the ion concentration of the species under investigation
but that the concentrations, net charges, and equivalent conductances of all ions in the solution must be taken into consideration.
Fig.2 shows the pH dependence of the sensitivity (normalized to
pH 7) calculated for a solution containing protons and ClO_4^- ions
at two values of the load resistor. It is seen that at reasonable
high ion concentrations (pH 2.5 - 3), where this method is also
limited by the increasing base line shift and base line fall-off,
the reduction of sensitivity is still rather small (10% or less)(8).

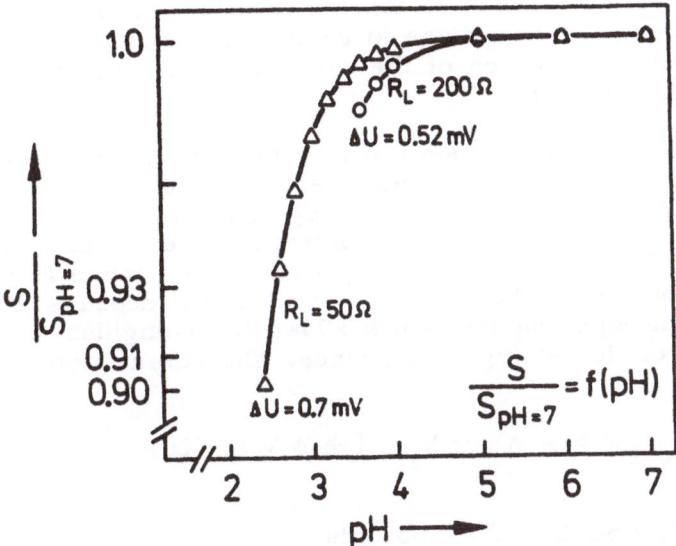

Figure 2. pH-Dependence of the sensitivity of conductivity measurements in aqueous solutions for two different load
resistors (200 and 50 Ω). ΔU represents the fall-off of
the base line over a period of 50 μs.

In summary, the parameters controlling sensitivity, time resolution and total ion concentration range are interrelated. Of particular importance is the load resistance, an increase of which yields a higher sensitivity but at the same time results in a decreased time resolution and a limitation of total ion concentration. For any practical applications the parameters will therefore have to be chosen according to the actual experimental requirements.

EXPERIMENTAL SET-UPS

In the following, some experimental set-ups for the measurement of transient conductance will be discussed.

Owing to polarization and electrolysis the DC-methods are limited for the micro- and submicrosecond time range. AC-conductivity methods are most often used to extend the time scale for conductivity measurements from the µs to ms and s time scale. For measurements in the ≥ 1 s time domain a train of short pulses of alternating polarity ($\bar{f} \leq 50$ Hz) is applied to the cell (10). The change in conductance is then given by the height of the individual response pulses. In addition, bridge configurations are reported utilizing one cell to which a sinewave of either a fixed frequency of 50 kHz (11) or a variable frequency of up to 500 kHz (12) are applied. The change in conductance is in these cases followed by the observation of the envelope of the nonrectified bridge output signal.

A two cell bridge system similar to that shown in Fig.1b is described by Lilie and Fessenden (4). The frequency of the applied voltage is 10 MHz and the bridge output signal is rectified by a synchroneous detector. Important features of this device are high sensitivity and good linearity over a wide range and the fact that the polarity of the detector output signal depends on the phase of the input signal, which allows to distinguish between an increasing or decreasing conductance. The output signal is

$$V_{out} = k \times \Delta V \times V_{ref} \left[\cos\phi + \cos(2\omega t + \phi) \right] \qquad (12a)$$

or after filtering the rf-components

$$V_{out} = k \times \Delta V \times V_{ref} \times \cos\phi \qquad (12b)$$

where k is an amplification factor, ΔV the bridge output signal, V_{ref} the amplitude of the reference voltage and ϕ the difference in phase of the input- and reference voltage. The scheme of an improved version (5) is shown in Fig.3. In this case the bridge

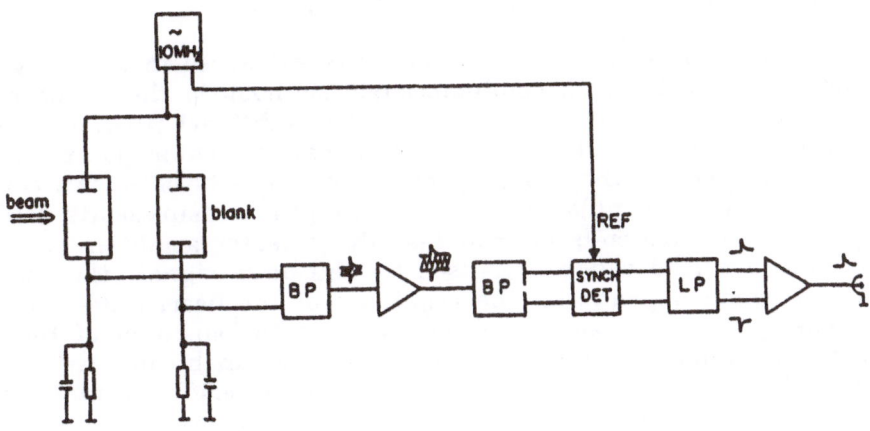

Figure 3. AC-conductivity set-up incorporating a two cell bridge

consists of two electrical, independent but closely mounted cells, load resistors and paralleled capacitances. For the purpose of balancing the bridge, load resistors and capacitors are variable. The AC-voltage ($10 \div 40$ V_{pp}) of 10 MHz as well as the reference voltage for the detector is delivered by a sinewave generator. The bridge output voltage passes a 10 MHz bandpass filter (balanced input, unbalanced output), followed by a low noise amplifier and passes again a bandpass filter (unbalanced input, balanced output). The amplified bridge signal is rectified by the synchroneous detector, the output is filtered from rf-components and again amplified. The overall voltage gain of the detector circuity is about 100. The observation time ranges from some microseconds to seconds.

A quite sophisticated set-up employing microwave techniques has been developed by Warman, de Haas, and corworkers (5), which allows AC-measurements also in nonpolar liquids on the nanosecond time scale. The operating frequency in this system is about 9 GHz.

During the last two decades, DC-conductivity measurements have been reported by a number of laboratories describing various methods for compensation of the initial offset voltage. The latter is due to the intrinsic conductance as it has been

mentioned already. Bridge configurations have been applied utilizing one cell (2,10,12,14) or two cells and a clamping circuit(15),
or an operational amplifier used as a differential amplifier (16),
as well as a voltage divider string followed by an AC-coupled
amplifier (17) or back-off circuits (14,18,19).

During the last years a special interest arose in studying
short-lived species with conductometric methods in the nanosecond
time domain. Within this time range some additional problems have
to be overcome: 1) a large spurious signal and subsequent "ringing" is created by the stopping of the electron beam in the cell.
The time after the pulse at which meaningful measurements could
be carried out depends on how fast the detector amplifier recover
from the overload which is caused by the beam signal, the amplitude of which depends on the amplitude of the beam pulse and on
the energy of the beam electrons. 2) Due to ionization of the
air surrounding the cell an additional signal can be present for
about 100 ns. 3) The metal electrodes, made mainly of platinum,
are sometimes "poisoned" by the solution under investigation,
which may result in another unwanted signal for the duration of
about one microsecond.

In a one cell arrangement incorporating a balanced to unbalanced transformer (6) and shown in Fig.4, the spurious signals
I_{sp} of both electrodes cancel at the output of this transformer
while the signals ΔI add if the beam is focussed properly to the
middle of the cell. With this set-up measurements could be done
from about 50 ns after the beginning of the beam pulse up to
some microseconds. Due to the low DC-resistance of the transformer measurements can be performed even at high ion concentrations without loosing sensitivity. In another set-up a limiting amplifier (8) is used which reduces the spurious signal (and only
this). This results in a considerably reduced recovery time of
the whole set-up. Measurements between 50 μs and 50 ps have
been carried out at pH values between 3 and 11.

The most recent development (Notre Dame Radiation Laboratory) is shown in Fig.5 (9). The voltage divider string consists
of the cell and the 50 Ω input resistance of a wideband DC-cou-
pled amplifier. A voltage pulse of plus or minus 200 V and 10 ms
duration is applied to the cell prior to the electron beam pulse.
The measuring cell is coaxial and glassy carbon is used for the
electrodes and outer shield. A 50 Ω connection to the cell is provided. The back-off circuit compensates currents of up to
± 50 mA. The recording of the transient signals is done by a
Biomation 6500 or 8100 digitizer connected to a PDP8 microcomputer. The final conductivity vs. time curve is composed of
two individual traces of opposite cell polarity. Each individual trace consists of the conductivity signal

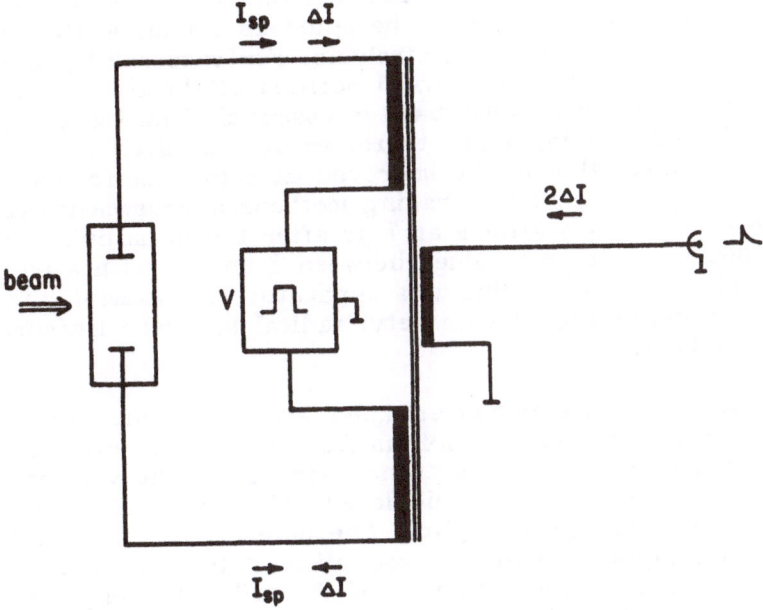

Figure 4. Submicrosecond conductivity set-up incorporating a balanced to unbalanced transformer.

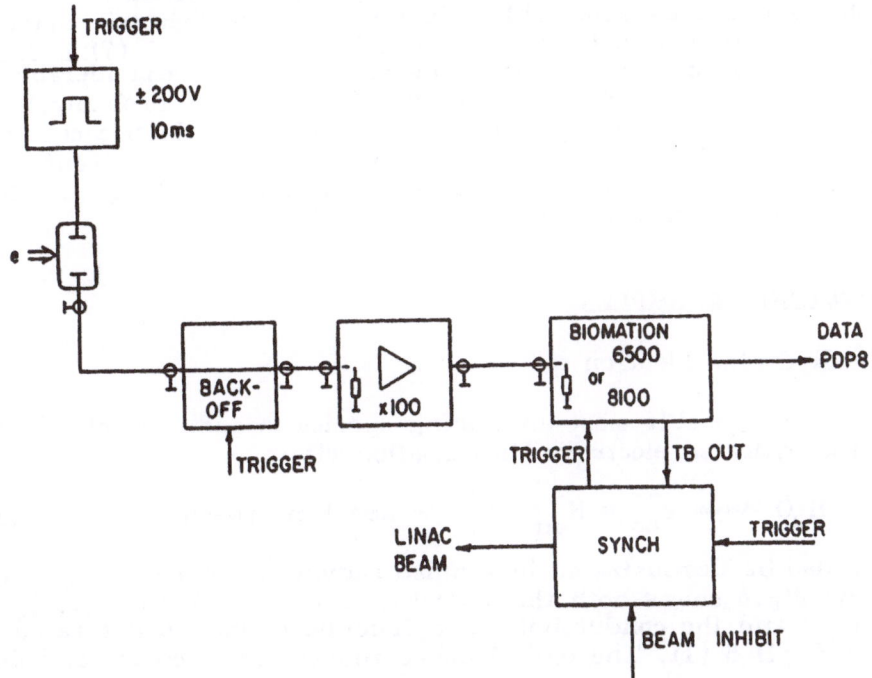

Figure 5. Scheme of conductivity set-up for the ns-time scale.

from which the base line - no radiation but voltage across the cell - is subtracted. By this, the spurious signal mostly cancels as well as any influence of electrolysis. For successful generation of the conductivity curve, a fixed position of the electron beam relative to the recorder time base is essential. The Linac beam and the Biomation trigger are therefore synchronized to the recorder time base. Due to the improved detector electronics, digital processing and signal avaraging methods, measurements could be done in aqueous solutions at 7 ns after the beam pulse up to some milliseconds for pH values between 3 and 11 with a day-to-day variation of $\leq$ 2%. Using this apparatus, for example, the protonation period of the acetone ketyl radical has been measured to 9.7 ± 5 ns (20).

All the above mentioned examples are - with one exception - generally valid for polar as well as for non-polar solutes. Differences in ion mobilities (equiv. conductances) in the various solvents have of course to be considered. This has a direct bearing on the observable signal heights. The lower dielectric constants of non-polar liquids result in a reduction of the capacitance and therefore in a lower time constant of the cell. The cell impedance is also influenced by the dielectric constant but such a change will not effect most of the experimental results since the cell is in most cases not matched to the transmission line. Such a match is required only for experiments in the subnanosecond time domain, and exactly achievable only for a certain dielectric constant. For example, if a cell with 50 Ω impedance for $\varepsilon_r \simeq 2$ (7) is filled with an aqueous solution, the impedance decreases considerably resulting in an observable mismatch. This effect is probably, however, not too serious for only small changes in dielectric constant, i.e. if going from one low-ε-solvent to another low-ε-solvent. Measurements in nonpolar liquids using 50 Ω coaxial techniques achieved step response times of about 100 ps (7).

CHEMICAL EXAMPLES

The Hydrated Electron

The probably most interesting species in radiation chemistry is the hydrated electron. Its formation via

$$H_2O \rightsquigarrow e_{aq}^- + H_{aq}^+ + OH^- + \text{non ionic species} \qquad (13)$$

can also be demonstrated in a pulse radiolysis conductivity experiment. Fig.6 shows both the optical absorption at 720 nm (λ_{max} of e_{aq}^-) and the conductivity as a function of time in deaerated H_2O at pH 5 (1). The optical curve represents formation and decay of e_{aq}^-.

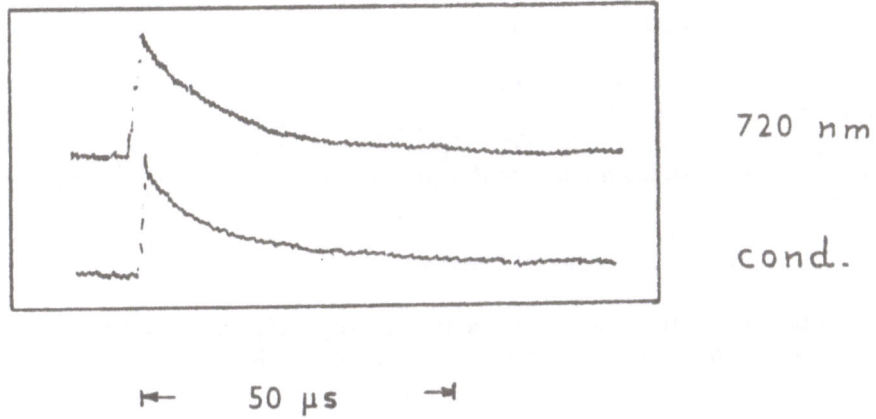

720 nm

cond.

|←   50 µs   →|

Figure 6. Optical absorption at 720 nm and conductivity vs.time
curves in pulse irradiated, deaerated water (pH 5).
Pulse width 0.3 µs.

The conductivity signal includes in principle the contributions of all ions present. The OH^- ions produced with a yield of
$G \approx 0.7$ (species per 100 eV absorbed energy) in the irradiation
process (21) do not appear, however, on the µs time scale experiment discussed here. At the prevalent pH of 5 they will be
neutralized by an equivalent amount of protons within a time of
less than 2 µs. These primary OH^- ions have, however, to be
taken into account for the interpretation of the conductivity signals (absolute yields, kinetics) when the neutralization process

$$H^+_{aq} + OH^- \longrightarrow H_2O \tag{14}$$

takes place on the time scale of investigation.

In the present case the observable conductivity signal on
the µs time scale is given by the sum of e^-_{aq} and excess H^+_{aq},
both of which are formed with an equal yield of $G \approx 2.7$. The
maximum change in conductance calculable via eqn.6 is given by

$$\sum_i c_i l_i = c(e^-_{aq}) \times l(e^-_{aq}) + c(H^+_{aq}) \times l(H^+_{aq}) \tag{15}$$

Since $c(e^-_{aq})$ can be calculated from the optical signal and
the known extinction coefficient of e^-_{aq}, and furthermore $c(e^-_{aq}) =$
$c(H^+_{aq})$ it is possible to evaluate the specific conductance of the

hydrated electron. It amounts to $l(e_{aq}^-) = 190 \pm 10\ \Omega^{-1}cm^2$ at 18°C. This value is almost equal to that of the OH^- ion, which reflects the fundamental structural similarity of the two species.

The decay of the conductivity signal is explained as follows: Some of the e_{aq}^- neutralize with H_{aq}^+ to yield H-atoms, alternatively e_{aq}^- are converted into OH^- (possible reactions are e_{aq}^- with $OH\cdot/e_{aq}^-/H_2O$) followed by fast neutralization of the OH_{aq}^- via reaction 14.

The Neutralization Process in Water

The neutralization process itself (equ.14) as it occurs in N_2O saturated water at pH 4.6 is shown in Fig.7 (22). The pulse

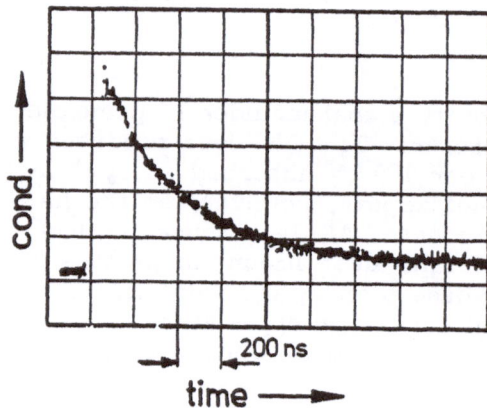

Figure 7. Conductivity vs. time curve in pulse irradiated, N_2O saturated aqueous solution of pH 4.6. Pulse length 5ns.

length was 5 ns and the absorbed dose around 10 Gy (1000 rad). Under these conditions the OH^- ion concentration formed during the pulse and via

$$N_2O + e_{aq}^- + H_2O \longrightarrow OH^- + \cdot OH + N_2 \qquad (16)$$

($t_{1/2} \approx 3$ ns for reaction 16) is relatively small ($\approx 3 \times 10^{-6}M$)

compared with the total H_{aq}^+ concentration of ca. 2.5×10^{-5}M. Accordingly, the decay of the conductivity signal occurs almost exponentially. From the first $t_{1/2} = 260$ ns a rate constant $k_{14} \approx 1 \times 10^{11}$M^{-1}s^{-1} is derived which is in reasonable agreement with the documented value of 1.4×10^{11}M^{-1}s^{-1} (23).

Oxidation of Sulphoxides

The reaction of sulphoxides with ·OH radicals (24) proceeds via the reaction sequence

$$R_2SO + \cdot OH \xrightarrow{\ k_{17}\ } R_2SO(OH) \cdot \xrightarrow{\ k'_{17}\ } R \cdot + RSO_2H \quad (17)$$

The first step is an addition of the ·OH radical to the sulphoxide function. Dissociation of this adduct leads then to the formation of alkyl radicals and sulphinic acid. The latter acid is of course subject to possible dissociation according to its acid-base equilibrium

$$RSO_2H \underset{k_{-18}}{\overset{k_{18}}{\rightleftharpoons}} RSO_2^- + H^+ \quad (18)$$

The pK_a values of sulfinic acids are usually quite low (e.g. $pK = 2.28$ for CH_3SO_2H). Accordingly, the reaction of an uncharged sulphoxide molecule with the also neutral ·OH radical leads to the formation of an anion/H_{aq}^+ ion pair at pH > pK_a. This is demonstrated inFig.8a, which shows a recording of the conductivity vs. time signal obtained from a pulse irradiated, N_2O saturated solution of 10^{-3}M$(CH_3)_2SO$ at pH 4.4. With rate constants of $k_{17} = 7 \times 10^9$M^{-1}s^{-1}, $k'_{17} = 1.5 \times 10^7$s^{-1} and an estimate of $k_{18} = 5.25 \times 10^7$ (based on $k_{18} = 5.25 \times 10^{-3}$ and $k_{-18} \approx 10^{10}$M^{-1}s^{-1}) the half-lives of the processes involved for the ion formation are 99 ns, 46 ns, and 13 ns, respectively, i.e. these processes are completed within the applied 1 μs pulse. The signal resembles a step function since the $CH_3SO_2^-$/H_{aq}^+ ion pair is stable over the detection period.

The yield of the conductivity signal, i.e. of $CH_3SO_2^-$/H_{aq}^+ ions relative to the yield of ·OH radicals is plotted as a function of pH in Fig.8b. The experimental data (limited to pH ≥ 2), are seen to fit a calculated pK-curve with $pK_a = 2.35$. This value is practically identical with the reported $pK_a = 2.28$ for the acid/base dissociation equilibrium of CH_3SO_2H (25).

Using equ.6 and assuming quantitative reaction of ·OH radicals via reaction 17, the specific conductance of $CH_3SO_2^-$ amounts

K.-D. ASMUS AND E. JANATA

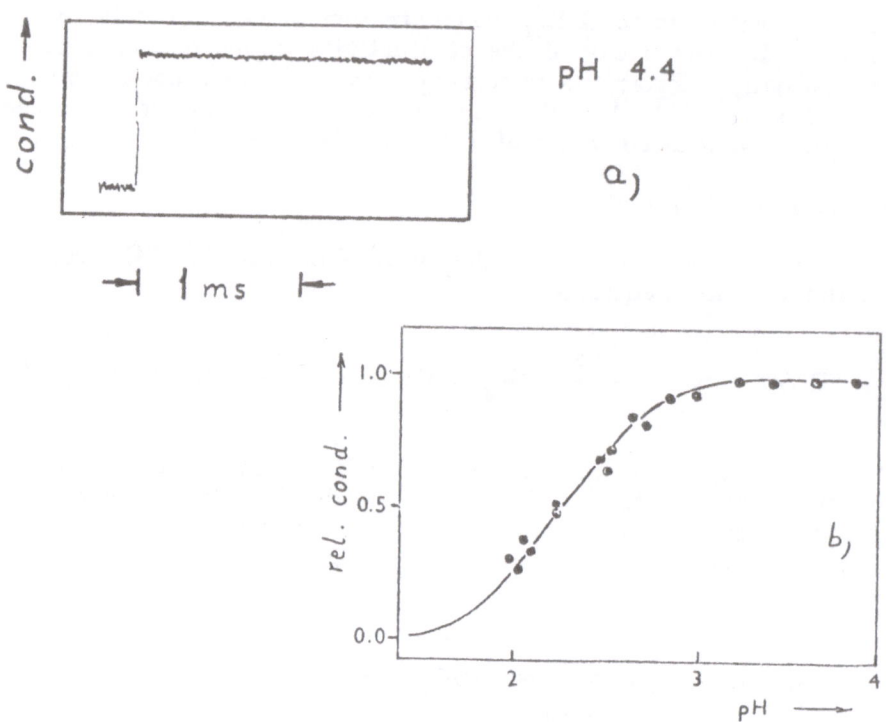

Figure 8.a) Conductivity vs. time curve in pulse irradiated, N_2O sat.
  solution of $10^{-3}M$ $(CH_3)_2SO$ at pH 4.4. Pulse width $1\mu s$.
  b) Relative yield of conductivity changes (normalized to actual $\cdot OH$ radical yield) vs. pH in pulse irradiated, N_2O sat.
  solutions of $10^{-3}M(CH_3)_2SO$.

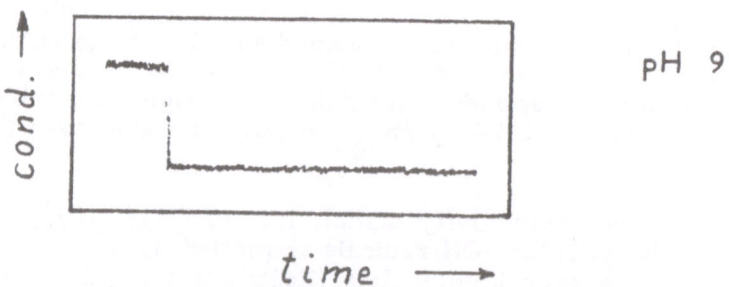

Figure 9. Conductivity vs. time curve in pulse irradiated, N_2O
  sat. solution of $10^{-3}M$ $(CH_3)_2SO$ at pH 9. Pulse width $1\mu s$

to $42 \ \Omega^{-1}cm^2$ which is of the right magnitude for normal anions.

A different conductivity signal from the same solution is obtained in basic environment as is seen in Fig.9. Under these conditions a decrease in conductance is observed. This results from the fact that the protons formed in reaction 18 are of course not stable in base and will be neutralized via reaction 14 which at $pH \geq 8$ proceeds with $t_{1/2} \leq 5 \ \mu s$. Stoichiometrically this results effectively in a replacement of an OH^- ion by a $CH_3SO_2^-$ ion(summation of reactions 17, 18, and 14), and since $l(OH^-)$ is higher than $l(CH_3SO_2^-)$ it explains the overall decrease in conductance.

The same change in conductance is also obtained if one would assume a single step mechanism

$$CH_3SO_2H + OH^- \longrightarrow CH_3SO_2^- + H_2O \qquad (19)$$

This is likely to occur, however, only at high pH when reaction 19 can compete directly with the fast proton elimination via reaction 18 $(t_{1/2}) = 13 \ ns)$.

Reduction of Tetranitromethane

Another example is provided by the reduction of tetranitromethane (26). Pulse irradiation of an N_2O saturated solution of $10^{-3}M \ C(NO_2)_4$ and $2 \times 10^{-1}M$ 2-propanol $(pH \approx 4)$ leads to the optical and conductivity signals shown in Fig.10. The underlying chemical reaction for the formation of the optically absorbing $(\lambda_{max} = 350 \ nm)$ nitroform anion $C(NO_2)_3^-$, is

$$(CH_3)_2\dot{C}OH + C(NO_2)_4 \longrightarrow C(NO_2)_3^- + NO_2 + H^+$$
$$+ (CH_3)_2CO \qquad (20)$$

Since $C(NO_2)_3^-$ is stable and the only absorbing species at 350 nm, and reaction 20 is completed within the duration of a 1 μs pulse under the above mentioned experimental conditions $(k_{20} = 5 \times 10^9 M^{-1}s^{-1})$ (27) the optical curve again shows up as a step function.

The corresponding conductivity curve exhibits a different picture and two processes are indicated. The initial fast step is due to the formation of the $C(NO_2)_3^-/H_{aq}^+$ ion pair formed in reaction 20. The secondary slow increase in conductance refers to the hydrolysis of the NO_2 which had been eliminated from the tetranitromethane in its dissociative reduction mechanism. The overall hydrolysis

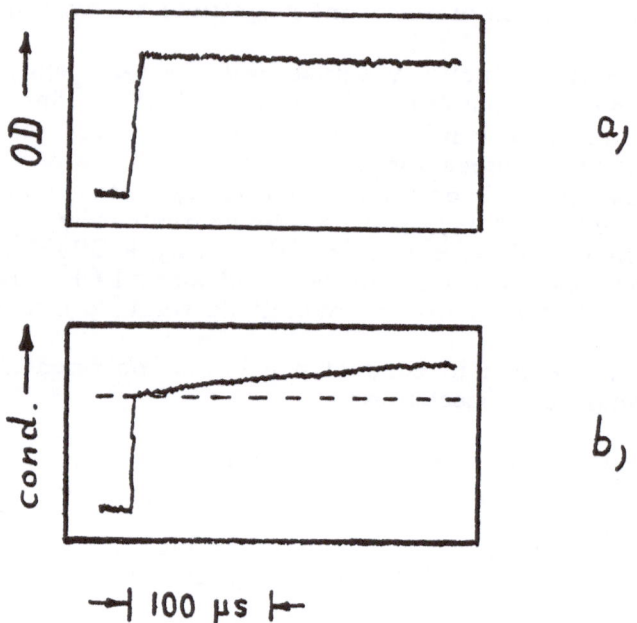

Figure 10. a) Optical absorption at 350 nm vs. time curve in pulse
irradiated, N_2O saturated solution of $10^{-3}M$ $C(NO_2)_4$
and $2 \times 10^{-1}M$ 2-propanol (pH 4-5). Pulse width 1μs
b) Conductivity vs.time signal in same solution

$$2\ NO_2 + H_2O \longrightarrow NO_2^- + NO_3^- + 2H^+ \tag{21}$$

also leads to the formation of protons and anions in a process
which is, however, considerably slower than reaction 20 (28).

This tetranitromethane example nicely demonstrates that conductivity is not as selective as optical measurements at particular
wavelengths could be, since conductivity always responds to all
ions produced or destroyed in an irradiated solution.

Oxidation of Dimethyldisulfide

An example of formation of positively charged species in a
radiation chemical process is the ·OH radical induced oxidation of
an organic disulfide (29). Pulse irradiation of an N_2O saturated
solution of $2 \times 10^{-4}M$ CH_3SSCH_3 at pH 4.7 and 8.0 leads to the
conductivity vs. time curves shown in Figs. 11a and b. The underlying process in this case is

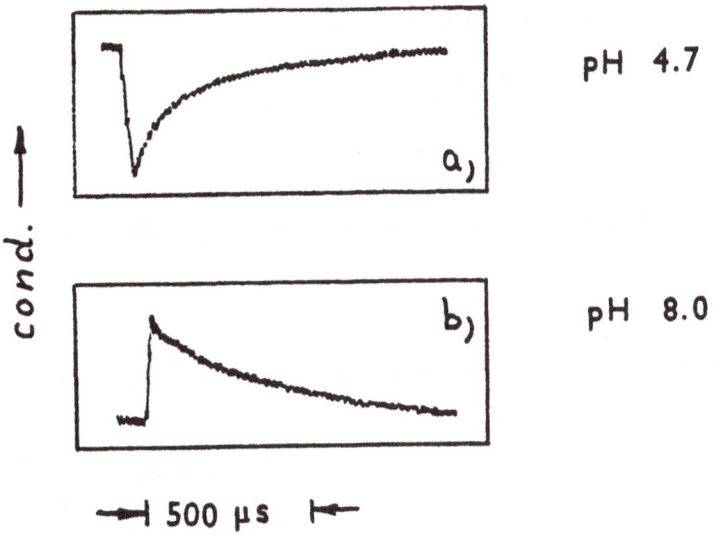

Figure 11. Conductivity <u>vs.</u> time curves in pulse irradiated, N_2O
saturated solutions of $2 \times 10^{-4}M$ CH_3SSCH_3 at pH 4.7
(a) and pH 8.0 (b). Pulse width $2\mu s$.

$$\cdot OH + CH_3SSCH_3 \longrightarrow CH_3SSCH_3^+\cdot + OH^- \qquad (22)$$

which in acid solution, where OH^- is not stable, is followed by the
neutralization reaction 14. Reaction 22 occurs with $k_{22} = 1.7 \times 10^{10}$
$M^{-1}s^{-1}$, i.e. is completed within the 2 μs pulse. The formation of
the radical cation/OH^- ion pair is reflected in the initial increase
in conductance in the basic solution. In acid solution, the OH^-
neutralizes an equivalent amount of H_{aq}^+, so that the net effect
(summation of equs.22 and 14) is a replacement of a proton by
the radical cation. Since the equivalent conductance of a proton
($1 = 315 \, \Omega^{-1}cm^2$) is much higher than that of a normal cation
($1 \approx 40 \pm 20 \, \Omega^{-1}cm^2$) an overall decrease in conductance is expected, and in fact experimentally found in the acid solution. The de-
cay of both curves at longer times is explained by chemical reactions which eventually lead to the liberation of a free proton
again from the radical cation.

Although the behaviour of the conductivity signal in the
above example is unambiguously indicative for the formation of a
positively charged species (relative to the zero charge of the original disulfide), associated with the replacement of a proton in
acid solution and formation of an OH^- ion in basic solution, an
exact identification of the radical cation solely on the basis of
conductivity remains ambiguous. The same conductivity signals

would, for example, also show up if the oxidation of the disulfide would proceed via an ·OH addition

$$\cdot OH + CH_3SSCH_3 \longrightarrow CH_3SSCH_3(OH)\cdot \qquad (23)$$

followed by protonation

$$CH_3SSCH_3(OH)\cdot + H^+ \longrightarrow CH_3SSCH_3(OH_2)^+\cdot \qquad (24a)$$

or

$$CH_3SSCH_3(OH)\cdot + H_2O \longrightarrow CH_3SSCH_3(OH_2)^{+}_{\cdot} + OH^- \qquad (24b)$$

where the resulting radical cation stoichiometrically would include the elements of a water molecule as compared with $CH_3SSCH_3^+$. Identification of the exact structure of the radical cation therefore requires additional investigations. In this particular case the radical cation could be identified as $CH_3SSCH_3^+$ although the ·OH radical reacts both via reaction 22 and 23 with about equal probability. The OH· adduct, $CH_3SSCH_3(OH)\cdot$, immediately dissociates, however, into the nonionic species $CH_3SH + CH_3SO\cdot$ or $CH_3SOH + CH_3S\cdot$ (depending on pH) (30).

Electron Reaction in Tetramethylsilane

Finally an example shall be presented in which Beck has demonstrated the applicability of conductivity methods in a low dielectric liquid, tetramethylsilane, at even the ps time scale (31). The possibility to do these experiments is based on the high mobility of excess electrons in this liquid (100 $cm^2V^{-1}s^{-1}$), the availability of very short pulses, and appropriate electronic equipment for the recording and analysis of the experimental data. Fig. 12a shows a trace of the conductivity as a function of time in a solution of tetramethylsilane irradiated with a train of 30 ps fine structure pulses of a linear accelerator. The increase in conductance is due to the formation of excess electrons. Fig. 12b-d show the signals in solutions where various increasing amounts of an electron scavenger, namely $CHCl_3$, has been added. The increasingly faster decay is caused by the reaction

$$e^- + CHCl_3 \longrightarrow CHCl_2^- + Cl^- \qquad (25)$$

which occurs with $k_{25} = 2.7 \times 10^{13} M^{-1}s^{-1}$ and leads to an ion, Cl^-, whose mobility is lower than that of e^- by orders of magnitude. A detailed analysis of these data reveals even more inter-

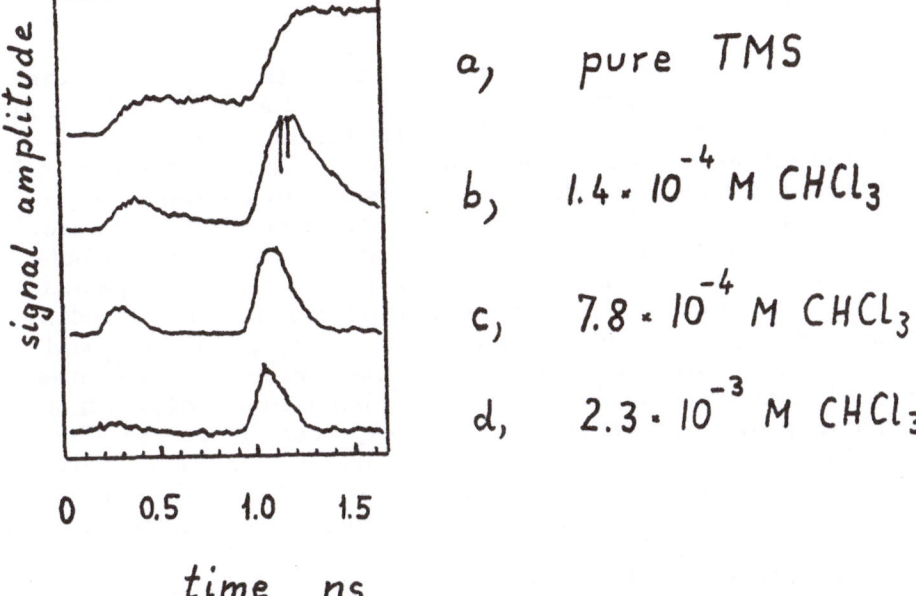

a) pure TMS

b) $1.4 \cdot 10^{-4}$ M $CHCl_3$

c) $7.8 \cdot 10^{-4}$ M $CHCl_3$

d) $2.3 \cdot 10^{-3}$ M $CHCl_3$

Figure 12. Conductivity signals obtained from tetramethylsilane solutions containing various amounts of $CHCl_3$, and pulse irradiated with a train of 30 ps pulses from a LINAC. Rel. vertical sensitivity: x 1 (a), x 2 (b), x 4 (c), x 8 (d).

esting aspects obtainable from such ps experiments but this shall not be subject of the present discussion.

CONCLUSION

The method of pulse radiolysis conductivity has proven to be a useful technique for obtaining qualitative and quantitative data on the formation or destruction of charged species, the associated kinetics and the determination of specific conductances and mobilities of transient charged species. It is of particular value if the detection of other physical properties it not possible. Conductivity finds also a most useful application in the investigation of acid/base equilibria which are a characteristic property of many radicals containing functional groups with hetero atoms (O, N, S, etc.). This is of particular relevance in aqueous solutions where free H_{aq}^+ and OH_{aq}^- exhibit significantly different specific conductances as compared with normal cations or anions. Another advantage of the conductivity method is its high sensitivity which allows detection of concentrations as low as 10^{-7}M or even less. Finally, the time resolution of $\leq$ 10 ns for aqueous systems and $\leq$ 1 ns for low polarity solvents opens a wide window on the time scale for the investigation of fast processes in radiation and associated general chemistry.

Considering all the advantages and positive aspects of the conductivity method, it is also necessary to be aware of its limitations. The parameters controlling time resolution and sensitivity, and the influence of total ion concentration as well as the associated hardware have been discussed in detail. In addition, one should always keep in mind that it is the sum of all ions, i.e. at least one ion pair which gives rise to the observable effects. In fact, most often, and this is in particular true for aqueous systems, the major fraction of conductivity signals is caused by species of high specific conductance, e.g. H^+_{aq}, OH^-, e^-_{aq} and only a small contribution arises from the species of actual interest. It is also necessary to remember that conductivity, although it characterizes the charge of a species, does not per se allow an unambiguous determination of the identity of a species. A simultaneous application of the conductivity technique with other methods (optical, polarographic, ESR, steady state experiments etc.) is therefore advised whereever possible. Finally, if dealing with thermodynamic equilibria such as acid-base equilibria, or neutralization processes the associated kinetics should be considered in order to avoid misinterpretation of both absolutely measured conductivity signals and their time dependence. This refers in particular to the submicrosecond time domain, but may also become of significance even at longer times as, for example in aqueous solutions near pH 7 or in buffered systems.

Being aware of all these advantages, specifications and limitations pulse radiolysis conductivity methods can be an excellent tool for the study of fast physico-chemical processes and chemical reaction mechanisms.

REFERENCES

1) Asmus,K.-D., Beck,G., Henglein,A., and Wigger,A.: 1966, Ber.Bunsenges.Phys.Chem. 70, pp. 869.
2) Schmidt,K.H., and Buck,W.L.: 1966, Science 171, pp. 70.
3) Barker,G.C., and Sammon,D.C.: 1967, Nature 213, pp. 65.
4) Lilie,J., and Fessenden,R.W.: 1973, J.Phys.Chem.77, pp.674.
5) Infelta,P.P., de Haas,M.P., and Warman,J.M.:1977, Radiat. Phys.Chem. 10, pp. 353.
6) Maughan,R.L., Michael,B.D., and Anderson,R.F.: 1978, Radiat.Phys.Chem.11, pp. 229.
7) Beck,G.: 1979, Rev.Sci.Instr.50, pp. 1147.
8) Janata,E.: 1980, Radiat.Phys.Chem. 16, pp. 37.
9) Janata,E.: 1981, Radiat.Phys.Chem. 17, pp. 0000.
10) Klever,H.G.: 1974, PhD Thesis, Ruhr-Univ.Bochum
11) Schöneshöfer,M., Beck,G., and Henglein,A.: 1970, Ber. Bunsenges.Phys.Chem. 74, pp. 1011.
12) Schmidt,K.H.: 1972, Int.J.Radiat.Phys.Chem.4, pp. 439.
13) Lilie,J., and Janata,E.: unpublished results.
14) Schmidt,K.H., Gordon,S., Thompson,R.C., and Sullivan,J. C.: 1980, J.Inorg.Nucl.Chem. 42, pp. 611.
15) Barker,G.C., Fowles,P., Sammon,D.C., and Stringer,B.: 1970, Trans.Faraday Soc.66, pp. 1488.
16) Erikson,T.E.: 1975, Chemica Scripta 7, pp. 193.
17) Beck,G.: 1969, Int.J.Radiat.Phys.Chem.1, 361.
18) Baxendale,J.H., Keene,J.P., and Rasburn,E.J.: 1974, J. Chem.Soc.Faraday Trans.I, 70, pp. 718.
19) Michaels,H.B., Rasburn,E.J., and Hunt,J.W.: 1976, Radiat. Res.65, pp. 250.
20) Janata,E., and Schuler,R.H.: 1980, J.Phys.Chem.84, pp. 3351.
21) Rabani,J., Grätzel,M., Chaudhri,S.A., Beck,G., and Henglein,A.: 1971, J.Phys.Chem. 75, pp. 1759.
22) Janata,E., Veltwisch,D., and Asmus,K.-D.: 1980, Radiat. Phys.Chem.16, pp. 43.
23) Eigen,M., and DeMaeyer,L.:1955, Z.Elektrochem. 59, pp. 986.
24) Veltwisch,D., Janata,E., and Asmus,K.-D.: 1980, J.Chem. Soc. Perkin II, pp. 146.
25) Wudl,F., Lighter,D.A., and Cram,D.J.: 1967, J.Amer.Chem. Soc. 89, pp. 4099.
26) Asmus,K.-D.: 1972, Int.J.Radiat.Phys.Chem.4, pp. 417.
27) Asmus,K.-D., Henglein,A., Ebert,M., and Keene,J.P.: 1964, Ber.Bunsenges.Phys.Chem. 68, pp. 657
28) Grätzel,M., Henglein,A., Lilie,J., and Beck,G.: 1969, Ber. Bunsenges.Phys.Chem.73, pp. 646.
29) Möckel,H., Bonifacić,M., and Asmus,K.-D.: 1974, J.Phys. Chem. 78, pp.282.
30) Bonifacić,M., Schäfer,K., Möckel,H., and Asmus,K.-D.:1975, J.Phys.Chem. 79, pp. 1496.
31) Beck,G.: 1979, Proc.VI.Int.Congr.Radiat.Res.Tokyo,pp. 279.

POLAROGRAPHY MONITORING TECHNIQUES

K.-D.Asmus and E.Janata

Hahn-Meitner-Institut für Kernforschung Berlin GmbH
Bereich Strahlenchemie, Glienicker Str. 100
D-1000 Berlin 39, Fed. Republic of Germany

ABSTRACT

The technique of pulse radiolysis polarography is described and some applications are discussed. The latter include the analysis of time dependent oxidation and reduction currents and the interpretation of polarograms.

INTRODUCTION

An important type of chemical reactions which free radicals are known to undergo are electron transfer processes. A quantitative determination of the redox properties of these, usually short lieved, species has therefore been the aim of many radical chemists. Significant contributions and data have been provided by electrochemical methods, e.g. cyclic voltametry. However, these techniques usually require lifetimes of the radicals in the order of milliseconds and longer, i.e. the respective studies are limited to relatively long lived radicals. Furthermore, electrochemically produced radicals have an inhomogeneous distribution in solution which makes any kinetic treatment of these radicals rather difficult. These difficulties can be overcome if free radicals are produced in the vicinity of an electrode by, for example, a single short pulse of radiation as in time resolved polarography. Flash photolysis experiments of such kind have been reported by Barker (1,2), Berg (3,4), and Perone (5,6). They could observe currents caused by the oxidation or reduction of radicals, depending on the applied electrode potentials.

A very suitable technique for such investigations has also

J. H. Baxendale and F. Busi (eds.),
The Study of Fast Processes and Transient Species by Electron Pulse Radiolysis, 115–128.
Copyright © 1982 by D. Reidel Publishing Company.

proven to be pulse radiolysis polarography which has been developed at the Hahn-Meitner-Institut in Berlin. This lecture will
be concerned with this technique. It includes some experimental
details and underlying principles and a discussion of a few examples which have been selected from the relevant literature (7-21).

EXPERIMENTAL SECTION

        The schematic of an experimental arrangement for simultane-
ous observation of the optical absorption and the polarographic
current is shown in Fig.1. The measuring cell consists of a rect-

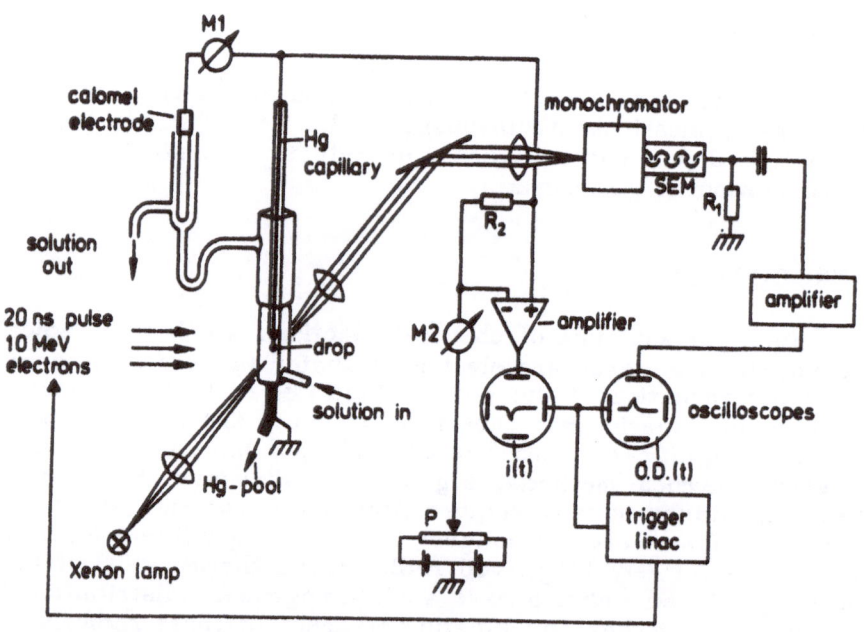

Figure 1. Schematic description of the experimental arrangement
          for simultaneous measurements of optical absorption and
          polarographic current of a short-lived radical

angular quartz vessel containing a hanging drop mercury electrode as the working electrode, a mercury pool as the counter electrode and a saturated calomel electrode (SCE) as a reference electrode. The solution under investigation flows continuously through
the cell passing the working and the reference electrodes. A li-

near accelerator (Linac) delivers electron pulses of 20 ns pulse width and 12 MeV of energy. The beam is focused around the mercury drop.

For the measurement of the optical absorption, the analyzing light is produced by a XBO 450 xenon lamp. The light is focussed through the cell just below the mercury drop, then passes several lenses and mirrors and finally hits the entrance slit of a monochromator. The conversion from light of the selected wavelength into an electrical signal is done by a photomultiplier (EMI 9558QB) and then recorded with a storage oscilloscope.

The polarographic current is recorded as a voltage drop across the working resistor R2 (20Ω) by means of another storage scope and a differential amplifier plug-in. The potential of the working electrode is set by potentiometer P and is measured with a high impedance voltmeter (M1) with respect to the reference electrode. The polarographic currents are in the order of 1 to 20 µA; the voltage drop across the working resistor is only a fraction of a millivolt. Potentiostatic methods are therefore not necessary.

All solutions contained electrolytes of about 0.3M of concentration which are inert with respect to radiation as well as electrochemically; sodiumsulfate is normally used.

The time constant for the recording of the polarographic current is simply given by

$$\tau = R \cdot C \tag{1}$$

where C is the differential capacity of the double layer around the working electrode and R is the overall resistance of the circuity, i.e. mainly the resistance of the solution between the working electrode and the mercury pool and the working resistor R2. The resistance of the solution is estimated to 50 to 70 Ω for a distance of 0.5 cm between the drop and the mercury pool. The double layer capacitance is about $10 \, \mu F \, cm^{-2}$. In order to keep this value low, a drop with a surface as small as $0.015 \, cm^2$ is normally used. These values give a time constant of about 10 µs. The differential double layer capacitance depend on the substances in solution, on the solvent and on the applied potential. Additional measurements of the double layer capacitance have been carried out therefore for each solution under investigation using an AC-bridge (22) similar to that described by Grahame(23).

With this apparatus the polarographic current is recorded at a given electrode potential as a function of time after the application of the irradiating beam pulse. Anodic currents have a negative sign and cathodic currents a positive one in accordance with pola-

rographic conventions. For radicals R·, for example, a cathodic
current is indicative for the reduction

$$R\cdot + e^- \rightarrow R^-$$ (2a)

and an anodic current refers to the oxidation

$$R\cdot \rightarrow R^+ + e^-$$ (2b)

at the electrode. Besides the interpretation of the received kinetic
data at a certain potential, the possible existance and position of
anodic or cathodic waves is of great interest for the identification
of a species. The polarogram is constructed from a number of experiments taken at different potentials by plotting the values of
the current at a certain time after the beam versus the potential.

EXAMPLES

Kinetics of Radical Reactions

     In pulse radiolysis experiments the radicals are produced ho-
mogeneously in the solution. Those radicals close to the electrode
surface can be oxidized or reduced.The rate of these reactions is
dependent on the rate constant of the electron transfer and the
actual radical concentration at the electrode surface. This concentration decreases with time due to the electrochemical process as
well as to other chemical reactions of the radicals. The electrode
process is maintained however as long as new radicals can be provided by diffusion from the bulk of the solution. In order to describe the polarographic current as a function of time, the corresponding diffusion problem has to be solved.

     Long-lived radicals. In the first example it will be assumed
that the radical of interest is very long lived with respect to the
measuring time and thus the current vs. time curves only reflect
the electron transfer and diffusion kinetics. Such a situation is
given for the radical from ascorbic acid. Fig.2a,b show the polarograms taken at two different electrode potentials (-0.5 and -0.8V
vs. SCE) in N_2O saturated solutions at pH = 5 containing ascorbic
acid and $2.5 \times 10^{-2}M$ KBr and $0.3M$ Na_2SO_4 as supporting electro-
lyte (10). Under these conditions the radical is formed via the oxidation of ascorbic acid by $Br_2^-\cdot$ and is stable over at least a millisecond. The latter can be seen in Fig.2c which shows a simultaneously recorded trace of the optical density at 360nm where the ab-
sorbic acid radical strongly absorbs. In contrast, the cathodic current decays to a significant extent over that period. A quantitative
descritpion of the curves is given by

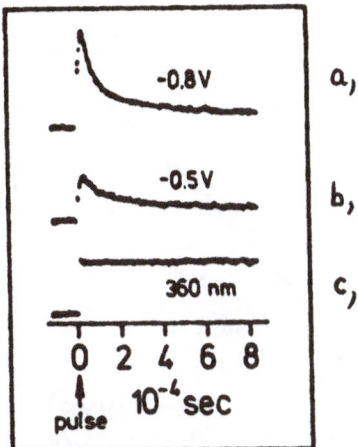

Figure 2. Cathodic current and optical absorption as functions of
time for the ascorbic acid radical. The solution contain-
ed 10^{-3}M ascorbic acid, 2.5×10^{-2}M KBr, 2.5×10^{-2}M
N_2O, and 0.35M Na_2SO_4. On the left side of the oscillo-
grams, the zero lines are shown. $c_o = 1.5 \times 10^{-5}$M,
A = 0.023 cm^2, pH = 10.

$$i(t) = \frac{n \cdot F \cdot A \cdot D^{1/2} \cdot c_o}{t^{1/2}} \, \lambda \, \exp(\lambda^2) \, \mathrm{erfc} \, (\lambda) \qquad (3)$$

with n = number of electrons transferred, F = Faraday constant,
A = surface area of electrode, D = diffusion constant, c_o = actual
concentration of radicals at the electrode, and $\lambda = k_f \cdot t^{1/2}/D^{1/2}$
with k_f = rate constant of the transfer reaction in the forward direction. The time dependence of the current is thus given by i =
$f(t^{-1/2})$, a plot of which is shown in Fig.3 for the electrode potentials of -0.5 and -0.8V.

Based on equ. (3) it is also possible to evaluate the rate
constants k_f which depend exponentially on the potential (12).
This relationship

$$k_{f,c} = k^o_{f,c} \exp \left(- \frac{\alpha \, nF}{RT}(E - E^o)\right) \qquad (4)$$

finally allows to calculate the transfer coefficient of the cathodic
process to $\alpha = 0.23$. (E_o = standard redox potential of the system)
(12).

The whole polarogram for the ascorbic acid radicals is shown
in Fig.4. It was constructed by plotting the currents observed at

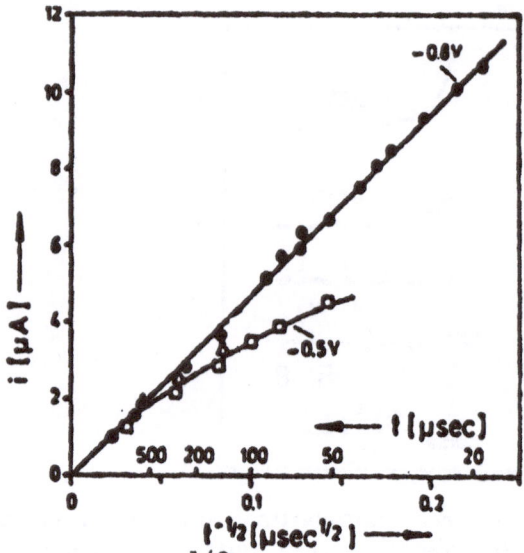

Figure 3. Plot of i vs. $t^{1/2}$ at different potentials for reduction of the ascorbic acid radical.

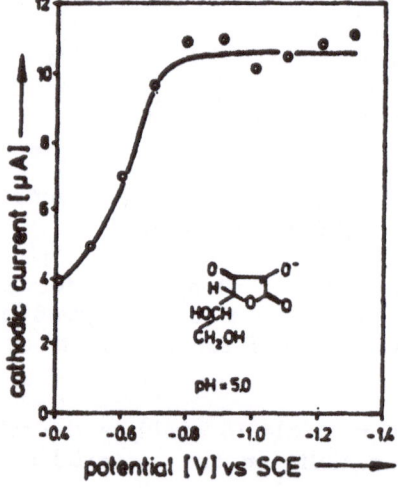

Figure 4.
Polarogram of the ascorbic acid radical at 15 μs after the pulse.

15 μs after the pulse vs. potential. It can be seen that the catho-dic reduction current is constant at potentials more negative than -0.8 V and decreases towards more positive potentials.

This example has demonstrated that pulse radiolysis polaro-graphy not only provides a possibility to identify a radical through its polarogram, i.e. its characteristic redox behaviour, but it also provides interesting quantitative information on electrode processes.

Short-lived radicals. The analysis of current vs. time curves is more complex, of course, if the radicals of interest are short-lived, i.e. their initial radical concentration in the bulk of the solutions decreases due to chemical reactions with periods comparable to the electron transfer and diffusion processes. In case the radicals disappear by a second order radical-radical reaction with a rate constant 2 k_2, the polarographic current vs. time curve is described by the relationship

$$\frac{1}{i(t) \cdot t^{1/2}} = \frac{\pi^{1/2}}{nFAD^{1/2}c_o} + \frac{2k_2\pi^{1/2}}{nFAD^{1/2}} t \tag{5}$$

but only at times shorter than the first half-life of the radical[7]. Examples are the radical anions $(C_6H_5COCH_3)^{\overline{\cdot}}$, $(CH_3COCH_3)^{\cdot}$ and $(CH_3COCOCH_3)^{\overline{\cdot}}$. Fig.5 shows the time dependence of the polarographic curves at particular potentials and traces of the optical density taken at a wavelength were these radical anions absorb. At a potential of -0.4 V both the $(C_6H_5COCH_3)^{\overline{\cdot}}$ and the $(CH_3COCH_3)^{\cdot}$ are oxidized at the electrode which is indicated by an anodic current. A cathodic current found at -0.8 V for the $(CH_3COCOCH_3)^{\cdot}$ anion radical means that this species is reduced at this potential. The analysis of the time dependence of these polarographic curves according to equ. 5 is shown in Fig.6 together with the first half-lives of the species which could be derived from the optical measurements. The linearity of the $1/t^{1/2}$ vs. t curves is particularly good for the long lived radical anion of acetophenone while in the other two cases the curves fall off at times which considerably exceed the first half-lives of the radicals.

Reversible waves. The examples discussed so far refer to a situation where the polarographic wave is completely irreversible, i.e. where back reactions at the electrode do not occur. This is in fact true for most of the carbon centered radicals. A reason for this may be that carbonium and carbanions, formed in the oxidation or reduction, respectively, have probably a too short lifetime in water to undergo further electrode reactions. For a polarographic wave to be reversible, it is necessary that the reduced or oxidized form of the radical is sufficiently stable and also that the back reaction at the electrode occurs with a high enough rate constant. Under these conditions the cathodic current i at a given time is related with the potential E by

$$E = E^o + \frac{0.058}{n} \log \frac{i_d - i}{i} \tag{6}$$

($i_d = n \cdot F \cdot A \cdot D^{1/2}/\pi^{1/2} \cdot t^{1/2}$, current due only to diffusion) [7].

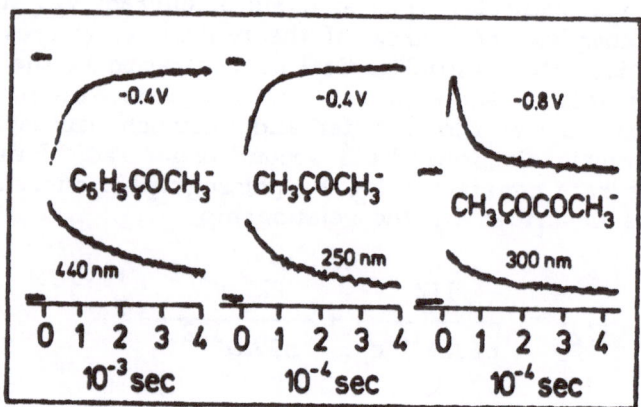

Figure 5. Current vs. time curves and absorption vs. time
curves for the oxidation of $C_6H_5COCH_3^-$, $CH_3COCH_3^-$
and for the reduction of $CH_3COCOCH_3$.
A: 0.0086 cm^2 ($C_6H_5COCH_3$), 0.015 cm^2 ($CH_3COCH_3^-$)
and 0.018 cm^2 ($CH_3COCOCH_3^-$). c_o = 1.5 x 10^{-5}M.

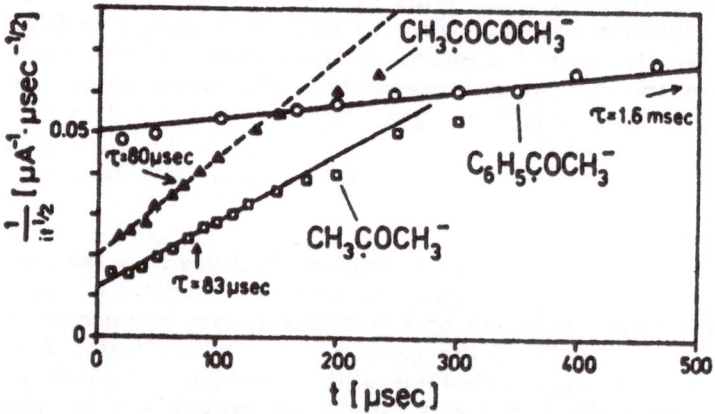

Figure 6. Plot of $1/it^{1/2}$ vs. t for the polarographic current
vs. time curves of Fig.5.

A reversible wave in the 10^{-5}s range has been observed for the oxidation of the ethanol radical anion (12). The latter is conveniently formed by $\cdot OH$ radical attack on ethanol

$$CH_3CH_2OH + \cdot OH \rightarrow CH_3\dot{C}HOH + H_2O \tag{7}$$

and subsequent dissociation in basic solution according to the equilibrium

$$CH_3\dot{C}HOH \rightleftharpoons CH_3\dot{C}HO^- + H^+ \tag{8}$$

(pK_a of 11.6) (24,25). Oxidation of this anion at the electrode, which is indicated by an anodic current, leads to acetaldehyde

$$CH_3\dot{C}HO^- \rightleftharpoons CH_3CHO + e^- \tag{9}$$

Since the back reaction is fast this process is reversible.

The reversibility also shows up in the polarograms shown in Fig.7 (12). At pH > pK ($CH_3\dot{C}HOH$) the waves are quite steep and exhibit a half-wave potential which agrees well with the conventionally measured values for the reduction of acetaldehyde around -1.8 V (vs. SEC) (26).

At lower pH where the radical exists in its neutralized form, $CH_3\dot{C}HOH$, the curves become less steep and the half-wave potentials are shifted towards more positive potentials. For reversible processes this latter shift can be described by the relation

$$E_{1/2} = E^o + \frac{RT}{nF} \ln \left(\frac{K_a + [H^+]}{K_a} \right) \tag{10}$$

where K_a represents the acid/base equilibrium constant of the radical. The corresponding theoretical curve is shown in Fig.8. The deviation of the actually measured half-wave potentials reflects the increasing irreversibility of the process with decreasing pH owing to the irreversibility of the oxidation of the neutral radical.

Some Selected Polarograms

Fig.9 shows the polarograms of the $CO_2^-\cdot$ and $(SCN)_2^-\cdot$ radicals constructed from the maximum changes in current after application of the pulse.

The polarogram of $CO_2^-\cdot$ (7) represents a steep wave ranging from oxidation directly into reduction with the polarographic current being zero at -1.5 V. CO_2 is known to be reduced

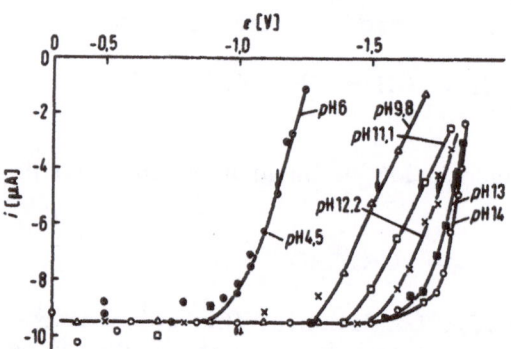

Figure 7. Anodic wave of the ethanol radical at various pH values (arrows point to the half-heights of the waves).

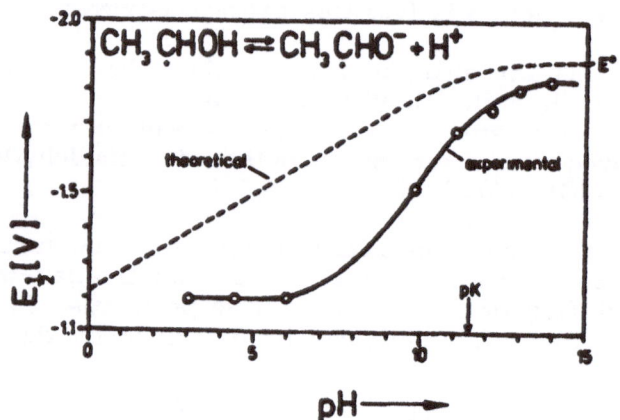

Figure 8. Half-wave potential vs. pHfor the oxidation of the ethanol radical.

only at rather high negative potentials (≈ 2.2 V) as has
been measured in conventional polarography (26). From the pulse
experiments it is now apparent that $CO_2^-\cdot$ is also and in fact very
rapidly reduced at these negative potentials via

$$CO_2^-\cdot + e^- \rightarrow CO_2^{2-} \qquad (11)$$

It can therefore be concluded that the electrochemical reduction of CO_2 itself at an electrode is a two-electron process

$$CO_2 + 2e^- \rightarrow CO_2^{2-} \qquad (12)$$

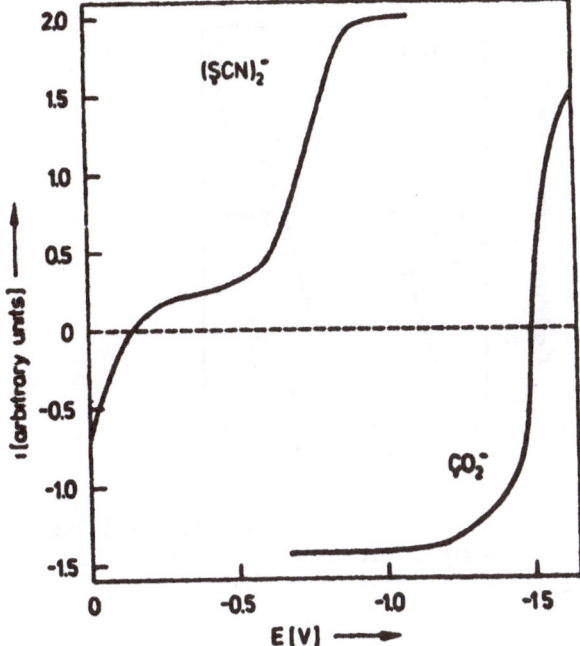

Figure 9. Polarogram of the carboxyl radical and of the $(\text{SCN})_2^-$
radical-anion complex at pH = 6.

The anodic current which can be observed at more positive potentials is due to the oxidation of the carboxyl radical in

$$CO_2^-\cdot \;\rightarrow\; CO_2 + e^- \tag{13}$$

The polarogram of $(\text{SCN})_2^-\cdot$ (27) also shows a cathodic reduct-
tion and an anodic oxidation. The cathodic wave breaks at -0.74V
and the anodic process occurs at potentials which are more positive than -0.15 V. This finding is somewhat surprising since
$(\text{SCN})_2^-$ is known to be a good oxidant with $E^\circ > 0$. Similar cathodic waves have also been observed for the halide radical anions,
$Cl_2^-\cdot$, $Br_2^-\cdot$, and $I_2^-\cdot$. To explain these data, it has been proposed (7) that mercury compounds, such as HgSCN, may be formed
at the surface of the mercury drop electrode which could more
easily be reduced or oxidized than the halide radical anions.

The polarogram of radicals from trimethylamine and N-methyl
acetamide are shown in Fig.10 (7). The anodic wave for the
$\dot{C}H_2N(CH_3)_2$ radical lies at rather negative potentials (-1.1 V) indicating that this radical is easily oxidized. In fact α-amino radicals are known to be excellent reductants (28, 29). The oxidation
wave does, however, immediately go over into a reduction wave
above -1.2 V, i.e. at very negative potentials these radicals can

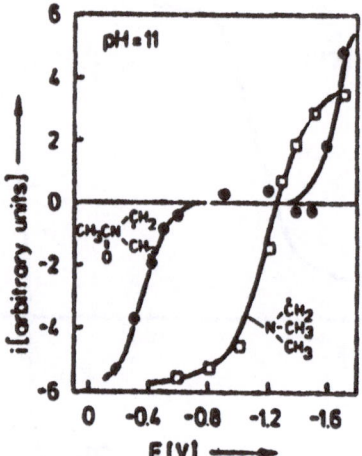

Figure 10. Polarograms of radicals obtained by H abstraction
from trimethylamine and N,N-dimethylacetamide.

also be reduced.

The $\dot{C}H_2NCH_3COCH_3$ radical also shows both an oxidation
and a reduction wave which, however, are well separated. The
oxidation wave breaks at about -0.35 V, showing that this radical is a much less powerfull reductant than the $\dot{C}H_2N(CH_3)_2$ radical.

Finally the polarograms of chlorinated methyl radicals shall
be considered (7). Fig.11 exhibits cathodic currents which have
been recorded from solutions in which $CH_2Cl\cdot$, $CHCl_2$ and CCl_3^-

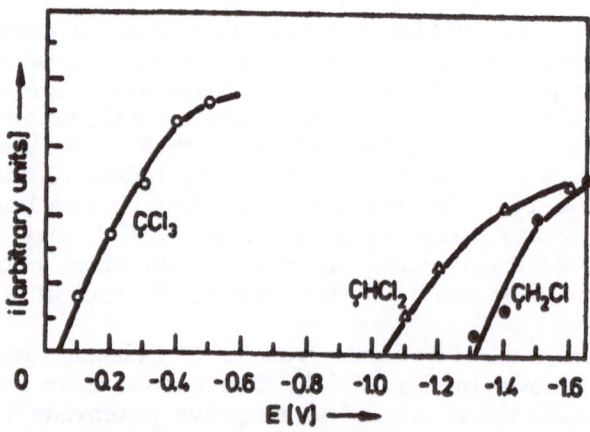

Figure 11. Cathodic waves of chlorine-substituted methyl radicals.

were produced. The waves are seen to be shifted toward less negative potentials with increasing number of chlorine atoms. This would mean that particularly the CCl_3 radical would be a relatively good oxidant since it can easily be reduced already at only slightly negative potentials. It has recently been shown, however, that oxidizing reactions which have been attributed to CCl_3^- have in fact to be attributed to the corresponding peroxy radical $CCl_3O_2^-$ (30). These species are easily formed in the presence of already small amounts of molecular oxygen and are reasonably good oxidants. The assignment of the polarographic waves may therefore remain ambiguous.

CONCLUSION

The examples presented in this lecture have been chosen to indicate possible applications of pulse radiolysis polarography techniques but also to point out some of the problems which may arise in the interpretation of the obtainable data. It should also be mentioned that it has not been attempted to cover all the aspects which have come up in the course of polarographic investigations. Thus the interesting reader has to be referred to the relevant literature. A comprehensive review of many more examples and a theoretical consideration of electrode kinetics and polarographic waves is found in an article by Henglein (7) and details may be studied in the associated original literature (8 - 21).

Although polarography is one of the more sophisticated time resolved techniques in pulse radiolysis its proper application not only helps to identify short-lived species through their redox properties but also opens a wide field for associated electrochemical studies.

REFERENCES

1) Barker,G.C., Gardner,A.W., and Sammon,D.C.: 1966, J. Electrochem.Soc. 113, pp.1182.
2) Barker,G.C.: 1968, Electrochimica Acta 13, pp.1221; Barker,G.C. and Bolzan,J.A.: 1974, Electroanal.Chem. and Interfac.Electrochem. 49, pp.227.
3) Berg,H.:1962, Naturwissenschaften 49, pp.11.
4) Berg,H. and Reissmann,P.: 1970, J.Electroanalyt.Chem.Interfacial Electrochem. 24, pp. 427.
5) Birk,J.R. and Perone,S.P.: 1968, Analytic.Chem. 40, pp.496.
6) Perone,S.P. and Drew,H.D.: 1971, in "Analytical Photochemistry and Photochemical Analysis", J.M.Fitzgerald (ed.), Marcel Dekker Inc., New York, chapter 7.

7) Henglein,A.: 1976 in "Electroanalytical Chemistry - a Series of Advances", Allan J.Bard (ed.), Vol.9, Marcel Dekker Inc., New York.

8) Lilie,J., Beck,G., and Henglein,A.: 1971, Ber.Bunsenges. phys.Chem. 75, pp.458.

9) Grätzel,M., Henglein,A., Lilie,J., and Scheffler,M.: 1972, Ber.Bunsenges. phys.Chem. 76, pp.67.

10) Grätzel,M. and Henglein,A.: 1973, Ber.Bunsenges.phys.Chem. 77, pp.2.

11) Grätzel,M., Henglein,A., and Bansal,K.M.: 1973, Ber.Bunsenges.phys.Chem. 77, pp.6.

12) Grätzel,M., Bansal,K.M., and Henglein,A.: 1973, Ber.Bunsenges.phys.Chem. 77, pp.11.

13) Henglein,A. and Grätzel,M.: 1973, Ber.Bunsenges.phys.Chem. 77, pp.17.

14) Bansal,K.M., Grätzel,M., Henglein,A., and Janata,E.: 1973, J.phys.Chem. 77, pp.16.

15) Bansal ,K.M., Henglein,A., Janata,E., and Sellers,R.M.:1973, Ber.Bunsenges.phys.Chem. 77, pp.1139

16) Bansal,K.M., Schöneshöfer,M., and Grätzel,M.: 1973, Z.Naturforschung 28b, pp.528.

17) Bansal,K.M. and Henglein,A.: 1974, J.phys.Chem. 78, pp.160.

18) Bansal,K.M., Henglein,A., and Sellers,R.M.: 1974, Ber.Bunsenges.phys.Chem. 78, pp.569.

19) Bansal,K.M. and Sellers, R.M.: 1975, in "Fast Processes in Radiation Chemistry and Biology", G.E.Adams, E.M.Fielden, and B.D.Michael (eds.), Wiley, New York, pp.259.

20) Sellers,R.M., Janata,E., Henglein,A., and Bansal,K.M.:1974, Ber.Bunsenges.phys.Chem. 78, pp.1085.

21) Grätzel,M., Bansal,K.M, and Henglein,A.: 1975, in "Radiat. Res.", O.F.Nygaard, H.I.Adler, and W.K.Sinclair (eds.), Academic Press, New York, pp.493.

22) Janata,E.: 1975, PhD-Thesis, Technical University Berlin,D83.

23) Grahame,D.C., 1949: J.Amer.Chem.Soc. 71, pp.2975.

24) Asmus,K.-D., Henglein,A., Wigger,A., and Beck,G.: 1966, Ber.Bunsenges.phys.Chem. 70, pp.756.

25) Asmus,K.-D., Möckel,H., and Henglein,A.: 1973, J.Phys. Chem. 77, pp.1218.

26) Meites,L.: 1967, Polarographic Techniques, Interscience Publ. New York.

27) Lilie,J.: 1972, J.phys.Chem. 76, pp.1487.

28) Griller,D. and Lossing,F.P.: 1981, J.Amer.Chem.Soc. 103, pp.1586.

29) Hiller,K.-O., Masloch,B., Göbl,M., and Asmus,K.-D.: 1981, J.Amer.Chem.Soc. 103, pp.2734.

30) Packer,J.E., Slater,T.F., and Willson,R.L.: 1978, Life Sciences, 23, pp.2617, and 1979, Nature, 278, pp.737.

THE MICROWAVE ABSORPTION TECHNIQUE FOR STUDYING IONS AND IONIC PROCESSES

John M. Warman

Interuniversity Reactor Institute
Mekelweg 15, 2629 JB Delft, The Netherlands

ABSTRACT

Microwaves propagating in a weakly conducting medium are attenuated. This effect can be used to study the yield and nature of
the primary ionic species and their subsequent reactions in pulse
ionised non polar media on a nanosecond timescale. Changes in
conductivity of 5×10^{-9} $ohm^{-1}cm^{-1}$ can be measured with a signal
to noise ratio of 10. This corresponds to a concentration of less
than 10^{-7} M of molecular ions (mobility $\underline{ca}$ 1×10^{-3} $cm^2V^{-1}s^{-1}$)
in a liquid of viscosity 1 cP. The experimental requirements and
theoretical background are discussed and a critical comparison
is made with other techniques. Examples of the application of the
method to some current problems in ionic processes are given.

J. H. Baxendale and F. Busi (eds.),
The Study of Fast Processes and Transient Species by Electron Pulse Radiolysis, 129–161.
Copyright © 1982 by D. Reidel Publishing Company.

INTRODUCTION

The application of microwave techniques to the study of ionic
processes following pulsed ionisation is by no means of recent
origin. Indeed, only a few years after the free availability of
microwave equipment following its wartime development, Biondi and
Brown illustrated the use of microwaves not only to probe the
electron concentration following pulsed ionisation of gases but
in addition to provide the ionising pulse itself [1a,b]. Since that
time the basic technique of Biondi and Brown, with slight modifications [2], has been used by several workers to investigate the
attachment, recombination, thermalisation and transport properties
of electrons in low pressure gases. Microwaves were also used at
quite an early date to study the conductivity resulting from
pulsed irradiation of semiconductor materials [3].

The method introduced by Biondi and Brown was based on the
phase shift or change in wavelength of electromagnetic waves which
can occur when the medium through which they propagate is ionised.
This effect however is expected to be extremely small in high
density, condensed media and one is forced to turn rather to
measurement of the change in amplitude of the probing waves as a
possible means of studying ionic processes. The reason for this
can best be understood by considering the motion of an ion of
mass M in an electric field.

In the presence of a field in the direction x, E_x, the normally random velocity distribution of the ion is perturbed by the
combined effects of an accelerating force eE_x in the field direction and a decelerating or frictional force due to interactions
of the ion with the medium. This results in the following equation
of motion for the ion

$$M \frac{dv_x}{dt} = eE_x - \nu_m M v_x \qquad (1)$$

where v_x is the net velocity component in the field direction superimposed on the otherwise isotropic velocity distribution of the ion. The friction parameter v_m may be seen as the rate at which any net momentum component of the ion in a given direction is isotropically randomised or "destroyed". In what follows v_m will be referred to as the momentum relaxation frequency and its reciprocal τ_m as the momentum relaxation time. For ions in moderate pressure ($\lesssim 1$ atm) gases, the momentum can be considered to be randomised by effective isotropic scattering collisions with individual molecules and v_m is referred to as the momentum transfer collision frequency.

Immediately following the application of a field, the net velocity component v_x will increase according to

$$v_x = \frac{eE_x}{Mv_m} [1 - \exp(-v_m t)] \tag{2}$$

For a constant field and times considerably longer than τ_m, a constant drift velocity is attained equal to eE_x/Mv_m. corresponding to a mobility of e/Mv_m. The conductivity of the medium due to the ion, present at a concentration N, is then

$$\sigma_{dc} = eN\mu \tag{3}$$

$$= e^2N/Mv_m \tag{4}$$

The total conductivity is of course the sum of all negative and positive ion contributions. The value of v_m to be used in (4) is an effective relaxation frequency obtained by an averaging procedure over the distribution of kinetic energies of the ion.

If the applied field varies sinusoidally with time with a radian frequency ω, then two extreme situations can occur. For $\omega \ll v_m$, the ion can at all times during the oscillation be considered to be moving in the field direction with the appropriate steady state value of the drift velocity in the same way as in the d.c. case above. If however ω is much greater than ν_m, no frictional damping of the motion of the ion occurs during a field cycle and it will move 90 degrees out of phase with the applied field.

For intermediate situations the motion of the ion is complex and can be considered to consist of an in phase or real component and an out of phase or imaginary component. This results in the a.c. conductivity being given by

$$\sigma_{ac} = \sigma_{dc}/(1 - i\omega/\nu_m)$$

$$= \frac{\sigma_{dc}}{1+(\frac{\omega}{\nu_m})^2} \left[1 + i\left(\frac{\omega}{\nu_m}\right)\right] \qquad (5)$$

As will be discussed in more detail later, for small conductivities the real component in (5) results in an attenuation of the applied voltage or a decrease in power of an electromagnetic wave and the imaginary component results in a phase shift. The conditions under which the real component begins to dominate can be roughly estimated for the case of electrons in a gas from the necessary condition $\nu_m/\omega > 1$. On making the substitution $\nu_m = e/m\mu$ this leads to

$$\mu < \frac{e}{m\omega} \qquad (6)$$

where m is now the electron mass. For X-band microwaves ($\omega \sim 5 \times 10^{10} \text{ s}^{-1}$), condition (6) is fulfilled if the electron mobility is less than $3 \times 10^4 \text{ cm}^2\text{V}^{-1}\text{s}^{-1}$. This corresponds to pressures greater than 400 torr in N_2 and above only 5 torr in the high collision frequency dipolar gas ammonia. It is apparent therefore that, even for moderately dense systems, attenuation measurements could prove to be more sensitive than the phase shift approach at least for the detection of electrons.

It was on the basis of the above considerations that it was decided in the early 70's to attempt to apply the microwave absorption technique to the study of excess electrons in pulse irradiated hydrocarbon liquids. At that time the first results using nanosecond time resolution d.c. conductivity methods were becoming available [4-6]. In addition infrared absorptions attributable to excess electrons had been discovered [7-10] and were being used to investigate the kinetic properties of electrons using nanosecond optical pulse radiolysis techniques. The first paper in which microwaves were used, to study excess electron kinetics in liquids, appeared in 1973 [11]. Apart from the presentation simply of results obtained using the method, two publications have appeared which concern more specifically the application of the technique to the study of ions in liquids [12, 13].

It was not initially apparent that the microwave conductivity method would have any significant advantages over the d.c. methods already available and being concurrently developed at the same laboratory. One large disadvantage was however immediately obvious. This was that only the product of ion yield times mobility could be determined using microwaves while with d.c. methods the mobility could be separately determined by measuring the drift time of ions between the electrodes.

Experience has shown however that microwave methods can provide information which is not obtainable using d.c. methods, for example in the study of geminate ion kinetics and ion pairs or investigations of conduction transients in heterogeneous or polycrystalline materials. Also, while d.c. techniques may be intrinsically more sensitive, they are much more prone to pick-up noise from the large transient currents associated with the ionising pulse delivered by particle accelerators and incidentally also by present-day pulsed light sources. This problem is alleviated considerably when waveguide transmission systems are used

since only a narrow band of frequencies can propagate in the circuitry.

In what follows a description of the requirements for microwave absorption experiments will be given followed by a discussion
of the interpretation of results and some examples of the application of the technique to real problems.

EXPERIMENTAL

General

While basically a conductivity technique, the experimental set up
used for microwave absorption measurements resembles more closely
that used for optical absorption studies. The fundamentals are
illustrated in figure 1. Microwaves are produced by the source and
propagate in waveguide to a circulator which "directs" the microwaves to the cell containing the medium of interest. The waves are
reflected back through the medium by a metal plate at the back of
the cell. The reflected wave is directed by the circulator to a
detector the output of which is measured.

Following this rather simplified description it is perhaps
only fair to point out immediately some of the problems which
help to make the experiment more "interesting". The first of these
is of a theoretical nature and will be dealt with as such later
but it has certain consequences as far as choice of equipment is
concerned and therefore is briefly mentioned here. Thus, apart
from a reflection occurring at the metal surface terminating the
cell, microwaves will undergo reflection at any dielectric dis-
continuity in the waveguide; for example the gas liquid interface
at the front of the cell. Due to the resulting interference effects
resulting from the coherent nature of the radiation and the similarity between practical cell dimensions and the wavelengths involved, a strong dependence of the measured absorption signal on

Electromagnetic waves propagating in a weakly
conducting medium are attenuated (absorbed)

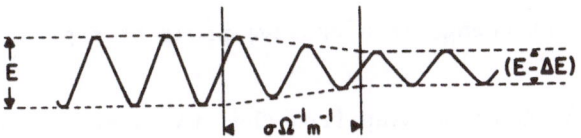

For small absorptions:

$$\frac{\text{Power absorbed}}{\text{Incident power}} = \frac{\Delta P}{P} \left(= \frac{2\Delta E}{E} \right)$$

$$= A\sigma$$

Measurements are made in reflection as illustrated below

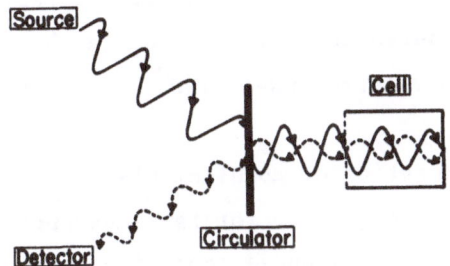

Figure 1. Illustration of the basic principles of
the microwave absorption method.

cell geometry as well as on the frequency of the radiation can
result for a given transient conductivity in the cell. For
maximum sensitivity and ease of quantitative measurements,
components with as large a bandwidth as possible are therefore
advisable. This includes source, detector and ferrite components
such as circulators and isolators.

A second problem lies in the nature of the point contact or
schottky barrier detector diodes used for fast time response
detection. The voltage output of these devices does not vary

linearly with either the power or field amplitude of the incident
radiation at the power levels used (ca 100 mW). Because of this,
either an accurate calibrated variable attenuator or a power meter
is therefore necessary in order to calibrate the change in detector
output for a small change in microwave power at a given incident
power level.

The third problem derives from the, in general, much higher
mobilities of ionic species (in particular electrons) in the gas
phase. Disproportionately large transient signals can therefore
result from small gas bubbles in the cell. This will be mentioned
further under cell design. In addition an appreciable transient
absorption may result from inadvertent ionisation of the air in
waveguide sections other than the cell. This results in a conductivity transient with a lifetime of approximately 15 ns caused
by electrons which undergo attachment to O_2 in the air. This can
be prevented for temperatures above $-120^{\circ}C$ by flushing the waveguide with SF_6.

A further consideration of considerable importance in the
design of the microwave circuitry for fast absorption measurements
is simplicity. In particular, phase sensitive components such as
"tees" and tuners which can convert frequency instability of the
source into amplitude noise should be excluded as far as possible.

Circuit Components

With the above comments in mind, a basic detection circuit is
shown in figure 2. It is assumed initially that this is for
operation in the X-band (8.2 - 12.4 GHz).

Beginning with the microwave source: present day Gunn diode
sources are advisable. Requiring usually only a 20 Volt, 2 ampere
power supply, they are capable of providing at least a 50 mW
output with noise figures comparable with the much more cumber-
some Klystron systems. Certain manufacturers provide relatively

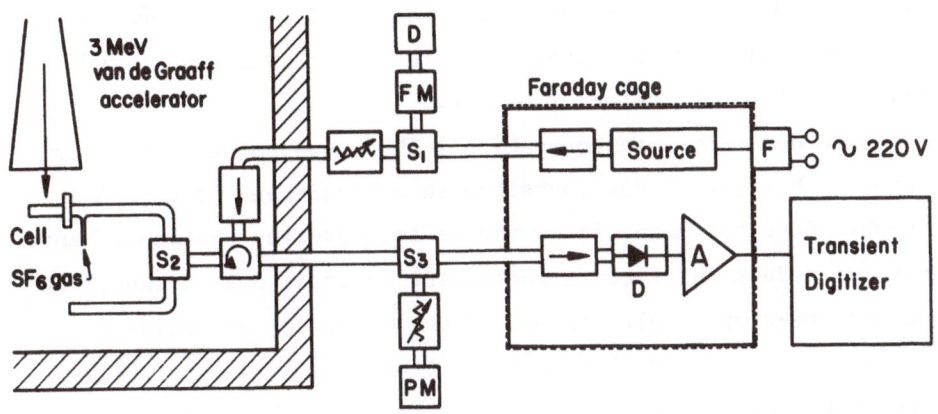

Figure 2. A schematic representation of the detection apparatus. The letters S, D, FM, PM, A
and F refer respectively to a microwave switch,
diode detector, frequency meter, power meter, amplifier and line filter.

cheap, cavity tunable Gunns capable of at least ± 1 GHz frequency
variation around a chosen centre frequency. The Philips (Sivers
Lab) model PM 7016X, used for several years by the author and
colleagues is unfortunately no longer available although the
lower power (10 mW) version, PM 7015X, is still catalogued. The
source presently used is a Hewlett-Packard sweep oscillator
mainframe, model 8620C, with a high power (> 50 mW) 86250C
(8.0 - 12.4 GHz) plug-in oscillator.

The ferrite isolators are incorporated to protect the source
and reduce effects due to reflections in the system, in particular
where extended lengths of waveguide are used. At present the
isolators and circulator used are Ramco Engineering 8.2 - 12.4 GHz
bandwidth models X200 and X615 respectively.

Electrically driven latching switches (Philips model PM 7299X)
are used for the following: S_1, frequency measurement using a

Marconi Sanders 6051 cavity meter; S_2, monitoring of the
incident power level at the cell (required for lossy samples and
resonant cavity applications); S_3, for power measurement and
calibration of the detector output.

The detectors are conventional IN23 type point contact or
schottky barrier diodes contained in a broad-band holder. A
tunable detector mount is advisable for covering the whole band
but can reduce the time response. The detector output should be
on the order of 1 volt, into a 50 ohm load, for an incident
power level of 50 mW. The response time is approximately 1 ns and
the output noise level is less than 50 μvolt. The calibration
factor n relating the fractional change in output voltage to a
small fractional change in incident power i.e. $\Delta P/P_o = n\Delta V/V_o$ is
usually in the range 1.6 ± 0.4.

The output of the detector is either connected via a 50 ohm
coaxial transformer [14] to the 50 ohm input of an oscilloscope
or digitizer, or the transient change is first amplified. The
Hewlett-Packard 8447A or 462A amplifiers which have risetimes of
1 ns and 4 ns respectively have been found to be suitable. Due
to the low frequency cut-off of these amplifiers a signal distortion of a few percent occurs for total times of 500 ns and
50 μs respectively. Due to the nonresistive a.c. coupling in
the input of the 8447A amplifier the output of the detector must
be connected to the input of this amplifier via a 50 ohm coaxial
transformer.

As can be seen in figure 2, the source, voltage supply,
detector and amplifiers are contained in a Faraday cage with
mains power being introduced through a high-frequency 60 dB line
filter. Pick-up in the coaxial cable from the Faraday cage to
the transient digitizer can be eliminated by surrounding the
cable with ferrite rings.

At present, measurement of the detector output is carried
out using a computer interfaced Tektronix R7912 transient digi-

tizer with a 7A19 vertical amplifier (risetime 0.7 ns). Using this,
together with the 8447A amplifier, an overal risetime of ca 1.5 ns
is found for the X-band system with a noise level after 100 x
amplification of approximately 5 mV. Defining as a detectable
signal the value of $\Delta P/P$ which would result in a change in output
equal to the noise level of 5 mV, one finds the detection
sensitivity to be one part in 10^4. Further reduction of noise and
random transients is accomplished by signal averaging.

A recently constructed Ka band (26.5 - 42 GHz) system has
been found to have a risetime of ca 0.6 ns using a 30 ps risetime,
sequential sampling system for measurements of transient changes
in the detector output.

Irradiation Cells

The cells are constructed of a length of metal waveguide (internal
dimensions 10 x 23 mm for X-band) with a section of one of the
broadwall sides reduced to 0.5 mm thickness to allow transmission
of the electron beam (see reference [12] for illustrations). One
of the ends of the waveguide is closed with a metal plate while
the other end is fitted with a flange for attachment to the
waveguide system and a high vacuum tight mica window which allows
microwaves to enter the cell.

Two "window" designs are currently used. The first, for
fastest time response, consists of a rectangular sheet of mica,
0.15 mm thick with dimensions approximately 1 mm larger than the
internal dimensions of the waveguide. This is sealed into a
0.5 mm deep recess in the flanged end of the cell using two-
component "Ultra Torr" cement. In the second design, used for
greater sensitivity at the expense of time resolution, a low-Q
resonant cavity construction is used with an iris coupling hole
of 8 mm diameter which is also sealed by attachment of a thin
mica sheet. The latter design reduces the time response of

detection to on the order of 10 ns.

The cells are connected through a 3 mm inlet hole in the front corner of the cell, via a metal to glass transition, to a glass bulb fitted with a greaseless stopcock. The interior of the cells is gold plated using Engelhard Industries "Atomex" gold-plating solution.

The problem, mentioned previously, of bubble formation in the cell can be overcome by enclosing the glass bulb in a heating mantle and ensuring that the temperature of the residual liquid in the bulb is slightly higher than that in the cell. The placing of the inlet hole of the cell in the front corner also facilitates the removal of bubbles.

Irradiation Conditions

The radiation facility used by the present author is a Van de Graaff accelerator which produces pulses of 3 MeV electrons. The pulse widths available are from 200 ps to 250 ns with a routinely available beam current of 2 amps. The total charge in the pulse (qnC) can be measured by deflecting the beam onto a 50 ohm coaxial collector. The cells are usually located at 10 cm from the exit port of the accelerator. The average dose at this location is approximately 20 rad per nanocoulomb beam charge or ca 10^{15} eVcm^{-3} nC^{-1} in a medium of density 1 gcm^{-3}. For a yield of 1 particle per 100 eV (G value = 1), the concentration formed would be 10^{13} cm^{-3}nC^{-1} or 1.6 x 10^{-8} MnC^{-1}.

DATA REDUCTION

The primary information obtained is the transient change in the output voltage of the detector which occurs when the sample in the cell is made conducting by a pulse of radiation. From this change the fractional change in power reflected by the cell can be derived using the calibration factor n.

$$\frac{\Delta P}{P_o} = n \frac{\Delta V}{V_o} \qquad (7)$$

For small (< 10%) changes in power level, $\Delta P/P_o$ is linearly
dependent on the radiation induced conductivity change in the
medium [12] i.e.

$$\frac{\Delta P}{P_o} = A\Delta\sigma \qquad (8)$$

which from (3) yields

$$\frac{\Delta P}{P_o} = AeN\mu \qquad (9)$$

It should be remembered that $N\mu$ is in fact the sum of the
products of the concentrations and mobilities of all ions present

$$N\mu = \sum_i N_i\mu_i \qquad (10)$$

It is this last parameter that one is basically interested in and
this is seen to be related to the output voltage change by

$$\sum_i N_i\mu_i = \frac{n}{Ae} \frac{\Delta V}{aV_o} \qquad (11)$$

where a is the factor by which the transient output change was
amplified.

For certain types of kinetic studies, for example first order
or pseudo first order reactions, it is unnecessary to have an
absolute measure of the concentration of the reactive species in
order to be able to derive kinetic parameters of interest.
Measurement of a parameter directly related to the concentration
is sufficient. If, using the present technique, one particular
ion has a mobility considerably greater than that of the other

ions present then the transient voltage change will be directly
proportional to the concentration of that species and even without
an absolute value of A or μ being known, useful kinetic data can
be obtained from the relaxation of the voltage transient. This
situation often applies to excess electrons which usually have
mobilities much larger than molecular ions. Also, if it can be
taken that only one positive and one negative ion is present
on the timescale of the investigations, then ΔV will be directly
proportional to the ion concentration and can give kinetic information on, for example, geminate ion kinetics.

    Since quantification of the method can yield extra informa-
tion on ion yields, mobilities and second order kinetics, it
is undoubtedly worthwhile. In a previous publication ⌈12⌉ the
derivations of the factor A in (8) were gone into in considerable
depth and those readers interested in this aspect are directed to
that paper. In what follows a summary will be given only of the
final conclusions and the analytical equations which can be used,
under certain conditions, to obtain absolute measurements of
the conductivity change.

    The data analysis can best be separated into experiments
using a straight forward reflection cell with mica window and
an iris coupled cavity cell.

Reflection Cell

As was mentioned above, interference effects due to multiple
reflections at dielectric interfaces result in the factor A in
(8) being dependent on cell geometry and on the frequency of the
microwaves. For a short circuited piece of waveguide with a
length d of uniform dielectric material adjacent to the short
circuit, the voltage reflection coefficient, Γ, in the air at
the air/dielectric interface is given by

$$\Gamma = \frac{1 - \gamma_1 \coth(\gamma_1 d)/\gamma_a}{1 + \gamma_1 \coth(\gamma_1 d)/\gamma_a} \tag{12}$$

The parameters γ_1 and γ_a are the propagation constants of microwaves in the dielectric and air filled waveguide respectively and describe the variation of the electric field amplitude in the propagation, or z, direction i.e.

$$E(x,y,z) = E_o(x,y)e^{-\gamma z} \tag{13}$$

The propagation constant is a complex number

$$\gamma = \alpha + i\beta \tag{14}$$

with the real part giving the change in amplitude due to attenuation and the imaginary part the variation in amplitude due to the wave nature of propagation. For propagation of the TE_{10} mode in rectangular waveguide of broad dimension "a", α and β are related to the real component of the relative dielectric constant of the medium ε_r and the high frequency conductivity σ (if small) by

$$\beta = \left[\left(\frac{\omega}{c}\right)^2 \varepsilon_r - \left(\frac{\pi}{a}\right)^2 \right]^{\frac{1}{2}} \tag{15}$$

$$\alpha = \frac{\omega\sigma}{2\varepsilon_o c^2 \beta} \tag{16}$$

The wavelength of propagation is related to β by

$$\lambda = 2\pi/\beta \tag{17}$$

The ratio of the power reflected by the dielectric plus short-circuit combination to the power incident is equal to the square of the absolute magnitude of the voltage reflection coefficient. The change in the reflected power due to a small induced conductivity, $\Delta\sigma$, in the otherwise nonconducting dielectric medium is therefore

$$\left(\frac{\Delta P_r}{P_r}\right)_{\Delta\sigma} = 1 - |\Gamma_{\Delta\sigma}|^2 \tag{18}$$

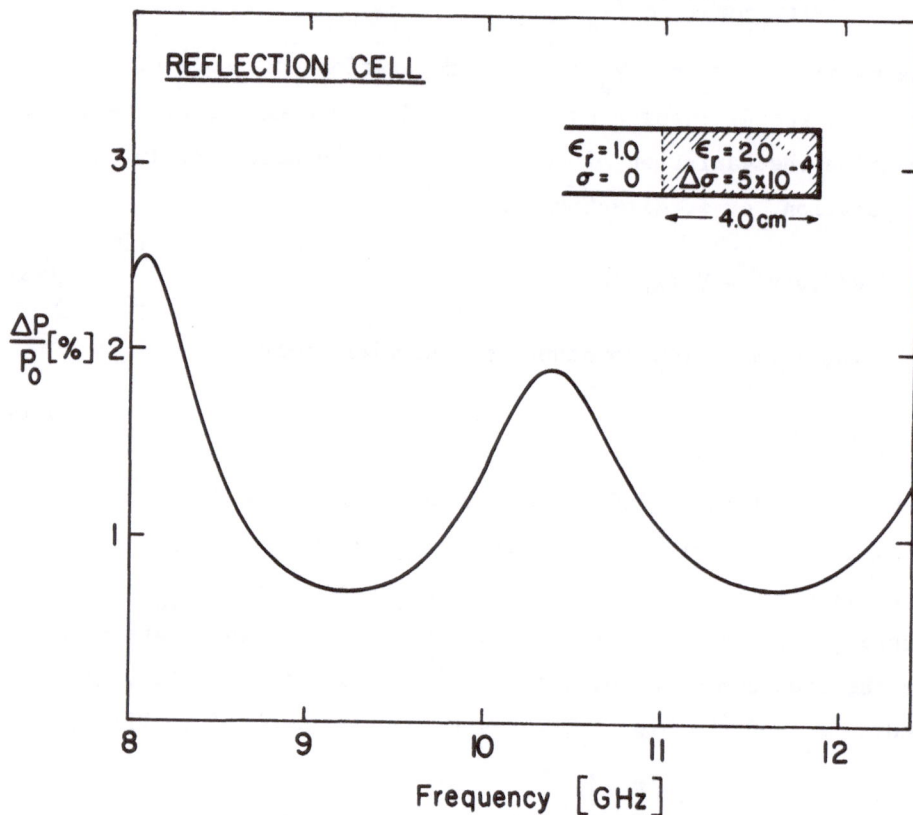

Figure 3. The calculated frequency dependence of
the absorption signal for a conductivity change
of 5×10^{-4} $ohm^{-1}m^{-1}$ in a 4.0 cm length of a
medium of dielectric constant 2 in a short-circuited piece of X-band waveguide.

In figure 3 is shown the change in power reflected from a medium
of dielectric constant 2 and length 4.0 cm as a function of
frequency for X-band waveguide for a uniform conductivity change
of 5×10^{-4} $ohm^{-1}m^{-1}$.

At the frequency for which the power change is a maximum, a

relatively simple analytical expression can be derived relating $\Delta P/P_o$ to $\Delta \sigma$ viz.

$$\left(\frac{\Delta P}{P_o}\right)^{max} = -\frac{2\lambda_a f_{max} d}{\varepsilon_o c^2} \Delta \sigma \tag{19}$$

In (19), λ_a is the wavelength of the microwaves in air, d is the total length of the dielectric medium, ε_o is the permittivity of free space and c the velocity of light in vacuum. An interesting aspect of (19) is that the sensitivity factor is independent of the dielectric constant of the irradiated medium. This is however important in determining the frequency of maximum absorption. This occurs when the medium is exactly an odd number of quarter wavelengths long i.e.

$$d = (2n + 1) \frac{\lambda_1}{4} \tag{20}$$

From (17) and (20) the frequencies at which maxima in the absorption will occur for a given length d are

$$f_{max} = \frac{c}{2\sqrt{\varepsilon_r}} \left[\left(\frac{2n+1}{2d}\right)^2 + \left(\frac{1}{a}\right)^2\right]^{\frac{1}{2}} \tag{21}$$

The value of λ_a corresponding to f_{max} is obtained from

$$\lambda_a = \frac{2\pi}{\left[\left(\frac{2\pi f_{max}}{c}\right)^2 - \left(\frac{\pi}{a}\right)^2\right]^{\frac{1}{2}}} \tag{22}$$

The parameters f_{max} and λ_a may then be substituted in (19) to obtained the sensitivity factor.

A practical example would be a cell of length 4×10^{-2} m and a medium of relative dielectric constant 2.0 with the experimental restriction of frequencies only available in the range 8 to 10 GHz. The values of f_{max} are given from (21) by

$$f_{max} = 1.06 \times 10^8 \left[\frac{(2n+1)^2}{6.4 \times 10^{-3}} - 1.89 \times 10^3 \right]^{\frac{1}{2}} \qquad (23)$$

The only frequency which fits into the available range is that at
8.07 GHz (n = 2) with the n = 1 and n = 3 possibilities lying at
6.09 and 10.36 GHz respectively. (It should be pointed out here
that a restriction in the bandwidth available can at least in part
be compensated for by increasing the length of the cell.) The
values of λ_a and λ_1 corresponding to 8.07 GHz are 6.31 x 10^{-2} and
3.20 x 10^{-2} m respectively. The relationship between the reflected
power and a change in the conductivity is then

$$\frac{\Delta P}{P_o} = -51.1 \, \Delta\sigma \qquad (24)$$

for conductivity in units of $ohm^{-1}m^{-1}$.

In practical cases, because of flanges, inlets, etc. the
liquid is not irradiated over its total length but rather over
a length d_i adjacent to the short-circuited end of the cell. To
take this into account the right hand side of (19) must be
multiplied by a correction factor $F(d_i/d)$ i.e.

$$\frac{\Delta P}{P_o} = - \frac{2\lambda_a f_{max} d}{\varepsilon_o c^2} F(d_i/d)\Delta\sigma \qquad (25)$$

with

$$F(d_i/d) = \int_o^{d_i} \sin^2 (2\pi z/\lambda_1) dz \Big/ \int_o^d \sin^2 (2\pi z/\lambda_1) dz \qquad (26)$$

which for the conditions of maximum absorption (d = $(2n + 1)\lambda_1/4$)
results in

$$F(d_i/d) = \frac{d_i}{d} - \frac{\sin[(2n+1)\pi d_i/d]}{(2n+1)\pi} \qquad (27)$$

If therefore only 2.5 cm of the cell were irradiated for the

above case of d = 4 cm and n = 2, then the value of $F(d_i/d)$ would
be 0.649 and the corresponding factor A would be 33 ohm m.

From the one part in 10^4 sensitivity to changes in $\Delta P/P_o$
mentioned previously in the experimental section, the detectable
change in conductivity using this system would then be 3×10^{-6}
$ohm^{-1} m^{-1}$.

Resonant Cavity Cell

Making an analogy with optical absorption techniques, the cavity
design acts effectively as a multiple reflection cell which
results in a longer pathlength through the medium. For a resonant
cavity, microwaves can only enter the medium behind the iris
coupling plate within a very narrow frequency band centered
around that corresponding to a voltage node at the iris. The
frequency must therefore be such as to result in the length of
the medium being an even number of quarter wavelengths, as opposed
to the odd number required for maximum absorption when using the
reflection cell.

For the reflection cell using non lossy liquids, the power
reflected is the same as the incident power. For a cavity, however
a sharp reduction in reflected power due to losses in the cavity
walls and the medium is observed as illustrated in figure (4).
This resonant absorption is characterised by a parameter R_o,
which is the ratio of the power reflected to the power incident
at the resonant frequency f_o, and by a frequency width Δf_o at
half absorption. The loaded quality factor of the cavity, Q_1, is
given by $f_o/\Delta f_o$.

In terms of the above parameters the following relationship
has been derived [12] between the fractional change in the power
reflected by the cavity at resonance and the conductivity change
induced in the medium

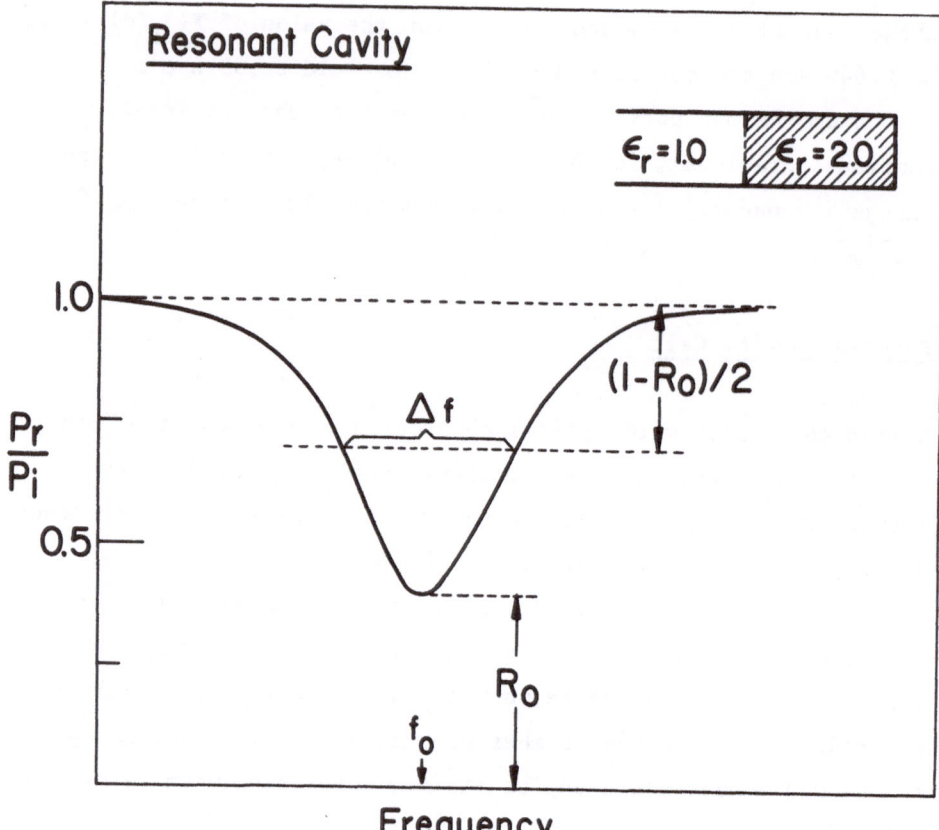

Figure 4. A representation of the frequency dependence of the ratio of the reflected, P_r, to incident, P_i, power for a cavity cell at resonance.

$$\frac{\Delta P}{P_r} = \mp \frac{(1/\sqrt{R_o} \pm 1)}{\pi \Delta f_o \varepsilon_o \varepsilon_r} F(d_i/d)\Delta\sigma \tag{28}$$

As can be seen, a small change in conductivity of the medium can result in either a decrease or an increase in the power reflected by the cavity. This can be understood by considering a gradual increase in the total background losses within the cavity to which the induced conductivity adds. Thus for very low loss R_o will be almost unity and will gradually decrease as the losses increase. At this stage a small further loss will result in a further de-

crease in R_o or an effective absorption. As the background losses
are increased R_o will eventually become zero which is a point known
as "critical coupling". Increasing the losses beyond this point
causes R_o to increase again and at this stage an additional loss
will give a further increase in R_o and hence effectively an emission.

While the quantitative relation (28) is quite simple, the
design of cavities and measurements of their properties for work
on dielectric liquids has its problems. The aim is to obtain an
optimum value of R_o, which is between 0.6 and 0.2, by the choice
of the diameter of the iris. However, for a given iris diameter,
R_o is very sensitive to the nature of the liquid and is for
example quite different for cyclohexane and isooctane due to the
slight dipole moment of the latter compound which results in a
higher background loss. The usual habit in spin resonance experiments of using tuning screws to adjust cavities to almost critical
coupling must be avoided in the present type of measurement since
for $R_o = 0$, $\Delta P = 0$. Tuning away from critical coupling is however
a possibility. Experience has shown that it is convenient to have
a set of cells with different iris diameters, from 6 to 9 mm in
X-band, with the choice for a given liquid being made on an
empirical basis. This choice can also depend on the solute used
particularly for compounds with high dipole moments which can
increase background losses considerably even in low concentration.

As mentioned previously, the time response of detection is
reduced by the use of resonant cavities due to their own intrinsic
relaxation time. The effect is the same as that of an RC circuit
with the exponential rise, or fall time being given by $Q/2\pi f_o$. A
Q of 500 results therefore in a response time of ca 6 ns. Concern
with producing liquid filled cavities with excessively high Q's
and hence response times much longer than 10 ns would seem on the
basis of experience to be unwarranted.

In order to illustrate the gain in sensitivity obtained using
a cavity, values of Q = 500 and $R_o = 0.5$ may be taken together

with the parameters d = 4 cm, d_i = 2.5 cm and ε_r = 2.0 used above
for the reflection cell. The resonant frequency within the hypo-
thetically available range 8.0 to 10 GHz is found using

$$f_o = \frac{c}{2\sqrt{\varepsilon_r}} \left[\left(\frac{n}{d}\right)^2 + \left(\frac{1}{a}\right)^2 \right]^{\frac{1}{2}}$$ (29)

to be 9.19 GHz for n = 3 with, for n = 2 and n = 4, 7.03 and
11.57 GHz respectively. Using (28) with the above parameters
gives $\Delta P/P_r$ = -1.53 x $10^3 \Delta\sigma$ for an absorption and +2.63 x $10^2 \Delta\sigma$
for an emission. Compared with the value of A = 33 $ohm^{-1}m^{-1}$ for
the reflection cell it can be seen that an increase in sensitivity
by between one and two orders of magnitude can be achieved by
using a resonant cavity with a decrease in time response from
ca 1 ns to ca 10 ns. The value of A = 1.53 x 10^3 given above
results in a detectable conductivity change of approximately
10^{-7} $ohm^{-1}m^{-1}$.

The Yield-Mobility Product

Having derived an absolute value of the sensitivity factor A, it
is possible to obtain at any time t, either during or following
the ionising pulse, the parameter $\Sigma N_i \mu_i$ from the voltage transient
using (11). It is then usual to normalize this quantity in some
way to the total energy absorbed by the sample during the pulse.
This is occasionally carried out by simply dividing the measured
conductivity by the nanocoulombs beam charge, q, as measured
using a coaxial collector

$$\frac{\Delta\sigma}{q} = \frac{n\Delta V}{AqaV_o}$$ (30)

If dosimetry has been carried out, then the average dose in
the sample per nanocoulomb beam charge, D $Gy.nC^{-1}$, is known.
(One Gray (Gy) is equal to an energy deposition of one joule per

kilogram or 6.2×10^{15} eV per gm.) Since the Gray is based on the
energy absorbed in water the total dose absorbed per m^3 in a
compound of density ρ_x gm cm^{-3} will be

$$\text{Dose (eV } m^{-3}) = 6.2 \times 10^{15} \, Dq\rho_x \times \frac{18}{10} \frac{\Sigma(\text{electrons})_x}{(\text{MWt})_x}$$

where $\Sigma(\text{electrons})_x$ is the total number of electrons per molecule
of the irradiated solvent and MWt_x is the molecular weight. A
yield times mobility product, $G\mu$, similar to the yield extinction
coefficient product, $G\epsilon$, determined in optical absorption experiments, can now be calculated from the conductivity

$$G\mu = \sum_i G_i \mu_i \tag{31}$$

$$= 5.6 \times 10^4 \frac{(\text{MWt})_x}{D\rho_x \Sigma(\text{electrons})_x} \frac{n}{Aa} \frac{\Delta V}{qV_o} \tag{32}$$

with the mobility μ being in units of $m^2 V^{-1} s^{-1}$.

APPLICATION AND COMPARISON WITH OTHER TECHNIQUES

Apart from the present technique, the two main fast detection
methods which have been used to study ions and ionic processes
in pulse ionized systems are the optical absorption and d.c.
conductivity methods. It was by use of the former method that the
first studies of ionic intermediates in pulse irradiated hydrocarbon liquids on a nanosecond timescale were carried out [15].
By measuring the absorption of the biphenyl anion in cyclohexane
solutions of biphenyl it was shown that the majority of ions
initially formed do not become homogeneously distributed in the
medium but rather undergo geminate recombination. The importance
of ionic processes in the chemistry of even low dielectric
constant media, which had previously been suggested on the basis
of steady state scavenging experiments, was therefore substantiated

and the impetus was provided for the further development of
techniques for the study of the nature and reaction kinetics of
the primary ionic species.

It would be rather pointless and misleading to intimate that
one particular technique is better suited to ion studies than
another. Optical, microwave and d.c. methods each have their own
specific areas in which they are superior and sometimes exclusively so. An obvious example is the separate absolute measurement of
the escaped yield, mobility and even the sign of the charge [16]
of an ionic species which can only be carried out using d.c.
conductivity methods. Optical methods come to the fore when
specific identification of a molecular ion is required or for the
study of geminate recombination which cannot be observed using
the d.c. method. Also the absorption spectrum of an electron or
ion allows studies to be made in very low mobility, high viscosity
media and in polar liquids or glasses. The microwave method is
particularly useful for studying geminate recombination processes
when the ions present have no optical absorption or when the
optical absorption of the ion overlaps with other species present.
It is also better suited to the study of the kinetics of homogeneous ion recombination processes than either optical or d.c.
methods since underlying absorptions for the former and distortion
of the kinetics due to separation of the ions for the latter often
present problems in data analysis.

The question as to which of the conductivity techniques, d.c.
or microwave, is the more sensitive is one which now and again
arises. The answer depends on how one defines the sensitivity of
a technique and how one estimates the limiting levels of noise or
interference.

Taking the sensitivity to be the lowest value of the conductivity change which can be measured, then a theoretical limit can
be derived for a fast d.c. conductivity system if it is taken that
measurements are made using a 50 ohm impedance, 400 MHz bandwidth

detection system. This results in a minimum, thermal noise
level of 20 μV. The voltage drop, ΔV, across the 50 ohm resistor
in the external circuit is related to the conductivity change,
Δσ, in a cell with electrodes of surface area S m^2 across which
a field strength of E Vm^{-1} is applied by

$$\Delta V = 50 \, ES\Delta\sigma \tag{33}$$

Taking E = 10^6 Vm^{-1} and S = 10^{-4} m^2, which are of the magnitude
normally used, then the value of Δσ required to produce a voltage
transient equal in magnitude to the thermal noise is 4 x 10^{-9}
$ohm^{-1}m^{-1}$. This value is more than an order of magnitude lower than
the detectable conductivity change using the microwave method
even when a resonant cavity cell is used.

The above calculation was based on the assumption that the
noise level was determined only by the thermal noise of the 50 ohm
resistor. In practice noise and/or interference levels considerably greater than this limiting value are normally observed in
d.c. measurements. This is due to ringing in the circuitry due to
lack of impedance matching of the various circuit components and
to signals induced by pick-up from the electromagnetic waves
emanting from the source of the ionising pulse. While the former
problem can be alleviated considerably by using impedance matched
coaxial or stripline cell designs [17], in general the practically
achievable sensitivity of the d.c. method will be approximately
an order of magnitude less than the limit given above and hence
comparable with that using a microwave cavity method.

For a mobility sum of 1 x 10^{-7} $m^2V^{-1}s^{-1}$, which is of the
magnitude found for molecular ions in a liquid of viscosity
1 cP, a sensitivity of 10^{-7} $ohm^{-1}m^{-1}$ corresponds to a detectable
ion concentration of approximately 2 x 10^{12} cm^{-3} (10^{-8} Moles per
litre). In order to attain this concentration, for a typical

free ion yield on the order of 0.1 (100 eV)$^{-1}$, a total dose of
approximately 1 Gy must be delivered to the sample in the pulse.
This dose cannot be obtained using X-ray sources. Therefore, for
the study of normal molecular ions or even ionic intermediates
with mobilities an order of magnitude larger than normal with
nanosecond time resolution one must resort to direct electron
irradiation of the sample. As pointed out in the experimental
section, this can be accomplished using the microwave system with
no significant increase in noise level. Direct irradiation of d.c.
cells with electrons however usually substantially increases the
noise level above that given above which was based on the use of
X-rays. The microwave absorption method therefore is better suited
than d.c. methods for fast, pulse radiolysis studies of low mobility charged species. It is perhaps worth pointing out that the
actual maximum field strengths involved in microwave measurements
for a 100 mW power level are only 5 and 100 Vcm^{-1} for the reflection and resonant cavity designs respectively.

An example of the use of the microwave technique to study
low mobility transients using electron radiation is given in
figure 5. The liquid used was trans-decalin for which even the
excess electron has a mobility of only approximately 10^{-2} cm^2
v^{-1}s^{-1}. The lower trace in the figure illustrates the use of the
technique to also study geminate ion kinetics which is responsible
for the initial rapid decay of the molecular ions in the fully
scavenged system.

The results in figure 5 were obtained using an X-band resonant cavity with a time response of 8 ns. In order to increase
the time response of molecular ion studies to 1 ns by use of a
reflection cell it was necessary to considerably increase the
radiation density which could only be achieved by decreasing the
dimensions of the electron beam (cell position closer to the exit
port of the accelerator) and reducing the cell dimensions
accordingly. This was one of the reasons for constructing a system

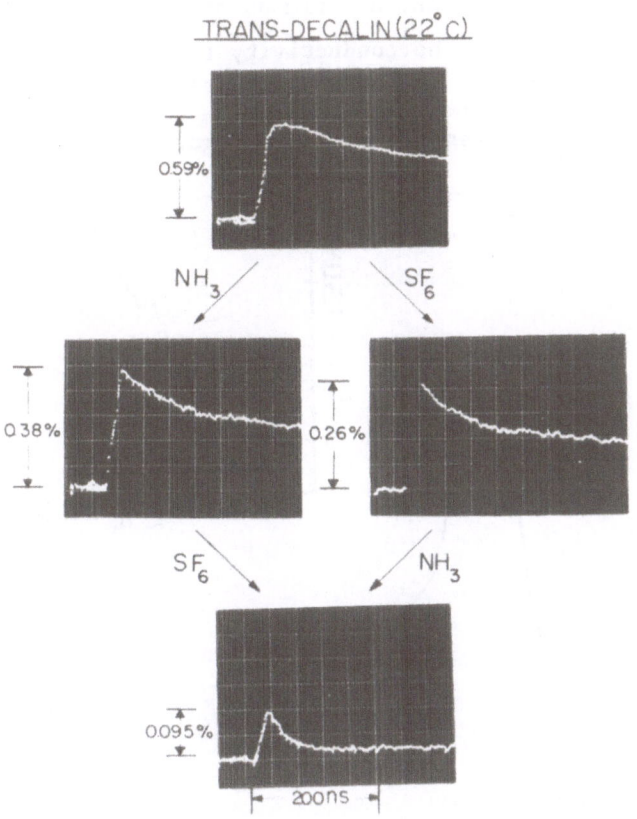

Figure 5. Changes in the voltage output of the microwave detector diode resulting from pulse ir-radiation (20 ns, 1A) of liquid trans-decalin con-tained in an X-band cavity cell. The ionic species responsible for the conductivity transients are: upper trace, excess electron plus radical cation; left centre, excess electron plus NH_4^+; right centre, SF_6^- plus radical cation; lower, SF_6^- plus NH_4^+.

at a frequency approximately a factor of three higher than X-band
and using waveguide of internal dimensions 0.71 cm by 0.36 cm.
Preliminary results on the conductivity transient due to molecular
ions (total mobility $\underline{ca}$ 1 x 10^{-3} $cm^2v^{-1}s^{-1}$) in liquid CCl_4 obtained using this system are shown in figure 6.

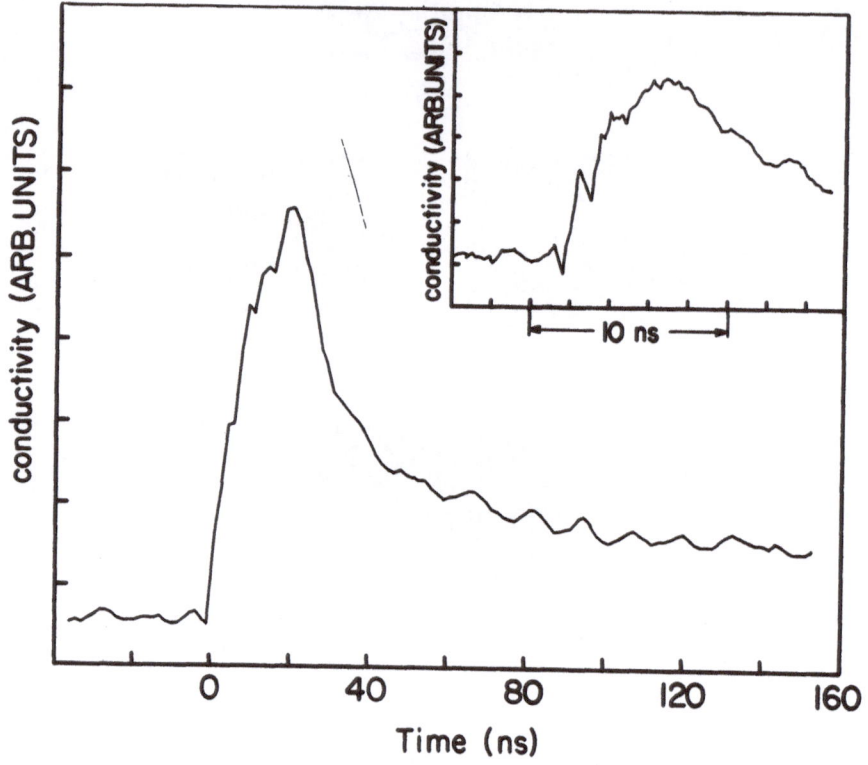

Figure 6. The conductivity induced in liquid CCl_4
by a 20 ns, 1A pulse of 3 MeV electrons, measured
using a Ka-band reflection cell. Insert: As for
the main figure but using a 5 ns pulse.

Particular individual contributions of the microwave method
to the general pool of information on ionic processes in dielectric liquids have been the measurement of the mobilities and
reaction kinetics of highly mobile radical cations (holes) in
certain hydrocarbon liquids [13, 18-20]. Also the finding of a

highly mobile radical anion in the electron attaching liquid perfluorobenzene [21-23]. Recent applications of microwave data to
the testing of theoretical solutions of the Smoluchowski equation
for geminate recombination [24,25] and to the measurement of the
times required for electron thermalisation in rare gas liquids
[26,27] may also be quoted. An illustration of the results obtained in the last named application is given in figure 7.

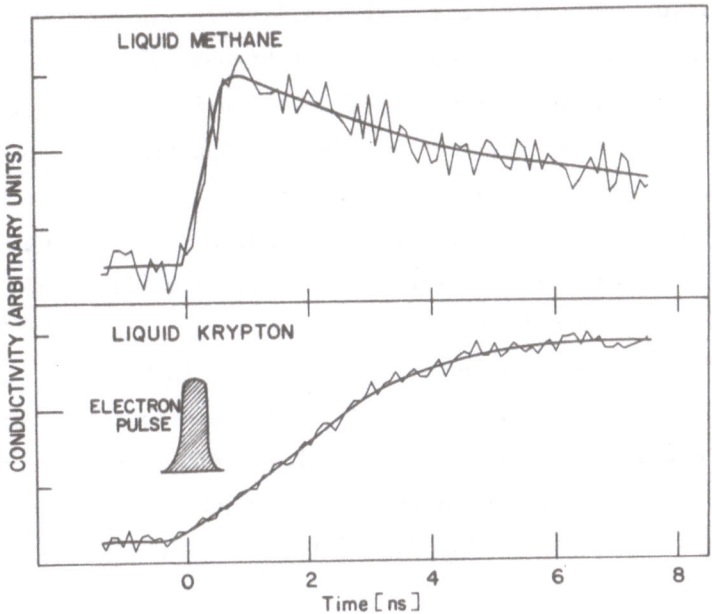

Figure 7. The conductivity change in liquid methane
and liquid Krypton due to excess electron formation on irradiation with a 400 ps, 2 A pulse of
3 MeV electrons. The signals were measured using
a Ka-band reflection cell capable of pressures up
to 250 atmospheres and a 30 ps risetime sampling
system to monitor the change in detector output.
The growth of the signal after the pulse in liquid
Krypton is due to the increase in electron mobility as the electrons cool down to thermal energies.

In conclusion it should be mentioned that the present technique is by no means limited to the study of ions in dielectric liquids although this has been the main topic considered here. As was pointed out in the discussion relaxation microwave techniques using phase shift measurements have been widely applied to low pressure gaseous systems. The absorption method is ideally suited to the study of electrons in high density or high collision frequency gaseous systems. This aspect has been used for the measurement of electron thermalisation times in N_2, He and Ar at pressures of an atmosphere or higher [26,28] and for studying the pressure dependence of electron ion recombination in gases such as H_2O, NH_3 and CO_2 [29,30]. Applications to solid systems where the advantages of an electrodeless conductivity method are significant (e.g. the use of polycrystalline samples) have also been made. In particular data on the mobility of electrons and protons in ice and frozen aqueous solutions [31-35] has been obtained and conductivity transients in hydrated biopolymer systems have been measured [36]. And, last but not least, the microwave conductivity method should provide and extremely useful tool in the study of high dipole moment excited states and charge transfer complexes formed in systems subjected to flash photolysis [37].

REFERENCES

[1a] M.A. Biondi and S.C. Brown, Phys. Rev., 75 (1949) 1700

[1b] M.A. Biondi, Rev. Sci. Instr., 22 (1951) 500

[2] see for example R.W. Fessenden and J.M. Warman, Adv. in Chem. Ser. Nr. 82(II) (1968) 222

[3] A.F. Gibson, Proc. Phys. Soc., B69 (1956) 488

[4] G. Bakale, E.C. Gregg and R.D. McCreary, J. Chem. Phys., 57 (1972) 4246

[5] G. Beck and J.K. Thomas, J. Chem. Phys., 57 (1972) 3649

[6] W.F. Schmidt and G. Bakale, Chem. Phys. Letters, 17 (1972)
     617

[7] I.A. Taub and H.A. Gillis, J. Amer. Chem. Soc., 91 (1971) 650

[8] J.T. Richards and J.K. Thomas, Chem. Phys. Letters, 10 (1971)
     317

[9] J.H. Baxendale, C. Bell and P. Wardman, Chem. Phys. Letters,
     12 (1971) 347; J. Chem. Soc. Farad. Trans, 69 (1973) 776

[10] H.A. Gillis, N.V. Klassen, G.G. Teather and K.H. Lokan, Chem.
     Phys. Letters, 10 (1971) 481

[11] J.M. Warman, M.P. de Haas and A. Hummel, Chem. Phys. Letters,
     22 (1973) 480

[12] P.P. Infelta, M.P. de Haas and J.M. Warman, Radiat. Phys.
     Chem., 10 (1977) 353

[13] J.M. Warman, P.P. Infelta, M.P. de Haas and A. Hummel, Can.
     J. Chem., 55 (1977) 2249

[14] L.H. Luthjens and A.M. Schmidt, Rev. Sci. Instrum., 44 (1973)
     567

[15] J.K. Thomas, K. Johnson, T. Klippert and R. Lowers, J. Chem.
     Phys., 48 (1968) 1608

[16] A.O. Allen, M.P. de Haas and A. Hummel, J. Chem. Phys., 64
     (1976) 2587

[17] M.P. de Haas, PhD Thesis, University of Leiden (1977)

[18] M.P. de Haas, J.M. Warman, P.P. Infelta and A. Hummel, Chem.
     Phys. Letters, 31 (1975) 382

[19] M.P. de Haas, J.M. Warman and A. Hummel, Proc. 5th Int.
     Conf. on Conduction and Breakdown in Dielectric Liquids,
     July 1975, ed. J.M. Goldschvartz, Delft University Press,
     Delft, the Netherlands

[20] J.M. Warman, P.P. Infelta, M.P. de Haas and A. Hummel, Chem.
     Phys. Letters, 43 (1976) 321

[21] L. Nyikos, C.A.M. van den Ende, J.M. Warman and A. Hummel,
     J. Phys. Chem., 84 (1980) 1154

[22] C.A.M. van den Ende, L. Nyikos, U. Sowada, J.M. Warman and

A. Hummel, Proceedings 7th Int. Conf. on Conduction and
Breakdown in Dielectric Liquids, July 1981, Berlin, West-
Germany

[23] C.A.M. van den Ende, L. Nyikos, J.M. Warman and A. Hummel,
submitted for publication

[24] C.A.M. van den Ende, L. Nyikos, J.M. Warman and A. Hummel,
Radiat. Phys. Chem., 15 (1980) 273

[25] C.A.M. van den Ende, L.H. Luthjens, J.M. Warman and A. Hummel,
submitted for publication

[26] U. Sowada and J.M. Warman, Proc. 7th Int. Conf. on Conduction
and Breakdown in Dielectric Liquids, July 1981, Berlin, West-
Germany

[27] U. Sowada, J.M. Warman and M.P. de Haas, submitted for
publication

[28] J.M. Warman and M.P. de Haas, J. Chem. Phys., 63 (1975) 2094

[29] J.M. Warman, E.S. Sennhauser and D.A. Armstrong, J. Chem.
Phys., 70 (1979) 995

[30] E.S. Sennhauser, D.A. Armstrong and J.M. Warman, Radiat.
Phys. Chem., 15 (1980) 479

[31] J.B. Verberne, H. Loman, J.M. Warman, M.P. de Haas, A. Hummel
and L. Prinsen, Nature, 272 (1978) 343

[32] J.M. Warman, M.P. de Haas and J.B. Verberne, J. Phys. Chem.,
84 (1980) 1240

[33] M. Kunst and J.M. Warman, Nature, 288 (1980) 465

[34] J.M. Warman, M. Kunst, M.P. de Haas and J.B. Verberne, Proc.
7th Int. Conf. on Conduction and Breakdown in Dielectric
Liquids, July 1981, Berlin, West-Germany

[35] J.M. Warman, M.P. de Haas, A. Hummel, D. van Lith, J.B.
Verberne and H. Loman, Int. J. Radiat. Biol., 38 (1980) 459

[36] M.P. de Haas, D. van Lith, J.M. Warman, A. Hummel, J.B.
Verberne and H. Loman, Radiat. and Environmental Biophys.,
17 (1980) 271

[37] R.W. Fessenden, P.M. Carton, H. Paul and H. Shimamori, J.

Phys. Chem., <u>83</u> (1979) 1676

Acknowledgement
===

The development and application of the microwave absorption
method has involved a considerable number of people other than
the author. Of particular importance are Andries Hummel for his
continuing support and encouragement, Pierre Infelta for his
important role in the development of the theoretical background
and of course Thijs de Haas who saw the technique through its
early teething troubles and has been to a large extent responsible
for any degree of adulthood and success which it could now be
said to have achieved.

EPR AND NMR DETECTION OF TRANSIENT RADICALS AND REACTION PRODUCTS

Alexander D. Trifunac

Chemistry Division
Argonne National Laboratory
Argonne, Illinois 60439
U.S.A.

ABSTRACT

   Magnetic resonance methods in radiation chemistry are illustrated. The most recent developments in pulsed EPR and NMR studies
in pulse radiolysis are outlined with emphasis on the study of
transient radicals and their reaction products.

INTRODUCTION

   Magnetic resonance techniques in radiation chemistry are relative newcomers. To be sure, people have for many years irradiated
solids or frozen solutions and studied the resulting paramagnetic
species by EPR. But, that sort of application of magnetic resonance, while obviously useful, can be used to observe only radicals
that live for an extended time and will not be of prime concern
to our discussion here. We focus here on the methods that allow
us to study transient radicals.

   In particular, we will focus on the most recent developments
that allow us to observe and study transient radicals in the microsecond and nanosecond time domains (1,2). Thus, an area of application of EPR which involves study of transient radicals by
continuous irradiation ("steady state") will not be covered here.
Everything we have to say about the magnetic resonance methods
would, of course, include those topics as well. But, my intention here is to tell you how to go to the present limits of technique. If you can do that, you will also have the capability to
do all the other types of "non-time resolved variety".

J. H. Baxendale and F. Busi (eds.),
The Study of Fast Processes and Transient Species by Electron Pulse Radiolysis, 163–178.
Copyright © 1982 by D. Reidel Publishing Company.

Implicit in all this discussion is that somehow the magnetic resonance method can provide some unique information, that in spite of its limitations, it has some superior attributes compared to the more conventional optical spectroscopy. Well, that obviously is determined by your point of view. Magnetic resonance can never equal the time resolution and only sometimes can approach the sensitivity of optical spectroscopy. But, where magnetic resonance excells is in the information content of the spectra; one can usually make a definitive assignment of the identity and structure of the transient radical on the basis of EPR and NMR; this is not so straightforward in optical spectra.

Before we plunge into the details of various magnetic resonance methods in radiation chemistry, let me, for the sake of those not so familiar with it, remind you of some very basic principles of magnetic resonance (3).

Magnetic Resonance

Any study of magnetic resonance involves the study of the spin energy levels and the populations of these levels in magnetic fields.

We concern ourselves either with electron spin in the study of EPR or with the nuclear spin in NMR. Here, we concern ourselves only with spin 1/2 systems, as this includes most of the cases of interest.

In a magnetic field H, the spin has two possibilities:

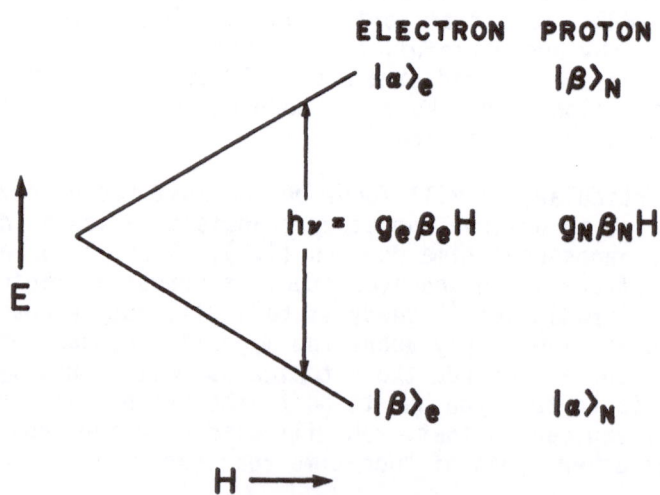

Figure 1. Energy levels of a spin 1/2 species in a magnetic field.

The steady state magnetic field, H, interacts with the magnetic moment of the spin μ. This is usually represented by a Hamiltonian

$$\mathscr{H}_N = -\mu_N \cdot H = -\gamma_N hH \cdot I = -g_N \beta_N H \cdot I \tag{1a}$$

$$\mathscr{H}_E = -\mu_E \cdot H = +\gamma hH \cdot S = +g_E \beta_E H \cdot S \tag{1b}$$

Since I and S have ± 1/2 values, the energy gap $h\nu = g\beta H$. The difference in the magnetogyric ratios of the electron and nucleus are $\sim 10^3$ so we observe NMR in the 10^6 Hz range and EPR in the 10^9 Hz range. Also, note that since at thermal equilibrium $N_\alpha/N_\beta = \exp^{-(g_N\beta_N H)/(kT)}$ according to the Boltzman law, there will be considerably larger population differences in EPR than in NMR by a factor of 10^3.

When we have a radical with electron spins coupled to nuclear spins, we get a more complicated set of energy levels. I will show such energy level schemes later. However, the basic picture is the same.

What will concern us quite a bit when we use time resolved magnetic resonance is that usually we will observe the NMR or EPR transitions before they have attained equilibrium, that is Boltzman population ratios. This non-equilibrium population (chemically induced magnetic polarization, CIDNP, and CIDEP) arises because transient radicals in solutions encounter and react or separate without reaction depending on their spin levels. For example, the reactive encounter (the bond making) requires spin phasing to be ↑↓- (this is singlet). Thus spin, while its energy gap is tiny in size when compared to the chemical energy, does play a very important role in controlling the outcome of the chemical reaction. The use of magnetic fields allows us to both study and, to some extent, influence the outcome of the chemical reaction. The effects are small, but if we can make the chemistry repeat itself in a sort of cyclical fashion, we can accomplish, for example, isotope separation between isotopes with different magnetic properties (4).

In order to gain some understanding of how spin enters into chemical reaction, we must consider pairs of reacting radicals in a magnetic field:

$$R - R \longrightarrow R \cdot \cdot R \overset{diff}{\rightleftharpoons} R \cdot + R \cdot$$
$$\downarrow \qquad\qquad\qquad EPR \tag{2}$$
$$R - R$$
$$NMR$$

In the magnetic field, the electron spin levels are singlet or triplet:

$$T_{+1} \quad\rule{1.5cm}{0.4pt}$$
$$\Big\updownarrow g_E \beta_E H$$
$$T_0 \quad\rule{1.5cm}{0.4pt} \quad\rule{0.8cm}{0.4pt}\; S \qquad S = \frac{1}{\sqrt{2}}(|\alpha\beta\rangle - |\beta\alpha\rangle): \quad T_1 = |\alpha\alpha\rangle$$
$$\Big\updownarrow g_E \beta_E H$$
$$T_{-1} \quad\rule{1.5cm}{0.4pt} \qquad\qquad T_0 = \frac{1}{\sqrt{2}}(|\alpha\beta\rangle + |\beta\alpha\rangle): \quad T_{-1} = |\beta\beta\rangle \qquad (3)$$

If we are doing chemistry in a magnetic field, which is of some magnitude H, which in the case of NMR will be ∿18 kg or EPR ∿3 kg, only the S and T_0 will be close in energy to each other.

After the pair of radicals is created in some spin phasing, e.g., $S \equiv R{\uparrow}{\downarrow}R$, say after breaking the bond, in time the interaction of the electron spin ↓ with the magnetic environment at R may dephase it to become to some extent $R{\uparrow}{\uparrow}R \equiv T_0$. This time evolution is summarized by the following time dependent Schrödinger equation:

$$[-J(2S_1 \cdot S_2 + 1/2) + \mathcal{H}]\psi(t) = i\frac{\partial \psi}{\partial t} \qquad (4)$$

$\mathcal{H}$ is the magnetic Hamiltonian:

$$\mathcal{H} = \beta(g_1 S_1 + g_2 S_2) \cdot H_0$$

$$+ \Sigma A_{1n} I_{1n} \cdot S_1 + \Sigma A_{2n} I_{2n} \cdot S_2 \qquad (5)$$

where β is the Bohr magnetron, H_0 is the external magnetic field, g_1 and g_2 are the electron g factors of the two radicals, A's are the hyperfine coupling constants and J is the electron exchange integral.

Chemically Induced Dynamic Nuclear Polarization (CIDNP) arises from the spin selective reactions of the radical pair. The word "polarization" is used to describe non-equilibrium spin population of nuclei (CIDNP) or electrons (CIDEP). The recombination probability can be assumed to be proportional to the S character and will thus depend on the $S-T_0$ mixing.

Electron Spin Polarization (CIDEP) originates at interradical separation where the exchange J is nonzero, as the interplay of $S-T_0$ mixing and J.

Without going into a much more detailed picture of CIDEP and CIDNP, one can see that the nature of the radicals, i.e., their g-factors and their hyperfine couplings will play a role in the patterns of polarization created by the interactions of such transient radicals.

By studying CIDEP and CIDNP, we can gain information on the past history of the radical or the radical reaction product. This is what we are after when we study reaction mechanisms.

TIME RESOLVED EPR

When very short-lived radicals are studied, nonstandard EPR techniques have to be employed. The conventional EPR spectrometer that one can buy from the manufacturer, uses 100 kHz field modulation and typically time constants are $\sim$1 sec. This improves the sensitivity. The modified commercial equipment can be used to study radicals that live longer than 20-100 μsec. To study shorter lived radicals, $\sim$1 μsec, one can utilize higher field modulation. Historically this was done first (5,6). At Argonne, B. Smaller constructed the 2 MHz field modulated EPR spectrometer (5). Smaller and co-workers were the first to observe the EPR of e_{aq} (7). However, modulation introduces problems in kinetic analysis, so the better choice is to not use it at all. That is the most popular course now and one can easily study radicals with time resolution of 0.2-0.3 μsec (8,9).

The main changes from the commercial instrument are a broad-band amplifier, a boxcar integrator and/or a fast transient recorder. The use of a broad-band amplifier reduces sensitivity. We are talking about $\sim$10^3 less sensitive instrument. This is to some extent made up by the use of boxcar averaging and the fact that nonequilibrium electron spin populations are almost always observed, giving considerable signal enhancement.

The most recent development is pulsed EPR, where two or three microwave pulses are utilized (2). With this approach, time resolution up to the frequency definition limit, which is several nanoseconds, is possible.

Pulsed EPR

The X band microwave bridge which can be utilized for direct detection is modified for use in the pulsed EPR experiment by incorporating three microwave switches, traveling wave tube amplifier (10 watt - TWT) and Ga-As FET amplifier (2). The instrument schematic is illustrated in Figure 2. The microwaves from the Gunn diode are switched by a fast microwave switch #1. This switch provides up to three microwave pulses of appropriate length

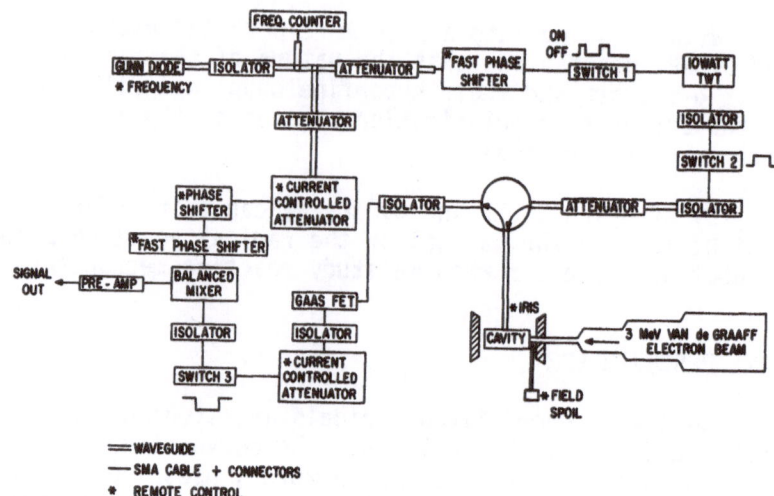

Figure 2. Schematic of microwave network used in pulsed EPR experiments with the Van de Graaff accelerator. The switching functions of the microwave switches are also indicated.

($\sim$10 nsec-1 μsec) and spacing (for example, 90-τ-180; 180-τ-90-τ-180; 90-τ-90-τ-90° sequences). A 90° pulse is typically 30-100 nsec. That means that the microwave field H_1 available at the sample is $\sim$1 gauss ($H_1 = \pi/\gamma\ t_p$).

Switches #2 and #3 are not as fast as switch #1, and are used to protect the detection network from noise and overloads during the time that the 10 watt TWT is amplifying. When switch #1 is providing pulses, switch #2 is open to allow them to get to the cavity and sample, but switch #3 is closed. At the time when the echo is detected, only switch #3 is open.

The microwave switching is controlled by a pulse sequencer, and timing can be automatically swept by the time-delay programmer. The whole system of pulse timing is controlled by 100 MHz clocks, where the smallest time increment is 10 nsec (one can get 1 nsec increments) and the time jitter is $\sim$2 nsec in our experiment.

In pulse radiolysis, the EPR magnet control, and the micro-wave bridge tuning must be performed remotely. Remote functions include Gunn diode frequency, phase, phase arm bias, microwave signal attenuation, TWT gain, iris tuning, and magnetic field setting and sweeping.

Additional and somewhat unusual features are the field spoil and the fast phase shifter. They are used to reduce free induc-tion decay interference as will be illustrated.

   Time Resolved Spectra (Field Sweep). After a transient radical is generated by a short electron beam pulse, we can examine
this magnetization by applying a two pulse microwave sequence
(90°-τ-180°) at a fixed time t after the electron beam pulse and
sweeping the magnetic field. Thus, the EPR spectrum at that instant of time is obtained. This timing of pulses is illustrated
in Figure 3. The action of the microwave pulse on the spin system
can be best understood if we use a rotating frame picture (Figures
4 and 5).

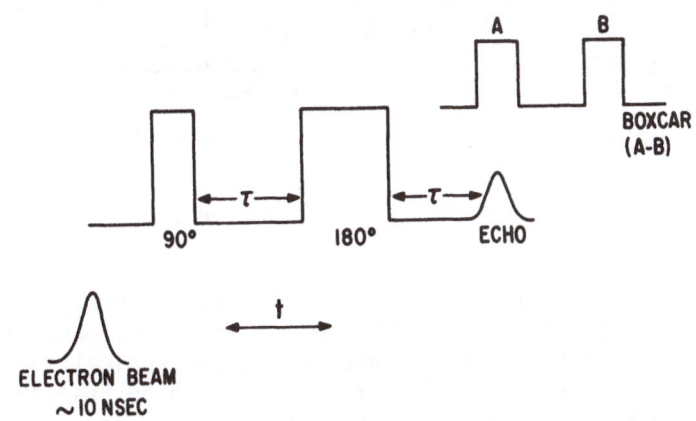

Figure 3. Timing sequence of the electron beam pulse,
microwave pulses and the observation (boxcar) gates.

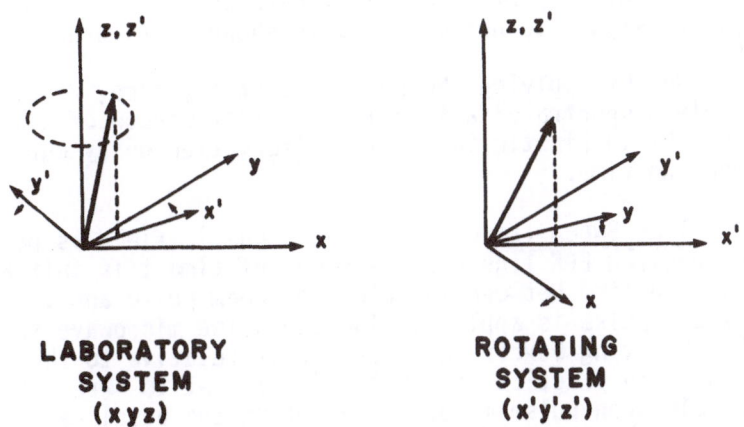

Figure 4. Precessing vector in laboratory and rotating
coordinate systems.

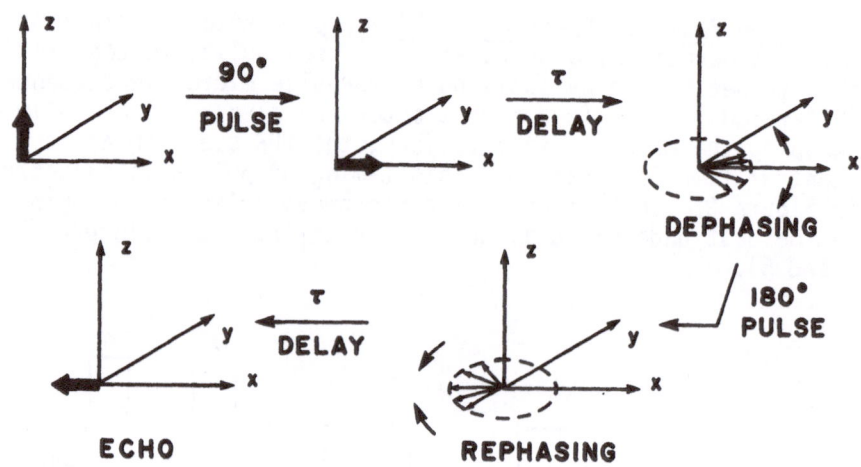

Figure 5.   Pulse sequence to observe a spin echo.

The main thing is to realize that only the magnetization
sampled during the time of the 90° pulse is refocused to give an
echo.   The EPR spectrum is obtained on the radical species present
during the 90° pulse application.   This can be transient or steady
state magnetization present at that time.

For example, if we look at the aqueous radicals of sodium
acetate, we can see the radical $\cdot CH_2COO^-$ in neutral or slightly
basic solution, as shown in Figure 6A.   The big signal in the
middle is from the suprasil walls of the EPR flat cell.   We can
reduce it somewhat by using phase shifted microwave pulse sequences.
This essentially provides light-dark subtraction.   However, we
lose $\sqrt{2}$ signal-to-noise.   This is shown in Figures 6A and 7.

So, by applying the 90° pulse at any time, we obtain time-
resolved spectra of a transient at that window of time.   We can
also obtain kinetic information (formation decay curves) by
sweeping time.

Time Sweep (Kinetics).   The magnetic field is positioned on
the desired EPR line and the sweep of time t is initiated.   Time
t is the time between the electron beam pulse and the time when
the 90° pulse is applied.   The two pulse microwave sequence and
the observing boxcar gates are swept relative to the electron beam
pulse.   In practice, the 90° pulse is set to sweep ∿200 nsec before
the electron pulse in order to obtain the baseline.   A baseline
can also be obtained when the signal decays to noise but requires
longer time sweep.

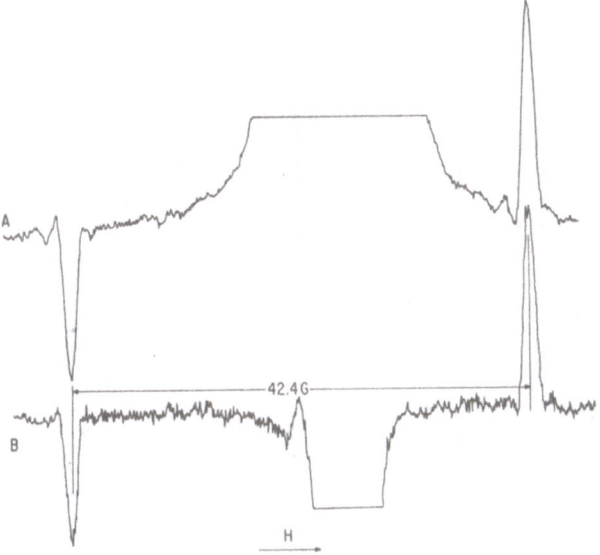

Figure 6. Field swept EPR spectrum (at 1 μsec) obtained
during radiolysis of aqueous 0.5 M potassium acetate at
pH ∿ 11 (N$_2$O sat); the middle line of the $\cdot CH_2CO_2^-$ radical can be seen (B) when the quartz signal is reduced
by beam-no beam subtraction.

The kinetic sweep of the two acetate radical lines are illustrated in Figure 8.

Free Induction Decay. The free induction decay (FID) following the 90° and 180° microwave pulses interferes with the echo.
For radicals in non-viscous liquids $T_2 \sim T_1$, thus in a homogeneous
magnetic field the FID may last for several microseconds. T_1 and
T_2 are the spin lattice and spin-spin relaxation times. Modern
NMR is based on the utilization of FID to obtain spectra. However, the EPR FID does not last seconds as in NMR and too much of
the EPR FID is lost at early times, so it is hard to apply the
Fourier transformation. Nevertheless, one can obtain "FID" spectra using a single microwave pulse as shown in Figure 9.

In the spin echo experiment, FID interference has to be minimized. We accomplish this by making the magnetic field artifi-
cially more inhomogeneous by the use of a field spoil and by phase
alternation of the 180° pulses.

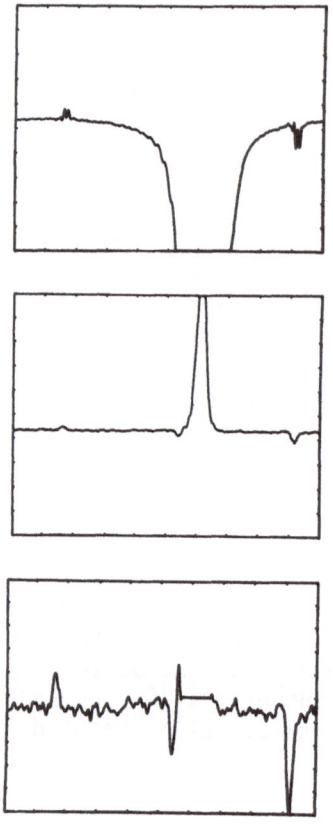

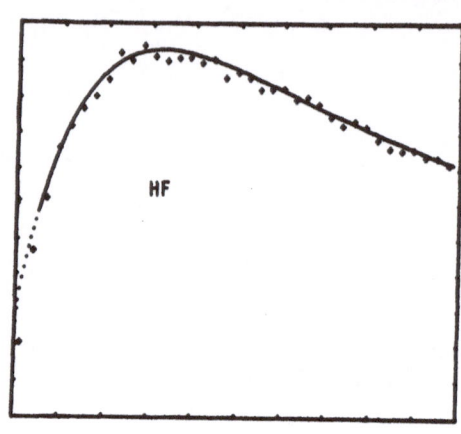

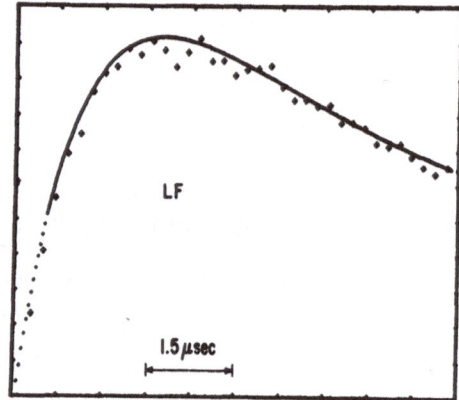

Figure 7. EPR spectra of
·CH$_2$CO$_2^-$ radical at 1 μsec.
Top-without and middle-with
the phase shifted pulse sequences, bottom-the y-enlarged
version of the middle spectrum.

Figure 8. Least squares fitting (solid line) of data
(crosses) of kinetic traces of
the two lines of the acetate
radical. Dots indicate the
calculated line through data
points that were not used in
the least squares fitting.
The plots show absolute, normalized intensities of the two
lines.

Data analysis and other varieties of pulsed EPR will be discussed later. We will also try to point out the advantages and
problems encountered in certain applications of pulsed EPR to
radiolytic systems.

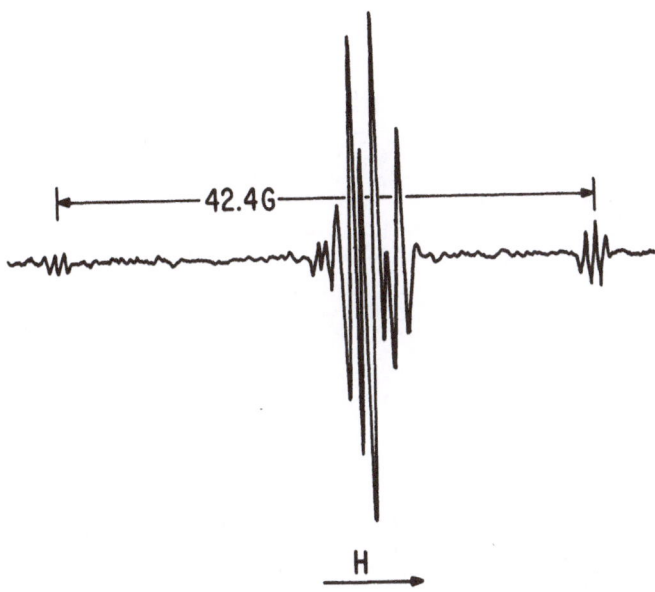

Figure 9. "FID" spectrum of the acetate radical (at 1 μsec) obtained by using only a 90° pulse.

NMR IN RADIATION CHEMISTRY

Over the last two decades, NMR has revolutionized several areas of chemistry. Organic chemists, especially, have found NMR an indispensable tool for analysis of reaction products, etc. Then, over the last ten years, pulsed NMR-Fourier Transfer NMR has made another quantum improvement in NMR, opening a whole new set of possibilities.

It is perhaps surprising that NMR has not become a tool in radiation chemistry. It appears that radiation chemists are a very conservative lot, even the EPR - especially time resolved EPR - is not very widespread. So far, we at Argonne are the only ones to use NMR in radiolysis (10-12).

Our approach aims at using NMR to study products of radical reactions in radiolysis and, as you will see, we have recently developed some new NMR based detection methods that allow us to obtain time resolved information on the reacting radicals.

The NMR magnet has to be very homogeneous since we are studying energy differences between the nuclear spin levels, and as you remember, they are 10^{-3} smaller than the electron spin energy difference observed by EPR. We cannot make axial holes in the NMR magnet and still do high resolution NMR. So, we use two magnets.

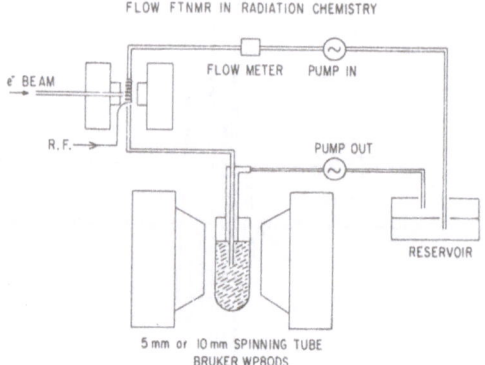

Figure 10.   The schematic of the flow NMR experiment
for pulse radiolysis.

The electron beam enters axially a small electromagnet (0-8 kg)
in which the sample irradiation is carried out (Figure 10).   A
fast flow system is used to transfer the irradiated solution to
the probe of the NMR spectrometer.   All the radical reactions are
over in ∿100 μsec and only the diamagnetic products are transferre
to the NMR probe for examination.   However, in the nuclear spin
population levels of these diamagnetic products is contained the
memory of the radical encounters of precursors of this product.
The nonequilibrium nuclear spin population - CIDNP - will be observable as long as the examination of the diamagnetic product is
carried out before nuclear spin relaxation (T_1) obliterates it.

In protons, $T_1 \sim$ 1-3 seconds and in ^{13}C T_1's are even longer;
so as long as the flow system is capable of transferring the irradiated solution to the NMR probe within 1-2 seconds, the experiment is feasible.

As it turns out, and I will illustrate this further, it is
very convenient that we can do chemistry (irradiation) in a variable magnetic field, because there is much information to be
gained from the field dependence of CIDNP.

To illustrate how we do the NMR experiments and what sort of
information is obtainable, I will use a simple system of aqueous
methanol radiolysis.   The principal radical intermediate is the
hydroxymethyl radical.   The principal products seen by NMR-CIDNP
are methanol and ethylene glycol.   Formaldehyde, another known
product, is not seen.   Apparently, either a nonradical mechanism

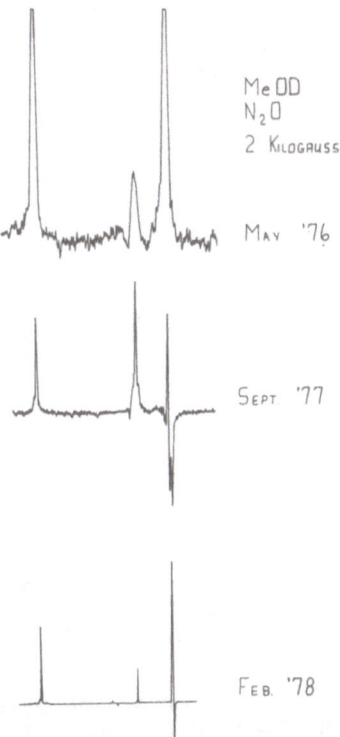

MeOD
N₂O
2 KILOGAUSS

MAY '76

SEPT. '77

FEB. '78

Figure 11. NMR spectra showing CIDNP of products observed in methanol radiolysis. Evolution of experimental technique is illustrated. Top and middle spectra were obtained using CW-NMR. Bottom spectrum was
obtained using FTNMR.

is responsible for formaldehyde formation, or there is no efficient polarization pathway.

The observed polarization (Figure 11) is explained by the
following scheme (* indicates polarization):

$$e^-_{aq} + \cdot CH_2OD \nearrow \begin{array}{l} CH_2OD\overline{*} \xrightarrow{D^+} DCH_2OD \ (E) \\[2em] \cdot CH_2OD* \xrightarrow{\cdot CH_2OD} (CH_2OD_2) \ (A) \end{array}$$

(6)

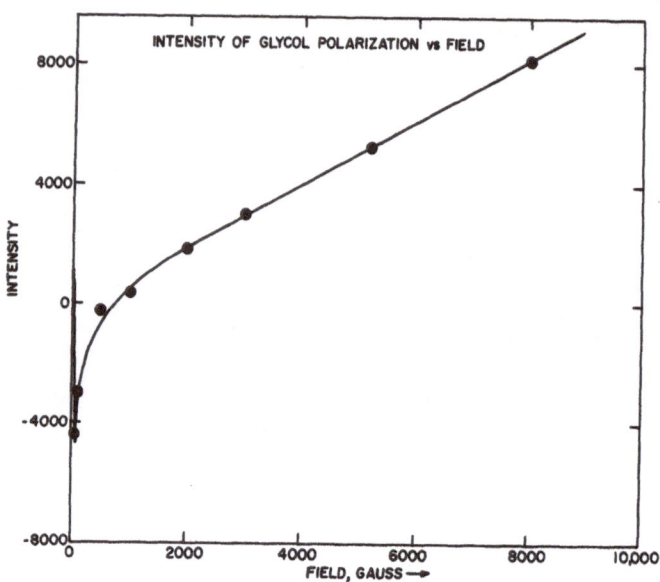

Figure 12. Field dependence of glycol polarization in pulse radiolysis of aqueous methanol (D_2O solution).

The e^-_{aq} is the important partner in producing polarization as substantiated by observing the field dependence of the ethylene glycol polarization. Figure 12, which clearly indicates that polarization is very much field dependent, i.e., g-factor difference between the radicals in the radical pair is substantial. Also, N_2O addition substantially reduces the polarization.

Radiolysis of several simple compounds was studied in some detail. Acetone, acetaldehyde, acetate, dimethylsulfoxide, and reactions involving H radicals were studied. In many radiolytic systems NMR is ideally suited to provide considerable information on the mechanism of radical reactions. Also, compared to EPR, the experiment is rather straightforward. Only the interpretation of the more complex NMR spectra can be difficult and time consuming. Also, one can do ^{13}C experiments using natural abundance or enriched samples, and we have done so (13).

So, in combination with just detecting and sorting out all the various products found in radiolysis of a given chemical system, flow NMR can provide details of the reaction mechanism, i.e., the history of a product or product fragment in terms of the reactions and encounters of its precursor radical.

Actually, the NMR experiment, as illustrated, is not a time resolved one. A rather simple modification provides us with a whole new approach to the study of transient radicals (14). In

that experiment, we observe nuclear resonance of transient radicals by NMR by applying a pulse of rf at the radical NMR frequency
during the radical lifetime. One gets a sort of ENDOR spectrum
of the transient radicals, and thus, one obtains hyperfine coupling constants with great precision. One can do kinetics and
get all sorts of other information, as will be illustrated later.

CONCLUSION

     State-of-the-art of magnetic resonance in radiation chemistry
has been illustrated. The experimental aspects of the technique
were outlined.

     Magnetic resonance, while lacking the sensitivity and the
time resolution of conventional optical methods, can provide much
more definitive information about the reactive radical intermediates in solution.

ACKNOWLEDGMENT

     Work supported by the Office of Basic Energy Sciences,
Division of Chemical Sciences, U. S. Department of Energy, under
Contract W-31-109-Eng-38.

REFERENCES

(1) Trifunac, A.D. and Thurnauer, M.C.: 1979, *Time Domain Elec-
     tron Spin Resonance*, ed. L. Kevan and R. N. Schwartz, John
     Wiley & Sons, Inc., New York.
(2) Trifunac, A.D., Norris, J.R., and Lawler, R.G.: 1979,
     J. Chem. Phys. 71, p. 4380.
(3) Carrington, A. and McLachlan, A.D.: 1967, *Introduction to
     Magnetic Resonance*, Harper and Row, New York.
(4) *Chemically Induced Magnetic Polarization*,: 1977, L. T. Muus
     et al., eds. (NATO ASI), D. Reidel, Publishers, Dordrecht.
(5) Smaller, B., Remko, J.R., and Avery E.C.: 1968, J. Chem.
     Phys. 78, p. 5174.
(6) Atkins, P.W., McLauchlan, K.A., and Simpson, A.F.: 1970,
     J. Phys. (E) 3, p. 547.
(7) Avery, E.C., Remko, J.R., and Smaller, B.: 1968, J. Chem.
     Phys. 79, p. 951.
(8) Trifunac, A.D., Johnson, K.W., Clifft, B.E., and Lowers,
     R.H.: 1975, Chem. Phys. Lett. 35, p. 566.
(9) Verma, N.C. and Fessenden, R.W.: 1976, J. Chem. Phys. 65,
     p. 2139.
(10) Trifunac, A.D., Johnson, K.W., and Lowers, R.H.: 1976,
     J. Am. Chem. Soc. 98, p. 6067.

(11) Trifunac, A.D. and Nelson, D.J.: 1977, J. Am. Chem. Soc. 99,
      p. 1745.
(12) Nelson, D.J., Trifunac, A.D., Thurnauer, M.C., and Norris,
      J.R.: 1979, Revs: Chem. Intermediates 3, p. 131.
(13) Lawler, R.G., Nelson, D.J., and Trifunac, A.D.: 1979,
      J. Phys. Chem. 83, p. 3444.
(14) Trifunac, A.D. and Evanochko, W.T.: 1980, J. Am. Chem. Soc.
      102, p. 4598.

RADICAL IONS AND EXCITED STATES IN RADIOLYSIS. OPTICALLY DETECTED TIME RESOLVED EPR

Alexander D. Trifunac and Joseph P. Smith

Chemistry Division
Argonne National Laboratory
Argonne, Illinois 60439
U.S.A.

ABSTRACT

Excited state production and radical ion recombination kinetics in pulse irradiated solutions of aromatic solutes in cyclohexane are studied by a new method of optical detection of time-resolved electron paramagnetic resonance (EPR) spectra.

INTRODUCTION

The EPR spectra of radical ion pairs that recombine to produce an excited singlet state can be detected optically *via* observation of a resonant change in the intensity of fluorescence arising from ion recombination. This not only permits the application of magnetic resonance to some very interesting problems, but also results in a major improvement in the specificity and sensitivity of time resolved EPR (1,2).

One area of current research (3,4) is the radiation chemistry of hydrocarbon solutions of aromatic molecules. Radical ion recombination is a major source of the excited states produced by pulse radiolysis. The recombination pathways for excited state production are summarized in the following scheme where S and Ar represent the solvent and aromatic solute, respectively.

$$S \xrightarrow{\text{e}^- \text{ beam}} S^+ + e^- \tag{1}$$

$$S^+ + Ar \longrightarrow Ar^+ + S \tag{2}$$

J. H. Baxendale and F. Busi (eds.),
The Study of Fast Processes and Transient Species by Electron Pulse Radiolysis, 179–187.

$$e^- + Ar \longrightarrow Ar^- \tag{3}$$

$$Ar^+ + e^- \longrightarrow Ar* \tag{4}$$

$$Ar^- + S^+ \longrightarrow Ar* + S \tag{5}$$

$$Ar^- + Ar^+ \longrightarrow Ar* + Ar \tag{6}$$

The primary products of ionization (1) are rapidly scavenged by solute molecules (2,3). Several ion recombination pathways (4-6) can then yield excited solute molecules. Experimentally, excited singlet state solutes(*) are observed *via* fluorescence and excited triplet states are most conveniently observed by triplet-triplet absorption.

The EPR spectra of radical ion pairs can be detected by observation of a resonant decrease in fluorescence following the application of a single (30-100 ns) microwave pulse. Here we will discuss several features of the fluorescence detected magnetic resonance (FDMR) technique.

METHOD, RESULTS, AND DISCUSSION

The experimental setup is illustrated in Figure 1. We use the same pulsed EPR spectrometer described previously. · The light from excited state emission escapes through a hole in the bottom of the cavity and is detected at a selected wavelength. At some time after irradiation, a single microwave pulse is applied and the fluorescence is sampled during a subsequent time window selected by a boxcar averager. An EPR spectrum corresponding to a superposition of the spectra of the isolated radical ions that recombined to produce the excited state is obtained by measuring the fluorescence intensity as a function of the applied magnetic field. For each spectrum, the positions of the microwave pulse and boxcar averager sampling window are fixed with respect to the time of irradiation. Typical FDMR spectra obtained for kinetic studies (Figure 2) consist of a single peak with no resolved hyperfine structure. In some cases, the hyperfine structure of the FDMR spectrum can be resolved through the use of long (250-500 ns), low power ($H_1 \leqslant 0.4$ g) microwave pulses (Figure 3). Although the hyperfine structure is resolved at the expense of some of the time resolution and sensitivity of the FDMR method, the spectra obtained in this way can provide useful data on the identity and structure of the radical ion precursors of excited states in more complex chemical systems. A plot of integrated FDMR intensity vs. time, obtained from a series of spectra similar to those in Figure 2, provides information about radical ion pair spin dynamics and chemical kinetics (Figure 4). The intensity of the FDMR signal increases for ∿70 ns after application of the microwave pulse and

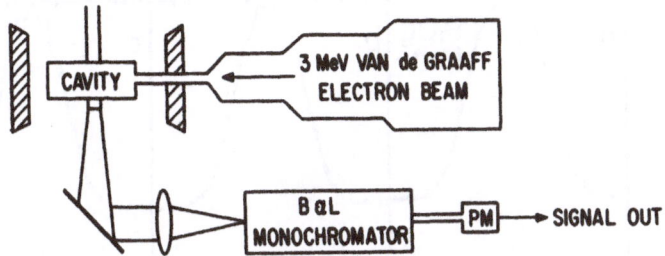

Figure 1. Experimental setup for optical detection of magnetic resonance showing the arrangement of the microwave cavity containing the sample cell; the electron beam and the light detection apparatus.

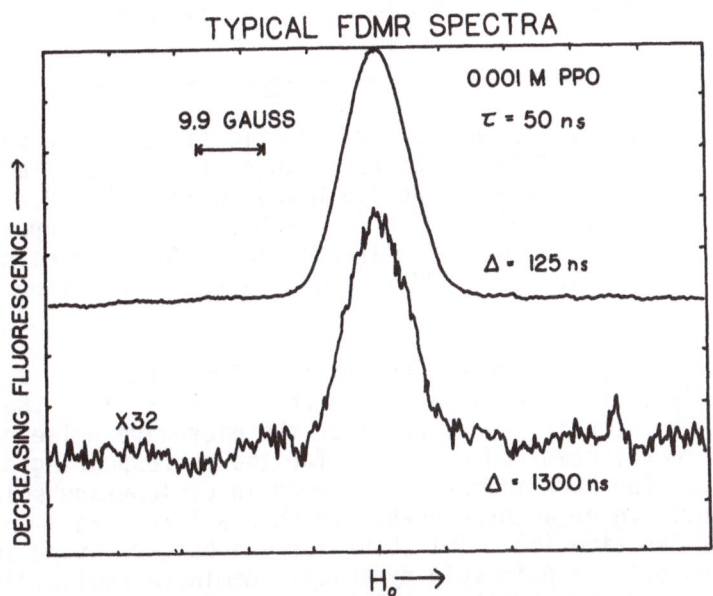

Figure 2. Typical FDMR spectra of 10^{-3} PPO (2,5 diphenyloxazole) in cyclohexane. The microwave pulse was applied from 0 to 100 ns after irradiation with a 5 ns electron beam pulse. The fluorescence was sampled from 75-175 ns (upper trace) or 1250-1350 ns (lower trace) after irradiation.

then decays. The time dependence of the FDMR intensity is dominated by spin dynamics in the time domain immediately following the microwave pulse and by the recombination kinetics of the radical ion pair population at later times.

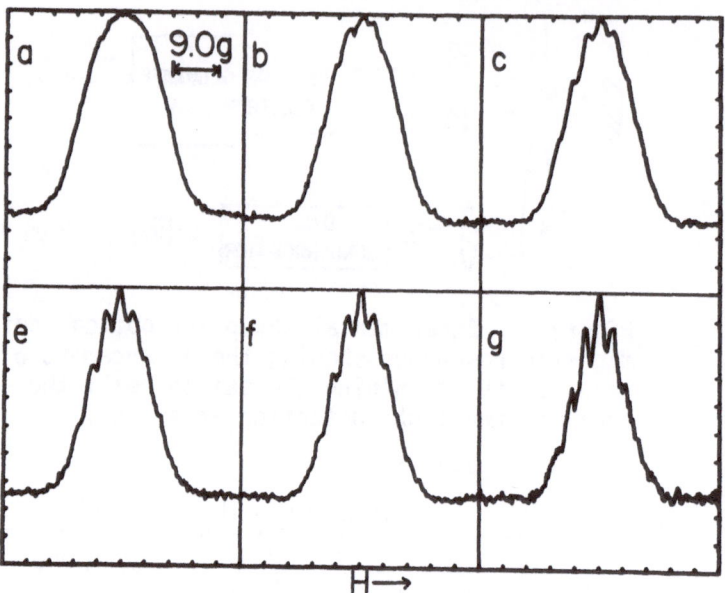

Figure 3. Microwave magnetic field (H_1) dependence of
FDMR of 10^{-3} M biphenyl in cyclohexane. A microwave
pulse of H_1 value 1.5 G (a), 0.74 G (b), 0.6 G (c),
0.47 G (d), 0.37 G (e) or 0.26 G (f) was applied from
0 to 500 ns after irradiation. The fluoresence was
sampled from 500 to 600 ns after irradiation.

When the time dependence of the FDMR intensity is measured
with higher time resolution (Figure 5), it is clear that most of
the FDMR intensity develops after the microwave pulse has ended.
Furthermore, compared with data for the corresponding deuterated
solutes, the FDMR intensity observed in cyclohexane solutions of
biphenyl-h_{10} or anthracene-h_{10} reaches a larger maximum value at
an earlier time (5). This behavior can be understood in terms of
the radical ion pair spin dynamics. Geminate radical ion pairs
are created with purely singlet orientation of the unpaired electron spins. In a static magnetic field, the electron spin wave-
function Ψ rapidly evolves from its initial state where $|\langle\Psi|S\rangle|^2 =$
1 to a quasi steady state where $|\langle\Psi|S\rangle|^2 = |\langle\Psi|T_0\rangle|^2$ (6). Micro-
wave radiation of the appropriate frequency connects states with
T_0 character with states of T_{+1} or T_{-1} character and a decrease
in $|\langle\Psi|T_0\rangle|^2$ results. After the microwave pulse, hyperfine in-
duced S-T_0 mixing acts to re-establish the steady state at a rate
determined by the magnitude of the hyperfine coupling constants.
The S-T_0 mixing rate is given by:

$$\omega_{S-T_0} = (2h^{-1})[(g_1 - g_2)\beta H_0 + \sum_i a_{1i} m_{1i} + \sum_j a_{2j} m_{2j}] \qquad (7)$$

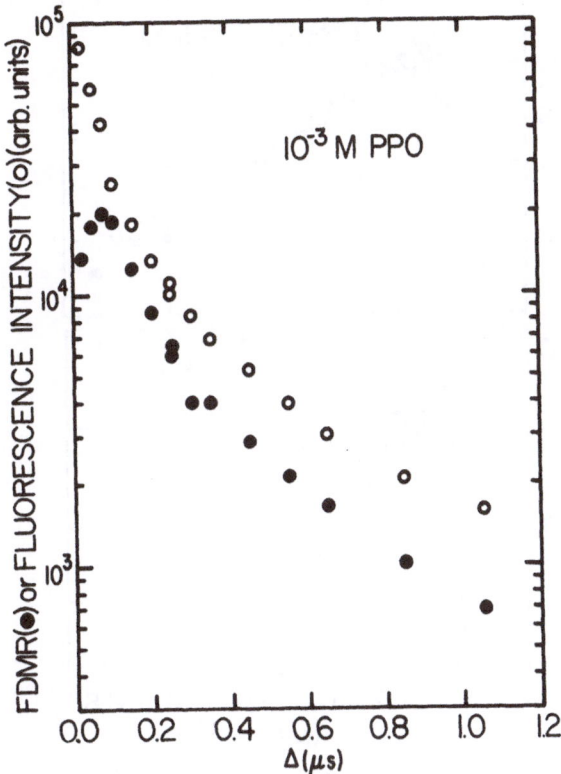

Figure 4. Comparison of fluorescence intensity (○) and
FDMR intensity (●) observed in 10^{-3} M PPO-cyclohexane.
A microwave pulse was applied from 0-100 ns after irradiation. The fluorescence was sampled at time Δ after
the center of the microwave pulse. For comparison, the
FDMR intensity has been scaled by an arbitrary multi-
plicative constant.

which is the familiar expression from treatments of CIDNP, CIDEP,
and the magnetic field effect (7). Since usually there is no
substantial g factor difference for aromatic radical cations and
anions, the hyperfine coupling constants determine the $S-T_0$ mixing
rate. Consequently, in the time domain immediately following the
microwave pulse, we observe the decay of $|<\Psi|S>|^2$ as the steady
state is approached. Due to the smaller hyperfine coupling constants of the deuterated radical ions, the rate of $S-T_0$ mixing,
and thus the approach to the steady state is slower for deuterated
solutes. After the steady state is established, the time dependence of the FDMR intensity is determined by the recombination
kinetics of the radical ion pairs.

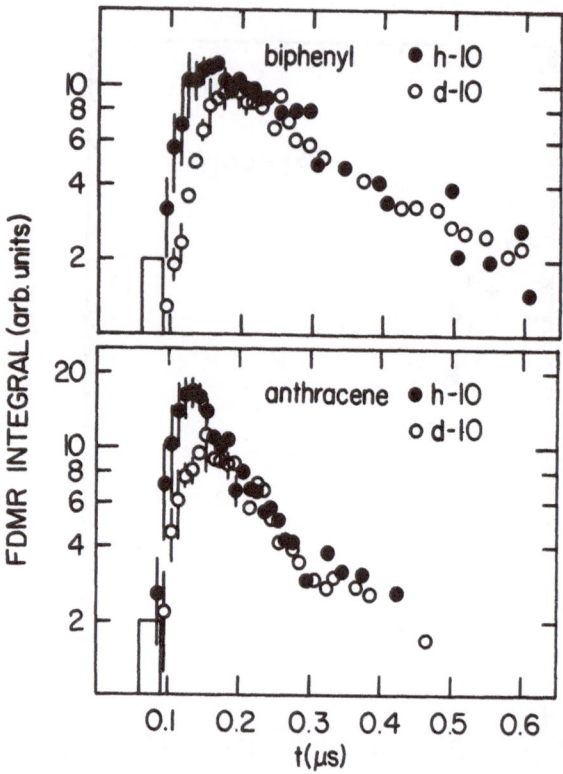

Figure 5. Time dependence of integrated FDMR intensity observed in cyclohexane solutions (0.001 M) of (a) biphenyl-h_{10} (●) and biphenyl-d_{10} (○) or (b) anthracene-h_{10} (●) and anthracene-d_{10} (○). Vertical bars represent standard deviation of the mean FDMR intensity for times where multiple spectra were measured. The rectangular box indicates the position of the 30 ns microwave pulse. The integrated area units of biphenyl and anthracene FDMR are not directly comparable.

The time and radiation dose dependence of the FDMR intensity establish that the FDMR phenomenon is a resonant change in the yield of excited singlets from *geminate* ion recombination. Time resolved absorption and conductivity measurements (8,9) have shown that the yield (G_{gi}) of geminate ion pairs decays according to

$$G_{gi}(t) = G_{fi}[1 + (\gamma/t)^{\frac{1}{2}}] \tag{8}$$

where G_{fi} is the free ion yield and γ is characteristic of the recombination reaction kinetics. If the FDMR effect is a resonant change in the yield of excited states arising from geminate recombination then the FDMR intensity should be proportional to $t^{-3/2}$

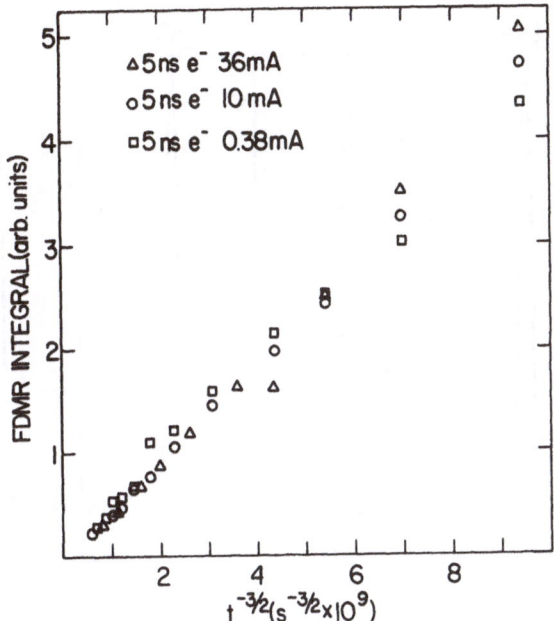

Figure 6. Radiation dose dependence of FDMR of 10^{-3} M PPO in cyclohexane plotted on a $t^{-3/2}$ scale. Samples were irradiated with 5 ns electron beam pulses with peak current values of 36 mA ($\triangle$), 10 mA ($\bigcirc$), and 0.38 m ($\square$). The microwave pulse was applied from 0 to 100 ns after irradiation and the fluorescence was sampled at time t after irradiation. Data for different doses were scaled so that all data sets had a common value at $t^{-3/2} = 4 \times 10^9$ s$^{-3/2}$.

and scale linearly with the absorbed dose, since the fluorescence intensity is proportional to the number of recombination events per unit time. This behavior is observed for a broad range of radiation doses (Figure 6). At higher doses, an increased rate of decay of the FDMR intensity suggests a possible contribution by a dose dependent rate of spin lattice relaxation due to Heisenberg exchange (2).

The FDMR spectrum is sensitive to the structure of the radical ion precursors of excited states. Consequently, the FDMR technique could be used to distinguish the contributions of the various recombination pathways (4-6) to excited state production. The FDMR spectrum of the PPO-cyclohexane system at a low solute concentration shows the presence of broad "wings" in addition to a narrow central peak (Figure 7). The decay of the radical species responsible for the wings is somewhat faster than the decay of the species responsible for the central peak (2). This behavior is

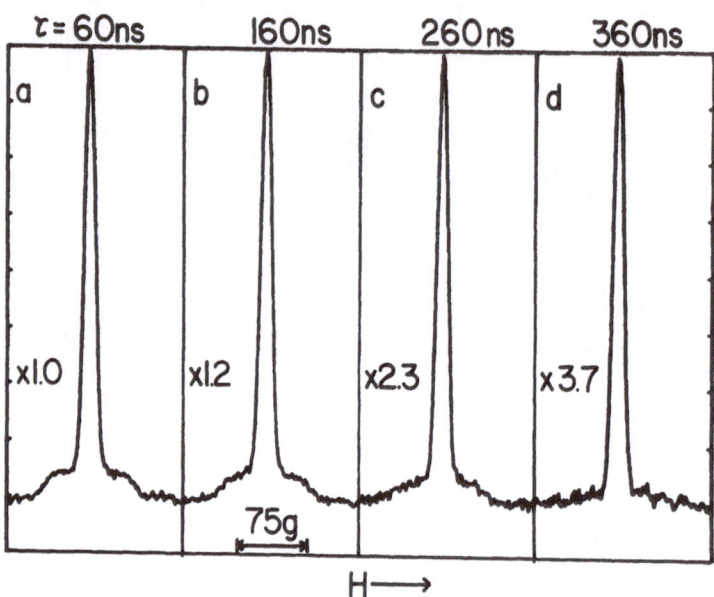

Figure 7. Dependence of the FDMR of 10^{-4} M PPO in cyclohexane wave pulse was started at 10 ns (a), 110 ns (b), 210 ns (c), and 310 ns (d) after irradiation. The fluorescence was sampled from 75 ns to 175 ns after the start of the microwave pulse. For comparison, the y axis values of the spectra have been multiplied by the factors indicated.

consistent with the idea that the wings are essentially the EPR spectrum of the cyclohexane radical cation or "hole." The kinetic properties of the species responsible for the wings are being actively investigated. If the FDMR method could be used to obtain quantitative data on the relative importance of solvent-solute or solute-solute recombination in excited state production, it could provide the first direct experimental check on theoretical calculations of the recombination kinetics (10,11).

Future applications of FDMR in radiation chemistry will explore excited state production in increasingly complex chemical systems where the sensitivity and specificity of the technique can be exploited to best advantage. Another potentially fruitful area of inquiry is the recombination of photoproduced ion pairs in liquid solution. Accordingly, we have recently undertaken an FDMR investigation of photoionization induced by an excimer laser excitation source.

SUMMARY

Time resolved pulsed EPR experiments utilizing optical detection were illustrated. The FDMR method provides a very sensitive and specific tool for the study of radical ion reactions that yield excited states.

ACKNOWLEDGMENT

Work supported by the Office of Basic Energy Sciences, Division of Chemical Sciences, U. S. Department of Energy, under Contract W-31-109-Eng-38.

REFERENCES

(1) Trifunac, A.D. and Smith, J.P.: 1980, Chem. Phys. Lett. 73, p. 94.
(2) Smith, J.P. and Trifunac, A.D.: 1981, J. Phys. Chem. 85, p. 1645.
(3) Jonah, C.D., Sauer, M.C.,Jr., Cooper, R., and Trifunac, A.D.: 1979, Chem. Phys. Lett. 63, p. 535.
(4) Katsumara, Y., Tagawa, S., and Tabata, Y.: 1980, J. Phys. Chem. 84, p. 833.
(5) Smith, J.P. and Trifunac, A.D.: 1981, Chem. Phys. Lett. in press.
(6) Brocklehurst, B.: 1976, J. Chem. Soc. Faraday Trans. II, 72, p. 1869.
(7) For a general reference see *Chemically Induced Magnetic Polarization*, L. T. Muus *et al.*, eds., D. Reidel Publishing Co., Dordrecht, 1977.
(8) Warman, J.M., Infelta, P.O., DeHaas, M.P., and Hummel, A.: 1977, Can. J. Chem. 55, p. 2249.
(9) Van Den Ende, C.A.M., Nyikos, L., Warman, J.M., and Hummel, A.: 1980, Radiat. Phys. Chem. 15, p. 273.
(10) Rzad, S.J.: 1972, J. Phys. Chem. 76, p. 3722.
(11) Sauer, M.C.,Jr. and Jonah, C.D.: 1980, J. Phys. Chem. 84, p. 2539.

LIGHT SCATTERING TECHNIQUES FOR INVESTIGATION OF TRANSIENTS PRODUCED IN ELECTRON PULSE RADIOLYSIS

Michael A. J. Rodgers

Center for Fast Kinetics Research, University of Texas, Patterson Bldg. Rm. 131, Austin, Texas 78712 U.S.A.

INTRODUCTION

In the two decades since electron pulse radiolysis was first developed several diagnostic techniques have been employed to detect and characterise the various species which are formed during the energy absorption stage. By far the commonest technique has been optical spectrophotometry (both emission and absorption) and the spectral and dynamic features of many radiation-induced reactions have been followed in this way. However, the characteristics of high sensitivity and amenability to high time resolution which have made kinetic spectrophotometry so attractive are also shown by other physical methods such as electrical conductivity, polarography, electron spin resonance and light scattering. The spectrophotometric, conductimetric, polarographic and magnetic resonance techniques are considered in detail elsewhere in this volume (1-4); here the use of light scattering spectroscopy as a diagnostic tool for pulse radiolysis studies is examined. Elastic (Rayleigh) and inelastic (Raman) scattering methods have been employed, leading to different kinds of information; both techniques are described.

RAYLEIGH SCATTERING

Elastic scattering is said to occur when part of the energy of a plane light wave travelling in a forward direction is taken up by a molecular electronic system and emitted in directions other than the incident one. Although attenuation of the main beam results, the total light emanating from the sample, integrated over all angles, is undiminished. Furthermore, the scattered

J. H. Baxendale and F. Busi (eds.),
The Study of Fast Processes and Transient Species by Electron Pulse Radiolysis, 189–197.

photons have the same frequency as the incident photons. The intensity of light scattered depends on the incident intensity, the observational angle, the observer distance and the incident frequency. For a single particle of small dimensions (compared to wavelength of light, λ) the intensity I_θ scattered through an angle θ is given by (5)

$$I_\theta = \frac{I_o}{r_s^2} \cdot \frac{8\pi^4\alpha^2 (1 + \cos^2\theta)}{\lambda^4} \qquad (1)$$

In equation (1) I_o is the incident light intensity, r_s is the distance between the scattering volume and the detector and α is the polarisability of the scatterer. Extensions of the Rayleigh theory have been made to solutions of macromolecules where both shapes and sizes of different parts of the same molecule are taken into account (5). These lead to the conclusion that I_θ depends on the number of scattering particles per unit volume and their molecular weight (weight average).

Using this principle, Schnabel and coworkers developed a technique for following the time profile of the scattered light intensity after a polymer solution had been subjected to a pulse of high energy electrons (6). Under these conditions chain scissions occur which lead to molecular weight changes in the affected polymer molecules. Hence the weight average molecular weight of the ensemble decreases and the scattered intensity falls. The time profile of the intensity change leads to rate constants for reaction causing scission (6). This technique was subsequently extended to monitoring the kinetics of untangling of DNA fragments after attack by hydroxyl radicals (7). In addition to monitoring molecular weight decreases, Rayleigh scattering could be applied to observation of the kinetics of macromolecule growth (e.g. colloid formation) after initiation by an electron pulse.

RAMAN SCATTERING

Background

While elastic scattering (Rayleigh) yields only kinetic information about the perturbations induced by electron beam irradiation, inelastic scattering (Raman) contains information concerning the nuclear motions of the scattering molecule. During an interaction between a photon and a molecule either inelastic or elastic scattering can occur. In the inelastic process the photon loses some of its energy if a transition is stimulated in the molecule. Since molecular transitions are quantized the frequency of the inelastically-scattered radiation will differ from that of the incident radiation by discrete amounts which are related to the vibrational or rotational characteristics of the molecule. Thus spectral dispersion of the scattered radiation shows a series of lines at lower frequencies than the Rayleigh component (Stokes lines). In addition a series of lines occurs at higher frequencies than the Rayleigh component. These anti-Stokes lines result from interactions in which molecules already in excited molecular states, transfer energy to the photons. Energy transfer between photons and quantized molecular states in which only a small amount (1-10%) of the photon energy is removed per interaction can be visualized as a consequence of photons possessing momentum. During a photon-molecule collision a transfer of momentum can occur between the quantized energy states of the nuclear motions and the photon resulting in scattering of the photon and a change in energy. In this way the Raman effect is the analog of the Compton effect in which high energy electromagnetic radiations (e.g. x-rays) are inelastically scattered by electrons.

    The intensity of the light scattered by the Raman effect is usually very much weaker than the elastically-scattered component and normal Raman spectroscopy measurements require high concentrations of scatterer and long observation times. Additionally, normal Raman scattering is non-specific and all components of a mixture will contribute to the observed spectrum. This situation is markedly improved by the phenomenon of resonance-enhancement. Here, the wavelength of the light-source (usually a laser) which interrogates the sample is selected to be within an electronic absorption band of a chromophoric group in the sample. This results in several advantages:
    (1) Only those vibrational modes which are coupled to the electronic transition lead to enhancement.
    (2) Scattering from molecules or parts of macromolecules not in resonance will be extremely weak.
    (3) Concentrations as low as 10^{-6} mol l^{-1} can be detected.
    (4) High time resolution is possible since enough photons scattered to allow analog photodetection methods to be used.

The origin of resonance enhancement

    The detailed theory of resonance Raman has been described (8-12) and no attempt is made to reproduce it here. The only aim is to demonstrate why enhancement occurs.

    In a system of randomly-oriented molecules the intensity of a Raman transition from state m to state n is given by:

$$I_{mn} = I_o \frac{2^7 \pi^5}{3^2} \left(\frac{\nu_s}{c}\right)^4 \sum_{\rho\sigma} |(\alpha_{\rho\sigma})_{mn}|^2 \qquad (2)$$

where I_0 is the incident beam intensity, ν_s is the frequency of
the scattered radiation and $\alpha_{\rho\sigma}$ is the transition polarizability
tensor (ρ and σ refer to the incident and scattered polarizations,
respectively). The elements of $\alpha_{\rho\sigma}$ are shown to be:

$$(\alpha_{\rho\sigma})_{mn} = \frac{1}{h} \sum_e \frac{<m|\mu_\rho|e><e|\mu_\sigma|n>}{\nu_e - \nu_m - \nu_0 + i\Gamma_e} + \frac{<m|\mu_\sigma|e><e|\mu_\rho|n>}{\nu_e - \nu_n + \nu_0 + i\Gamma_e} \qquad (3)$$

where ν_0 is the incident laser frequency, μ_ρ and μ_σ are dipole
moment operators, Γ_e is a damping factor and $|e>$ is the wave
function for an intermediate state of energy equivalent to ν_e.
The denominator of the first term is the factor which determines
resonance: when $\nu_0 << \nu_e - \nu_m$ both terms are independent of the
laser frequency and normal Raman scattering results with I_{mn}
proportional to ν_s^4. As ν_0 approaches $\nu_e - \nu_m$ for a state e that
has a nonzero value of μ, the eth term becomes dominant and the
denominator of the first term becomes very small (limited by Γ_e).
Under such conditions the second term of equation (3) is
negligible.

The use of the Born-Oppenheimer approximation to separate
electronic and nuclear wave functions allows the Raman polarisability (α_{ij}) for a vibrational transition to be written as a sum
of two terms

$$(\alpha_{\rho\sigma})_{ij} = A + B \qquad (4)$$

where

$$A = \frac{M_e^\rho M_e^\sigma}{h} \sum_k \frac{<i|k><k|j>}{\nu_{ik} - \nu_0 + i\Gamma_k} \qquad (5)$$

and

$$B = \frac{M_e^\rho (\delta M_e^\sigma|\delta Q)^0}{h} \sum_k \frac{<i|k><k|Q|j>}{\nu_{ik} - \nu_0 + i\Gamma_k} \qquad (6)$$

where ν_{ik} is the energy gap between the lowest vibrational level
of the ground state and the kth vibrational level of the resonant-
ly-enhanced excited electronic state e, $|i>$, $|k>$ and $|j>$ are the
initial, intermediate and final vibrational wave functions, M_e is
the electronic transition moment from ground to e^{th} electronic
state and Q is the normal coordinate.

For resonance scattering by allowed transitions, M_e is
larger than $\delta M_e|\delta Q$ and the A term dominates. Vibrational Raman

intensities depend on the $<i|k>$, $<k|j>$ overlap integrals in
equation (5) (Franck–Condon factors). Vibrations will undergo
marked resonance enhancement if electronic excitation affects
the binding between atoms whose motions are important constituents of that vibrational mode. Totally symmetric vibrational
modes are enhanced under the A term whereas nontotally symmetric,
weak transitions are B term dependent. The damping factor, Γ
corresponds to the limiting line width of the transition.

Experimental considerations

Both normal and resonance–enhanced Raman spectra yield the
same information - viz. vibrational frequencies of scattering
species. The advantage of resonance enhancement is that it
offers a ca 10^6 times increase in intensity thus making it more
useful for species in low concentration. Further its adaptability to time resolution makes it an extremely useful technique for
observing short-lived transient entities as produced, for example,
in electron pulse radiolysis experiments. Therefore, time-
resolved resonance Raman (TR^3) spectroscopy yields vibrational
spectra and therefore structural information about transient
species - a characteristic not conveyed by the more commonly used
diagnostic methods.

Like all physical techniques however, TR^3 suffers disadvantages. One concerns fluorescence which can often result from
interacting a photon with a molecule which absorbs the photon.
Typically, fluorescence emission occurs with a probability in the
10^{-3} - 10^0 range whereas resonance enhanced scattering probabilities are nearer 10^{-6}. Also, both fluorescence and Raman scattering occur most significantly in the Stokes region, near to the
incident frequency. In many cases fluorescence overloads photodetectors optimized for observing resonance Raman intensities.
In situations where the transient of interest is present at times
much longer than the emitting state, the fluorescence can be
electronically gated out. A second disadvantage lies in the existence of poor Franck–Condon factors which prevent significant
resonance enhancement by even highly allowed electronic transitions.

Pulse radiolysis + TR^3 detection

The combination of pulse radiolysis and TR^3 detection has to
date been demonstrated in three laboratories. It was pioneered
by Wilbrandt, Pagsberg and their colleagues at Risø, Denmark who
obtained the RR spectrum of thiocyanate radical ions in aqueous
solutions (13). Subsequently Dallinger, Woodruff and Rodgers at
Texas used a similar method to obtain the RR spectrum of the
triplet state of β–carotene in benzene solution (14). Recently,
Lee, Schmidt, Gordon and Meisel at Argonne assembled TR^3 equipment for observing the resonance Raman spectrum of the stable
free radical formed by hydrated electron reduction of paraquat

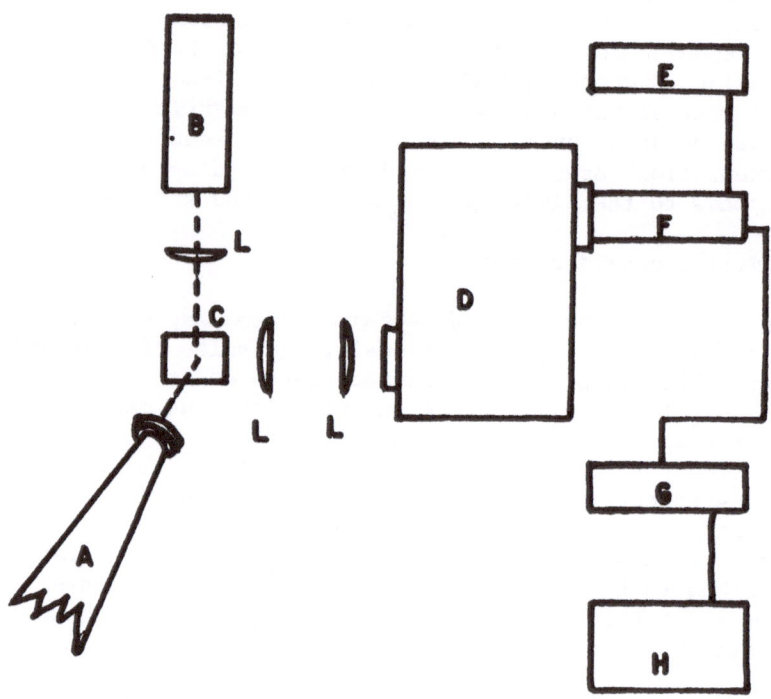

Figure 1. Schematic representation of pulse radiolysis + TR^3
apparatus at CFKR. A - Van der Graaff (downwards), B - Nd:YAG
laser, C- flow cuvette, D - 0.5 m spectrograph, E - gating pulse
generator, F - ISIT detector, G - OMA controller, H - plotter,
L - lenses.

(15). A schematic view of the experimental arrangement at the
Center for Fast Kinetics Research at the University of Texas at
Austin is shown in Figure 1.

     Electron pulses of 800 ns duration from a vertically-mounted
5.5 MeV Van der Graaff (HVEC) were incident upon the target solution in a specially-constructed quartz flow cell. These caused
excitation of some benzene solvent molecule into the triplet
manifold. Subsequent energy transfer to β-carotene solute molecules populates the β-carotene triplet state which has a strong
absorption at 530 nm. This process is complete in less than 1 μs
after which the β-carotene state decays with a natural lifetime
of ca 6 μs. At 1 μs after the electron pulse an 8 ns pulse of
532 nm light from a frequency-doubled Nd:YAG laser (Quanta Ray)
was sent into the cuvette via a focussing lens. Light scattered

at right angles was picked up by a fast lens system and focussed
onto the input slit of a Spex 0.5 m spectrograph and the dispersed spectrum recorded at the focal plane on a PAR 1205I ISIT
vidicon. The vidicon signal was processed by an optical multichannel analyzer and recorded on an X-Y plotter. The vidicon
was gated off during the electron pulse in order to avoid saturating the detector with Cerenkov light. To improve signal-to-
noise ratio the Van der Graaff and the laser were run at 6 Hz and
signals occumulated in the OMA memory. The sample was flowed
through the cuvette continuously during the irradiation period.

Spectra obtained in this way contained the peaks arising
from solvent, β-carotene (ground state) and β-carotene (triplet
state). Ground state and solvent spectra (normal Raman) were
collected in channel B of the OMA under identical conditions except that the electron beam was disabled. The contents of
channel B were subtracted from channel A (laser + electron beam)
to obtain the true triplet spectrum. The RR spectrum of β-carotene (T_1) obtained in this way is shown in Figure 2.

Departures from the equipment described here are, of course,
possible. Different electron beam generators can be employed
(the Risø group has used a field emission device). Tunable dye
lasers can advantageously replace the fixed-frequency Nd:YAG
system thereby allowing interrogation of transients absorbing in
different spectral regions. Again, new generation multi-diode
arrays can replace the vidicon detector. Time resolution of
these experiments is governed by the widths of the exciting and
interrogation pulses. Sub-nanosecond resolution is possible
with current electron gun and laser technology. The response
time of the photodetectors are unimportant in the experiment
since they act only as integrating devices. It must be noted
that generation of transients is not restricted to electron excitation. Several workers are currently using pulsed lasers of
various kinds for producing unstable species (16-19) and obtaining their vibrational spectra by TR[3].

Current activity

At this time only a few systems have been investigated.
Some free radical vibrational spectra have been obtained at Risø
and at Argonne (13, 15). The most detailed studies have concerned the vibrational spectra of triplet states of carotenoid and
polyene pigments (14, 20-22). The technique is currently at an
early stage of development; accumulating spectra, logging vibrational frequencies of transients, assigning normal modes, comparing with related stable species and rationalizing line shifts.
Clearly, as more research effort is put into this area, necessary
experience will be built up and resonance Raman spectroscopy of

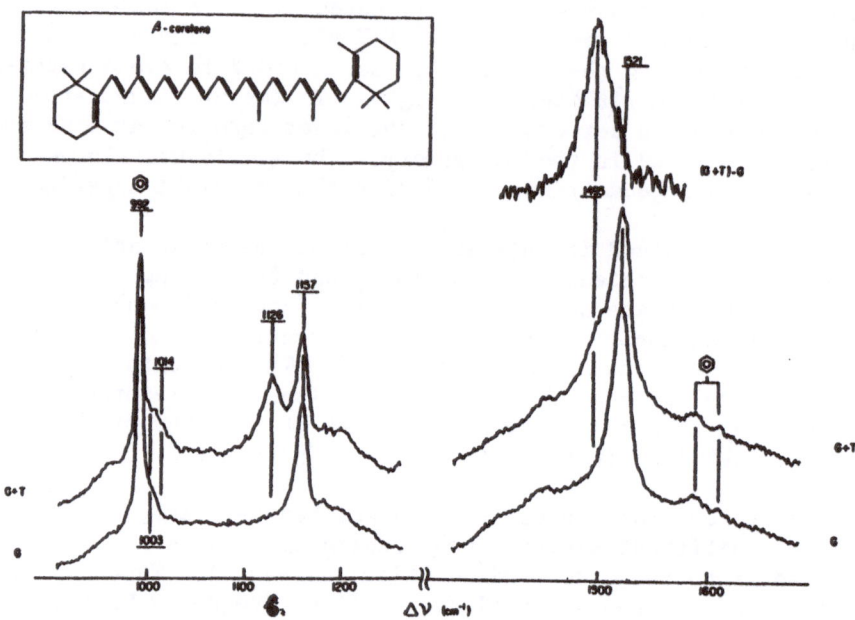

Figure 2. Transient RR spectrum in a pulse-irradiated solution
of β–carotene (inset) in benzene. Two vidicon frames are placed
side-by-side. Lower spectrum (G) taken with electron beam disabled (ground state solute and solvent peaks). Upper full
spectrum (G + T) taken with both electron and laser beams incident upon sample. Top right: difference spectrum showing T_1 band
at 1495 cm^{-1}. (Reproduced with permission of American Chemical
Society.)

short-lived transient species will eventually offer a real contribution to the assignment of structural features of rapidly-
decaying molecules.

ACKNOWLEDGEMENTS

The expertise of Drs. W. H. Woodruff and R. F. Dallinger
was crucial to the coupling of pulse radiolysis and TR^3 at CFKR.
Work at CFKR is supported by the Biotechnology Branch of the
Division of Research Resources of NIH (RR00886) and by the
University of Texas at Austin.

REFERENCES

1. Roffi, G.: this volume, p. 63.
2. Asmus, K.-D. and Janata, E.: this volume, p. 91.
3. Asmus, K.-D. and Janata, E.: this volume, p. 115.
4. Trifunac, A.: this volume p. 163.
5. Moore, W. J.: 1972, "Physical Chemistry", Prentice-Hall, Inc.,
    New Jersey, p. 935.
6. Beck, G., Kiwi, J., Lindenau, D., and Schnabel, W.: 1974,
    European Polymer J., 10, p. 1069.
7. Lindenau, D., Hagen, U., and Schnabel, W.: 1976, Z.
    Naturforsch., 31c, p. 484.
8. Albrecht, A. C., and Hutley, M. C.: 1971, J. Chem. Phys.,
    55, p. 4438.
9. Peticolas, W. L., Nafie, L., Stem, P., and Fanconi, B.:
    1970, J. Chem. Phys., 52, p. 1576.
10. Warshel, A., and Dauber, P.: 1977, J. Chem. Phys., 66,
    p. 5477.
11. Spiro, T. G., and Stein, P.: 1977, Ann. Rev. Phys. Chem.,
    28, p. 501.
12. Mathies, R.: 1979, in "Chemical and Biochemical Applications
    of Lasers", Vol IV (C. B. Moore, ed.), Academic Press, New
    York, Chapter 3.
13. Pagsberg, P., Wilbrandt, R., Hansen, K. B., and Weisberg,
    C. V.: 1976, Chem. Phys. Letters, 39, p. 538.
14. Dallinger, R. F., Guanci, J. J., Woodruff, W. H., and
    Rodgers, M. A. J.: 1979, J. Amer. Chem. Soc., 101, p. 1355.
15. Lee, P. C., Schmidt, K., Gordon, S., and Meisel, D.: 1981,
    Chem. Phys. Letters, 80, p. 242.
16. Dallinger, R. F., and Woodruff, W. H.: 1979, J. Amer. Chem.
    Soc., 101, p. 4391.
17. Atkinson, G. H., and Dosser, L. R.: 1980, J. Chem. Phys.,
    72, p. 2195.
18. Terner, J., Spiro, T. G., Nagumo, M., Nicol, M. F., and El-
    Sayed, M. A.: 1980, J. Amer. Chem. Soc., 102, p. 3238.
19. Friedman, J. M., and Lyons, K. B.: 1980, Nature, 284, p. 570.
20. Dallinger, R. F., Farquharson, S., Woodruff, W. H., and
    Rodgers, M. A. J.: 1981, J. Amer. Chem. Soc., in press.
21. Jensen, N-H, Wilbrandt, R., Pagsburg, P. B., Silleson, A-H,
    and Hansen, K. P.: 1980, J. Amer. Chem. Soc., 102, p. 7441.
22. Wilbrandt, R., and Jensen, N-H: 1981, J. Amer. Chem. Soc.,
    103, p. 1036.

DATA ACQUISITION AND ANALYSIS IN PULSE RADIOLYSIS
PART I: CONTROL, DIGITIZATION, AND ANALYSIS

David C. Foyt*

Center for Fast Kinetics Research,
Paterson Laboratory,
University of Texas, Austin,
Texas, U.S.A.

Aspects of data acquisition that are common to many different
detection schemes and chemical systems are surveyed. These aspects
include timing and control, digitization, and data processing and
analysis.

1.0 INTRODUCTION

Other papers in this collection have discussed techniques
for generating pulses of ionizing radiation, for measuring the
radiation dose delivered to the sample, and for monitoring the
resulting transients by a variety of analytical methods. The
present paper and the one following are concerned with overall
strategies for control of the fast kinetic apparatus, for digitization of the transient signal, for processing and storing the
resulting data, and for extracting from it the kinetic information
of chemical and biological interest.

In keeping with the subject of the NATO Advanced Study
Institute, these topics will be treated from the viewpoint of
their application to the study of fast process induced by ionizing
radiation. However, it should also be recognized that the
techniques discussed here are equally applicable to the study of
fast processes induced by pulsed lasers or flash lamps, rapid
mixing of reactants, temperature jump, electromagnetic pulse, and
a variety of other methods.

No attempt will be made to survey all of the different
techniques, such as optical absorption, fluorescence emission,
electrical conductivity, microwave absorption, Raman scattering,

J. H. Baxendale and F. Busi (eds.),
The Study of Fast Processes and Transient Species by Electron Pulse Radiolysis, 199–212.

EPR, NMR, polarography, and others that might be employed for the
detection of transients produced by pulses of ionizing radiation;
several of these are discussed elsewhere in this collection by
specialists in the respective fields. Instead, those aspects of
data acquisition and analysis that are common to the field as a
whole will be emphasized, drawing primarily upon the optical
absorption technique when concrete examples are required.

The present paper deals with three main topics in pulse
radiolysis: general considerations of sequencing and timing of
events, conversion of the fast transient signal to a permanent
form for storage and analysis, and data processing and analysis
techniques. The following paper surveys the history of the use
of laboratory computers in this field and discusses some of the
practical considerations involved in the selection and implementation of dedicated laboratory computer systems.

2.0 TIMING AND CONTROL

Even the simplest pulse radiolysis system requires that
several events be initiated at precisely controlled intervals.
To fix ideas, the sequence of events in a typical optical
absorption system is given below:

1. Fresh sample is flushed into the sample cell.

2. The monochromator is advanced to the desired wavelength.

3. The light intensity is adjusted to fall within the linea
    range of the detector.

4. The digitizer is set for the desired time base,
    sensitivity, and vertical offset, and any required arm
    or pre-trigger functions are executed.

5. The shutters, which protect the sample from continuous
    irradiation by the analyzing light, are opened.

6. The light source may be pulsed briefly if high intensity
    is required over the time span of the transient.

7. The quiescent light level is measured and recorded, and
    an offset (backoff) may be applied to the detector
    output, so that the transient signal will span the full
    dynamic range of the digitizer.

8. The actual digitizer sweep is triggered.

9. The pulsed radiation source is triggered, and the dose
    is measured.

10.   After completion of the trace, the shutters are closed.

11.   The entire sequence may be repeated at a series of
       wavelengths, radiation doses, time scales, temperatures,
       pH's, or other variables.

The timing of at least some of these events is crucial to
the success of the experiment. The most critical is the interval
between the digitizer trigger and the radiation pulse. If this
interval is too short, the initial baseline of the trace will not
be recorded; if it is too long, the conclusion of the trace will
not be seen; and of course this interval must be adjusted as the
time scale of the digitizer is varied. Other devices whose timing
is especially crucial include the lamp pulser, the backoff, and
the radiation dose monitor.

For detection techniques other than optical spectrometry, a
sequence of operations of comparable complexity will generally be
required. In fast conductivity experiments, for example, the
bias may be applied to the cell several times in rapid sequence
for the purpose of obtaining baselines and determining the effect
of a calibration resistor, and these events must be coordinated
with the trigger of the radiolytic pulse.

It is evident that a data acquisition apparatus for pulse
radiolysis must be able to initiate events at precisely spaced
intervals on time scale comparable to that of the transient of
interest.

In experiments that do not require a high level of automation,
the fast oscilloscope or other device employed for digitization
of the transient signal can also perform the most critical timing
functions, particularly if it is equipped with a dual time base.
On sub-microsecond time scales, delays can also be generated by
passing a control pulse or the transient signal itself through
variable lengths of cable, or by changing the optical path length.
And of course, the timing of such events as sample flushing,
monochromator setting, digitizer adjustment, and shutter control
is not often critical, and these operations can be performed
manually.

A greater degree of flexibility and convenience can be
attained by the use of electronic delays, constructed from digital
one-shots and controlled by variable RC circuits. The relative
timing of each event in the data acquisition sequence can be pre-
determined by adjusting the corresponding potentiometer. The
delays generated by this technique are reproducible enough for
all but the fastest time scales, for which optical or cable path
delays are preferred.

The most efficient use of the fast kinetic apparatus requires, however, that these methods be supplemented by the high-level control capabilities of a digital computer, which can initiate and control the timing of the events of the data acquisition sequence, implement over-all strategies for a sequence of wavelengths or other parameters, perform validity checks on the data sets, and carry out on-line data analysis and provide graphical displays. Since computer instruction times are typically of the order of a few microseconds, the fastest timing functions must still be performed by separate pulse sequence generators, although these can also be made programmable.

The topic of laboratory computerization is treated more fully in the following paper. However, to give one example of the advantages of computerized systems, it is estimated that the data collection, assembly, and plotting of a time-resolved spectrum over the ultraviolet and visible wavelengths, at ten-nanometer intervals and with five to ten time values included, requires more than a day in a system without computer/facilities; in a computer-based laboratory, several such spectra can be obtained within an hour.

3.0 DIGITIZATION OF THE TRANSIENT SIGNAL

Several different techniques are available for recording transient signals on the time scales of interest in pulse radiolysis. Some of these are listed below. In some cases, the transient signal has already been converted to a voltage by an appropriate detector, and it is this voltage which is to be digitized; in others, an optical signal is to be digitized directly.

3.1 Oscilloscopes

Routine, convenient measurement of events on the microsecond and nanosecond time scales was made possible by the development of fast oscilloscopes, which can be coupled with a camera to capture and preserve the electronic transient produced by the detector. Oscilloscopes are widely used for this purpose, and they have several advantages over other types of digitizers:

1. The data are recorded immediately, in permanent form, as a visual image that is readily assimilated by the experimenter.

2. A wide selection of scopes is available from a large number of vendors, making it easy to select one tailored to a particular application.

3. The scope represents an established technology and can

be made relatively reliable.

4. Users may already be acquainted with the operation of an oscilloscope, and thus no learning is involved for this part of the procedure.

5. Many laboratories already have an oscilloscope, or must acquire one for other purposes.

The use of oscilloscopes for digitization of the transient signals produced in pulse radiolysis also has two major drawbacks:

1. Polaroid film is expensive.

2. Data reduction and analysis are tedious. Quantitative treatment requires that the data be extracted by hand from the photographs (a painstaking and error-prone procedure), and both the number of points and the number of significant figures obtained will be fewer than in automated methods.

The time required for processing such data can be considerably reduced if a computer equipped with a tablet digitizer is available. The operator selects each point on the photograph by means of a special pen, and the tablet reads and transmits the coordinates to a computer. Several digital plotters, such as the Tektronix 4662, include a graphical input mode that can be used for this purpose, providing both the capabilities of digitization and of processed data display in a single unit.

3.2 Electronic Digitizers

The popular workhorse of pulse radiolysis laboratories, the Biomation 8100, typifies a class of digitizers employing primarily the techniques of digital electronics. Consisting of an analog-to-digital converter and a digital memory, it is capable of recording data points as closely spaced as a few nanoseconds or as widely spaced as hours. Its eight-bit resolution is somewhat coarse, providing only 256 distinguishable signal levels, but the wide range of sensitivities, together with the large number of points recorded (2048), make it adequate for almost any application on these time scales. The 8100 has the further advantage that all of its controls, including time base, sensitivity, vertical offset, and arm and trigger functions, are programmable via a parallel interface. The digitizer's memory can also be read across this interface at a rate that is typically limited only by the speed of the computer in use. Like the oscilloscopes, the digitizers in this category are based on well-established technology and are relatively reliable and trouble-free.

The Datalab digitizer, which is especially popular in Britain
falls in the same category as the Biomation. Several other
manufacturers, including Tektronix and LeCroy, have recently
offered digitizers in this time regime. Although more expensive
than the Datalab and the 8100, they provide certain advantages
that may be important in some specialized applications. In
particular, the increments between the different selectable
settings of the time base are smaller, and the time base can be
changed many times during a single sweep (as opposed to once on
the 8100). These devices are also designed to interface to
computers by means of such standard I/O buses as the GPIB and
CAMAC. The relative merits of such buses, as contrasted with
the simple parallel interface of the 8100, are discussed in the
following paper.

3.3   The Diode Matrix Technique

The Tektronix R7912 and 7912AD digitizers achieve significantly faster data rates than those in the previous category.
They are essentially computer-readable oscilloscopes, in which
the phosphor screen of the scope is replaced by a matrix of diodes
that are discharged by the writing beam. The state of these
diodes can then be read, and an array of points representing the
trace can be constructed. The R7912 is the older of the two, and
it lacks the programmability of the 7912AD. Its front panel
settings are computer-readable, however, and it is capable of
using a faster time base for relatively low-voltage signals. The
picosecond resolution attained by the R7912 is somewhat deceptive,
since the limits of its usefulness for fast kinetic measurements
are not set by its resolution, but rather by the 500 MHz bandwidth
of its amplifier (unless the detector in use produces signals as
large as several volts). Nevertheless, transient halftimes as
small as 0.5 nanoseconds can be measured, a 100-fold improvement
over the Biomation 8100. The R7912 requires a special controller
for computer interfacing (available from Tektronix for some
computers), whereas the 7912AD employs the GPIB bus.

3.4   Streak Cameras

The digitization of an amplified voltage signal, as in all
the above techniques, has the intrinsic disadvantage that the
electronic amplifiers and/or filters required have bandwidth
characteristics that limit the time resolution to transient halftimes of approximately 0.5 nanoseconds or greater. The streak
camera is based on the relatively new vidicon technology, which
effectively circumvents this difficulty. In this method, the
light signal is directly incident through a slit onto a photocatho
and the electrons produced are accelerated and swept across a
phosphor screen to produce a glowing streak, whose intensity at
each point along the direction of sweep is proportional to the

incident light intensity at the corresponding time. The streak
is digitized in its two dimensions, and the resulting data are
integrated across the width of the streak to give a representation
of the trace. Resolutions of a few picoseconds can be attained,
although the signal-to-noise ratio is significantly worse than
for the slower methods.

Techniques have also been described in which the light is
dispersed by a spectrograph before striking the slit. The
digitized streak then provides both spectral and kinetic information
from a single pulse.(1)

3.5 Pulse-Probe (Stroboscopic) Technique

The earliest measurements of fast radiation-induced transients
were carried out by the stroboscopic method first introduced by
Norrish and Porter(2) for flash photolysis. In this method, the
analyzing lamp is pulsed at a selected time after the radiolytic
pulse, and the transmitted light is passed through a spectrograph
and recorded on film. Repeated measurements are required to
construct a kinetic trace. The advent of fast photomultiplier
tubes and oscilloscopes has largely displaced this technique on
the time scale for which it was originally developed, although
sampling scopes may still be gated at varying times after the
radiolytic pulse to obtain kinetic traces in a similar way. The
pulse-probe technique is also employed for optical absorption
measurements on the picosecond time scale, utilizing a pulse of
light produced by the pulsed radiation source (the Cerenkov
light, for example). The probe pulse is routed along a variable-
length optical delay path and passed through the sample at a
time that is a function of the path length used. The attenuated
probe pulse may be electronically stretched and integrated by a
relatively slow digitizer(3), or digitized directly by a vidicon.(4)
A series of such measurements at different path lengths can be
used to construct a kinetic trace of the transient being produced
in the sample, or a wavelength spectrum at a fixed time following
the pulse can be assembled.

3.6 Counting Methods

The photon counting technique of detection and digitization
should also be mentioned, although its restriction to the
monitoring of low intensity, repetitive light signals makes it
more suited to fluorescence experiments employing pulsed light
excitation than to pulse radiolysis work. In this method, the
first photon emitted after each excitation pulse is detected and
counted in a channel corresponding to the elapsed time. Better
statistics are obtained than with the analog methods, especially
for weak signals, but the data collection times are significantly
longer. The most stable systems can measure signals as fast as

100 or 200 picoseconds, with careful adjustment. Counting
methods have also been used for the acquisition of kinetic traces
from fluorescence produced by low-intensity radioactive decay;
each initiating event is detected separately and matched with the
corresponding photon produced.

3.7 The Computer as Digitizer

If a digital computer is to be used for control and data
processing functions, it may be equipped with a relatively
inexpensive analog-to-digital (A/D) converter that can digitize
the transient signal as well. This method is limited, however,
by the time required for the computer to store the digitized data
points, and data rates faster than microseconds are not generally
possible. One advantage of this method of digitization is that
continuously variable (e.g. logarithmic) spacing of the points
in time can be achieved. Laboratory computers have also been
used as integral parts of pulse-probe(3) and photon counting(5)
data collection systems.

4.0 DATA PROCESSING AND ANALYSIS

The particular processing and analysis procedures employed
will depend both on the method of detection of the fast transient
and on the nature of the physical and chemical processes being
observed. The following topics are of general interest for many
kinds of experiments.

4.1 Initial Data Processing

The raw data obtained from a scope or a digitizer must
typically be converted to a variable, such as optical density or
conductivity, that is proportional to the concentration of the
species to be observed. It may also be desirable to correct for
digitizer astigmatism, to subtract a baseline light level, to
normalize by radiation dose, to average several traces in order
to improve the signal-to-noise ratio, or to apply a smoothing
algorithm to the data. Successive sets of data points within a
trace may be averaged together to lessen high-frequency noise and
to reduce the number of points that must be stored in computer
memory(6), although this procedure introduces some error in rate
constants if applied on a time scale over which the signal is
changing rapidly. Alternative smoothing techniques are available
in the form of linear filtering algorithms, which leave the total
number of points unchanged.(7)

The order of performing such preliminary processing operations
may be significant. For example, the result obtained by averaging
several raw digitizer traces and converting them to optical density

using an averaged value of 10 will be different from the result
obtained by converting each trace to optical density and averaging
the transforms. Normalization by radiation dose prior to
averaging will increase the weight given to low-dose traces, and
traces taken at significantly different doses should not generally
be averaged together at all, except in the case of first-order
kinetics. Deconvolutions and other procedures that assume a
linear detector response function should be performed before any
non-linear transformations are made.

The asymptotic baseline of the trace should be determined,
if possible, and subtracted before fitting is attempted. A zero-
time point should also be chosen, usually at the time of the
radiolytic pulse, for generation and extrapolation of fits. Here
and elsewhere, the importance of interactive video graphics for
manual input of such selections cannot be overemphasized.

4.2 Simple Linear Fitting

Simple first and second order kinetics may be fitted by
means of traditional linearization and linear regression techniques,
employing the logarithmic and reciprocal transforms for first
order and second order cases, respectively. A correct prior
choice of the asymptotic baseline is essential if these methods
are to succeed; second order traces, in particular, approach their
asymptotes slowly, so that it may be difficult to obtain good
asymptotes directly. In such situations, and in other situations
involving unusually noisy data, the iterative method discussed in
a subsequent section may be preferable even for simple kinetics.

There are two important difficulties in the use of the
linear transformation method:

1.   A least-squares fit to a linear transform of the
      trace does not, in general, correspond to a least-
      squares fit of the back transform to the original
      data. In particular, the transformation introduces
      a weighting function that places too much importance
      on points near the asymptote. Corrective weighting
      functions may be applied to compensate for this
      effect, or data near the asymptote may be excluded
      from the fitting sums. In any case, the goodness of
      fit should not be judged from the appearance of
      the transform fit alone, since small deviations
      near the asymptote will appear disproportionately
      large.

2.   Linear regression only ensures that a minimum in
      the residual has been found. For noisy data with
      a small number of points, the minimum obtained may

depend more on the details of the noise than on the
kinetics of the trace. In such cases, the experimenter
may prefer to specify the linear fit directly, by
means of interactive video graphics.

4.3 Iterative Linear Regression

Even for simple kinetic models, it is sometimes desirable
to include the asymptotic baseline among the parameters to be
fitted. Furthermore, the transients produced by ionizing
radiation frequently involve mixed or competing kinetics, permanent
products (on the time scale in use), simultaneous bleaching and
growth of signals from different species, and other complex
situations. Experimental parameters such as wavelength, dose,
concentration, or choice of sensitizer can often be adjusted to
simplify the kinetics. But when it is necessary to fit complex
traces by numerical methods, the iterative linearized least-
squares procedure provides a general technique that is suitable
for a wide range of cases, especially on computer systems equipped
with interactive video graphics.(8,9,10) To understand this
method, let $F(p,t)$ be the desired fit, which is in general non-
linear in the parameters p. Let p represent a set of starting
parameters for the current iteration, and let p'represent the
improved parameters at the end of this iteration. Then expand
F in a Taylor series and truncate after first order:

$$F(p,t) = F(p',t) + \sum_i (p_i - p_i') \frac{F}{p_i}$$

This function is fitted to the data by conventional multiple
linear regression techniques, to obtain the differences (p-p').
The improved parameters are calculated by adding these differences
to the starting parameters, and the procedure is repeated until
the fractional change in the parameters becomes insignificant.
The derivatives required by the above equation can often be
obtained analytically, although the secant approximation can be
used equally well in all cases.

Some of the types of kinetics to which this method has
been applied are the following: biexponential processes,
simultaneous independent first and second order processes, competing
first and second order processes, and first or second order kinetics
with the asymptote also to be fitted.(9,10)

If fits such as these, employing three or four parameters
for a single trace, are to be used productively in the analysis
of fast transient data, it is essential to recognize that two or
more models will frequently fit the data equally well by all the
usual statistical criteria. In such cases, any method designed
to distinguish these different types of kinetics(11) will also fail.

Therefore, chemical criteria must be employed to determine the appropriate kinetic model. If the analysis of the rate constants remains consistent as dose, concentration, or wavelength is changed, the model is confirmed much more reliably than would be possible on the basis of a single trace. It may also be possible to determine one or more of the parameters independently and constrain it during fitting, or to separate the model equation into two parts, one of which can be applied independently to the long-time part of the data and subtracted, leaving a relatively simple kinetic trace for fitting at earlier times. Clearly, the speed and convenience of a computerized data acquisition and analysis system will make it possible to carry out tests and procedures of this kind, which otherwise could not be routinely employed.

4.4 Direct Solution of Kinetic Equations

In cases of even greater complexity, it may be possible to obtain kinetic parameters by comparing the experimental trace to the result of numerical integration of the appropriate set of coupled differential equations. However, the number of unknown rate constants is often too great to make this approach useful.

4.5 Deconvolution

Regardless of how fast the ultimate time scale of a pulse radiolysis system may be, there will always be applications that push its resolution to the limit. This limit is determined primarily by the width of the radiolytic pulse itself and by the response functions of the detector and the digitizer. Careful application of deconvolution techniques makes it possible, under favourable conditions, to extract rate constants on the order of five to ten times faster than these characteristic parameters of the apparatus.

In a linear system, the observed trace $D(t)$ is given by the convolution integral

$$D(t) = \int_0^t L(t') \, R(t-t') \, dt'$$

where $L(t)$ is itself the convolution of the pulse shape with the detector/digitizer response function. $L(t)$ can be measured by passing a signal proportional to the pulse through the detector. Thus $D(t)$ and $L(t)$ are known, and $R(t)$, the ideal response function of the chemical system to a pulse of negligible width, is to be determined by deconvolution.

O'Connor et al (12) have compared six deconvolution methods on the basis of their ability to resolve fast or closely spaced

decays and to deal with distortions in the data. They find that
the iterative linearized least-squares technique is generally
superior to the others provided, of course, that a parameterized
form for the deconvolved result is known beforehand. The method
is rather slow because a reconvolution must be performed each
time a derivative is calculated; the Marquardt algorithm (13)
may be employed, however, to speed up the search. If the
functional form of the desired result is not known, it may be
approximated by a sum of about ten exponentials with fixed lifetimes, the pre-exponential coefficients being determined by
ordinary linear least-squares fitting. No physical significance
is attached to the resulting parameters, but the deconvolved
curve can then be analyzed by ordinary fitting techniques. The
Fourier transform method also avoids any prior assumptions about
the form of the deconvolved curve, but it is apparently unsuitable
for deconvolution of very short lifetimes, and it may produce
serious oscillations if the observed decay contains non-linear
instrumental distortions. Unlike the exponential sum method,
the Fourier transform approach is not easily modified to analyze
for scattered light.

Considerable care must be taken in the application of
deconvolution techniques and in the interpretation of the results,
and reports of physically meaningful extraction of more then
two exponentials should be regarded with scepticism. It is
especially important that the functions $L(t)$ and $D(t)$ be aligned
in time, not on the rising edges of the two signals, but at the
time of the radiolytic pulse.

4.6 Statistical Considerations

Statistical criteria of goodness of fit, such as the residual
and the number of crossings, are more useful for data acquired
by counting methods than for analog data; the latter often
contain systematic noise which invalidates the assumptions used
in the derivation of the statistical equations. Furthermore,
since an increase in the number of fitting parameters will
typically improve the fit, regardless of whether the kinetic
model is correct, these statistical criteria should be used
only to compare fits having the same number of parameters. A
careful visual examination of the fit superimposed on the data
is often more decisive in selecting the best fit than the
residual is, and the autocorrelation of the difference between
the data and the fit provides an especially revealing visual
display of systematic deviations between the two curves. Of
course the most important fitting criterion is the consistency
of the interpretation of the parameters as dose, wavelength, or
other experimental variables are changed.

Once a fit has been selected, estimates of the standard

errors in the fitting parameters can be obtained by conventional techniques from the linear regression matrix. However, these estimates are unreliable for analog data, since they are based on the assumption of random statistical noise. Even for statistically sound data, such techniques generally underestimate the error in the parameters for two reasons: First, they assume that the overall uncertainty of the fit is distributed statistically among all of the parameters involved; and second, they do not take into account systematic errors in the calibration and measurement techniques employed.

5.0 CONCLUSION

This brief survey of data acquisition and analysis techniques for fast radiation-induced transients cannot be exhaustive. The topic is vast and diverse, touching every aspect of every kind of scientific research performed in the field. Nevertheless, several areas of widespread common interest have been described, and a unifying perspective has been developed for the acquisition and treatment of a considerable diversity of experimental data.

This work was carried out at the Center for Fast Kinetics Research in Austin, Texas. The support of the Biotechnology Branch of the Division of Research Resources of the National Institutes of Health, Grant RR-00886, and of the University of Texas at Austin is gratefully acknowledged.

REFERENCES

*   Present address SRI International, 333 Rowenswood Avenue, Menlo Park, California, U.S.A.

1.   S. Gordon, K.H. Schmidt and J.E. Martin, Rev. Sci. Instrum. 45, 552 (1974); K.H. Schmidt, S. Gordon and W.A. Mulac, Rev. Sci. Instrum. 47, 356 (1976).
2.   R.G.W. Norrish and G. Porter, Nature 164, 658 (1950); G. Porter, Proc. Roy. Soc. A 200, 284 (1950).
3.   M.J. Bronskill, W.B. Taylor, R.K. Wolff and J.W. Hunt, Rev. Sci. Instrum. 41, 333 (1970); J.E. Aldrich, P. Foldvary, J.W. Hunt, W.B. Taylor and R.K. Wolff, Rev. Sci. Instrum. 43, 991 (1972).
4.   P.M. Rentzepis, Biophys. J. 24, 272 (1978).
5.   L.K. Patterson, personal communication.
6.   L.K. Patterson and J. Lilie, Int. J. Radiat. Phys. Chem. 6, 129 (1974).

7.   A. Savitzky and M.J.E. Golay, Anal. Chem. 36, 1627 (1964).

8.   D.F. DeTar, Computer Programs for Chemistry, New York,
     Academic Press, Vol. 1-4 (1972).

9.   D.S. Gorman and J.S. Connolly, Int. J. Chem. Kinetics 5, 977
     (1973).

10. D.C. Foyt, Computers & Chemistry 5, 49 (1981).

11. D.C. Foyt and J.S. Connolly, Biophys. J. 24, 60 (1978).

12. D.V. O'Connor, W.R. Ware and J.C. Andre, J. Phys. Chem. 83,
     1333 (1979).

13. D.W. Marquardt, J. Soc. Ind. Appl. Math. 11 (2), 431 (1963);
     P.R. Bevington, "Data Reduction and Error Analysis for the
     Physical Sciences", McGraw-Hill, New York, 1969.

DATA ACQUISITION AND ANALYSIS IN PULSE RADIOLYSIS
PART II: COMPUTERIZATION

David C. Foyt*

Center for Fast Kinetics Research,
Paterson Laboratory,
University of Texas, Austin,
Texas, U.S.A.

A historical survey is given of the development of digital computers and of their use in the acquisition and analysis of data from fast radiation-induced transients. The practical considerations involved in the choice of computer hardware and software for dedicated laboratory applications are discussed, and some more general aspects of the design and implementation of such systems are presented.

1.0 INTRODUCTION

It is certainly possible to carry out experimental studies of fast radiation-induced transients without the use of a dedicated laboratory computer. However, the considerable advantages of computerized systems, together with the increasing availability of sophisticated computer hardware at a reasonable cost, have led to the widespread use of computers both for automation of the pulse radiolysis apparatus and for on-line analysis of the data thus acquired. The principal advantages of computerization in this field are as follows:

1. Overall strategies for the data acquisition session, such as a series of wavelengths, time bases, or radiation doses, are readily implemented and modified in computer software.

2. The lengths of the various timing delays in the data acquisition sequence can be adjusted automatically when the time base is changed.

213

J. H. Baxendale and F. Busi (eds.),
The Study of Fast Processes and Transient Species by Electron Pulse Radiolysis, 213–225.
Copyright © 1982 by D. Reidel Publishing Company.

3. Corrections can be made automatically for variations
    in the quiescent light level, the dose, or other
    parameters.

4. Validity checks can be performed on each data set,
    and the necessary adjustments to keep the next set in
    the sequence on scale can be estimated in advance.

5. Many useful data processing and analysis techniques
    are difficult, if not impossible, without a computer.
    Timesharing computer services are often used for this
    purpose, but they may be expensive and inconvenient,
    and they cannot generally provide the control and on-
    line analysis capabilities enjoyed with a smaller
    computer dedicated to the use of a single laboratory
    and configured specifically for its applications.

6. The course of the experiment can be guided interactively
    by the investigator, on the basis of the on-line
    analysis and graphics provided by the computer. The
    speed and efficiency of a computerized system make it
    possible to test new ideas and to alter the course of
    an experiment in progress, in response to preliminary
    results, while the original sample is still in the
    apparatus.

On the other hand, the computerization of pulse radiolysis
laboratories also has some negative aspects. The sheer volume
of data that can be produced in a short time may actually
obscure rather than clarify the significant results, unless
the experiments are guided carefully and sensibly. Furthermore,
although automation eliminates many opportunities for operator
error, an error on the part of the programmer can have far-
reaching and disastrous consequences, and subtle programmer errors
may go undetected for some time. Equally disastrous results can
occur if the user misunderstands the functioning of the program.
And finally, users of the computerized system may be tempted to
rely too much upon the computer, instead of understanding the
details of the experiment, assessing the validity of the results
and making the kinds of judgements that only the scientist who
is deeply involved in the particular study at hand can properly
make. It is essential, therefore, in designing, implementing,
and administering laboratory computer systems, to promote the
intelligent, critical use of such apparatus. The computer
cannot be a substitute for creative and painstaking human guidance
of the experiment and evaluation of the data that are produced.

2.0 HISTORICAL SURVEY

2.1 The Computer Revolution

It is customary to begin a historical survey of the development of computers with the anonymous invention of the abacus in the third or fourth millennium, BC. This device employs the technology of beads and strings, and in the hands of a skilled operator, it can achieve times of a few seconds for a typical operation.

The next major advance in computer technology occurred in the 17th century AD, with the invention of the mechanical adding machine (Pascal) and the desk-top calculator (Leibnitz). These devices performed on the same time scale as the abacus, but they provided an operator interface that was more readily accepted by Western users.

The 19th century British inventor, Charles Babbage, extended and refined the mechanical technology pioneered by Pascal and Leibnitz to produce a Difference Engine, which automated practically the entire sequence of operations required to solve ordinary difference equations. Babbage's ambitions were much greater, however, and he spent the last several decades of his life promoting the idea of a complete Analytical Engine, a mechanical device, driven by steam power, which encompassed all the essential elements of what we now call a computer. The engine was never constructed; shortly after Babbage's death, a commission of the British government concluded that the technology of the time was incapable of producing a machine containing over 10,000 gears, all of which must work together to a very close tolerance.

Babbage's dream was finally fulfilled in the 1940's, when several groups of inventors, including Stibitz and collaborators at Bell Laboratories, Aiken's group at Harvard, Eckert and Mauchley, and von Neumann, applied the new technologies of telephone switching relays and vacuum tubes to the problem of computer design. The ENIAC, completed in 1945, contained some 18,000 vacuum tubes and 70,000 resistors, occupied 1400 square meters of floor space, consumed 150 KW of power, and weighed 27 tons.

In the intervening 40 years, the growth of computer technology has accelerated tremendously. The invention of transistors and the development of ever higher levels of circuit integration have led to an object measuring only a few centimeters and weighing a few grams, but having considerably greater capabilities than the ENIAC. And the price of such a "computer on a chip" is usually less than the cost of the printed circuit board on which it will be mounted.

2.2 Computers in Pulse Radiolysis

Needless to say, the development of computer software and
of computer applications in all fields of science and engineering
has been unable to keep up with the rapid progress in computer
hardware, and most laboratory applications are obsolete by the
time they come completely into use.

The earliest applications of computers to fast kinetics
employed the computer (often a PDP-8) only for storage and analysi
of the collected data.(1,2,3,4) In 1971, Lilie (5) reported a
polarographic system in which the computer sensed the presence
of a forming mercury drop, triggered the radiolytic pulse, operate
the A/D converter, and provided video and hard-copy graphics.
Patterson and Perone (6) described a combined electrochemical and
photometric monitoring system in which a computer performed some
of the selection and control functions.

The first fully automated pulse radiolysis facility was
described by Patterson and Lilie (7) in 1974. This system was
based on a Biomation 8100 digitizer and a DEC PDP-8 computer,
and was designed for optical absorption measurements. It included
computer control of all the events of the data acquisition sequenc
and required 8 seconds for acquisition, initial processing, and
storage of each kinetic trace. Complete spectra could be taken
without operator intervention. Each kinetic trace in the series
was verified as being on scale, and appropriate changes in the
digitizer settings from one trace to the next were anticipated
by the program. Kinetic fitting and 3-dimensional video displays
of assembled spectra were provided, and hard copy graphical output
was made available.

Subsequent reports of computerized pulse radiolysis systems
have elaborated and adapted the basic approach of Patterson and
Lilie. A single system supporting both optical and conductivity
measurements has been described,(8) as well as a facility supporti
both pulse radiolysis and flash photolysis with two complete
computer systems.(9) Most systems have employed hardware modules
separately interfaced to the bus of the computer, although at
least one (10) has made use of the uniform CAMAC standard. Many
computerized laboratories are now in existence, and the details
of most of them have not been separately published.

Recent advances in computer hardware and software have been
reflected by two major directions of current development in
computerized fast kinetic systems:

1.   The relatively new multi-user minicomputer operating
       systems, such as Bell's UNIX and DEC's RSX-11M, have
       been exploited to support several separate data acquisi-

tion systems from a single computer, with concurrent
data acquisition and analysis from different terminals.
Successive generations of systems developed by Patterson
at Michael Reese Hospital and at Notre Dame exemplify
this trend. And the system developed by Foyt (11) at
the Center for Fast Kinetics Research employs RSX-11M
on a PDP-11/34 to support one electron pulse radiolysis
and three laser flash photolysis units, including the
Biomation 8100, Tektronix R7912, and streak camera
digitizers, and it provides for optical absorption,
fluorescence emission, and electrical conductivity
detection techniques. This system also furnishes over
100 interactively selected software options, including
15 different types of kinetic analysis.

2.   The phenomenal growth of the personal computer industry
has reduced the cost of a single-user, microcomputer-
based laboratory system to significantly less than the
cost of a good oscilloscope. Most of the work with
microcomputers in pulse radiolysis laboratories is still
in progress, but at least one has developed an operational
data collection and analysis system based on a PET
microcomputer. (12)

3.0   TECHNIQUES OF LABORATORY COMPUTERIZATION

The selection and implementation of a laboratory automation
scheme involves four major choices: computer hardware, interfacing techniques, operating system software, and programming
languages. A detailed treatment of the considerations involved
in making these choices is beyond the scope of the present paper,
but a survey of the options available in each area is given below.

3.1   Computer Hardware

Digital computers are conventionally classified in three
categories, determined by their word size, speed, available
memory, and degree of integration of electronic components:

3.1.1 Mainframe computers - Large, fast computers with at least
32-bit words and several magabytes of digital memory. These are
usually operated in a timesharing mode by a central computer
facility, and therefore are not often suitable for real-time
laboratory automation. But they may be used for off-line data
analysis, especially when demanding numerical procedures, such
as the solution of coupled differential equations or the
diagonalization of large matrices, are required. Some new 32-bit
machines, such as the DEC VAX-11/750, might be considered for
dedicated use by a laboratory in which several independent pieces

of apparatus are to be controlled concurrently.

3.1.2 Minicomputers - Medium-sized computers, having 16-bit words
and usually no more than 64 K bytes of digital memory available
to each running program. Many automated pulse radiolysis systems
to date have used machines in this category. DEC's growing LSI-11
series, and similar products on the borderline between micro and
minicomputers, have brought 16-bit hardware and minicomputer operating system software into a price range only slightly above the
best-equipped 8-bit computers.

3.1.3 Microcomputers - The recent development of large scale integration (LSI) of electronic components has made it possible to produce
an 8-bit computer processor (and more recently, an entire computer,
including memory) on a single chip. Several complete systems,
incorporating a video monitor, cassette tape storage, and the BASIC
language, are now available for under $1000, most of them employing
the second-generation 6502 or Z-80 processor. The addition of
more digital memory and a floppy disk system makes these computers
suitable for fairly ambitious single-user laboratory applications.
The Cromemco, North Star, and Apple systems are perhaps better
suited to laboratory use than the PET and the TRS-80, although the
larger volume of commercial software available for the TRS-80
should also be taken into account, and the rapid growth of the
field precludes any definite long-term recommendations. Complete
systems (excluding disks) for under $200 have now begun to appear
(the Sinclair ZX-80, for example), although these may have small,
awkward keyboards and only the most rudimentary hardware interfacing
capabilities. With the recent entry of IBM into the personal
computer market, and with the imminent arrival of Japanese products
in this price range, it is safe to assume that more powerful systems
will continue to be offered at a very reasonable price.

3.2 Hardware Interfacing

     In most cases, commercial computer interfaces will be available
for the hardware devices to be controlled and the signals to be
measured in pulse radiolysis experiments. Analog-to-digital
converters are useful for measurement of dose, initial light level,
temperature, and other parameters. Parallel digital interfaces
are appropriate for digital voltmeters and for many digitizers,
and may also be used for generating trigger pulses. A crystal
clock is useful for precise timing and control of events, and a
programmable pulse sequence generator can be used for timing
requirements that are faster than the instruction time of the
computer (usually a few microseconds). Inexpensive graphics may
be furnished by a dual digital-to-analog converter, which can
drive either an oscilloscope or an x-y plotter. More sophisticated
graphics systems are available commercially, usually employing
serial interfaces like those used for terminals.

Several interfacing standards have been developed for laboratory and business applications. Their advantages include uniformity of software interface, interchangeability of modules from different vendors, and ease of hardware connection. Their principal disadvantages are that they are somewhat slower and more expensive than simple parallel interfaces. Also, in some cases, they have not been correctly or thoroughly implemented. Four of these standards are considered below:

3.2.1 CAMAC -

The CAMAC standard is the oldest and strictest of those listed here. A controller module, unique to the bus structure of the particular computer, connects the computer to a CAMAC crate. The crate, which is independent of the type of computer used, contains plug-in modules that implement the various detection, control, timing, and other functions that are desired. A large variety of modules is available, and the standard is sufficiently strict that modules from different vendors may be mixed with relative impunity. Several crates may be connected in branching dataways for complex interfacing requirements. CAMAC systems are generally somewhat more expensive than other standard interfacing systems.

3.2.2 GPIB -

The GPIB bus (IEEE-488) is standard equipment on several commercial microcomputer systems. Newer instruments from Tektronix are usually designed with GPIB interfaces, as are many products from Hewlett-Packard and from other vendors. The GPIB standard is less expensive, less versatile, and slower than CAMAC, and it permits a greater degree of diversity (hence incompatibility) among different vendors. Furthermore, it has sometimes been implemented incorrectly or incompletely; a notorious example is the failure of the PET microcomputer to support some of the hardware and software features of the standard, making the unmodified PET incompatible with certain Tektronix instruments.(13)

3.2.3 S-100 Bus -

This microcomputer interfacing standard, although quite popular for disks and other peripheral interfacing applications, is perhaps less widely used for scientific instrumentation than the GPIB. It is found primarily in systems based on the 8080 processor and its descendants. The S-100 standard has also been plagued by variations among different manufacturers, although this situation is said to be improving.

3.2.4 Ethernet -

The interfacing standards discussed above all require
parallel cabling for the address and data lines, making inter-
connection of modules difficult and, compared to the cost of
microcomputers, relatively expensive. Intel, DEC, and Xerox
have joined forces to implement a new local networking standard,
called Ethernet, which employs only a single coaxial cable.
Data rates on the order of several MHz are achieved, and inter-
connections can be made, by means of a simple clamp, while the
network is in use. Considerable intelligence is required of
each module, since all modules share equal responsibility with
the computer(s) for control and integrity of the network. It is
too early to determine how useful this standard will become for
laboratory applications, although it is generating a great deal
of interest in its preliminary stages.

3.3 Operating Systems

The operating system of a computer provides all the necessary
services for entering and editing programs, translating them to
machine language, linking them with subroutine libraries, running
them, and reading and writing files on tape or disk. It may
also provide real-time device drivers, multi-programming, timing,
and a variety of other system services.

Minicomputer operating systems may be divided into two
categories:

1. Single-user systems, usually with foreground/background
    capability. These systems are suitable for dedicated
    control of one application at a time. They are
    smaller and easier to use than multi-programming systems,
    and they may permit program development or data analysis
    to proceed concurrently with a program that controls
    the scientific apparatus in real time. (Example: RT-11)

2. Multi-programming systems, which allow concurrent
    execution of several different applications from
    different terminals. These require more memory than
    single-user systems, and some expertise in systems
    programming is usually needed in order to maintain them
    and to make full use of their capabilities. (Examples:
    RSX-11M, UNIX)

Microcomputer operating systems are often limited to the
BASIC language, implemented in read-only memory (ROM) and
augmented with file control (and possibly also interface control)
commands. Systems with hard disks may provide disk-based operating
systems that support other languages. More sophisticated operating

systems such as CP/M, and now even UNIX or subsets of it, are also available for some microcomputers.

3.4 Programming Languages

The selection of a programming language has far-reaching consequences for the development of an automated laboratory. Once a major committment has been made to one programming language, it will be very difficult to change to another and to merge new programs and data structures with old ones. Furthermore, since computer software may continue to be used, modified, and enhanced long after its original authors and their colleagues have gone elsewhere, it is important to choose a language that encourages the writing of well-organized, documented, and readable code. The cost of software development will easily exceed the cost of hardware, over the lifetime of a computer system, and a small amount of time devoted to organising the development of software from the beginning will result in substantial savings overall.

Unfortunately, the languages that are most readily available and most popular with scientific programmers are not always the best languages by the above criteria. Many microcomputer systems provide only BASIC and assembly language. The latter is preferred for hardware device drivers, which must be compact and fast, and must set and clear register bits directly. The longer, more complex programs used for control logic and data analysis, however, should be written in a higher-level language if at all possible. BASIC is such a language, but it was originally designed as an instructional language for beginners, and its extremely slow execution times, rudimentary subroutine capabilities, and lack of sophisticated control and data structures make it tedious and difficult to avoid errors in the writing of large, complex programs. Its one advantage is that program changes can be made and run immediately, without the intermediate steps of compiling and linking. A few microcomputer manufacturers are now offering the FORTH language, which retains this feature of BASIC, avoids most of BASIC's faults, and provides for unusually compact and well-organised programs.

Minicomputers and microcomputers with hard disks are capable, in principle, of supporting any of the many compiled languages that have been devised. By far the most popular of these for scientific purposes is FORTRAN, although ALGOL has been more widely used in Europe than in the USA. FORTRAN is widely understood and reasonably well standardized, so that it is more easily transported from one installation to another and more readily modified and adapted by different programmers. On the other hand, computer scientists are generally in agreement that structured programming techniques, employing such constructs as

IF-THEN-ELSE, WHILE, and UNTIL, are much better suited to the
production of readable, easily modified code than are the
primitive GO TO and DO statements of FORTRAN.

The most popular structured languages are PASCAL, a descendant
of ALGOL with a powerful repertoire of data structures, and C,
the programming language of the UNIX operating system developed
at Bell Laboratories. Neither of these is as readily available
as FORTRAN, and software developed in these languages for one
computer system is likely to be more difficult to transport to
another. For many users, the best compromise seems to be the
RATFOR language, which is implemented as a FORTRAN pre-processor.
RATFOR furnishes the structured programming constructs mentioned
above, together with conditional compilation parameters and
several cosmetic features that are lacking in standard FORTRAN.
A program written in RATFOR is first run through the pre-processor,
which translates it into ordinary FORTRAN that can be compiled,
linked, and run in the usual way. The generated FORTRAN can be
transported to other sites, and programmers unfamiliar with
RATFOR can still read and modify the software at the FORTRAN level.

4.0 Some General Aspects of Design and Implementation

A dedicated laboratory computer system should not only
provide for control of the apparatus and analysis of the data,
but it should do so in a way that encourages innovation and
creativity, and that liberates the experimenter to think about
the research rather than about the computer. These goals can
only be achieved by experience and the exercise of good judgement,
but some suggestions can at least be given, based on the experience
of one computerized fast kinetics laboratory:

4.1 Flexibility

The control and analysis software should provide for
considerable flexibility in the interaction between people and
the computer. Frequently used sequences of operations should
be concatenated and made available as single commands; however,
low-level entry should also be provided to the individual parts
of these sequences, so that users can carry out non-standard
procedures as well. For example, it should be possible either
to acquire a sequence of kinetic traces for assembly of a complete
spectrum, or at the opposite extreme, to advance the monochromator
or pulse the radiation source apart from such a sequence.

Special software features intended for ease or elegance of
operation should be made optional, since an enhancement provided
for one experiment might prove to be a hinderance in a different
context. For example, automatic elimination of the Cerenkov

radiation might be introduced to avoid compressing the vertical scale on graphical output; however, unless this feature can be disabled, it will play strange tricks when the transient trace results from bleaching of an initial signal rather than creation of a new one. Another example is the automatic generation of the various delays of the data acquisition sequence as a function of the total digitizer sweep time; the software should not prevent experimenters from specifying innovative timing sequences as the need may arise.

4.2 Ease of Use

The computerized system should be easy to learn and to use routinely. A small, easily remembered set of mnemonics should provide for the most common sequences, even though a much larger number of options may be available for use in special situations. Menu displays of options, with the desired one selected by means of interactive graphics cursors or a light pen, can be used to simplify the choice of commonly-used procedures. Separate buttons can be interfaced to the computer for initiating the pulse sequence and for replying to YES/NO questions, if users prefer them to the typing of commands at the keyboard. The questions, complaints, and errors of users should be monitored carefully to determine how the human/machine interface can be improved. User errors should be trapped and processed whenever possible, instead of allowing the program to abort or to produce mysterious system error messages. Many operating systems provide facilities for programs to link to system error trapping routines, which can be used or modified for this purpose.

4.3 Manual Control Options

One of the most important design decisions for a computerized laboratory is the degree to which manual control options are to be provided, to supplement or replace automated control by the computer. Experimenters who are not accustomed to computerized facilities may prefer to maintain a complete manual mode of operation, so that data can be taken and processed even when the computer is inoperative. This may be a wise decision, especially in cases where the laboratory does not have exclusive use or control of the computer, or where it lacks adequate electronics and programming staff to maintain and upgrade the system as required.

On the other hand, provision of full manual operation requires considerable duplication of effort, and persons who have become accustomed to the speed and convenience of a computerized system may choose to wait for the computer to be repaired anyway, instead of taking and analyzing data manually. A maintenance contract from the computer vendor, although usually

expensive, will go a long way toward making manual control
options unnecessary. Manual hardware for timing and control
typically requires frequent modifications as the apparatus is put
to new uses, and frequent hardware alterations are tedious and
conductive to eventual hardware failure. Computer software, on
the other hand, is ideal for constructing flexible, easily-
modified control procedures.

4.4 How Much Computerization Is Enough?

Beyond a certain point, the further refinement and enhancement
of a laboratory automation system becomes unproductive, or even
counterproductive. The optimal degree of computerization will
differ from one laboratory to another, depending on such factors
as the following:

1.   The number of persons using the system and their rate
     of turnover and degree of experience, both with computers
     and with fast radiation-induced transients

2.   The extent to which experienced staff are available to
     assist inexperienced users

3.   The continuing availability of staff for program and
     interfacing enhancements

4.   The complexity of the apparatus and of the kinds of
     experiments being performed

5.   The capacity of the computer

6.   The relative merits of human and computerized decision
     making.

To illustrate the last point, it would be counterproductive in
most cases to write computer programs to take data, determine
rate constants, and then decide what kind of experiments to
perform next, all without operator intervention. At least at
the present stage of laboratory computer development, such
strategic decisions are better left to the experimenter.

5.0 CONCLUSION

The field of laboratory automation is changing rapidly as
the technology of digital computers continues to expand and
improve, seemingly without limit. Much of what has been said
above will be out of date very quickly, and any laboratory
automation project should draw upon the advice of persons who are
experienced in the field and who are in touch with the latest

developments in digital electronics and system software.

This work was carried out at the Center for Fast Kinetics Research in Austin, Texas. The support of the Biotechnology Branch of the Division of Research Resources of the National Institutes of Health, Grant RR-00886, and of the University of Texas at Austin is gratefully acknowledged.

REFERENCES

*   Present address SRI International, Menlo Park, California, U.S.A.

1. D.S. Gorman and J.S. Connolly, Rev. Sci. Instrum. 43, 1175 (1972).
2. J.E. Aldrich, P. Foldvary, J.W. Hunt, W.B. Taylor and R.K. Wolff, Rev. Sci. Instrum. 43, 991 (1972).
3. S. Gordon, K.H. Schmidt and J.E. Martin, Rev. Sci. Instrum 45, 552 (1974).
4. C.D. Jonah, Rev. Sci. Instrum. 46, 62 (1975).
5. J. Lilie, J. Phys. Chem. 76, 1487 (1972).
6. J.I.H. Patterson and S.P. Perone, Anal. Chem. 44, 1978 (1972).
7. L.K. Patterson and J. Lilie, Int. J. Radiat. Phys. Chem. 6, 129 (1974).
8. T.E. Eriksen, J. Lind and T. Reitberger, Chemica Scripta 10, 5 (1976).
9. A.T. Thornton and G.S. Laurence, Radiat. Phys. Chem. 11, 311 (1978).
10. C.K. Ross, K.H. Lokan and G.G. Teather, Computers & Chemistry 3, 89 (1979).
11. D.C. Foyt, Computers & Chemistry 5, 49 (1981).
12. J.H. Baxendale, personal communication.
13. T. DeSantis, Electronic Engineering Times, Jan. 7, 1980, p.28.

RAPID TECHNIQUES FOR CORRECTING NANOSECOND KINETIC TRACES FOR
CONVOLUTION ERROR

David C. Foyt

Center for Fast Kinetics Research,
Paterson Laboratory,
University of Texas, Austin,
Texas, U.S.A.

1.0 INTRODUCTION

In the analysis of data on fast transients, such as are taken
in pulse radiolysis or flash photolysis, significant errors may
be introduced by neglect of convolution of the chemical or photophysical transient with the excitation pulse shape and with the
time response function of the detection apparatus. Due to the
commutative and associative properties of the convolution integral,
the observed signal $S(t)$ in a linear system is given by

$$S(t) = \int_0^t I(t') \, G(t-t') \, dt' = \int_0^t G(t') \, I(t-t') \, dt' \qquad (1)$$

where $I(t)$ is the ideal signal that would be obtained with a pulse
of negligible width and a detector of infinite bandwidth, and
where $G(t)$ is itself given by the convolution of the excitation
pulse shape $E(t)$ with the detector response function $R(t)$:

$$G(t) = \int_0^t E(t') \, R(t-t') \, dt' \qquad (2)$$

$G(t)$ is ordinarily measured directly by passing the excitation
pulse (or a signal proportional to it) through the detector,
although variations in the detector response function with wavelength may also have to be taken into account (1). The formalism
applies equally to the photon counting technique and to single
shot, analog experiments employing intense pulsed sources, photomultiplier tube detectors, and fast transient digitizers.

Duddell(2) has pointed out that neglect of convolution introduces much more serious errors into the measurement of the
fluorescence intensity than it does into the measurement of the

227

J. H. Baxendale and F. Busi (eds.),
The Study of Fast Processes and Transient Species by Electron Pulse Radiolysis, 227–240.
Copyright © 1982 by D. Reidel Publishing Company.

halflife of the transient species. Specifically, the halflife
can ordinarily be obtained without deconvolution within less
than 1%, within the limits of random statistical error, whenever
it exceeds the full width at half maximum (FWHM) of G(t). The
peak maximum (for monoexponential transients), on the other hand,
is not obtained within 1% until the halflife reaches almost 200
times the FWHM. The same point has been made by Michael(3) in his
treatment of convolution corrections in chemical dosimetry.

Thus there exists a rather large range of cases in which
reliable pre-exponential coefficients could be obtained for
dosimetry, the construction of spectra, or other purposes, without
the necessity of deconvolution, provided that suitable corrections
for the convolution error can be derived on the basis of the half-
lives of the transient processes. Such an approach would be
especially helpful for microcomputer-based instruments, for which
the deconvolution of even a few kinetic traces might require an
unacceptably long time.

Duddell(2) has suggested an empirical approach to the desired
corrections, based on computer-synthesized data. This approach
is restricted to the case of monoexponential fluorescence, it
applies only when G(t) has a particular functional form, and it
assumes (correctly, but without proof) that the necessary
corrections depend only on a single parameter, which is the ratio
of the FWHM of G(t) to the transient halflife.

The present treatment avoids all these difficulties by
providing analytic expressions for the corrections, either in
terms of the commonly-used analytic form for G(t), or in terms of
the measured G(t). Like that of Duddell, it applies only to those
(numerous) cases for which reliable kinetic parameters can be
obtained without deconvolution, or are otherwise known with
confidence. Both treatments also assume exponential kinetics;
therefore, in the case of optical absorption measurements, they
are restricted to the case of weak absorptions, for which the
transmitted light intensity change is proportional to the change
in optical density.

2.0 SEPARATION OF THE CONVOLUTION INTEGRAL

For arbitrary forms of the photophysical function I(t), the
convolution integral (1) approaches the form of I(t) at sufficientl
long times, provided that I(t) decays more slowly, in the limit of
long times, than G(t). To see this, observe that the integrand
of (1) is significant in such cases only for small values of
t-t', i.e. for t'≈t. But then t may be substituted for t' in I(t')
and

$$S(t) \overset{\sim}{=} I(t) \int_{0}^{t} G(t-t') \, dt' \tag{3}$$

where the remaining integral is independent of t for large t.

On the other hand, the convolution integral can be separated exactly in the case for which $I(t)$ is given by a finite sum of n exponentials, i.e.

$$I(t) = \sum_{i=1}^{n} a_i \, e^{-k_i t} \tag{4}$$

Then the convolution integral (1) becomes

$$S(t) = \sum_{i=1}^{n} a_i \int_{0}^{t} e^{-k_i(t-t')} G(t') \, dt' \tag{5}$$

and may be factored to yield

$$S(t) = \sum_{i=1}^{n} a_i \, q_i \, (t) \, e^{-k_i t} \tag{6}$$

where

$$q_i \, (t) = \int_{0}^{t} e^{k_i t'} G(t') \, dt' \tag{7}$$

The integrals $q_i(t)$ become independent of t for large t, provided that G(t) decays more rapidly than $\exp(k_i t)$ rises, i.e. that

$$\lim_{t \to \infty} e^{k_i t} G(t) = 0 \tag{8}$$

Then defining

$$\overline{q}_i = \lim_{t \to \infty} g_i \, (t) \tag{9}$$

we have, at long times,

$$S(t) = \sum_{i=1}^{n} a_i \, q_i \, e^{-k_i t} \tag{10}$$

This is, even if the rate constants are obtained correctly by fitting at long times without deconvolution, the measured intensities (which we shall call c_i) are in error by the factors $\bar{q}_i$, which are functions of the rate constants k_i. These errors can lead to serious distortions of the measured pre-exponential coefficients. However, since $G(t)$ and the k_i are known, the $\bar{q}_i$ may be calculated from (7), with $G(t)$ taken to long enough times so that (9) is satisfied, and the results of the calculations may be used to correct the pre-exponential coefficients c_i obtained in the fitting process. The details of performing these corrections are discussed in Section IV.

3.0 MODEL EXCITATION/RESPONSE FUNCTION

The excitation function

$$g(t) = \alpha t^2 e^{-\beta t} \tag{11}$$

has been widely used in the synthesis of model convolution data (2,4), and it frequently represents the experimental $G(t)$ quite well, both for photon counting and for single shot experiments. In such cases, the equations of Section II can be evaluated analytically, and certain interesting properties of the convolution errors can be derived.

First, in order to fit (11) conveniently to measured $G(t)$ functions, we require the peak height and FWHM of (11) in terms of α and β. The time at which (11) is maximum is given by the extremum condition

$$\frac{\partial g}{\partial t} = \alpha t (2 - \beta t) e^{-\beta t} = 0 \tag{12}$$

which yields the non-trivial solution

$$t_{max} = 2/\beta \tag{13}$$

The corresponding maximum value is

$$g(t_{max}) = \frac{4\alpha}{\beta^2 e^2} \tag{14}$$

and the FWHM is the difference Δt_h between the two positive solutions to the transcendental equation.

$$(\beta t_h)^2 = 2e^{\beta t_h - 2} \tag{15}$$

Equation (15) was solved numerically to yield

$$\Delta t_h = 3.395/\beta \tag{16}$$

Thus trial fits of (11) to experimental functions $G(t)$ can readily be obtained from the measured maximum and FWHM, via (14) and (16).

With $G(t)$ given by (11), the error corrections (7) become

$$q_i(t) = \alpha \int_0^t t'^2 \, e^{(k_i - \beta)t'} \, dt'$$

$$= \frac{\alpha}{(k_i - \beta)^3} \{ e^{(k_i - \beta)t} \, | (k_i - \beta)^2 t^2 - 2(k_i - \beta)t + 2 | -2 \} \tag{17}$$

and $\bar{q}_i$ is simply

$$\bar{q}_i = \frac{-2\alpha}{(k_i - \beta)^3} \tag{18}$$

The practical aspects of applying these analytic corrections are discussed in Section IV.

Finally, a derivation will be given of the expression for the error incurred by neglect of convolution when the observed maximum is used to approximate the initial deconvolved intensity. Substituting (17) into (6) yields

$$S(t) = \sum_{i=1}^{n} \frac{a_i}{(k_i - \beta)^3} \, e^{-k_i t} \{ e^{(k_i - \beta)t} [(k_i - \beta)^2 t^2 - 2(k_i - \beta)t + 2] - 2 \} \tag{19}$$

Now we simplify to monoexponential kinetics ($n = 1$) and apply the extremum condition $\frac{\partial s}{\partial t} = 0$ to obtain the transcendental equation

$$e^{(\beta - k)t_m} = \frac{\beta(\beta - k)^2 t_m^2}{2k} + (\beta - k) t_m + 1 \tag{20}$$

for t_m, the time at which the observed signal is maximum. Then $s(t_m)$, the maximum value of $s(t)$, is to be compared to the undistorted maximum s_m. The latter is simply

$$S_m = a \int_0^\infty g(t') \, dt' = \frac{2 \, \alpha \, a}{\beta^3} \tag{21}$$

and the ratio of uncorrected to corrected intensity is given by

$$r \left(\frac{\beta}{k} \right) = \frac{S \, (t_m)}{S_m} \tag{22}$$

To demonstrate that (22) depends only on the ratio of β and k, make the following substitutions:

$$\gamma = \beta/k \tag{23}$$

$$\xi = (\beta - k) \, t_m \tag{24}$$

Then

$$k \, t_m = \frac{\xi}{(\gamma - 1)} \tag{25}$$

and by (19), (21), and (22), r becomes a function of γ and ξ only:

$$r(\gamma, \xi) = \frac{\gamma^3}{2(1-\gamma)^3} e^{-\xi/(\gamma-1)} \{ e^{-\xi} [\xi^2 + 2\xi + 2] - 2 \} \tag{26}$$

Since (20) is also a function of γ and ξ alone, it follows that (26) is uniquely determined by (20) when γ is specified.

Equation (20) was solved numerically to obtain ξ for selected values of γ, and corresponding values of r were calculated from (26). The results, which are displayed in Fig. 1, are in agreement with the empirical results of Duddell.(2)

4.0 DISCUSSION

The treatment of convolution error developed above is useful primarily for cases in which the correct rate constants can be obtained directly from the transient trace without deconvolution, by fitting exponentials to the portion of the trace after the excitation/instrument response function has decayed to zero. As noted in the Introduction, there exists a wide range of such cases for which the peak maximum is still greatly distorted by convolutio Therefore, if the trace contains more than one exponential componen and if the relative intensities of these components change with wavelength, then it becomes impossible to obtain satisfactory spectra directly from the maxima of the individual decay curves. The same problem arises if the signal maximum is required for

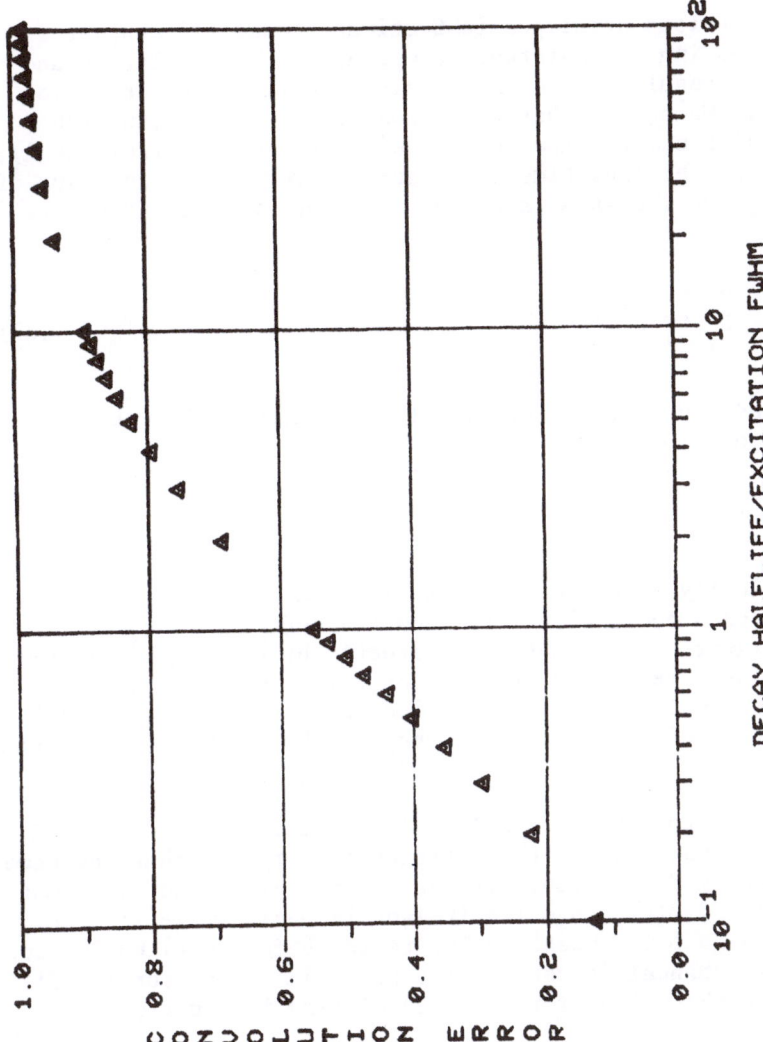

Fig.1. The ratio of uncorrected to corrected peak maximum as a function of $\gamma^1 = 4.9\ \beta/k$, which is the ratio of decay halflife to the full width at half maximum of the excitation function of equation (11).

dosimetry, quenching measurements, or other purposes. And the corrected maximum cannot be obtained simply by extrapolating a fit made at long times to a pre-selected zero-time point, since the resulting intensities would depend on the arbitrary point selected for the extrapolation.

The expressions developed in Sections II and III suggest a means of choosing a consistent end-point for extrapolation and correcting the resulting pre-exponential coefficients for convolution error. We assume that the deconvolved kinetic function $I(t)$ is a finite sum of exponentials with unknown coefficients A_i, as in (4); then the long-time fit, extrapolated to t_o, is a sum of exponentials with the same rate constants, but with coefficients C_i:

$$F(t) = \sum_{i=1}^{n} C_i \, e^{-k_i t} \tag{27}$$

Since $F(t)$ and the observed signal $S(t)$ coincide at long times, it follows from (10) that

$$A_i = C_i / \overline{q}_i \tag{28}$$

and the physically significant coefficients A_i can be obtained from the measured coefficients C_i once the corrections $\overline{q}_i$ are known. The latter can be obtained directly by summing the integrand of (7) over the entire region for which the measured $G(t)$ is non-zero and multiplying by the time per channel; or, if $G(t)$ is approximately of the form (11), they can be obtained analytically from (18).

Implicit in the above formalism is the fact that both the C_i and the $\overline{q}_i$ are functions of the starting time t_o, so that the same t_o must be employed for both. If the $\overline{q}_i$ are calculated by summation over the measured $G(t)$, this requirement is easily fulfilled; a single data channel is chosen in the region before $G(t)$ begins to rise, and this channel is used as the zero-time point for the fit which produces the C_i and for the zero of time in equation (7).

If the $\overline{q}_i$ are calculated analytically from (18), care must be taken to ensure that the t_o employed for the extrapolated fit which yields the C_i is chosen in such a way that (11) is fulfilled. The following procedure is recommended:

1.   Obtain the maximum and FWHM of $g(t)$ by inspection.

2.   Solve (16) for β and (14) for α.

3. The difference $t_{max} - t_0$ is given by (13), from which t_0 can be calculated.

4. Use this t_0 for the starting time of the exponentials in the fit that gives the coefficients C_i.

5. Obtain the $\bar{q}_i$ from (18) and the A_i from (28).

This procedure ensures that the zero-time point for extrapolation is consistent with the use of (11) to calculate $\bar{q}_i$ (eqs. 17-18).

5.0 EXAMPLES

In order to illustrate the correction technique presented here, two experimental excitation/response functions were obtained:

1. A single 30 ps pulse of 353 nm light was applied to a photomultiplier tube. The resulting analog signal was digitized by the Tektronix R7912 digitizer and read into a PDP-11/34 computer. Because of the very narrow excitation pulse, this signal contains primarily the response function of the photomultiplier tube and the digitizer.

2. The profile of a nanosecond nitrogen flash lamp was collected by means of a photon counting spectrofluorometer (Photochemical Research Associates). This profile is dominated by the shape of the lamp pulse.

The experimental profiles are shown in Fig. 2. Each of them was convolved numerically with a biexponential decay law having equal pre-exponential coefficients, with lifetimes of 3 and 12 ns. Each $g(t)$ was fitted to equation (11) by the method described in Section IV. These fits are reasonably good (Fig. 2). In both cases the analytic function rises more steeply than the experimental one, and in case (1), it decays more slowly.

The synthetic fluorescence curves were fitted to equation (27) by means of the iterative linearized least-squares technique, employing the software system previously described.(5) The results of these fits are given in Table I. In all cases, the rate constants obtained were correct within 5%; however, the ratio of the pre-exponential coefficients differed from 1 by about 100%, illustrating the importance of correcting such results for convolution error. Comparable errors were found when the end-point for extrapolation was chosen at the maximum of the signal. Use of the maximum of $g(t)$ gave a better ratio, but the signal maximum of $g(t)$ gave a better ratio, but the signal maximum was still grossly in error.

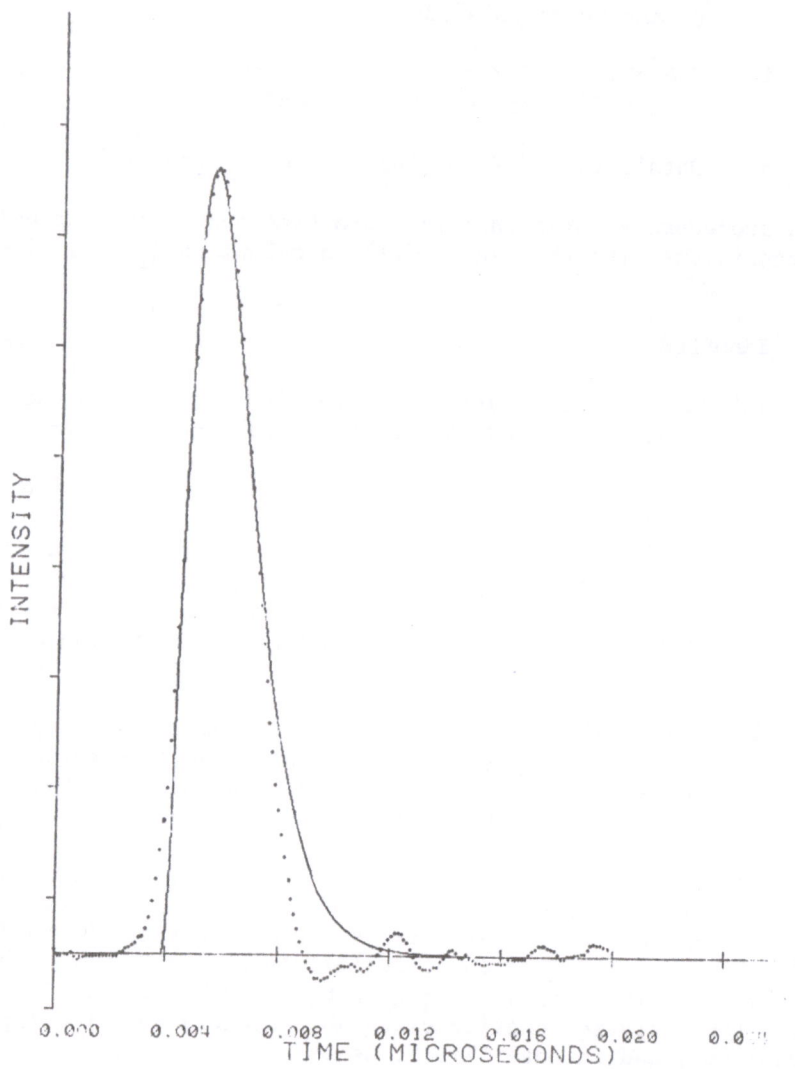

Fig. 2(a) Experimental excitation profiles with fits of g(t) determined by the method of Section 4. 30 ps pulse of 353 nm light, detected by a photomultiplier tube and recorded by a Tektronix R7912 digitizer.

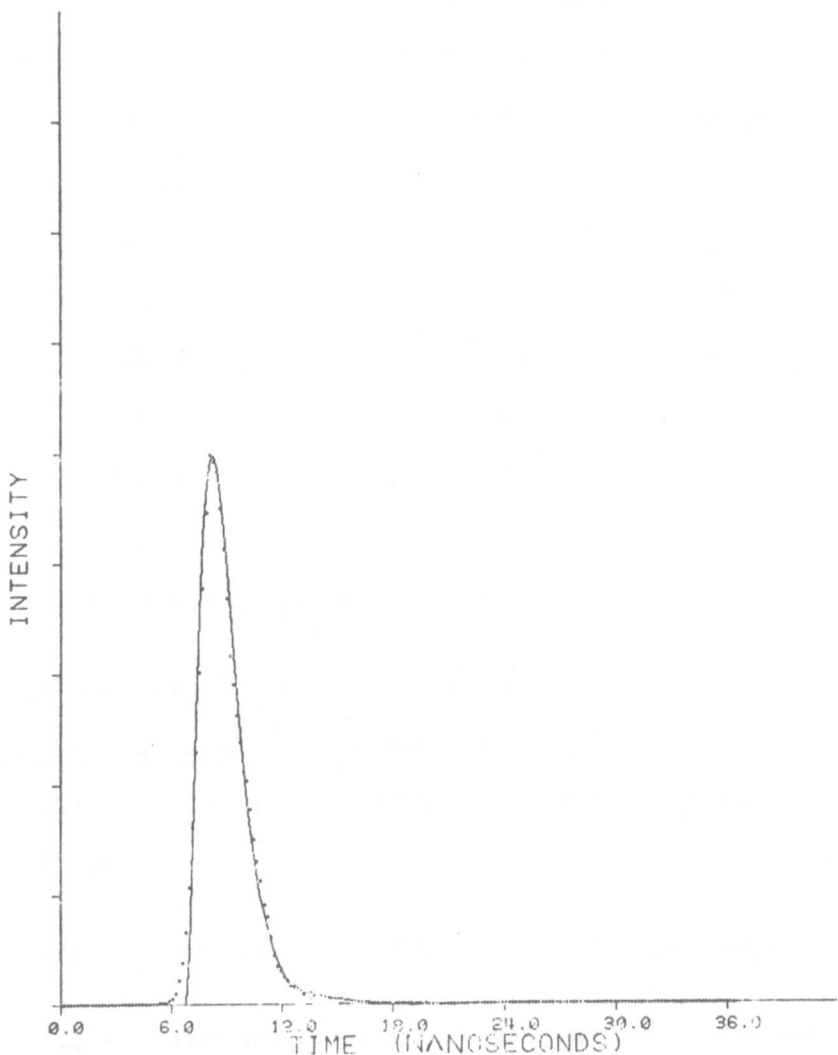

Fig. 2(b) Experimental excitation profiles with fits of g(t) determined by the method of Section 4. Nanosecond flash lamp, recorded by a photon counting spectrofluorometer.

Table I. Starting Parameters and Results of Fits to Synthesized
Fluorescence Data.

Used	(a) Analog q(t)	(b) Photon Counting q(t)
a_1/a_2	1.000	1.000
k_1	0.333	0.333
k_2	0.0833	0.0833
Found		
c_1/c_2	2.076	1.807
k_1	0.347	0.327
k_2	0.0798	0.0831

Table II. Results of Applying Correction Procedure to Data of
Table I.

	(a) Analog q(t)		(b) Photon Counting q(t)	
	Numerical	Analytic	Numerical	Analytic
$c_1/\bar{q}_1 a_1$	0.807	0.853	1.104	1.007
$c_2/\bar{q}_2 a_2$	0.958	0.841	1.008	1.019
$c_1\bar{q}_2/c_2\bar{q}_1$	0.842	1.014	1.095	0.988

Note: The first rows give the ratios of the corrected pre-
exponential factors to the ones originally used. The last r
gives the ratio of the two corrected factors.

The pre-exponential coefficients were corrected by both the
numerical and the analytic techniques discussed above. Table II
gives the ratio of the corrected coefficients to the original
ones used in the convolution. The last line gives the ratio of
the corrected coefficients. The deviation from unity is a measure
of the error incurred in the fitting and correction procedures.
These results display the strengths and weaknesses of the method
and suggest guidelines for its use:

1. The corrected coefficients for the analog case are consistently low, that is, $\bar{q}_i$ is consistently high. For the analytic correction, this is due to the fact that the fit of (11) decays more slowly than the experimental g(t) (Fig. 2a). For the numerical correction, the small upward drift in the baseline at long times is greatly amplified in (7) by the exponential term.

2. The final ratio of coefficients for the analog case is given quite well by the analytic correction, but badly by the numerical one. As noted in (1) above, small deviations at long times are amplified by the numerical procedure. It is evident from (7) that these errors will be larger for the faster of the two exponential processes. Therefore, the analytic correction technique should be employed whenever possible. The performance of the numerical procedure can be improved considerably by terminating the sum early, immediately after g(t) has effectively decayed to zero.

3. The photon counting data, which is relatively free of noise compared to the analog data, is corrected well by both techniques. The analytic correction is better for the faster component and for the final ratio, due to the factors discussed in (2). Although it is not evident from the table, the numerical correction is highly sensitive to small errors in the asymptotic baseline of g(t).

6.0 CONCLUSION

Techniques have been described for extracting component intensities from fast kinetic transient data without deconvolution. Good results are achieved when the rate constants can first be obtained without deconvolution. The method is most successful when the excitation/response function can be fitted by a simple analytic expression. Both the total intensity and its resolution into exponential components are obtained much more rapidly than by the usual technique of deconvolving each trace.

This work was carried out at the Center for Fast Kinetics Research in Austin, Texas. The support of the Biotechnology Branch of the Division of Research Resources of the National Institutes of Health, Grant RR-00886, and of the University of Texas at Austin is gratefully acknowledged.

REFERENCES

1. Ph. Wahl, J.C. Auchet and B. Donzel, Rev. Sci. Instrum. 45,
    28 (1974).
2. D.A. Duddell, Photochem. Photobiol. 31, 121 (1980).
3. B. Michael, Doctoral Dissertation.
4. A.E. McKinnon, A.G. Szabo and D.R. Miller, J. Phys. Chem. 81,
    1564 (1977).
5. D.C. Foyt, Computers & Chemistry 5, 49 (1981).

BASIC RADIATION CHEMISTRY OF LIQUID WATER

G. V. Buxton

University of Leeds

Abstract. – Water radiolysis generates approximately equal numbers of highly reactive reducing (e^-_{aq} and H) and oxidising species (OH), thus providing chemists with a powerful tool for generating in aqueous solution a host of inorganic and organic free radicals and unusual oxidation states of many metal ions.

This review focusses attention on the two features of water radiolysis which are of major importance in its use as a chemical tool. The first is the yields of e^-_{aq}, H and OH and how these yields depend on time, pH, radiation quality (LET) and scavenger concentration. The second concerns the properties of the primary species and the ways in which these may be utilised to produce totally oxidising or reducing conditions.

1. INTRODUCTION.

The radiation chemistry of water is very well characterised and, although there are still a few unanswered questions concerning the role of excited states and the exact details of track structure, it has for some time provided chemists with a powerful, and often unique, method of studying the reactions of unstable species in aqueous solution. Detailed accounts of the subject are to be found in standard texts (1-4) and in reviews by Thomas (5) and Hunt (6). It is the aim of this article to provide the non-specialist with the basic information needed to use the technique as a chemical tool.

There are two distinguishing features of radiation chemistry. The first is the non-selective absorption of energy, so that in

241

J. H. Baxendale and F. Busi (eds.),
The Study of Fast Processes and Transient Species by Electron Pulse Radiolysis, 241–266.

dilute aqueous solution energy is absorbed predominantly by the water. The second is the large energy involved, leading to the formation of ions and highly excited molecules in isolated volume elements, called spurs, along the track of the ionising particle. The yields of these species are measured as G-values, defined by

$$G(X) = \text{no. of molecules of } X/100 \text{ eV}$$

Although the precise structure of the spur and its distribution in size and space along the primary particle track is not known the energy loss per spur in liquid water is thought to be in the range 6 - 100 eV (7). In water vapour G(ion pairs) is 3.1 so that a typical spur in the liquid is expected to contain more than one ion pair in close proximity. There is abundant experimental evidence in support of the spur model, and together they provide a clear picture of the radiation chemistry of liquid water.

2. PRIMARY EVENTS

The sequence of events that is initiated by the passage of the ionising particle through water is shown in the scheme opposite.

Here H_2O^* represents an excited water molecule which may ionise, dissociate or simply return to the ground state. There is no clear evidence as yet that H_2O^* plays a significant role in water radiolysis, although it has been invoked to explain some observations in strongly alkaline water (see section 4.2).

The positive ion H_2O^+ is generally believed to be very short lived ($<10^{-14}$ s) on the grounds that the ion-molecule reaction in the gas phase has a rate constant of 8×10^{12} dm^3 mol^{-1} s^{-1} (8). However, Hamill (9) has proposed that H_2O^+ can be scavenged by high concentrations of halide ions, based on the idea that H_2O^+ can migrate rapidly by resonance electron transfer with a neighbouring water molecule.

The ejected electron loses energy via further ionisation and excitation events and finally becomes thermalised and solvated. Recent experimental evidence indicates that the solvation time is less than 10^{-12} s (10), which is much shorter than the dielectric relaxation time of water (10^{-11} s). This may mean that the thermal electron finds a preformed site in the medium that largely satisfies its solvation requirements. The electron is often termed 'dry' before it becomes solvated, and there is good evidence that it can be scavenged in this state (6) at high concentrations ($\sim$1 mol dm^{-3}) of solute.

Scheme for the Radiolysis of Water

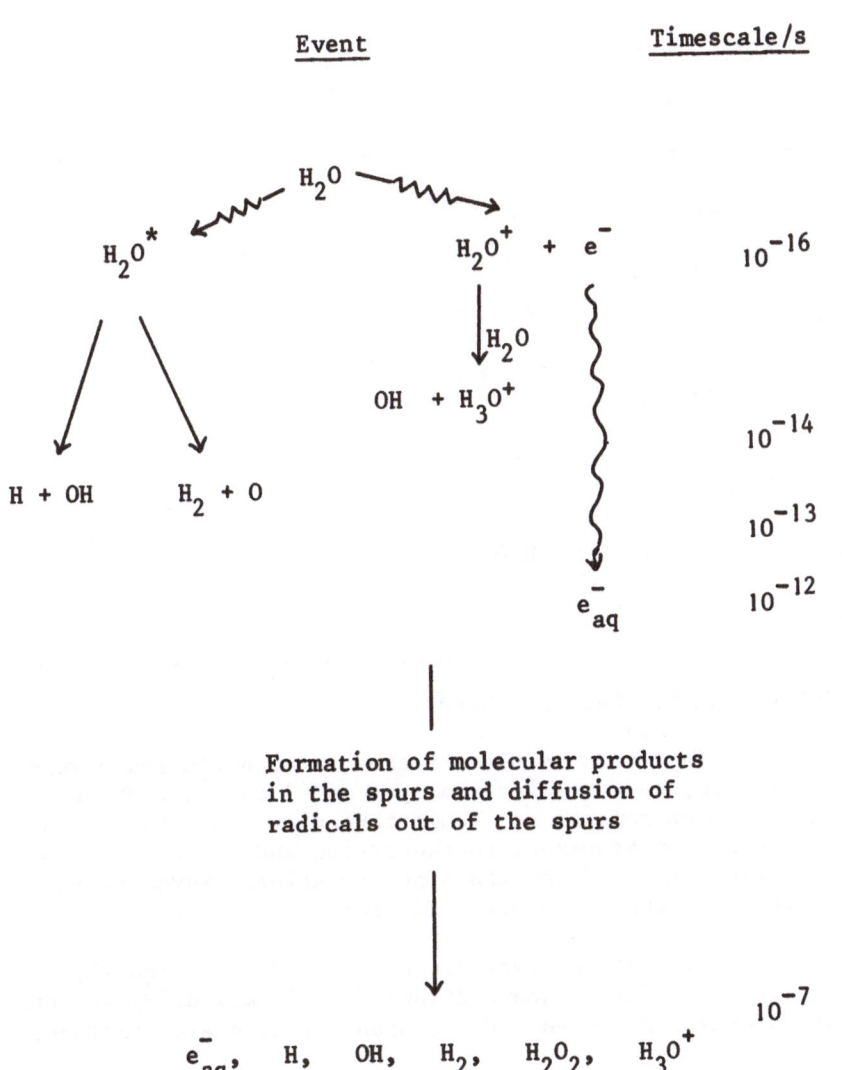

Event   Timescale/s

According to the generally accepted view, about 10^{-12} s after the initial ionisation event the species e_{aq}^-, OH and H_3O^+, together with any dissociation products of H_2O^*, are located in spurs. Subsequently these species begin to diffuse with the result that a fraction of them undergo intraspur reactions while the remainder escape into the bulk medium and become distributed homogeneously. These spur processes are complete by about 10^{-7} s. Reactions that are likely to occur in the spurs are listed in Table 1

Table 1. Spur Reactions in Water (1)

	Reaction	$10^{-10}k/dm^3$ mol^{-1} s^{-1}
(1)	$e_{aq}^- + e_{aq}^- \rightarrow H_2 + 2OH^-$	0.54
(2)	$e_{aq}^- + OH \rightarrow OH^-$	3.0
(3)	$e_{aq}^- + H_3O^+ \rightarrow H + H_2O$	2.3
(4)	$e_{aq}^- + H \rightarrow H_2 + OH^-$	2.5
(5)	$H + H \rightarrow H_2$	1.3
(6)	$OH + OH \rightarrow H_2O_2$	0.53
(7)	$OH + H \rightarrow H_2O$	3.2
(8)	$H_3O^+ + OH^- \rightarrow 2H_2O$	14.3

3. EXPERIMENTAL EVIDENCE FOR SPURS

Until the fast time-resolved technique of pulse radiolysis became available, evidence for the spur model outlined above was obtained from measurements of stable radiation products, including their dependence on scavenger concentration and the mean linear energy transfer (LET) of the ionising radiation. Three characteristic features of these measurements are:

1) A fraction of the radicals e_{aq}^-, H and OH are easily scavenged at low solute concentrations ($\leqslant 10^{-3}$ mol dm^{-3}) and the yield of molecular hydrogen and hydrogen peroxide are constant.

2) At higher solute concentrations the yields of scavenged radicals progressively increase and the yields of H_2 and H_2O_2 decrease.

3) As the LET increases the yields of e_{aq}^- and OH decrease while those of H_2 and H_2O_2 increase.

It is easy to see that these features of water radiolysis are consistent with the spur model. Firstly, at low solute concentrations only those species which diffuse out of the spurs are

scavenged and all the spur reactions leading to molecular
products are complete (see Table 1). Secondly, at sufficiently
high solute concentrations reactions (9) and (10) begin to occur

$$e_{aq}^- + S_1 \rightarrow P_1 \qquad\qquad (9)$$

$$OH + S_2 \rightarrow P_2 \qquad\qquad (10)$$

in the spurs. Hence $G(P_1)$ and $G(P_2)$, which are measures of
scavenged e_{aq}^- and OH, increase and the yields of H_2 and H_2O_2 fall.
Thirdly, as the LET increases the spurs are formed closer together
and eventually coalesce into a cylindrical track. As a result
the local concentration of radicals is increased so that reactions
leading to molecular products account for a higher proportion of
the initial radicals. Some data which illustrate this effect very
clearly are given in Table 2.

Table 2. Dependence of the Yields of Radiolysis Products from
Neutral Water on LET (1,11)

LET/eV nm^{-1}	$-H_2O$	e_{aq}^-	OH	G H	H_2	H_2O_2	HO_2
0.23	4.08	2.63	2.72	0.55	0.45	0.68	0.008
12.3	3.46	1.48	1.78	0.62	0.68	0.84	
61	3.01	0.72	0.91	0.42	0.96	1.00	0.05
108	2.84	0.42	0.54	0.27	1.11	1.08	0.07

These G-values also provide an explanation for the fact that
when pure deaerated water is irradiated, essentially no decomposition occurs with low LET radiation but H_2, H_2O_2 and O_2 are
produced at high LET. This is because the reactions in Table 3
take place in the bulk medium after the spur reactions are complete,
resulting in decomposition of the molecular products by the
radicals. At high LET net formation of molecular products is
observed because the radical yields are much lower than the molecular yields.

The development in the last decade of pulse radiolysis facilities capable of subnanosecond time resolution finally led to
the direct observation of spur reactions (6,12-14). Fig. 1 shows
the time profiles of e_{aq}^- and OH for 20 MeV electron radiation (13)
and Fig. 2 shows data for e_{aq}^- for 3 MeV proton radiation when the
LET is 100 times higher (14). In both cases addition of

Table 3.   Reactions Between Radical and Molecular Products

Reaction	$10^7 k/dm^3 \, mol^{-1} \, s^{-1}$
(11) $e_{aq}^- + H_2O_2 \rightarrow OH + OH^-$	1200
(12) $OH + H_2 \rightarrow H + H_2O$	4.9
(13) $OH + H_2O_2 \rightarrow HO_2 + H_2O$	2.7
(14) $HO_2 + HO_2 \rightarrow H_2O_2 + O_2$	0.25
(15) $O_2^- + HO_2 \rightarrow HO_2^- + O_2$	4.4
(16) $H + O_2 \rightarrow HO_2$	1900
(17) $e_{aq}^- + O_2 \rightarrow O_2^-$	1900
(18) $HO_2 \rightleftharpoons H^+ + O_2^-$	pK = 4.9

scavengers for OH and H_3O^+ slows down the rate of decay of e_{aq}^-, confirming that reactions (2) and (3) in Table 1 occur in the spurs. Fig. 3 illustrates this point for 20 MeV electron radiation (12).

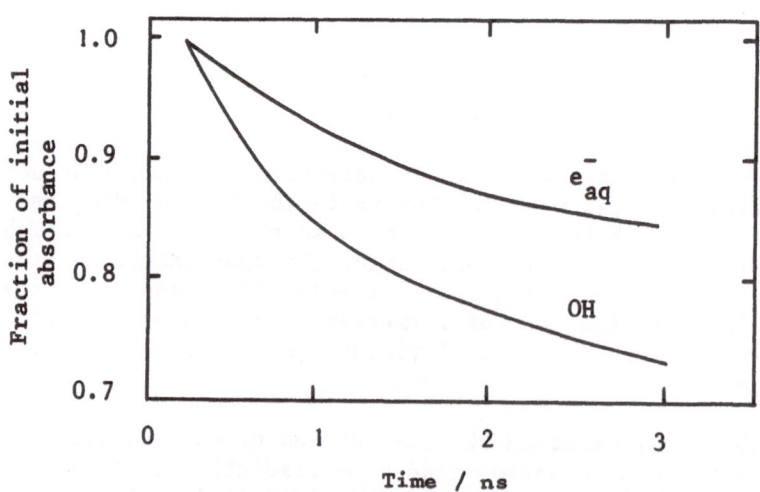

Figure 1.   Decay of e_{aq}^- and OH produced by a 30 ps pulse of 20 MeV electrons in water. Reprinted in part from (12) with permission. Copyright 1976 American Chemical Society.

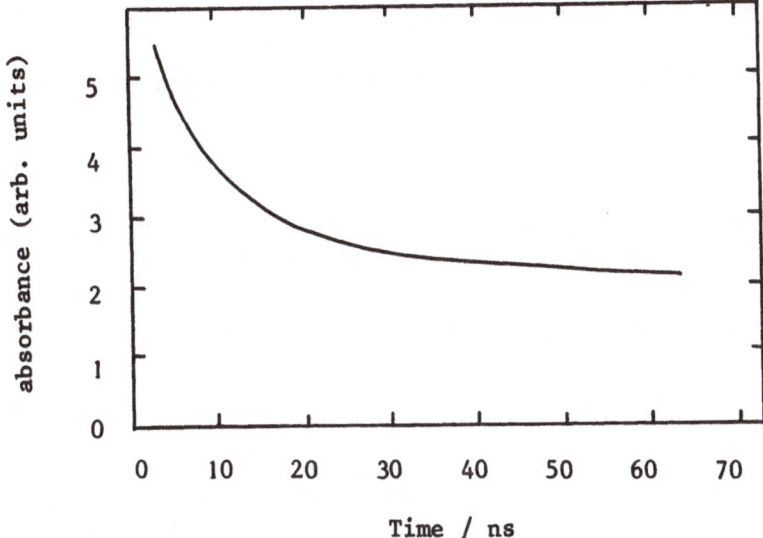

Figure 2.   Decay of e_{aq}^{-} produced by a 1 ns pulse of 3 MeV protons (Redrawn from (14)).

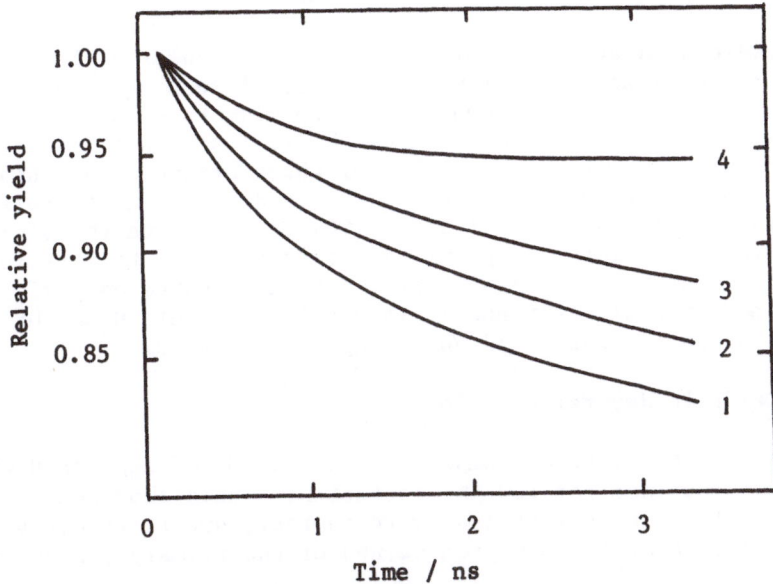

Figure 3.   Decay of e_{aq}^{-} produced by a 30 ps pulse of 20 MeV electrons. 1, water; 2, 3 mol dm^{-3} ethanol; 3, 1 mol dm^{-3} NaOH; 4, 3 mol dm^{-3} ethanol + 1 mol dm^{-3} NaOH. Reprinted in part from (12) with permission. Copyright 1976 American Chemical Society.

4. YIELDS OF THE PRIMARY SPECIES

Reaction (18) represents the radiolytic change in water at 10^{-7} s after the initial ionisation event,

$$H_2O \rightarrow e_{aq}^-, H, OH, HO_2, H_2O_2, H_3O^+ \tag{18}$$

and eqn. (19), or (20) if HO_2 is neglected at low LET (see Table 2), follow from material balance requirements,

$$G_{-H_2O} = 2G_{H_2} + G_H + G_{e_{aq}^-} - G_{HO_2} = 2G_{H_2O_2} + G_{OH} + 2G_{HO_2} \tag{19}$$

$$G_{-H_2O} = 2G_{H_2} + G_H + G_{e_{aq}^-} = 2G_{H_2O_2} + G_{OH} \tag{20}$$

These G-values represent the yields of radiolysis products present when the spur processes are complete. For historical reasons these yields are often termed primary yields and the products are described as primary products. This terminology is a little confusing because it is now known that the initial yields of e_{aq}^- and OH are much larger than their primary yields. When quoting G-values for aqueous solutions the accepted convention is to use G(X) when X is the measured product, and G_X or $g(x)$ when calculating the primary yield of X.

Numerous measurements of radical and molecular yields have been made, especially for low LET radiations such as ^{60}Co γ-rays and fast electrons. In steady-state experiments solutes must be added which scavenge one or more of the radicals to form identifiable products. The yields of these products and those of hydrogen and hydrogen peroxide, which are directly measurable, are then substituted into eqn. (19) or (20) to calculate the yield of each primary species. In principle it is possible to measure radical yields by pulse radiolysis, but in practice only e_{aq}^- is easily measured directly and it is usual to measure H and OH by adding suitable solutes and observing the products.

4.1 Yields in Neutral Solution

Table 4 shows some examples of the yields of e_{aq}^- and H which have been obtained for low LET radiation. These show some variation from one solute system to another, but it is not very large. The generally accepted values of the primary yields are

$$G_{e_{aq}^-} = G_{OH} = G_{H_3O^+} = 2.7$$

$$G_H = 0.6, \quad G_{H_2} = 0.45, \quad G_{H_2O_2} = 0.7$$

As a general rule these values will apply to dilute solutions in the pH range 3-11 for low LET radiation.

Table 4. Some Values of $G_{e_{aq}^-}$ and G_H in Neutral Solution (15)

System	$G_{e_{aq}^-}$	G_H	$G_{e_{aq}^-} + G_H$
2-propanol + acetone	2.65	0.55	3.2
2-propanol + N_2O	2.8	0.6	3.4
N_2O + NO_2^-	2.6	-	-
HCO_2^- + O_2	2.3	0.75	3.05
NO_3^- + HPO_3^-	2.65	0.55	3.2
NO_3^- + 2-propanol	2.8	0.6	3.4
$C(NO_2)_4$, pulse radiolysis	2.6	-	-

Some care must be taken in the choice of G-values in more concentrated solutions where radicals are scavenged in the spurs. Fig. 4 shows that widely differing concentration effects are observed which depend on the nature of the solute.

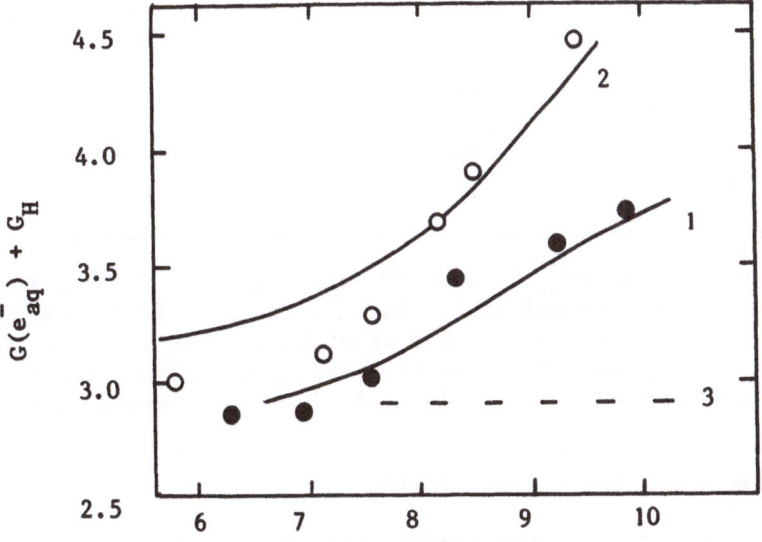

Figure 4. Dependence of $G(e_{aq}^-)$ on $k[S]$. O, N_2O; ●, NO_3^-; 1, H_3O^+; 2, eqn. (27); 3, no net spur reaction.

These effects can be rationalised in terms of the spur model as follows. The spur contains approximately equal numbers of oxidising and reducing radicals. Hence, if P_1 formed in reaction (9) is unreactive with OH, $G(P_1)$ will be larger than $G_{e_{aq}^-}$. On the other hand, if P_1 reacts efficiently with OH in reaction (21)

$$e_{aq}^- + S_1 \rightarrow P_1 \tag{9}$$

$$OH + P_1 \rightarrow OH^- + S_1 \tag{21}$$

there may be little or no enhancement in $G(P_1)$ at high concentrations of S_1. The analogous situation can occur with OH scavengers

$$OH + S_2 \rightarrow P_2 + OH^- \tag{10}$$

$$e_{aq}^- + P_2 \rightarrow S_2 \tag{22}$$

Examples of solutes which increase the yields of scavenged radicals at high concentration are listed in Table 5.

Table 5. Solutes which Increase Radical Yields at High Concentrations (15)

Solute	Scavenged Radical	Product
NO_3^-	e_{aq}^-	NO_3^{2-} ($NO_2^- + 2OH^-$)
N_2O	e_{aq}^-	$N_2 + O^-$
$ClCH_2CO_2^-$	e_{aq}^-	$Cl^- + CH_2CO_2^-$
HCO_2^-	OH	CO_2^-

Solutes such as Br^-, I^- and CNS^- are examples of those that do not enhance the radical yields because reaction (22) is as efficient as reaction (2). On the other hand, whenever spur scavenging takes place the molecular yields are always diminished because reaction (1) and (4-6) in Table 1 are suppressed (see Fig. 5).

4.2 Dependence of the Primary Yields on pH

The effect of pH on the radical and molecular yields for dilute solutions and low LET radiation is summarised in Fig. 6. At low pH e_{aq}^- is rapidly converted into the hydrogen atom (see Table 1), and the larger yield of H which escape from the spur is

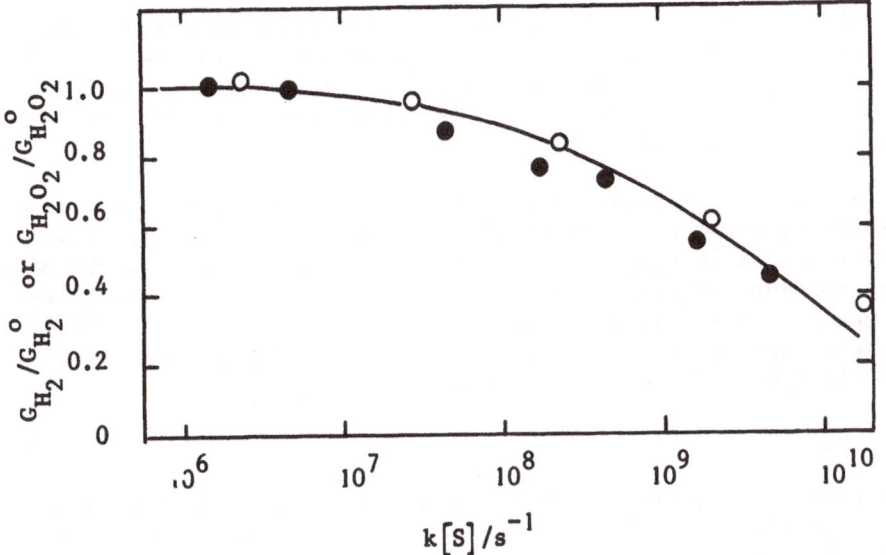

Figure 5. Effect of scavengers on the yields of molecular hydrogen and hydrogen peroxide. ●, Ethanol; O, Hydrogen peroxide; the solid line is the average for several scavengers (Redrawn from (3)).

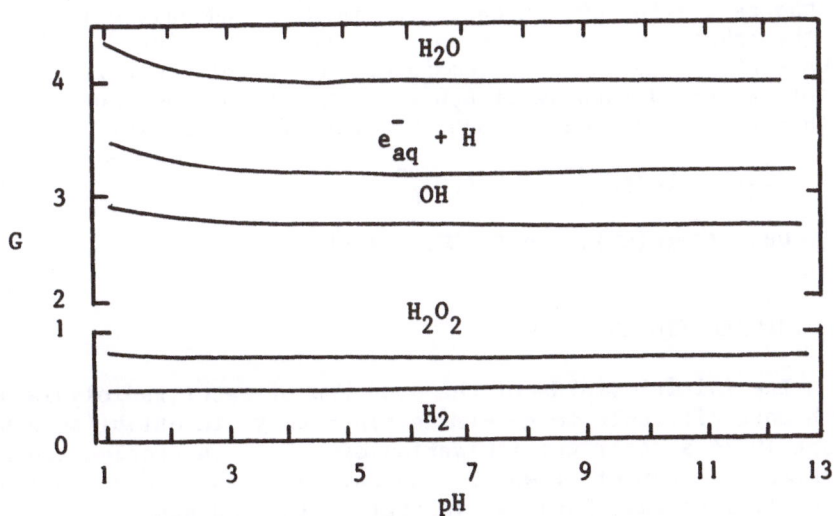

Figure 6. Dependence of radical and molecular yields on pH. (Redrawn from (3)).

probably the result of H having a higher diffusion coefficient than e_{aq}^-. It is significant that G_{OH} does not increase as much as $G_{e_{aq}^-} + G_H$ at low pH and that there is a small increase in $G_{H_2O_2}$. This suggests that reaction (6) is relatively more efficient when e_{aq}^- is replaced by H.

In alkaline solution (pH > 11) there is some doubt as to whether or not the yields of the radicals increase (15). There is a decrease in $G_{H_2O_2}$ at very high pH (15), but this is probably a consequence of OH(pK = 11.9) dissociating to O^- and the latter recombining more slowly in the analogue of reaction (6). Those systems that do show an increase in $G_{e_{aq}^-}$ at high pH generally contain N_2O as the electron scavenger,

$$e_{aq}^- + N_2O \rightarrow N_2 + O^- \tag{23}$$

and this may provide a clue as to why other scavengers such as ferricyanide ion indicate that $G_{e_{aq}^-}$ is independent of pH. It has been suggested (16,17) that reaction (24) is a source of radicals at high pH.

$$H_2O^* + OH^- \rightarrow e_{aq}^- + OH \tag{24}$$

In this reaction it is proposed that an excited water molecule accepts an electron from OH^- to form e_{aq}^- and OH close together. If one partner of this pair is scavenged irreversibly, as would be the case with N_2O, an enhanced yield of radicals will result. On the other hand, if the product of the scavenging reaction reacts efficiently with the remaining radical of the pair the net result is the quenching of H_2O^*. Such a situation could arise with ferricyanide ion through reactions (25) and (26)

$$e_{aq}^- + Fe(CN)_6^{3-} \rightarrow Fe(CN)_6^{4-} \tag{25}$$

$$OH + Fe(CN)_4^- \rightarrow Fe(CN)_6^{3-} + OH^- \tag{26}$$

5. INITIAL YIELDS

The initial yields of the products of water radiolysis are much more difficult to determine since they are established by about 10^{-12} s after the ionisation event. This problem can be tackled in a number of ways. One way is to use the diffusion model (18) to calculate the initial yields and spur dimensions which account for the primary yields. Another is to make direct observations of the time-profiles of the radicals using pulse radiolysis with high time-resolution. A third way is to transform the measured dependence of product yields on solute concentration into a time-profile for the scavenged entity. These three methods are outlined below.

The most important feature of the diffusion model is the requirement that OH, H and H_3O^+ should be in the core of the spur with e_{aq}^- distributed mainly outside this core. It is usually assumed that these species have a Gaussian distribution. Schwarz (18), for example, obtained good agreement between calculated and experimental primary yields using the parameters in Table 6. These parameters include initial yields of atomic and molecular hydrogen formed from excited water molecules. The initial yield of H is similar to G_H, suggesting that there is a balance between the formation and loss of H as the spurs expand.

Table 6. Spur Parameters Used to Fit Primary Yields (18)

Species	Initial G-value	Mean Initial Distribution Radius/nm
e_{aq}^-	4.78	2.3
H	0.62	0.75
H_2	0.15	-
OH	5.6	0.75
H_3O^+	4.78	0.75

Direct measurements by pulse radiolysis (Fig 1) show that $G(e_{aq}^-) = 4.6 \pm 0.2$ at 10^{-10} s and $G(OH) = 5.9 \pm 0.2$ at 2×10^{-10} s (although there could be a systematic error in G(OH) (13)). Thus the agreement between model calculations and experiment is encouraging. Moreover, Fig. 1 shows that OH decays faster than e_{aq}^-, which is consistent with the proposal that OH has a tighter spatial distribution than e_{aq}^-. However, there is some discrepancy between the time-profiles of e_{aq}^- obtained from model calculations and experiment. Fig. 7 shows that this descrepancy is more marked for higher LET radiation.

The discrepancy may be removed for low LET radiation by skewing the distribution of e_{aq}^- away from the centre of the spur (19), or by choosing larger spur radii and more energy loss per spur. Kuppermann (20), for example, obtained good agreement with the data for e_{aq}^- in Fig. 1 by choosing 2.25 nm as the radius for OH, H and H_3O^+ and 6.75 nm for e_{aq}^-.

The concentration dependence observed for electron scavenging by nitrous oxide and methyl chloride has been shown (21) to be

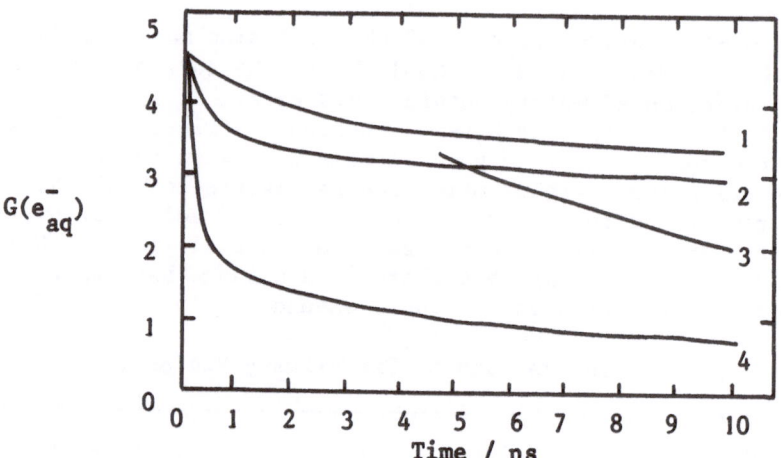

Figure 7. Observed (1,3) and calculated (2,4) decays of e_{aq}^- for
20 MeV electrons (1,2) and 3 MeV protons (3,4) (Redrawn from (14)).

described by eqn. (27) which is empirical (21),

$$G(P_1) = 2.55 + \frac{2.23(k[S_1]/\lambda)^{\frac{1}{2}}}{1 + (k[S_1]/\lambda)^{\frac{1}{2}}} \tag{27}$$

Here P_1 is the product of reaction (9), k is the rate constant for
reaction (9), λ is a constant estimated to be 8×10^8 s^{-1}, and
the initial and primary yields of e_{aq}^- are taken as 4.78 and 2.55
respectively. The initial yield was chosen to agree with Schwarz's
value (18). The Laplace transform of eqn. (27) is eqn. (28) from
which the time-profile of e_{aq}^- can be calculated.

$$G(e_{aq}^-) = 2.55 + 2.23e^{\lambda t}erfc(\lambda t)^{\frac{1}{2}} \tag{28}$$

The fit of eqn. (27) to the data for N_2O solutions is shown
in Fig. 4, from which it is seen that eqn. (17) may be used to
calculate the scavenged yield of e_{aq}^- for values of $k_9[S_1]$ up to
2×10^9 s^{-1} with an accuracy that is sufficient for most chemical
applications. It should be remembered, however, that eqn. (28)
will only apply to scavengers which form products that do not react with OH in the spur.

The time-profiles of e_{aq}^-, determined as described above, are
compared in Fig. 8. Because of its broader distribution e_{aq}^- will
be more easily scavenged than OH so that, in general, higher concentrations of OH scavengers are needed to increase the yield of
the scavenged radical.

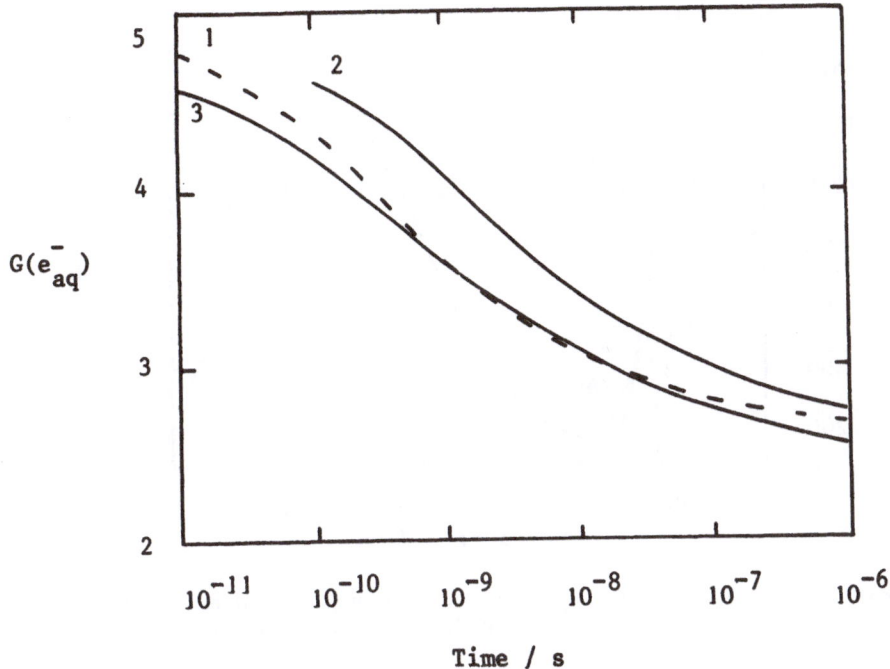

Figure 8. Time-profile of e_{aq}^-. 1, Calculated (Schwarz model
(18)); 2, Pulse radiolysis data (12); 3, Calculated using
eqn. (28) (21).

At very high solute concentrations (~ 1 mol dm^{-3}) it has been
demonstrated (6) that $G(P_1) > G(e_{aq}^-)$ at 3×10^{-11} s when S_1 is
Cd^{2+} or cystamine in reaction (9). The difference amounts to
G = 0.8 and has been attributed (6) to scavenging of 'dry'
electrons. On this basis it is concluded (12) that the initial
yield of electrons is 5.4 and that 4.6 of these become solvated.

6. PROPERTIES OF THE PRIMARY RADICALS

6.1 The Hydrated Electron

     The absorption spectrum of e_{aq}^- is shown in Fig. 9 and some
of its properties are listed in Table 7. The intense absorption
band in the visible region of the spectrum makes it very easy to
follow the reactions of e_{aq}^- using pulse radiolysis combined with
kinetic spectroscopy (22). Several hundred rate constants have
been measured by this method (23).

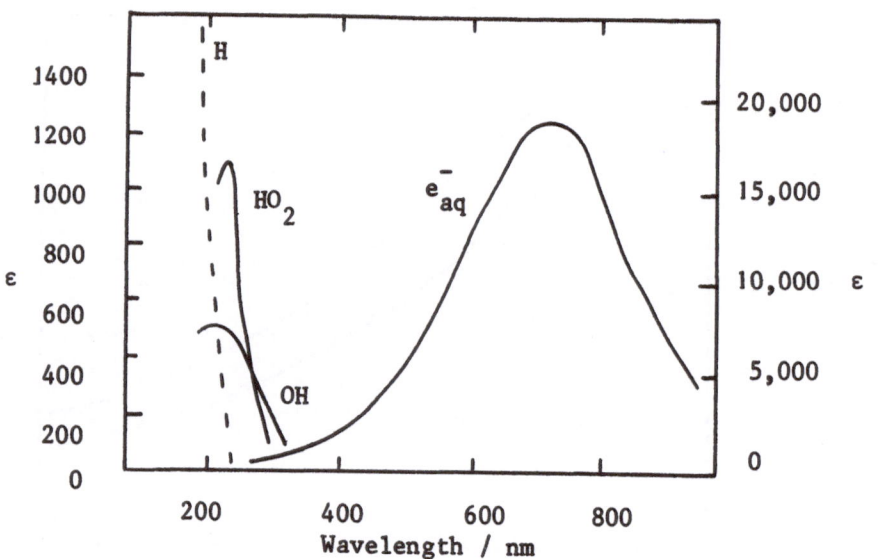

Figure 9. Spectra of e_{aq}^-, H. OH and HO_2. Redrawn from (1) and in part from (32) with permission. Copyright 1976 American Chemical Society.

Table 7. Properties of the Hydrated Electron

Radius of charge distribution	0.25 - 0.3 nm
Diffusion constant	4.9×10^{-5} cm^2 s^{-1}
Wavelength of maximum absorption	715 nm
Extinction coefficient at 715 nm	1.85×10^4 dm^3 mol^{-1} cm^{-1}
Standard reduction potential	- 2.9 V
Half-life in neutral water	2.1×10^{-4} s

The hydrated electron may be visualised as an electron surrounded by a small number of oriented water molecules and behaving in many ways like a single charge anion of about the same size as the iodide ion. It is a powerful reducing agent and its reactions are one-electron transfer processes represented by

$$e_{aq}^- + S^n \rightarrow S^{n-1} \qquad\qquad (29)$$

where n is the charge on the solute. The rate constants for these reactions range from 16 dm^3 mol^{-1} s^{-1} (reaction with water) up to the diffusion controlled limit, but the activation energy is invariably small (6 - 30 kJ mol^{-1} K^{-1}). This suggests that the magnitude of the rate constant is governed mainly by the entropy of activation, and the controlling factor here appears to be the availability of a suitable vacant orbital in the acceptor molecule. Thus molecules with no vacant low-lying orbitals, such as water, simple alcohols, amines and ethers, react slowly.

Some features of the reactivity of e_{aq}^- are illustrated in Table 9.

Table 9. Rate Constants for Some Reactions of the Hydrated Electron (23)

Solute	$10^{-10}k/dm^3 mol^{-1} s^{-1}$	Solute	$10^{-10}k/dm^3 mol^{-1} s^{-1}$
Inorganic		**Organic**	
O_2	1.9	C_6H_6	1.2×10^{-3}
H_3O^+	2.3	C_6H_5Cl	5×10^{-2}
NH_4^+	$< 2 \times 10^{-4}$	C_6H_5I	1.2
Ag^+	3.6	$CH_2=CH_2$	$< 2.5 \times 10^{-4}$
Cd^{2+}	5.0	$CH_2=CCl_2$	2.3
In^{3+}	5.6	$CH_2=CHCONH_2$	1.8
$Fe(CN)_6^{3-}$	0.3		
CrO_4^{2-}	1.8	CH_4	$< 10^{-3}$
NO_3^-	1.0	CH_3I	1.7
N_2O	0.87	CH_3OH	$< 10^{-6}$
H_2O	1.6×10^{-9}		

As might be expected, e_{aq}^- reacts rapidly with many species having reduction potentials more positive than - 2.9 V. In some cases, e.g. CrO_4^{2-}, the rate constant is higher than predicted by the Debye-Smoluchowski eqn. for diffusion controlled reactions, and it is suggested (4) that the reaction radius is bigger than normal because the electron tunnels several Ångstroms from its solvent trap to the acceptor. Tunnelling between solvent traps may also explain the mobility of e_{aq}^- since it is much higher than expected for a singly charged anion of radius 0.3 nm.

In its reaction with organic molecules e_{aq}^- behaves as a nucleophile. Thus electron withdrawing substituents adjacent to alkene double bonds or attached to aromatic rings greatly enhance the reactivity of the molecule. Increased reactivity is also observed when halogen atoms are substituted into aliphatic and aromatic hydrocarbons. In this case the negative ion formed in reaction (30) rapidly eliminates halide ion.

$$e_{aq}^- + RX \rightarrow RX^- \rightarrow R\cdot + X^- \tag{30}$$

6.2 The Hydrogen Atom

Although the hydrogen atom is the minor reducing radical in neutral and alkaline solution, it is the major one in acidic solution through reaction (3) (Table 1). Its properties are not so well documented as those of e_{aq}^-, principally because it does not absorb in an accessible part of the spectrum (see Fig. 9). Nevertheless, many of its rate constants have been measured using competition kinetics (24).

The hydrogen atom has a standard reduction potential of -2.3 V and is a slightly less powerful reducing agent than e_{aq}^-. It readily reduces cations having more positive reduction potentials than itself, but often at slower rates than the hydrated electron. In some cases it effectively reacts as an oxidant forming a hydride intermediate, e.g.

$$H + Fe^{2+} \rightarrow Fe^{3+}.H^- \xrightarrow{H^+} Fe^{3+} + H_2 \tag{31}$$

$$H + I^- \rightarrow (HI)^- \xrightarrow{H^+} H_2 + I \tag{32}$$

In its reactions with organic compounds the hydrogen atom generally abstracts H from saturated molecules and adds to centres of unsaturation,

$$H + CH_3OH \rightarrow H_2 + \cdot CH_2OH \tag{33}$$

$$H + CH_2=CH_2 \rightarrow CH_3CH_2\cdot \tag{34}$$

The abstraction reaction produces molecular hydrogen and this is a particularly useful way of separating yields of H and e_{aq}^-. A good illustration of this is provided by chloroacetic acid (or its anion) which reacts with e_{aq}^- and H as follows:-

$$e^-_{aq} + ClCH_2CO_2H \rightarrow Cl^- + {}^{\cdot}CH_2CO_2H \tag{35}$$

$$H + ClCH_2CO_2H \rightarrow H_2 + Cl\dot{C}HCO_2H \tag{36}$$

The yields of chloride ion and hydrogen vary with pH (Fig. 10) because of reaction (3)

$$e^-_{aq} + H_3O^+ \rightarrow H + H_2O \tag{3}$$

Reactions such as (35) and (36) also revealed that H is converted into e^-_{aq} in alkaline solution

$$H + OH^- \rightarrow e^-_{aq} \tag{37}$$

The rate constant for reaction (37) is $2.3 \times 10^7 \text{ dm}^3 \text{ mol}^{-1} \text{ s}^{-1}$ so that it may compete with reactions of H with other solutes in alkaline solution.

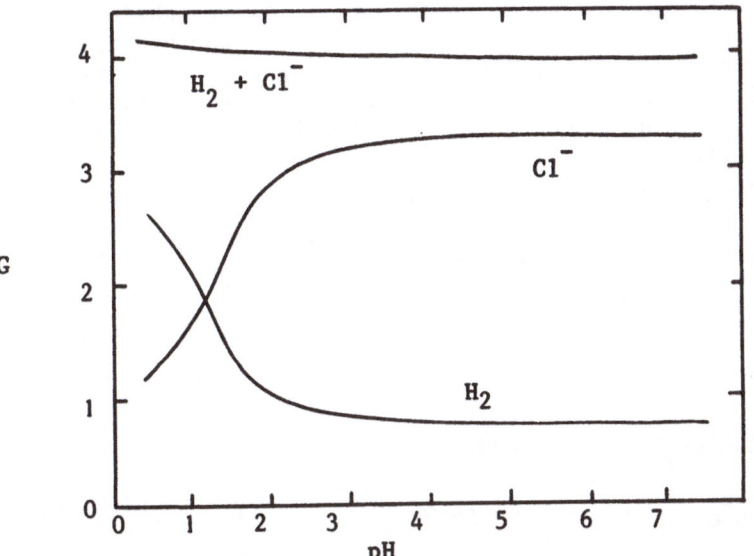

Figure 10. Effect of pH on $G(Cl^-)$ and $G(H_2)$ (Redrawn from (1)).

6.3 The Hydroxyl Radical

The hydroxyl radical has a standard reduction potential of + 2.8 V in acidic solution and is therefore a strong oxidant. It absorbs weakly in the ultraviolet (see Fig. 9) and so rate constants for its reactions are obtained using competition methods or by observing the formation of products (25). It readily oxidises inorganic ions, the reaction usually being re- presented as an electron transfer process

$$OH + S^n \rightarrow S^{n+1} + OH^- \tag{38}$$

although this is an oversimplified view in many instances. Reaction with halide ions, for example, proceeds via intermediate adducts which can be observed by pulse radiolysis

$$OH + X^- \rightarrow HOX^- \tag{39}$$

In several cases the rate of reaction of OH with anions, including negatively charged metal complexes, approaches the diffusion controlled limit. In contrast to this the rate constants for its reactions with aquated metal cations seems to be limited to $\sim 3 \times 10^8$ dm^3 mol^{-1} s^{-1}, i.e. some two orders of magnitude less than the diffusion controlled limit. A suggested explanation (26) for this is that OH abstracts a hydrogen atom from a water molecule in the coordination sphere of the metal ion which then transfers an electron to the oxidised ligand, e.g.

$$OH + M^{n+}.H_2O \rightarrow M^{n+}.OH \rightarrow M^{(n+1)+}.H_2O \tag{40}$$

The hydroxyl radical behaves as an electrophile in its reactions with organic molecules and, like the hydrogen atom, abstracts H from saturated molecules and adds to centres of unsaturation. However, it is less selective and more reactive than the hydrogen atom in its abstraction reactions. The reactivities of the two radicals with organic molecules are compared in Table 10.

Table 10. Rate Constants for Selected Reactions of OH and H

Solute	Reaction Type	$10^{-7}k$/dm^3 mol^{-1} s^{-1}	
		OH	H
$CH_2=CHCONH_2$	Addition	450	1800
C_6H_6	Addition	530	53
$C_6H_5NO_2$	Addition	340	170
C_2H_5OH	Abstraction	180	1.7
CH_3OH	Abstraction	84	0.16

In strongly alkaline solution OH is rapidly converted into its basic form O$^-$ through reaction (41) for which $k_{41} = 1.2 \times 10^{10}$ dm^3 mol^{-1} s^{-1}, $k_{-41} = 9.3 \times 10^7$ s^{-1} and p$K_{OH} = 11.9$

$$OH + OH^- \rightarrow O^- + H_2O \tag{41}$$

The standard reduction potential of the hydroxyl radical in basic solution is $+ 1.4$ V and so O^- is a less powerful oxidant than OH. This is reflected in its reactions with inorganic ions which tend to be much slower than those of OH. Notable examples are $Fe(CN)_6^{4-}$, CO_3^{2-} and Br^- whose rate of reaction with O^- is too slow to be measured although they all react rapidly with OH. On the other hand, O^- reacts rapidly with oxygen to form the ozonide ion whereas OH is unreactive.

$$O^- + O_2 \rightarrow O_3^- \tag{42}$$

Hydrogen atom abstraction by O^- from organic molecules is almost as rapid as that by OH, but addition to centres of unsaturation is generally much slower, reflecting the nucleophilic character of the radical ion. In fact if an aromatic ring carries an aliphatic side-chain, O^- preferentially attacks there whereas OH adds to the ring. Thus different products may be obtained from aromatic solutes at high pH.

6.4 The Perhydroxyl Radical

Although the perhydroxyl radical is not a significant primary radical at low LET, it is an important secondary species in oxygenated solution (27) as a result of the following reactions

$$e_{aq}^- + O_2 \rightarrow O_2^- \tag{43}$$

$$H + O_2 \rightarrow HO_2 \tag{44}$$

$$HO_2 \rightarrow O_2^- + H^+ \tag{45}$$

The pK of HO_2 is 4.9. Both forms can act as mild oxidants and reductants, although HO_2 tends to be the stronger oxidant and O_2^- the stronger reductant. The perhydroxyl radical is unreactive with most organic compounds unless they contain weakly bonded hydrogen, e.g.

$$HO_2 + RSH \rightarrow H_2O_2 + RS\cdot \tag{46}$$

7. WATER RADIOLYSIS AS A CHEMICAL TOOL

The radiolysis of water generates approximately equal numbers of powerful oxidising and reducing radicals, but in chemical applications it is desirable to achieve either totally oxidising or totally reducing conditions. Some useful systems that fulfil this requirement are described in brief below.

7.1 Oxidising Conditions

The most effective way of obtaining an almost totally oxidising system is to saturate the solution with N_2O which converts e_{aq}^- to OH

$$e_{aq}^- + N_2O \rightarrow N_2 + O^- \xrightarrow{H_2O} OH + OH^- \qquad (23)$$

Under these conditions the concentration of N_2O is 2.5×10^{-2} mol dm^{-3} and $G(N_2)$ is 3.2 (Fig. 4). Pulse radiolysis measurements on this system (28) show that $G(OH) = 6.2$, so it is clear that when e_{aq}^- is scavenged in the spur by N_2O there is a nearly equal increase in the yield of OH which escape from the spur, i.e. replacement of e_{aq}^- by OH through reaction (23) does not increase the extent of reaction (6) (Table 1). This is consistent with the conclusion that OH is located in the core of the spur.

Nitrous oxide does not react rapidly with the hydrogen atom ($k = 2.3 \times 10^6$ dm^3 mol^{-1} s^{-1} (24)), so that approximately 10% of the total radicals available in N_2O saturated solutions in the pH range 3 - 11 are H atoms and 90% are OH radicals. At pH > 11 $G(OH)$ can increase by up to 0.6 (i.e. G_H) through reactions (37) and (23), the extent of the increase depending on whether the solute competes with hydroxide ion for H. Below pH 3 reaction (3) competes with reaction (23) and N_2O is no longer a good solute for achieving predominantly oxidising conditions.

It should be borne in mind that the product of reaction (23) is O^- rather than OH (29). Thus if the solute reacts with O^- in competition with the protonation reaction (-41), i.e. $k[S] > 9 \times 10^7$ s^{-1}, then because O^- can react differently from OH, different products may be obtained in N_2O saturated solutions at high and low concentrations of OH scavengers.

7.2 Reducing Conditions

To achieve reducing conditions it is desirable to convert OH into a reducing radical. Whilst it is possible to convert it into e_{aq}^- through the reaction sequence

$$OH + H_2 \rightarrow H + H_2O \qquad (12)$$

$$H + OH^- \rightarrow e_{aq}^- \qquad (37)$$

this can only be achieved under the rather severe conditions of high pH and hydrogen at a pressure of 100 atm. The method most commonly used is to add an organic solute such as an alcohol or formate ion which reacts with OH as in (45)

$$OH + RH \rightarrow R\cdot + H_2O \tag{45}$$

Unfortunately the radicals $R\cdot$, although they are reductants, are not so powerful as e_{aq}^- or H and are generally much less reactive.

When RH is HCO_2^- or HCO_2H the radical is CO_2^- which has a reduction potential of -2.0 V (30). Because of the high acid dissociation constant of CO_2H (pK = 1.4 (31)) CO_2^- can be employed as an electron transfer reducing agent under conditions where e_{aq}^- would react predominantly with H_3O^+. All the primary radicals from water can be converted to CO_2^- through reactions (46) - (48)

$$e_{aq}^- + CO_2 \rightarrow CO_2^- \tag{46}$$

$$OH + HCO_2^- \rightarrow H_2O + CO_2^- \tag{47}$$

$$H + HCO_2^- \rightarrow H_2 + CO_2^- \tag{48}$$

By replacing CO_2 with O_2 in this system all the primary radicals are readily converted to O_2^- (or HO_2) through reactions (42 and (49)

$$CO_2^- + O_2 \rightarrow CO_2 + O_2^- \tag{49}$$

Alcohol radicals are less powerful reductants than CO_2^- and are generally much less reactive than e_{aq}^-. This property can be exploited in pulse radiolysis studies because here the hydroxyl radical can be converted to a radical which does not react on the same timescale as e_{aq}^-, leaving only the reactions of e_{aq}^- to be observed at short times. Tertiary butanol is a particularly good solute for this purpose because the radical $\cdot CH_2(CH_3)_2COH$ (reaction (45)) is rather unreactive. It has the added advantage of absorbing only very weakly in the ultraviolet and hence interferes to a minimal extent in the measurement of absorption spectra of other products.

8. CONCLUDING REMARKS

The radiolysis of liquid water is sufficiently well understood for it to be used routinely as a very effective, and often unique, way of generating and studying quantitatively the chemistry of inorganic and organic free radicals and metal ions in unusual oxidation states. In this respect pulse radiolysis is a particularly powerful technique because it allows direct observations of properties of the transient species such as their reaction rates and optical absorption spectra, and helps to elucidate reaction mechanisms. Steady state radiolysis also plays an important role, especially in mechanistic studies when the yields and identities of stable products are required.

REFERENCES

1. Spinks, J. W. T., and Woods, R. J., "An Introduction to
    Radiation Chemistry", 2nd ed., Wiley-Interscience, 1976, ch. 7.

2. Swallow, A. J., "Radiation Chemistry, An Introduction",
    Longman, 1973, ch. 7.

3. Draganic, I. G., and Draganic, Z. D., "The Radiation Chemistry
    of Water", Academic Press 1971.

4. Hart, E. J., and Anbar, M., "The Hydrated Electron", Wiley-
    Interscience, 1970.

5. Thomas, J. K., in "Advances in Radiation Chemistry", eds.
    Burton, M., and Magee, J. L., Wiley-Interscience, Vol. 1,
    1969, pp. 103-198.

6. Hunt, J. W., in "Advances in Radiation Chemistry", eds.
    Burton, M., and Magee, J. L., Wiley-Interscience, Vol. 5, 1976,
    pp. 185-315.

7. Mozumder, A., in "Advances in Radiation Chemistry", eds.
    Burton, M., and Magee, J. L., Wiley-Interscience, Vol. 1, 1969,
    pp. 1-102.

8. Lampe, F. W., Field, F. H., and Franklin, J. L., 1957, J. Am.
    Chem. Soc. 79, pp. 6132-6135.

9. Hamill, W. H., 1969, J. Phys. Chem. 73, pp. 1341-1347.

10. Wiesenfeld, J. M., and Ippen, E. P., 1980, Chem. Phys. Letters
    73, pp. 47-50.

11. Appleby, A., and Schwarz, H. A., 1969, J. Phys. Chem. 73, pp.
    1937-1941.

12. Jonah, C. D., Matheson, M. S., Miller, J. R., and Hart, E. J.,
    1976, J. Phys. Chem. 80, pp. 1267-1270.

13. Jonah, C. D., and Miller, J. R., 1977, J. Phys. Chem. 81, pp.
    1974-1976.

14. Burns, W. G., May, R., Buxton, G. V., and Wilkinson-Tough, G.
    S., 1981, J. Chem. Soc., Faraday Trans. I 77 pp. 1543 - 1551.

15. Buxton, G. V., 1968, Radiat. Res. Rev. 1, pp. 209-222.

16. Dainton, F. S., and Watt, W. S., 1963, Proc. Roy. Soc. A 275,
    pp. 447-464.

17. Singh, A., Chase, W. J., and Hunt, J. W., 1977, Faraday
     Discussions 63, pp. 28-37.

18. Schwarz, H. A., 1969, J. Phys. Chem 73, pp. 1928-1936.

19. Trumbore, C. N., Short, D. R., Fanning, J. E., and Olson, J.
     H., 1978, J. Phys. Chem. 82, pp. 2762-2767.

20. Kuppermann, A., in "Physical Mechanisms in Radiation Biology",
     Technical Information Centre, Office of Information Services,
     U. S. Atomic Energy Commission, Washington, D. C., 1974, p.
     155.

21. Balkas, T. I., Fendler, J. H., and Schuler, R. H., 1970, J.
     Phys. Chem. 74, pp. 4497-4505.

22. Matheson, M. S., and Dorfman, L. M., "Pulse Radiolysis",
     M. I. T. Press, 1969.

23. Anbar, M., Bambenek, M., and Ross, A. M., 1973, "Selected
     Specific Rates of Reactions of Transients From Water in
     Aqueous Solution. I. Hydrated Electron", NSRDS-NBS 43, U.S.
     Department of Commerce, National Bureau of Standards,
     Washington, D. C.

24. Anbar, M., Farhataziz, and Ross, A. B., 1975, "Selected
     Specific Rates of Reactions of Transients From Water in
     Aqueous Solution. II. Hydrogen Atom", NSRDS-NBS 51, U.S.
     Department of Commerce, National Bureau of Standards,
     Washington, D. C.

25. Dorfman, L. M., and Adams, G. E., 1973, "Reactivity of the
     Hydroxyl Radical in Aqueous Solutions", NSRDS-NBS 46, U.S.
     Department of Commerce, National Bureau of Standards,
     Washington, D. C.

26. Berdnikov, V. M., 1973, Russ. J. Phys. Chem. 47, pp. 1547-
     1552.

27. Bielski, B. H. J., and Gebicki, J. M., in "Advances in Rad-
     iation Chemistry", eds. Burton, M., and Magee, J. L., Wiley-
     Interscience, Vol. 2, 1970, pp. 177-280.

28. Buxton, G. V., 1969, Trans. Faraday Soc. 65, pp. 2150-2158.

29. Buxton, G. V., 1970, Trans. Faraday Soc. 66, pp. 1656-1660.

30. Breitenkamp, M., Henglein, A., and Lilie, J., 1976, Ber.
     Bunsenges. Phys. Chem. 80, pp. 973-978.

31. Buxton, G. V., and Sellers, R. M., 1973, J. Chem. Soc., Faraday Trans. I 69, pp. 555-559.

32. Nielson, S. O., Michael, B. D., and Hart, E. J., 1976, J. Phys. Chem. 80, pp. 2482-2488.

APPLICATIONS OF WATER RADIOLYSIS IN INORGANIC CHEMISTRY

G. V. Buxton.

University of Leeds.

Abstract. – The radiolysis of water produces short-lived
oxidising (OH) and reducing free radicals (e_{aq}^- and H) which
react rapidly with solutes to form other short-lived intermediates.
This review illustrates how the techniques of pulsed and continous
radiolysis have been exploited to obtain detailed kinetic and mechanistic information about fast reactions which occur in inorganic
systems in aqueous solution. Topics covered are: formation and
reactions of inorganic free radicals containing halogen, pseudo-
halogen, nitrogen, phosphorus, sulphur etc.; formation and reaction
of metal ions and their complexes in unstable valency states,
including inter- and intramolecular electron transfer, ligand
transfer and metal-carbon bond formation.

1. INTRODUCTION

The techniques of radiation chemistry provide the chemist with
powerful methods for generating and observing previously unknown
species, measuring very fast reaction rates, and elucidating the
role of short-lived intermediates in reaction mechanisms. In inorganic chemistry these techniques have so far been applied mainly
to redox chemistry in aqueous solution because the radiation
chemistry of this solvent is fairly well known.

The aim of this review is to provide examples of the broad
range of topics in inorganic chemistry to which radiation chemical
techniques have been and are being applied. A considerable literature already exists in this field, and continues to grow, so
that the examples chosen comprise only a very small fraction of
the work that has been done. The topics which have been widely

J. H. Baxendale and F. Busi (eds.),
The Study of Fast Processes and Transient Species by Electron Pulse Radiolysis, 267–287.
Copyright © 1982 by D. Reidel Publishing Company.

studied include inorganic free radicals, systems involving non-metallic compounds, aquo-metal ions in unusual oxidation states, and reactions of coordination complexes of transition-metal ions.

2. INORGANIC FREE RADICALS

These radicals are formed readily by oxidation of inorganic anions with the hydroxyl radical, e.g.

$$\cdot OH + CO_3^{2-} \rightarrow CO_3^{-} + OH^{-} \tag{1}$$

or by reduction of peroxo-compounds, e.g.

$$e_{aq}^{-} + S_2O_8^{2-} \rightarrow SO_4^{-} + SO_4^{2-} \tag{2}$$

Several of the radicals that have been investigated are listed in Table 1. Other, less studied, radicals include SO_2^{-}, SO_3^{-}, SeO_3^{-}, SeO_4^{-}, $NO_2^{\cdot}$, $NO_3^{\cdot}$ etc. These species are all oxidants but differ in their reactivities. This means that the hydroxyl radical can be replaced by one which is more selective in its reactions with solutes. A comprehensive list of rate constants of these radicals is given in ref. 1.

For inorganic compounds that contain abstractable H-atoms, e.g. H_2PO_3, HPO_3^{2-}, N_2H_4, NH_2OH etc., the order of reactivity is $\cdot OH > SO_4^{-} \sim H_2PO_4^{-} > HPO_4^{-} \sim PO_4^{2-}$; and HPO_4^{-} is slightly more selective than SO_4^{-} (2). The pattern of reactivity with inorganic ions which do not contain abstractable hydrogen is quite different. In this case the redox potentials of both radical and solute appear to be important. Thus the radicals react rapidly with strongly reducing anions such as SO_3^{2-}, N_3^{-} and I^{-}, but the rate constants show much more variation with weaker reductants like Br^{-} and Cl^{-}, and indicate that their oxidation potentials decrease in the order $SO_4^{-} > Cl_2^{-}$ $> HPO_4^{-} > PO_4^{2-}$. The reactivities of the halide and pseudohalide radicals generally decrease in the order $Cl_2^{-} > Br_2^{-} > (SCN)_2^{-} > I_2^{-}$.

These radicals generally react by electron transfer unlike the hydroxyl radical which is known to react by addition in several cases. Clear examples of this are the oxidation of halide and pseudohalide ions (3-6) where there is kinetic and spectral evidence for the formation of XOH^{-} intermediates which react further to form X_2^{-}, e.g.

$$\cdot OH + Br^{-} \rightleftharpoons BrOH^{-} \tag{3}$$

$$BrOH^{-} \rightleftharpoons Br^{\cdot} + OH^{-} \tag{4}$$

$$BrOH^{-} + Br^{-} \rightleftharpoons Br_2^{-} + OH^{-} \tag{5}$$

Table 1. Some Inorganic Free Radicals

Radical	pK	λ_{max}/nm	ε_{max}/dm^3 mol^{-1} cm^{-1}
$CO_3^{\bar{\cdot}}$		600	1880
$\overset{\cdot}{C}O_3H$	9.6	600	1880
$SO_4^{\bar{\cdot}}$		450	1100
$PO_4^{2\bar{\cdot}}$		530	2150
$HPO_4^{\bar{\cdot}}$	8.9	510	1550
$H_2PO_4^{\cdot}$	5.7	520	1850
$N_3^{\cdot}$		275	1400
$Cl_2^{\bar{\cdot}}$		340	12500
$Br_2^{\bar{\cdot}}$		360	8560
$I_2^{\bar{\cdot}}$		380	10000
$(SCN)_2^{\bar{\cdot}}$		480	7580

$$Br^{\cdot} + Br^- \rightleftharpoons Br_2^{\bar{\cdot}} \qquad (6)$$

Both $ClOH^{\bar{\cdot}}$ and $BrOH^{\bar{\cdot}}$ absorb in the same spectral region as their related radicals $Cl_2^{\bar{\cdot}}$ and $Br_2^{\bar{\cdot}}$ but with smaller extinction co-efficients, whereas $HOSCN^{\bar{\cdot}}$ has a different spectrum from $(SCN)_2^{\bar{\cdot}}$ as shown in Fig. 1.

Chloride ion is only oxidised by ·OH in acidic solution, showing that acid-base equilibria are involved as well as the analogues of reactions (3) - (6). However, Cl^- is oxidised (7) by $SO_4^{\bar{\cdot}}$ in neutral and alkaline solution

$$SO_4^{\bar{\cdot}} + Cl^- \rightarrow Cl^{\cdot} + SO_4^{2-} \qquad (7)$$

so it seems unlikely that ·OH reacts with SO_4^{2-} by simple electron transfer. Although the reactions of ·OH with inorganic anions are often represented as electron transfer reactions, it seems more likely that adduct formation occurs. This is because of the large reorganisation energy involved in the simple electron transfer process because the hydroxide ion is strongly solvated.

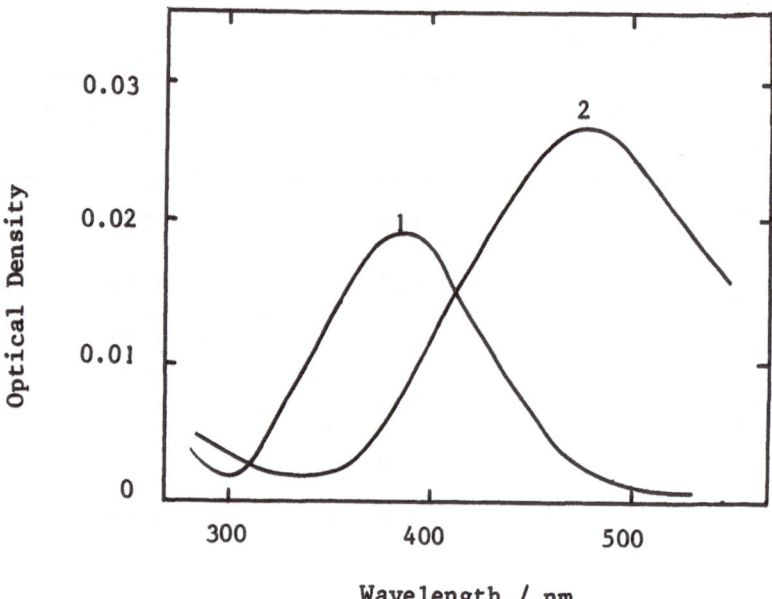

Figure 1. Absorption spectra of 1, $HOSCN^-$; 2, $(SCN)_2^-$. Reprinted in part from (4) with permission. Copyright 1972 American Chemical Society.

3. NON-METALLIC COMPOUNDS

3.1 Oxyhalogen Ions

The oxyhalogen ions provide classic examples of systems where water radiolysis can be applied to investigate one-electron processes when the normal modes of reaction involve two-electron changes in oxidation level, in this case through transfer of oxygen atoms, e.g.

$$ClO^- + ClO^- \rightarrow ClO_2^- + Cl^- \tag{8}$$

The oxyanions of chlorine, bromine and iodine have all been studied extensively (8-11). The principle features are summarise in Table 2 which shows the oxidation states that have been formed for each series of oxyanions.

The halogen oxides (XO_{n-1} or XO_n^{2-}) all absorb in accessible regions of the spectrum (Fig. 2) so that their reactions can be followed using kinetic spectroscopy combined with pulse radiolysi In the chlorine system $ClO\cdot$ dimerises to Cl_2O_2 which then is hydrolysed or reacts with chlorite ion.

$$Cl_2O_2 + H_2O \rightarrow ClO^- + ClO_2^- + 2H^+ \tag{9}$$

Table 2. Products of One-Electron Reduction and Oxidation of Oxyhalogen Ions in Aqueous Solution

Reaction	X	n
$e^-_{aq} + XO^-_n \rightarrow XO^-_{n-1} + O^{\overline{\cdot}}$	Cl	1,2
	Br	1
	I	1
$e^-_{aq} + XO^-_n \rightarrow XO^{2-}_n \xrightarrow{H_2O} XO^-_{n-1} + 2OH^-$	Br	2,3
	I	2,3,4
$\cdot OH(O^{\overline{\cdot}}) + XO^-_n \rightarrow XO\cdot_n + OH^-(O^{2-})$	Cl	1,2
	Br	1,2
	I	1(?),3,4

$$Cl_2O_2 + H_2O \rightarrow ClO^- + Cl^- + O_2 + 2H^+ \tag{10}$$

$$Cl_2O_2 + ClO^-_2 \rightarrow ClO^-_3 + Cl_2O \tag{11}$$

The bromine analogue BrO$\cdot$ reacts with itself to form BrO$^-$ and BrO$^-_2$, presumably via Br$_2$O$_2$, and also oxidises BrO$^-_2$ to BrO$_2$.

$$BrO\cdot + BrO^-_2 \rightarrow BrO^- + BrO^{\cdot}_2 \tag{12}$$

This oxide is also hydrolysed via dimer formation in a pH dependent reaction.

$$Br_2O_4 + OH^- \rightarrow BrO^-_2 + BrO^-_3 + H^+ \tag{13}$$

Some of the kinetic data for these reactions are given in Table 3.

3.2 Borohydride Ion

This ion is a strong reducing agent and is relatively stable in alkaline solution. The radiation chemistry of its aqueous solution was investigated because it was expected to react rapidly with the hydroxyl radical to form a reducing species. (12).

$$\cdot OH + BH^-_4 \rightarrow H_2O + BH^{\cdot}_4 \text{ or } BH_3 + H\cdot \tag{14}$$

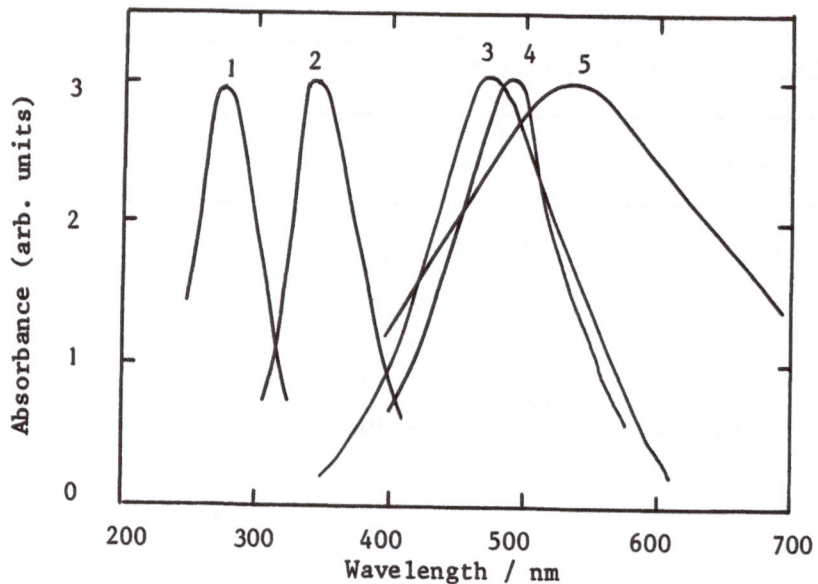

Figure 2. Absorption spectra of oxyhalogen radicals. 1, ClO$\cdot$; 2, BrO$\cdot$; 3, BrO$_2^\cdot$; 4, IO$_2^\cdot$; 5, IO$_4^\cdot$.

Table 3. Rate Constants for Reactions of Halogen Oxides in Aqueous Solution. ^a2k.

Reaction	$10^{-8}k/dm^3\ mol^{-1}\ s^{-3}$
ClO$\cdot$ + ClO$\cdot$ → Cl$_2$O$_2$	150[a]
BrO$\cdot$ + BrO$\cdot$ → Br$_2$O$_2$	49[a]
BrO$\cdot$ + BrO$_2^-$ → BrO$^-$ + BrO$_2^\cdot$	34
BrO$_2^\cdot$ + BrO$_2^\cdot$ $\rightleftharpoons$ Br$_2$O$_4$	28[a] (K = 1.9 x 10^4 dm^3 mol^{-1})
Br$_2$O$_4$ + OH$^-$ → BrO$_2^-$ + BrO$_3^-$ + H$^+$	7

γ- Radiolysis experiments showed that H$_2$ is produced in two stages a rapid one with G(H$_2$) = 6.8 and a slow one (hours) with G(H$_2$) = 6.3. A transient species absorbing in the UV was observed by pulse radiolysis and assigned to BH$_4^-$ formed in reaction (14) with k_{14} = 1.2 x 10^{10} dm^3 mol^{-1} s^{-1}. The following mechanism was proposed (12) to account for the two-stage hydrogen production

$$e_{aq}^- + e_{aq}^- \rightarrow H_2 + 2OH^- \tag{15}$$

$$BH_4^{\cdot} + BH_4^{\cdot} \rightarrow B_2H_6 + H_2 \tag{16}$$

$$B_2H_6 \rightleftharpoons 2BH_3 \tag{17}$$

$$BH_3 + H_2O \rightarrow H_2 + BH_2(OH) \tag{18}$$

$$BH_2(OH) + 2H_2O \rightarrow 2H_2 + B(OH)_3 \tag{19}$$

in which reactions (15) - (18) comprise the rapid stage and reaction (19) the slow stage. An interesting feature of this system
is the occurrence of a chain reaction in the presence of N_2O
through reactions (14) and (20)

$$BH_4^{\cdot} + N_2O \rightarrow N_2 + BH_3 + \cdot OH \tag{20}$$

4. AQUO-METAL IONS IN UNUSUAL OXIDATION STATES

The monovalent states Cd^+, Co^+, Ni^+ and Zn^+ were among the
first hyper-reduced metal ions to be studied by radiation chemical
methods in a series of experiments by Baxendale and co-workers (13).
These ions are readily formed by reduction with the hydrated
electron

$$e_{aq}^- + M_{aq}^{2+} \rightarrow M_{aq}^+ \tag{21}$$

and have intense optical absorption bands in the UV (Fig. 3.)
which makes their reactions easy to follow using pulse radiolysis
methods. It is not surprising, therefore, that these hyper-
reduced ions have been studied in some depth (14).

They are strong reductants and the general order of reactivity
with inorganic oxidants is $Zn^+ \sim Co^+ \gtrsim Cd^+ > Ni^+$. Although most
of the reactions of M^+ involve electron transfer (see section 6.1),
other types of reaction have been observed (14), e.g.

$$M^+ + N_2O \rightarrow MO^+ + N_2 \tag{22}$$

$$M^+ + O_2 \rightarrow MO_2^+ \tag{23}$$

$$M^+ + olefin \rightarrow (M - olefin)^+ \tag{24}$$

$$M^+ + R^{\cdot} \rightarrow MR^+ \tag{25}$$

Reaction (22) is interesting because it involves oxygen atom
transfer and formation of the +3 oxidation state of the metal.
Of the four metals mentioned above, nickel is the only case where
it is certain that reaction (23) occurs rather than electron

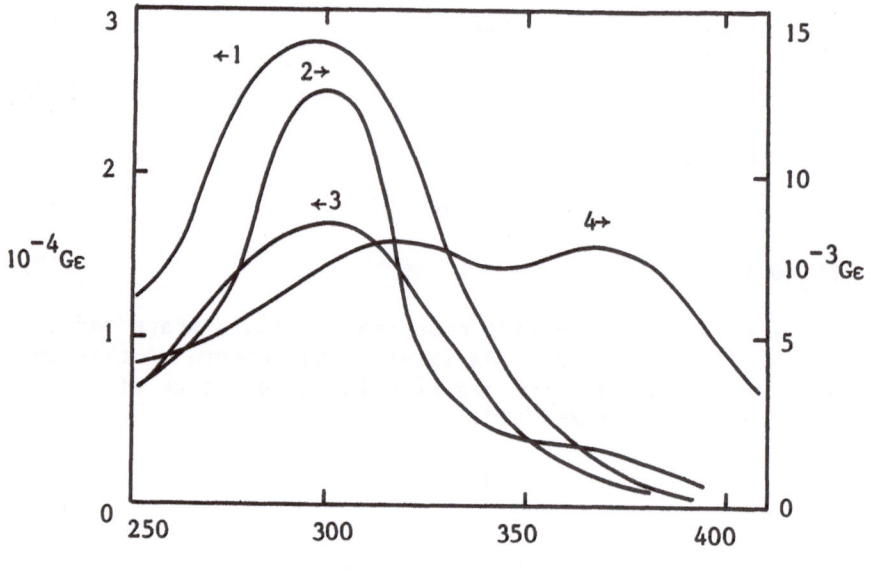

Figure 3. Absorption spectra of M^+. 1, Cd^+; 2, Ni^+; 3, Zn^+; 4, Co^+.

transfer, but it has also been observed with Cr^{2+} under conditions where $[Cr^{2+}] \ll [O_2]$ (14).

Reaction (25) is a general example of the formation of a metal-carbon bond between the metal and a carbon-centred free radical generated via ·OH reactions, e.g.

$$\cdot OH + (CH_3)_2 CHOH \rightarrow (CH_3)_2 \overset{\cdot}{C}OH + H_2O \qquad (26)$$

Several specific examples of reaction (25) are known involving simple organic radicals and metal ions such as Cr^{II}, Ni^I, Cd^I and Cu^I.

Hydrolysis constants of metal ions in unstable oxidation stat can be obtained by measuring conductance changes following pulse radiolytic oxidation of, for example, Tl^+, Ag^+, Cu^{2+} or Sn^{2+} by ·OH which proceeds via the adduct (15) in reaction (27)

$$\cdot OH + M^{n+} \rightarrow M(OH)^{n+} \qquad (27)$$

Hydrolysis equilibria are established and can be monitored through the number of protons released or taken up,

$$M^{(n+1)+} \underset{pK_1}{\rightleftharpoons} M(OH)^{n+} \underset{pK_2}{\rightleftharpoons} M(OH)_2^{(n-1)+} \underset{pK_3}{\rightleftharpoons} \ldots\ldots$$

Polarographic measurements (15) show that the oxidative properties of these unstable metal ions decrease as the ions become more hydrolysed.

The combination of pulse radiolysis and conductance measurements can also be used to measure hydrolysis constants of stable reducible ions. An advantage of this method is that it yields data for low concentrations and ionic strengths. It works as follows (16). Equilibrium (28) is disturbed by a sudden decrease in the concentration of M^{n+} brought about by reaction (29)

$$M^{n+} + H_2O \rightleftharpoons M(OH)^{(n-1)+} + H^+ \qquad (28)$$

$$e_{aq}^- + M^{n+} \rightarrow M^{(n-1)+} \qquad (29)$$

and this causes a post-pulse decrease in conductance as some of the radiation produced protons are consumed in restoring the hydrolysis equilibrium. This method has been used to obtain hydrolysis constants for $Yb(ClO_4)_3$, pK = 8.7; $Eu(ClO_4)_3$, pK = 8.7; $UO_2(ClO_4)_2$, pK = 5.2 (16).

5. LANTHANIDES AND ACTINIDES

The lanthanides and actinides are characterised by their 4f and 5f electrons respectively and the role these play in the chemistry of the ions. Pulse radiolysis provides a powerful method of probing this role using e_{aq}^- and ·OH to create unusual valency states.

5.1 Lanthanides

The stable valency state of the lanthanides is +3 but they can be reduced to the +2 state by e_{aq}^-. Rate constants for this reaction are listed in Table 4 and fall roughly into three groups.

The high rate constants for Eu^{3+}, Yb^{3+} and Sm^{3+} provide some support for the view that f^0, f^7 and f^{14} electron configurations have special stability. In this connection it is noteworthy that Ce^{3+} (f^1) and Pr^{3+} (f^2) are readily oxidised by ·OH to the +4 level whereas there is no evidence that Ce^{3+} is reduced by e_{aq}^-. However, there is clearly no systematic trend in the rate constants for the other lanthanide ions to support the idea that stability is favoured even if there is only an approach to f^7 or f^{14} configuration.

The divalent cations are powerful reductants and their reactivity with inorganic oxidants is in the order $Sm^{2+} > Yb^{2+} > Eu^{2+}$ which correlates with the order of the Ln^{III}/Ln^{II} redox potentials (Fig. 4).

Table 4. Rate Constants for Reduction of Ln^{3+} by e_{aq}^- (17)

Ion	No. of 4f electrons	$10^{-7}k/dm^3\ mol^{-1}\ s^{-1}$
Eu^{3+}	6	6500
Yb^{3+}	13	4300
Sm^{3+}	5	1960
Tm^{3+}	12	32.9
Gd^{3+}	7	9.1
Ho^{3+}	10	1.4
Er^{3+}	11	1.0
Lu^{3+}	14	0.49
Dy^{3+}	9	0.35
Nd^{3+}	3	0.32
Pr^{3+}	2	0.23
Tb^{3+}	8	<0.1

Making the assumption that the oxidation of Ln^{II} by nitrite ion follows an outer-sphere mechanism, Tendler and Faraggi (18) obtained the relationship

$$\log(k_{act}) = A + BE^o$$

between the activation controlled rate constant and the standard reduction potential E^o for Sm^{2+}, Yb^{2+} and Eu^{2+} which enabled them to calculate E^o for the other lanthanides. In this expression A and B are constants and k_{act} is given by

$$\frac{1}{k_o} = \frac{1}{k_{act}} + \frac{1}{k_{diff}}$$

where k_o is the measured rate constant extrapolated to zero ionic strength and k_{diff} is the rate constant calculated for diffusion control. Values were obtained for Pr, Nd, Tb, Dy, Ho, Er and Tm.

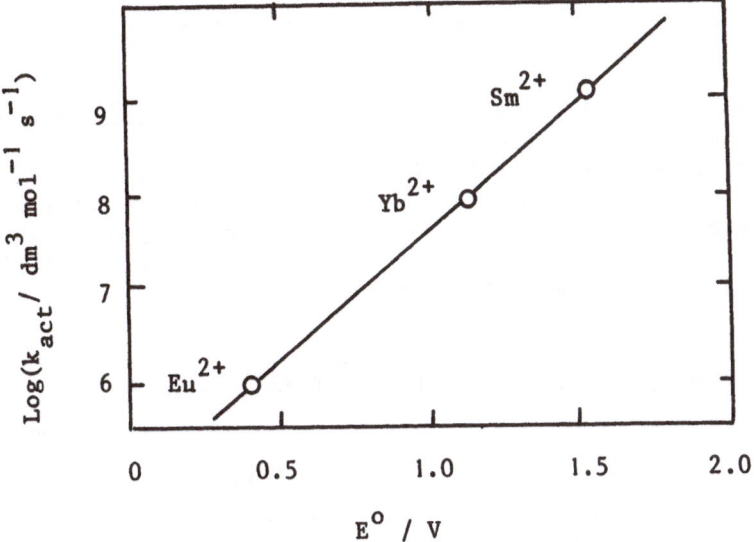

Figure 4. Correlation between the standard redox potential of Ln^{III}/Ln^{II} and the rate constant of $Ln^{II} + NO_2^-$. Data taken from ref. 18.

In each case E^o was close to -1.7 V and agreed quite well with values calculated using the Marcus theory for outer-sphere electron transfer.

5.2 Actinides

The radiation chemistry of the actinides has received considerable attention, mainly because of its importance in nuclear reactor fuel technology. Pulse radiolysis studies have been aimed at preparing unstable oxidation states and also comparing actinides with lanthanides in their reactions with e_{aq}.

The hydrated electron reacts at diffusion rates with An^V and An^{VI} for An = Am, Pu, Np, and U in the pH range 5.3 - 6.8 (17). In contrast to the lanthanides there is no correlation between the rate constant for reduction by e_{aq}^- and the oxidation potential of An^V or An^{VI}, and it is possible that an electron tunnelling mechanism is involved. On the other hand, Am^{III} is 100 times less reactive than Sm^{III} with e_{aq}^- although the two ions have similar radii and formal potential and the same charge. It has been suggested (19) that this difference in reactivity reflects the greater overlap of 5f orbitals, compared with 4f orbitals, with ligands in the first coordination shell resulting in a larger rearrangement free energy in the formation of the activated complex of Am^{III}.

Pulse radiolysis of aqueous perchlorate solutions of americium (III) and curium (III) provided the first evidence for the existence of Am^{II}, Am^{IV}, Cm^{II} under these conditions (19-21). Only Am^{IV} has so far been studied in any detail; some of the kinetic data are collected in Table 5 and its spectrum is shown in Fig. 5. Am^{IV} can also be generated by reduction of Am^{V} with e_{aq}^{-} (k = 3.2 x 10^{10} dm^3 mol^{-1} s^{-1}).

Table 5. Kinetic Data for Reactions of Am^{IV} at pH 4 (21)

Reaction	pH k/dm^3 mol^{-1} s^{-1}
Am^{IV} + Am^{IV} → Am^{III} + Am^{V}	6 x 10^6 (2k)
Am^{IV} + $HO_2^{\cdot}$ → Am^{III} + O_2 + H^+	6 x 10^7
Am^{IV} + H_2O_2 → Am^{III} + $HO_2^{\cdot}$ + H^+	1.4 x 10^6

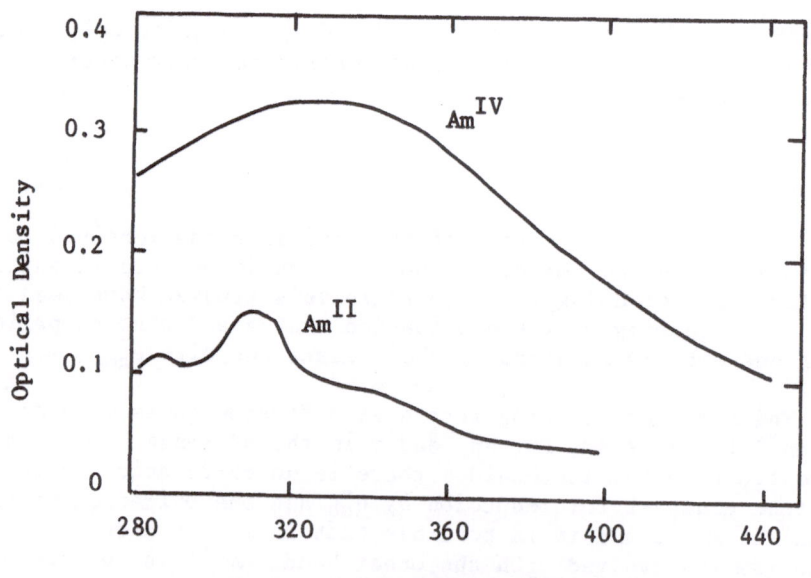

Figure 5. Absorption spectra of Am^{II} and Am^{IV}. Reprinted from (19) with permission. Copyright 1978 American Chemical Society.

In alkaline solution it has been found that Np^{VI}, Pu^{VI} and Am^{VI} are rapidly oxidised to the +7 state by O^{-}, e.g. $k \sim 6 \times 10^{7}$ dm^{3} mol^{-1} s^{-1} for Np^{VI} (22).

Technetium (VI) has also been generated in alkaline solution by reduction of TcO_{4}^{-} with e_{aq}^{-}. Tc^{VI} has a weak absorption spectrum in the visible region ($\varepsilon \sim 250$ dm^{3} mol^{-1} s^{-1} at 500-600 nm) and a lifetime of at least a few milliseconds, making it a viable chemical entity (23).

6. TRANSITION-METAL COMPLEXES

Pulse radiolysis has been widely applied to the study of the redox chemistry of complexes in unstable oxidation states formed by one-electron oxidation or reduction. Fast processes that have been investigated include electron transfer, formation of co-ordinated free radicals and ligand degradation, aquation, and changes in stereochemistry and coordination number. Illustrative examples of these processes are given below.

6.1 Electron Transfer

Although the oxidation and reduction of many transition-metal complexes by one-electron agents has been investigated it is not clear in most cases whether electron transfer occurs by an inner- or outer-sphere mechanism.

In an attempt to clarify the mechanism of reduction by the hyper-reduced cations Cd^{+}, Ni^{+} and Zn^{+}, Meyerstein and Mulac (24) measured their rates of reaction with a series of complexes $Co^{III}(NH_{3})_{5}X$ and $Co^{III}(en)_{2}YZ$, where X, Y and Z are typically NH_{3}, $H_{2}O$, halides or pseudohalides (Table 6).

The order of reactivity is $Zn^{+} \geqslant Cd^{+} > Ni^{+}$ for $Co^{III}(NH_{3})_{5}X$ and $Ru(NH_{3})_{6}^{3+}$ and $Cd^{+} \geqslant Zn^{+} > Ni^{+}$ for the $Co^{III}(en)_{2}YZ$ complexes. From these orders it was concluded (24) that Zn^{+} reacts mainly by outer-sphere, Ni^{+} mainly by inner-sphere, and Cd^{+} by both mechanisms.

Both mechanisms have been observed when transition-metal complexes are oxidised by dihalogen radical ions (X_{2}^{-}). For example, $Cr(H_{2}O)_{6}^{2+}$ reacts (26) with I_{2}^{-} and Br_{2}^{-} entirely by inner-sphere mechanisms (30), whereas Cl_{2}^{-} reacts with equal probability by inner- and outer-sphere paths (30) and (31)

$$X_{2}^{-} + Cr(H_{2}O)_{6}^{2+} \rightarrow Cr(H_{2}O)_{5}X^{2+} + H_{2}O + X^{-} \qquad (30)$$

$$X_{2}^{-} + Cr(H_{2}O)_{6}^{2+} \rightarrow Cr(H_{2}O)_{6}^{3+} + 2X^{-} \qquad (31)$$

Table 6. Specific Rates of Reaction of M^+ with Some Transition-Metal Complexes (24)

Complex	10^{-9} k/dm^3 mol^{-1} s^{-1}		
	Cd^+	Ni^+	Zn^+
$Ru(NH_3)_6^{3+}$ [a]	2.2	0.4	2.2
$Co(NH_3)_6^{3+}$	0.17	<0.005	0.84
$Co(NH_3)_5OH_2^{3+}$	0.62	<0.005	1.56
$Co(NH_3)_5OH^{2+}$	0.9	0.013	1.1
$Co(NH_3)_5Cl^{2+}$	2.2	0.65	2.2
cis-$Co(en)_2NH_3Cl^{2+}$	1.75	0.47	1.47
cis-$Co(en)_2Cl_2^{2+}$	2.3	0.59	1.91

[a] From ref. 25.

No evidence was found (26) for $Cr(H_2O)_4X_2^{2+}$, indicating that the X-X bond is broken in the electron transfer.

6.2   Coordinated Free Radicals

    When the ligands of the transition-metal complexes contain abstractable hydrogen atoms, reaction with ·OH can result in the formation of a free radical coordinated to the metal ion as an alternative to direct oxidation of the metal.

    Conductivity measurements combined with pulse radiolysis (27) show that the Co^{III}- polyamine complexes $Co(dien)_2^{3+}$, $CoCl_2(en)_2^+$ and $CoCl_2(trien)^+$, which all react with ·OH with k = $(3 \pm 0.2) \times 10^8$ dm^3 mol^{-1} s^{-1}, lose H from a ligand. The ligand radical then transfers an electron to the Co^{III} centre to form Co^{II} and the degraded ligand dissociates from the complex (see also section 6.3).

    Both types of oxidation are reported (28) for oxidation by ·OH of the first-row bivalent transition-metal complexes containing ethylenediaminetetraacetate (EDTA) and nitriloacetate (NTA) ligands. For M = Ni^{II}, Cu^{II} or Fe^{II} the metal centre is oxidised directly, but for M = Zn^{II}, Mn^{II} or Co^{II} abstraction of H from the ligand takes place. The results show no obvious pattern and it is suggested (28) that the first step in the reaction is the formation of an intermediate where ·OH interacts strongly with the complex.

The reorganisation energy involved in the oxidation of the metal centre may then determine whether direct electron transfer or H-atom abstraction from a ligand follows.

The results of pulse radiolysis experiments by themselves are often open to more than one interpretation. It is generally desirable, therefore, to obtain supporting information from the measurement of products using steady-state radiolysis. For example, it was concluded from pulse radiolysis studies alone that $\cdot OH$ oxidises Ni^{II} to Ni^{III} in nickel-glycine complexes (29). On the other hand comparative studies of the reactions of Br_2^- and $\cdot OH$ with nickel-glycine show that the two radicals form different products (30). An example is given in Table 7. The conclusion drawn from these results is that Br_2^- oxidises the metal centre while $\cdot OH$ mainly attacks the ligand

$$\cdot OH + Ni^{II}(NH_2CH_2CO_2^-) \rightarrow Ni^{II}(NH_2\overset{\cdot}{C}HCO_2^-) + H_2O \qquad (32)$$

and the radical intermediate disproportionates

$$2Ni^{II}(NH_2\overset{\cdot}{C}HCO_2^-) \rightarrow Ni^{II}\text{-gly} + Ni^{II}NH_3 + CHOCO_2 \qquad (33)$$

to form glyoxalate. The oxidation by Br_2^- may be inner- or outer-sphere. The Ni^{III} complex undergoes decarboxylation followed by disproportionation or oxidation of the radical and hydrolysis to formaldehyde

$$Ni^{III}(NH_2CH_2CO_2^-) \rightarrow Ni^{II}(NH_2\overset{\cdot}{C}H_2) + CO_2 \qquad (34)$$

$$\cdot CH_2NH_2 \overset{OX}{\rightarrow} CH_2 = NH \overset{H_2O}{\rightarrow} CH_2O + NH_3 \qquad (35)$$

Table 7. Product Yields in the γ-Radiolysis of Nickel-Glycine Complexes in N_2O-saturated Solution at pH 7 (30)

$|Ni^{II}|$	$|Gly|^{-3}$	$|Br^-|$	$G(CHOCO_2^-)$	$G(CH_2O)$
10^{-2}	3×10^{-2}	-	1.5	0.4
10^{-2}	3×10^{-2}	8×10^{-1}	0.8	3.8

Coordinated free radicals are also produced in the reactions of reducing radicals with complexes having aromatic ligands. Identification of the coordinated radical is based on the similarity of its absorption spectrum with that of the radical from the free ligand. For example, the absorption spectra of the transient species

formed in the reduction by e_{aq}^- of pentamminecobalt(III) complexes
containing a mono- or dinitrobenzoato ligand are identical to the
free ligand radical anions $^-O_2CC_6H_5NO_2^-$ $(31)_{2+}$. These transients
decay unimolecularly with the release of Co_{aq}^{2+}, showing that the
ligand radical transfers an electron to the Co^{III} centre. Rate
constants for this electron transfer for a number of complexes
are listed in Table 8. Their variation with the position of the
nitro-group in the aromatic ring led to the conclusion that the
rate is governed by the electron spin density at or adjacent to
the carboxylate group (31).

Table 8. Rate Constants for Intramolecular Electron Transfer in
$(NH_3)_5Co^{III}(^-O_2CC_6H_5NO_2^-)$ at pH 7 (31)

Position of NO_2	$k \ / \ s^{-1}$
ortho	4.0×10^5
meta	1.5×10^2
para	2.6×10^3
2, 4 dinitro	3.5×10^4
3, 5 dinitro	1.3×10^2

The rate constants in Table 8 are low and indicate that the
carboxylate group is not very permeable to electron transfer from
ligand to metal. This led Miller et al to measure the temperature
dependence of the transfer from the para isomer coordinated to one
and two Co^{III} centres (32) to see if the transfer involved electron
tunnelling at low temperatures. It was found that the transfer
remained activation controlled even when the rate constant was
reduced by a factor of 10^4 by lowering the temperature, and there
is no evidence of non-Arrhenius behaviour.

6.3 Aquation of Transition-Metal Complexes.

The process of ligand exchange with water molecules in labile
complexes can be readily followed using pulse radiolysis combined
with conductance measurements. Cobalt(III) complexes have been
investigated in some detail because the Co^{III} centre is rapidly
reduced to Co^{II} which has a labile d^7 configuration and readily

loses its ligands sequentially to form high-spin Co_{aq}^{2+}. Several
Co^{III}-ammine and Co^{III}-amine complexes have been studied (33).
The ammine complexes $Co^{III}(NH_3)_5X$, where X is NH_3, Cl^-, F^- or
H_2O all lose the first three ligands (including X) too rapidly
to be measured ($t_{\frac{1}{2}} < 2\ \mu s$) but the last three are detached at
slower rates (see Table 9).

Table 9. Half-Lives for the Detachment of the Last Three
Ammonias from Labile Cobalt(II) Complexes (33)

Complex	$t_{\frac{1}{2}}$ for Ligand Detachment / μs		
	4th	5th	6th
$Co(NH_3)_6^{2+}$	14	90	700
$Co(NH_3)_5Cl^+$	10	65	540
$Co(NH_3)_5F^+$	8	92	740
$Co(NH_3)_4Cl_2$	14	84	640
$Co(NH_3)_4ClH_2O^+$	7	75	550

In contrast to NH_3 and X, the bidentate ligand ethylenediamine
is released much more slowly and the rate is pH dependent (33).
It is suggested (33) that one end of the chelate comes off and is
protonated which leads to complete detachment. Degraded ligands
resulting from ·OH attack and electron transfer to the metal centres
(see section 6.2) tend to dissociate more rapidly than the normal
ligand (27).

6.4 Change in Symmetry

It has been suggested (34) that in some cases free radicals
may react with transition-metal complexes via intermediates in
which the coordination sphere is expanded. Plausible examples
are the oxidation of Ti^{III} and Fe^{II} by H-atoms and some aliphatic
radicals because the observed rate is faster than the exchange of
coordinated water.

Oxidation of the square planar complex $PtCl_4^{2-}$ by ·OH is
thought to result in a change of symmetry to trigonal bypyramidal
(35). The reaction generates a transient species which absorbs

at 450 nm and changes by a first-order process ($k = 2 \times 10^5 s^{-1}$)
to a second species with λ_{max} at 410 nm. Oxidation of $PtCl_4^{2-}$ by
Cl_2^- or Br_2^- does not produce the 450 nm band (36) and new
absorptions appear in the UV. On the other hand, reduction of
octahedral $PtCl_6^{4-}$ also produces the 410 nm band. The following
mechanism was proposed (35) to explain these observations

7. CONCLUDING REMARKS

The application of the techniques of water radiolysis to
inorganic chemistry has provided quantitative information on
the chemical and physical properties of a wide variety of unstable intermediates and an insight into their role in reaction
mechanisms where their involvement is accepted but where they
cannot be directly observed.

Pulse radiolysis studies alone are not always capable of
unravelling mechanistic complexities and product measurements from
steady state radiolysis can provide valuable complementary
information.

REFERENCES

1. Ross, A. B., and Neta, P., "Rate Constants for Reactions of
    Inorganic Radicals in Aqueous Solution," NSRDS-NBS 65, U. S.
    Department of Commerce, National Bureau of Standards,
    Washington D. C., 1979.

2. Maruthamuthu, P., and Neta, P., 1978, J. Phys. Chem. 82,
    pp. 710 - 713.

3. Zehavi, D., and Rabani, J., 1972, J. Phys. Chem. 76, pp.
    312 - 319.

4. Behar, D., Bevan, P. L. T., and Scholes, G., 1972, J. Phys.
    Chem. 76, pp. 1537 - 1542.

5. Jayson, G. G., Parsons, B. J., and Swallow, A. J., 1973, J.
    Chem. Soc., Faraday Trans. I 69, pp. 1597 - 1607.

6. Ellison, D. E., Salmon, G. A., and Wilkinson, F., 1972, Proc.
    Roy. Soc. A 328, pp. 23-36.

7. Chawla, O. P., and Fessenden, R. W., 1975, J. Phys. Chem. 79,
    pp. 2693 - 2700.

8. Buxton, G. V., and Subhani, M. S., 1972, J. Chem. Soc.,
    Faraday Trans. I 68, pp. 947 - 957.

9. Buxton, G. V., and Dainton, F. S., 1968, Proc. Roy. Soc. A
    304, pp. 427 - 439.

10. Barat, F., Gilles, L., Hickel, B., and Lesigne, B., 1972, J.
    Phys. Chem. 76, pp. 302 - 307.

11. Tendler, Y., and Faraggi, M., 1973, J. Chem. Phys. 58, pp.
    848 - 853.

12. Baxendale, J. H., Breccia, A., and Ward, M. D., 1970, Int. J.
    Radiat. Phys. Chem. 2, pp. 167 - 176.

13. Baxendale, J. H., Fielden, E. M., and Keene, J. P., 1965, Proc.
    Roy. Soc. A 286, pp. 320 - 336.

14. Buxton, G. V., and Sellers, R. M., 1977, Coord. Chem. Rev.,
    22, pp. 195 - 274.

15. Asmus, K.-D., Bonifacic, M., Toffel, P., O'Neill, P.,
    Schulte-Frohlinde, D., and Steenken, S., 1978, J. Chem. Soc.
    Faraday Trans. I 74, pp. 1820 - 1826.

16. Schmidt, K. H., Sullivan, J. C., Gordon, S., and Thompson,
     R. C., 1978, Inorg. Nucl. Chem. Lett. 14, pp. 429 - 434.

17. Gordon, S., Sullivan, J. C., Mulac, W. A., Cohen, D., and
     Schmidt, K. H., 1976, in "Proceedings of the Fourth Tihany
     Symposium on Radiation Chemistry" eds. Hedrig, P., and
     Schiller, R., Akademiai Kiado, Budapest, pp. 753 - 759.

18. Tendler, Y., and Faraggi, M., 1972, J. Chem. Phys. 57, pp.
     1358 - 1359.

19. Gordon, S., Mulac, W. A., Schmidt, K. H., Sjoblom, R. K.,
     and Sullivan, J. C., 1978, Inorg. Chem. 17, pp. 294 - 296.

20. Sullivan, J. C., Gordon, S., Mulac, W. A., Schmidt, K. H.,
     Cohen, D., and Sjoblom, R., 1976, Inorg. Nucl. Chem. Lett.
     12, pp. 599 - 601.

21. Pikaev, A. K., Shilov, V. P., and Spitsyn, V. I., 1977,
     Dokl. Akad, Nauk. SSSR 232, pp. 387 - 390.

22. Pikaev, A. K., and Shilov, V. P., 1978, Izvest. Akad. Nauk
     SSSR Ser. Khim. No. 9, pp. 2136 - 2139.

23. Dentsch, E., Heineman, W. R., ᴜurst, R., Sullivan, J. C.,
     Mulac, W. A., and Gordon, S., 1978, J. Chem. Soc. Chem. Comm.,
     p. 1038.

24. Meyerstein, D., and Mulac, W. A., 1969, J. Phys. Chem. 73,
     pp. 1091 - 1095.

25. Navon, G., and Meyerstein, D., 1970, J. Phys. Chem. 74,
     pp. 4067 - 4070.

26. Laurence, G. S., and Thornton, A. T., 1974, J. Chem. Soc.
     Dalton Trans., pp. 1142 - 1148.

27. Shinohara, N., and Lilie, J., 1979, Inorg. Chem. 18, pp.
     434 - 438.

28. Laiti, J., and Meyerstein, D., 1978, J. Chem. Soc., Dalton
     Trans., pp. 1105 - 1118.

29. Laiti, J., and Meyerstein, D., 1972, Inorg. Chem. 11, pp.
     2397 - 2401.

30. Battacharyya, S. N., and Neta, P., 1981, J. Phys. Chem. 85,
     pp. 1527 - 1529.

31. Simic, M. G., Hoffman, M. Z., and Brezniak, N. V., 1977, J.

Am. Chem. Soc. 99, pp. 2166-2172.

32. Beitz, J. V., Miller, J. R., Cohen, H., Wieghardt, K., and
     Meyerstein, D., 1980, Inorg. Chem. 19, pp. 966 - 968.

33. Lilie, J., Shinohara, N., and Simic, M. G., 1976, J. Am.
     Chem. Soc. 98, pp. 6516 - 6520.

34. Meyerstein, D., 1978, Acc. Chem. Res. 11, pp. 43 - 48.

35. Adams, G. E., Broszkiewicz, R. K., and Michael, B. D., 1968,
     Trans. Faraday Soc. 64, pp. 1256 - 1264.

36. Broszkiewicz, R. K., and Grodowski, J., 1976, Int. J. Radiat.
     Phys. Chem. 8, pp. 359 - 365.

APPLICATION OF PULSE RADIOLYSIS TO THE STUDY OF AQUEOUS ORGANIC SYSTEMS

A.J. Swallow

Paterson Laboratories
Christie Hospital & Holt Radium Institute
Manchester, England.

Organic free radicals can be produced pulse radiolytically in aqueous solution through the action of hydroxyl radicals, hydrated electrons or hydrogen atoms on the parent compound. Their properties can be studied by means of optical absorption. Experiments with suitably constructed systems enable their reactions with added substances to be observed. Different acid-base forms can often be distinguished. Discussion is given of the radicals formed by oxidation or reduction of simple organic compounds such as aliphatic and aromatic hydrocarbons, halides, alcohols, carbonyl compounds, phenols and sulphur compounds.

INTRODUCTION

Regardless of what the basic processes might be, the ionisation and excitation of water by fast electrons are known to result in the formation of free radical and molecular species, as well as H^+ and OH^-:

$$H_2O \longrightarrow OH, e_{aq}^-, H, H_2, H_2O_2, H^+, OH^- \tag{1}$$

These species are available to interact with any solutes which may be present in the water. The H_2, H_2O_2, H^+ and OH^- are not especially reactive, and in any case their chemistry has been adequately studied without using radiation. Hydroxyl radicals, hydrated electrons and hydrogen atoms on the other hand are highly active, generally reacting with organic compounds to give organic free radicals. Providing the solute concentrations is less than about 1-10% by weight, such interactions normally produce far more effect than can be produced by direct ionisation or excitation of

J. H. Baxendale and F. Busi (eds.),
The Study of Fast Processes and Transient Species by Electron Pulse Radiolysis, 289–315.

the solute.

In this contribution it will be shown how pulse radiolysis, mainly using optical detection, is used to study the formation, properties and reactions of organic free radicals. Free radicals form an important class of chemicals, characterised by high reactivity and therefore low abundance in nature. An adequate understanding of their reactions is essential to the understanding of the effects of radiation on aqueous organic systems as encountered for example in the processing of nuclear fuels, in radiobiology and in the radiation preservation of foodstuffs. They are also important in numerous other situations, for instance in polymerisation, in normal biochemistry and in oxidative degradation.

REACTIONS OF OH, e_{aq}^- AND H

Hydroxyl Radicals (1)

Hydroxyl radicals add to olefinic and aromatic compounds often reacting at every collision. They abstract hydrogen atoms from carbon-hydrogen bonds in compounds such as alcohols, aldehydes and ketones, carboxylic acids, amines etc. They readily abstract hydrogen atoms from -SH groups. The radicals formed in such reactions often have characteristic absorption spectra.

Amongst other reactions of hydroxyl radicals is their ionisation, which may be represented:-

$$OH + OH^- \rightleftharpoons O^- + H_2O, \quad pK = 11.9 \tag{2}$$

The significance of this reaction is that in alkaline solution hydroxyl radicals are converted into O^- which does not readily add to olefinic or aromatic groups, although it abstracts hydrogen atoms at a comparable rate to OH. Other reactions which are important in the present context include those which can be used to convert OH into species which act as useful electron accepting agents, e.g. the overall reaction:-

$$OH + 2Br^- \longrightarrow OH^- + Br_2^- \tag{3}$$

Typical rate constants for reactions of OH, O^- and Br_2^- are given in Table 1. Further values for OH and O^- are to be found in NSRDS-NBS 59(2) and for Br_2^- and certain other inorganic radicals in NSRDS-NBS 65(3). The rate constants in Table 1 are the values of k defined by:-

$$-\frac{d[OH]}{dt} = \frac{d[R\cdot]}{dt} = k[OH][RH] \tag{4}$$

Table 1

Rate constants for reactions of OH, O^- and Br_2^- ($M^{-1}s^{-1}$)

	OH	O^-	Br_2^-
Br^-	1.0×10^9	-	-
SCN^-	1.1×10^{10}	1.0×10^9	-
Fe^{2+}	2.3×10^8	-	3.6×10^6
$Fe(CN)_6^{4-}$	9.3×10^9	$\leqslant 3 \times 10^7$	-
Acetate	9.0×10^7	5×10^7	-
Acetone	1.0×10^8	-	-
Benzoate	5.7×10^9	4×10^7	-
Methanol	9×10^8	5.8×10^8	-
Ethanol	1.85×10^9	1.1×10^9	-
Isopropanol	2.2×10^9	1.5×10^9	-
Cysteine	1.2×10^{10}	-	1.8×10^8 (pH 6.6)
Formate	3.5×10^9	1.0×10^9	$< 10^3$
Ribose	1.5×10^9	-	-
Thymine	5.4×10^9	4×10^8	2×10^8 (pH 12)
Tryptophan	1.2×10^{10}	-	7.7×10^8

- where $[R \cdot]$ is the total concentration of radical species formed by action of OH on RH. The rate constant, k, for mutual reaction of hydroxyl radicals defined by:-

$$- \frac{d[OH]}{dt} = \frac{2d[H_2O_2]}{dt} = 2k[OH]^2 \tag{5}$$

is $6 \times 10^9 M^{-1}s^{-1}$, i.e. the rate constant $2k = 1.2 \times 10^{10} M^{-1}s^{-1}$. It may be noted that statements such as "the rate constant for the reaction is x" are ambiguous when dealing with the mutual reaction of like species. There is no possibility of ambiguity where unlike species are reacting with each other.

Hydroxyl radicals and O^- absorb weakly below 300nm (λ_{max} = 230 and 240nm respectively, ε_{max} = 600 and 240 $M^{-1}cm^{-1}$ respectively). The low extinction in a difficult region of the spectrum means that OH and O^- can rarely be observed in aqueous organic systems. However, the product radicals can usually be seen. Competition methods, e.g. competition with the easily observable reaction:-

$$OH + Fe(CN)_6^{4-} \longrightarrow OH^- + Fe(CN)_6^{3-} \tag{6}$$

- can be used to measure relative rate constants. Br_2^- is easily observable, λ_{max} = 360nm, ε_{max} = 9900 $M^{-1}cm^{-1}$.

Hydrated Electrons (4)

Hydrated electrons add rapidly to compounds containing low-lying vacant orbitals, for example most aromatics, halides, aldehydes and ketones, thiols, disulphides and nitro compounds. In the first instance the reaction is a straightforward addition, but in some cases (e.g. with many halides) bond breakage occurs within the time-scale of the addition, so the reaction can be considered as a dissociative electron capture. Hydrated electrons do not react at significant rates with water itself, or with aliphatic hydrocarbons, alcohols, ethers, carboxylic acids in their anionic form or the uncharged forms of amines. Another reaction of importance is the conversion of hydrated electrons into hydrogen atoms:-

$$e_{aq}^- + H^+ \longrightarrow H \tag{7}$$

- which can also be effected by the acidic form of buffers, but with a lower rate constant. Reaction with N_2O is important as it replaces hydrated electrons by hydroxyl radicals:-

$$e_{aq}^- + N_2O \xrightarrow{+H_2O} N_2 + OH + OH^- \tag{8}$$

Hydrated electrons react with oxygen:-

$$e_{aq}^- + O_2 \longrightarrow O_2^- \tag{9}$$

- the superoxide radical being the anionic form of the HO_2 radical, pK = 4.7. Selected rate constants, nearly all determined by observing the disappearance of the absorption due to the hydrated electron, are given in Table 2. Others are to be found in NSRDS publications (5,6). The rate constant for mutual reaction of hydrated electrons (to produce molecular hydrogen) is about $2k = 1 \times 10^{10} M^{-1}s^{-1}$.

The hydrated electron has an absorption spectrum extending through the entire visible range, λ_{max} = 715nm, ε_{max} = 18,500 $M^{-1}cm^{-1}$. The superoxide anion radical has λ_{max} = 240nm, ε_{max} =

Table 2

Rate constants for reactions of e_{aq}^- and H ($M^{-1}s^{-1}$)

	e_{aq}^-	H
H_3O^+	2.3×10^{10}	$< 10^4$
$Fe(CN)_6^{3-}$	3.0×10^9	7×10^9
N_2O	8.7×10^9	$< 10^6$
O_2	1.9×10^{10}	2×10^{10}
OH	3×10^{10}	2×10^{10}
Acetate	$< 10^6$	3×10^5
Acetic acid	1.8×10^8	1×10^5
Acetone	6×10^9	2.8×10^6
Benzoate	3.0×10^9	1×10^9
Ethanol	$< 10^5$	2.6×10^7
Benzoquinone	2.7×10^{10}	1×10^{10}
Cysteine	8.7×10^9 (pH 6.3)	1.1×10^9 (pH 6)
Formate	$< 10^4$	1.5×10^8
Formic acid	1.4×10^8	7.5×10^5
Ribose	$< 10^7$	5×10^7
Thymine	1.7×10^{10}	8×10^8 (pH 1)
Tryptophan	3.0×10^8	2×10^9

$2350 \text{ M}^{-1}\text{cm}^{-1}$ and HO_2 has $\lambda_{max} = 225$ nm, $\varepsilon_{max} = 1400 \text{ M}^{-1}\text{cm}^{-1}$.

Hydrogen Atoms (7)

Although they resemble hydrated electrons in being reducing agents, the reactions of hydrogen atoms with organic compounds are closer to those of OH. They add to olefins and aromatic compounds giving radicals whose absorption spectrum is similar to that produced by OH addition. They add to the disulphide group with accompanying scission of the -S-S- bond. They abstract hydrogen atoms from carbon-hydrogen and sulphur-hydrogen bonds. Other reactions of importance in the present context are that with hydroxyl ions:-

$$H + OH^- \rightleftharpoons e_{aq}^- + H_2O \tag{10}$$

(cf. Reaction 7), the hydrogen atom being a weak acid, pK = 9.6, and that with oxygen:-

$$H + O_2 \longrightarrow HO_2 \tag{11}$$

Some rate constants are included in Table 2. Other values are to be found in NSRDS-NB5 51 (8). Where the reaction resembles that produced by OH, the rate constants are generally smaller. Hydrogen atoms react with each other with $2k = 2 \times 10^{10} \text{M}^{-1}\text{s}^{-1}$.

DESIGN OF EXPERIMENTS

Radicals Formed Through Action of OH

To prepare organic radicals by action of hydroxyl radicals it is usual to saturate the solution with nitrous oxide (N_2O concentration at room temperature about 25mM) so as to replace hydrated electrons with OH (Reaction 8). This procedure also removes dissolved oxygen. At a concentration of 25mM, nitrous oxide interferes with spur reactions, and thus increases the yield of available active species (Equation 1) above the values found in pure water or very dilute solution. Under the present conditions it may be taken that the yield of hydroxyl radicals (including those formed from e_{aq}^-) is G(OH) = 6.0, although slightly different values have been adopted at various times in the past. The yield of hydrogen atoms in the solution is only about G(H) = 0.55. It often introduces little error if these are treated as if they were OH radicals, so that a total of about G = 6.5 radicals are available to react with the organic compounds.

In order to ensure that the majority of the hydrated electrons in fact react with N_2O rather than hydrogen ions, the solution must not be too acid. Less than 5% of hydrated electrons react with hydrogen ions if the hydrogen ion concentration is less than about:-

$$\frac{8.7 \times 10^9 \times 0.025}{20 \times 2.3 \times 10^{10}} = 4.7 \times 10^{-4}M \tag{12}$$

- i.e. the solution should have pH > 3.3. Organic buffers should not be used to control the pH as OH radicals generally react with them.

To prevent hydrated electrons reacting significantly with the solute of interest, its concentration should be small compared with $8.7 \times 10^9 \times 0.025/k_1$, where k_1 is the rate constant for reaction of e_{aq}^- with solute. For $k_1 = 2 \times 10^{10}M^{-1}s^{-1}$, less than 5% of the hydrated electrons will react in this way if the solute concentration is kept below about $5 \times 10^{-4}M$. For lower values of k_1, higher concentrations of solute are permissible.

To prevent secondary reactions of OH radicals with organic radicals and to ensure first order (pseudo unimolecular) reaction of OH with solute, the pulse dose should be small enough to convert no more than about 10% of the solute into radicals. It is also necessary to restrict the dose so as to ensure that most hydroxyl radicals react with the solute of interest rather than with each other. For this reason the concentration produced by the pulse should be small compared with $k_2[S]/1.2 \times 10^{10}$, where k_2 is the rate constant for reaction of OH with solute S. For a solution of concentration $5 \times 10^{-4}M$ and $k_2 = 3 \times 10^9M^{-1}s^{-1}$, the initial rate of loss of OH radicals by reaction with each other would be less than 10% of the rate of loss by reaction with S if the concentration of OH produced by the pulse is less than $1.25 \times 10^{-5}M$, i.e. for G(OH) = 6.0 the dose should be less than 200Gy.

Having prepared the solution, and chosen an appropriate dose, the optical path length of the cell should be selected such that the absorbance of the unpulsed solution at wavelengths of interest is less than about 0.9 (i.e. 10% transmission). Different cells could be used for different wavelength ranges if necessary. Delivery of the pulse will result in any changes in absorbance which correspond to conversion of some of the solute molecules into radicals:-

$$OH + S \longrightarrow radical \tag{13}$$

The half-life of Reaction 13 in seconds will be $0.693/k_2[S]$. For $k_2 = 3 \times 10^9M^{-1}s^{-1}$ and $[S] = 5 \times 10^{-4}M$, the half-life would be $4.6 \times 10^{-7}s$. It is common for organic radicals to react with each other. For the simple case where a single radical species is produced rapidly and then disappears more slowly by mutual reaction with a rate constant of $2k_3$, the first half-life is equal to $1/2k_3 a$, where a is the concentration of radical species produced by the pulse. For a typical rate constant of $2k_3 =$ $1 \times 10^9 M^{-1} s^{-1}$ and a radical concentration of $1.25 \times 10^{-5} M$ the first half-life would be 80μs. It is usually easiest to work out the kinetics if the pulse duration is short compared with the time-scale of the reaction of interest, so that production of radicals can be assumed to be instantaneous.

It is often desired to examine the reaction of an organic radical with an added solute, S'. Like the principal solute S, its concentration must be kept low so that hydrated electrons will not react with it significantly. To prevent OH radicals reacting with it significantly its concentration should also be small compared with $k_2[S]/k_4$ where k_4 is the rate constant for reaction of OH radicals with S'. These two requirements generally, but not invariably, dictate that the concentration of S' be appreciably less than that of S. For the straightforward reaction of S radicals (S·) with S', the kinetics will be simple first order if [S·] is less than about 10% of [S'] so that [S'] is essentially constant, and also small compared with $k_4[S']/2k_3$ so that most S· react with S' instead of with each other. These requirements usual dictate the use of a dose which is substantially less than may be needed when examining S· radicals in the absence of S'.

The above considerations apply to simple reactions of OH radicals with solutes, and to simple reactions of the organic radicals formed. Similar principles are applied to the choice of conditions when examining radicals formed by action of O^-, Br_2^- etc. In practice it is usually necessary to adapt and extend the considerations according to the known or suspected reactions of the radicals under investigation. Parameters such as pH, concentration of main and added solutes and dose can be varied to provide tests for proposed mechanisms. Reaction kinetics other than the simplest require more advanced treatment (9) and computers could be needed to solve equations.

Radicals Formed Through Action of e_{aq}^-

There is no practicable method for converting hydroxyl radicals into hydrated electrons. An alternative approach is to add t-butanol to the solution at a concentration which enables it to compete successfully with the solute of interest for the hydroxyl radicals:-

$$OH + (CH_3)_3COH \rightarrow H_2O + \cdot CH_2C(CH_3)_2OH, \quad k = 5.2 \times 10^8 M^{-1} s^{-1} \quad (14)$$

The radicals formed from t-butanol are rather inert, and do not usually react with organic solutes at a significant rate. Absence of reaction can be verified by saturating the t-butanol-containing solution with N_2O, whereupon t-butanol radicals become the only significant source of attack apart perhaps from hydrogen atoms (see below). If solutions containing t-butanol are deoxygenated, for example by bubbling with an inert gas such as argon or nitrogen (oxygen-free), the solutes are thus exposed only to the action of hydrated electrons, together with a small proportion of hydrogen atoms (the rate constant for abstraction of hydrogen from t-butanol by hydrogen atoms is only about $1 \times 10^5 M^{-1}s^{-1}$). In a typical experiment using 1M t-butanol, and a $10^{-3}M$ concentration of a solute where the rate constants for OH and H attack are 3×10^9 and $5 \times 10^8 M^{-1}s^{-1}$ respectively, more than 99% of the hydroxyl radicals and 80% of the hydrogen atoms are taken up by the t-butanol. The yield of hydrated electrons in solutions containing 0.1 - 1M t-butanol may be taken to be $G \sim 3$. In the example just given, the attack on the solute by hydrogen atoms would correspond only to $G = 0.1$.

A disadvantage of t-butanol is that although the radicals formed from it are usually unreactive to organic solutes, they might well react with the organic radicals under investigation, so interfering with more interesting reactions. Also acid solutions where Reaction 7 is significant cannot be used. Another method of eliminating hydroxyl radicals is to allow them to react with isopropanol present in adequate excess over the solute of interest. Abstraction of hydrogen atoms from the α-position:-

$$OH + (CH_3)_2CHOH \longrightarrow H_2O + (CH_3)_2\dot{C}OH \qquad\qquad (15)$$

accounts for all but 15% of the attack of OH on isopropanol, the remainder being mainly abstraction of hydrogen atoms from the β position to give radicals as unreactive as that formed from t-butanol. Hydrogen atoms, including any produced by Reaction 7, may also be taken to react predominantly at the α-position, giving the same radical as Reaction 15 ($k \sim 10^8 M^{-1}s^{-1}$). The α-hydroxyisopropyl radical is quite a powerful reducing agent, often able to transfer an electron with a high rate constant, giving the same reduced radical as formed by hydrated electron attack, although sometimes it reacts by addition or hydrogen abstraction. In alkaline solution the isopropanol radical dissociates at the hydroxyl group, $pK = 12.1$, giving a radical which is a stronger reducing agent than the neutral radical. In 0.1 or 1M isopropanol the total yield of reducing radicals may be taken to be about 5.5, while the $\cdot CH_2CH(CH_3)OH$ radicals can often be ignored as their yield is only about $G = 1$.

A third method of eliminating hydroxyl radicals is to allow them to react with formate:-

$$OH + HCO_2^- \longrightarrow H_2O + CO_2^- \qquad\qquad (16)$$

Hydrogen atoms also react with formate, giving the same radical:-

$$H + HCO_2^- \longrightarrow H_2 + CO_2^-$$ (17)

The free carboxylate radical is a reducing agent of comparable reducing power to the α-hydroxyisopropyl radical. In acid solutions where formate exists as formic acid (pK = 3.75) the rate constants for OH and H attack are 1.5×10^8 and $7.5 \times 10^5 M^{-1}s^{-1}$ respectively so that higher formate concentrations are required in acid solution than in neutral solution. As the solution becomes more acid, hydrated electrons become replaced by CO_2^- through Reactions 7, 17 and:-

$$H + HCO_2H \longrightarrow H_2 + CO_2^- + H^+$$ (18)

The carboxyl radical has pK = 1.4. In 0.1 or 1M formate the total yield of reducing radicals may be taken to be G ∿ 6.5.

Whatever the method used to eliminate interference from OH radicals, doses, optical path lengths etc. must be chosen according to principles like those discussed for the radicals formed by action of OH.

t-Butanol radicals have λ_{max} at 225nm, $\varepsilon_{260} = 100\ M^{-1}cm^{-1}$, α-hydroxyisopropyl radicals have $\lambda_{max} < 210$nm, $\varepsilon_{280} = 700$ in their neutral form and $\lambda_{max} < 230$nm, $\varepsilon_{300} = 1500$ in their anionic form. CO_2^- has λ_{max} at 240nm, $\varepsilon_{max} = 3000\ M^{-1}cm^{-1}$ and CO_2H has $\lambda_{max} = 250$nm, $\varepsilon_{max} = 1200\ M^{-1}cm^{-1}$.

Radicals Formed Through Action of H

Acid solutions are employed, of pH low enough to enable most of the hydrated electrons to react with hydrogen ions according to Reaction 7. A carefully chosen concentration of t-butanol may be added, sufficient to scavenge hydroxyl radicals (k = $5.2 \times 10^8 M^{-1}s^{-}$ but not hydrogen atoms ($10^5 M^{-1}s^{-1}$). For the example of a 10^{-4}M concentration of a solute which reacted with hydroxyl radicals, hydrated electrons and hydrogen atoms with k = 3×10^9, 2×10^{10} and $5 \times 10^8 M^{-1}s^{-1}$ respectively, it can readily be calculated that use of > 10^{-3}M H^+ and $1-2 \times 10^{-2}$M t-butanol would expose it to predominant H atom attack. An alternative method of examining radicals formed by H atom attack is to use simple acid solutions containing no added solute such as N_2O, t-butanol etc. The radicals will then be formed together with those resulting from OH attack. If the spectrum of the radicals produced by OH is determined separately, the spectrum of the radicals formed from H can be obtained by subtraction.

Acid-Base Properties (10,11)

Organic free radicals often exhibit acid-base properties which differ from those of the parent material. Now acidic and

basic forms of radicals often differ in absorption spectra, it
being quite common for the acid form to absorb at shorter wavelengths than the basic. This enables the pK of the radicals to be
determined spectroscopically. To do this it is necessary to
measure the spectrum at a range of pH values bracketing the
expected pK. The absorbances found should then correspond to
those of two forms of the radical at concentrations varying with
pK in agreement with dissociation of the acid form. The interpretation is most simple if the attacking species remains constant
over the pH range (i.e. if OH does not become replaced by O^- or
H by e_{aq}^-) and if the parent has no pK in the region. If the
attacking species or the parent does change with pH the possibility
has to be considered that any change in radical spectrum is due to
the formation of different types of radical rather than to the
formation of a constant radical in different acid-base forms.

    Another condition for the spectroscopic determination of the
pK of a radical is that the acid-base equilibrium should be attained
before the radicals react, whether by mutual reaction or in other
ways. The rate of attainment of an equilibrium is limited by the
faster of the forward and reverse rates. Equilibrium is expected
to be attained quickly in acid solution where the dissociation is:-

$$HA \rightleftharpoons H^+ + A^- \tag{19}$$

and in alkaline solution where it is:-

$$HA + OH^- \rightleftharpoons H_2O + A^- \tag{20}$$

- the back reaction of 19 and the forward reaction of 20 both
having rate constants in the region of $10^{10}M^{-1}s^{-1}$. In solutions
in the region of neutrality, buffers can speed up the process by
providing a proton donor and acceptor. Buffer is also necessary
to enable pH values in the region of neutrality to be measured
reliably, especially since a neutral unbuffered solution which has
been saturated with N_2O or deoxygenated by bubbling with argon
or nitrogen will acquire a different pH if exposed to air
(containing CO_2) during a pH measurement. Phosphate buffer is
suitable, but it is necessary to be aware of the reaction:-

$$e_{aq}^- + H_2PO_4^- \longrightarrow H + HPO_4^{2-}, \quad k = 10^7 M^{-1}s^{-1} \tag{21}$$

Borate buffer is suitable for use on the alkaline side of
neutrality. Formate itself provides an acceptable buffer on the
acid side when used as an OH scavenger. Other organic buffers
should generally be avoided, owing to the possibility of their
participating in radical reactions.

    For radicals capable of acting as polybasic acids, experimental
limitations may prevent the determination of every pK. In such

cases it becomes necessary to identify the state of protonation
of the species observed (e.g. if only one pK is seen for the
species AH_2^+, it would have to be decided whether this was the pK
for the dissociation $AH_2^+/AH^{\cdot-}$ or for $AH^{\cdot-}/A^{\cdot 2-}$). For sufficiently
stable radicals, e.s.r. can sometimes identify one of the
partners. For unstable radicals, determination of the effect of
ionic strength on rates of reaction can be of assistance since
rates of reaction do not depend on ionic strength when one of the
partners is uncharged, but for small molecules and at ionic
strength less than about $5 \times 10^{-2}M$, the rate constant k for
reactions between charged species at ionic strength μ is related
to the rate constant at zero ionic strength, k_0, by the expression

$$\log(k/k_o) = 1.02 \ Z_a Z_b \mu^{\frac{1}{2}}/(1 + \underline{a}\mu^{\frac{1}{2}}) \tag{22}$$

where Z_a and Z_b are the charges on the reactants and $\underline{a}$ is a
constant which for small molecules could be taken to have the
numerical value of 2 (when μ has units of M). According to this
equation, reactions between species possessing opposite charges
decrease in rate as the ionic strength increases, while reactions
between species of like charge increase in rate.

Errors

The accuracy of an absolute measurement of a rate constant
for the reaction of OH, Br_2^-, e_{aq}^-, CO_2^-, H etc. with an organic
compound depends on the extent to which the following criteria
are met:-

1. The composition of the solution, the dose and other condition
should be chosen in accordance with principles like those given
above, correction being made for any side reactions unavoidably
taking place. Measurements should be made at a number of
solute concentrations and doses as a check.

2. Allowance should be made for any impurities in the solute of
interest. This is particularly important when the reaction of
interest is a slow one (e.g. if the true rate constant for reactio
of e_{aq}^- with a solute were to be $10^7 M^{-1}s^{-1}$, 1% of an impurity
reacting with $10^{10} M^{-1}s^{-1}$ would result in an apparent rate constant
of $1.1 \times 10^8 M^{-1}s^{-1}$.

3. Allowance should be made for any change in solute concentration during the conduct of the experiment, e.g. due to instability
volatility during bubbling procedures, photosensitivity etc.

4. If both of the reactants are charged, the ionic strength of
the solution should be specified (cf. Equation 22).

5. The temperature of measurement (normally 15-25°C) should be
specified since activation energies cannot always be taken to be

negligible.

With reasonable care it is not difficult to attain an accuracy of ± 10% or even 5%. The physical capabilities of present day pulse radiolysis apparatus would in themselves permit accuracies of better than 1%.

When determining a rate constant by competition with a reference reaction of known rate constant (such as in the determination of an OH rate by competition between substrate and ferrocyanide) it is necessary that neither the organic radical nor the species produced in the reference reaction enters into further reactions during the measurement. The accuracy of the rate constant determined is, of course, also dependent on the accuracy of the reference reaction.

For the simplest case of radicals of a single species, R·, disappearing by mutual reaction to give a product P (2R· $\longrightarrow$ P), the rate constant is obtained from measured optical absorptions by application of Equation 23:-

$$- \frac{d[R\cdot]}{dt} = \frac{2d[P]}{dt} = 2k[R\cdot]^2 \qquad (23)$$

(cf. Equation 5 above) together with Beer's law. If neither the original substrate nor the product absorbs light at the chosen wavelength, the absorbance of the solution at any given time, OD, is given by:-

$$OD = \varepsilon[R\cdot]d \qquad (24)$$

- where ε is the extinction coefficient of the radical and d is the optical path length. According to Equations 23 and 24, measurements of optical absorbance as a function of time are able to yield values of $2k/\varepsilon$. The extinction coefficient of radicals is obtained by measuring the absorbance produced by delivery of a known pulse dose which gives radicals with a known G-value. The accuracy of k is therefore dependent on the accuracy of the dose and the validity of the assumed G-values as well as on criteria such as 1-5 above. Regarding the dose, it is only for dilute solutions that it can be assumed that the dose in the solution of interest would be the same as that which would be received by the dosimeter pulsed in the same way. Errors can also arise if the dose and hence the concentration of active species is non-uniform across the light path (12). In the ordinary way, accuracies in rate constant for reactions observed second order cannot be expected to be better than about ± 20%, and are often not so good.

Where radicals of more than one species are disappearing in

mutual reactions (e.g. when OH attack on a solute has produced
more than one type of radical, or when radicals formed from
hydrated electrons are present simultaneously with the radicals
formed by OH attack on t-butanol) proper treatment of the kinetics
is quite complex, more so than often assumed. Full treatment
from first principles should be attempted.

FREE RADICALS FORMED FROM ORGANIC COMPOUNDS

     There are hundreds of papers concerned with pulse radiolytic
studies of organic free radicals in aqueous solution. Among
other reviews (11,13) one article contains an almost exhaustive
review of literature published between October 1971 and December
1976, covering more than 99% of relevant papers published during
the period. It includes comprehensive tabulations of rate
constants, values of pK and one-electron reduction potentials
(14). A compilation of rate constants for reaction of aliphatic
carbon-centered radicals is in preparation (15). Selected
aspects of the chemistry of certain organic free radicals in
aqueous solution will now be touched upon for discussion. Pulse
radiolysis using optical detection has not been the sole source
of the new information described but it has been the most important one.

Hydrocarbons

     Simple aliphatic hydrocarbon radicals can be prepared from
hydrocarbons themselves by abstraction of hydrogen atoms by OH
radicals or from compounds such as alkyl halides by dissociative
electron capture resulting from the action of hydrated electrons.
They react together with rate constants close to $2k = 2-3 \times 10^9 M^{-1}s^{-1}$, typical for diffusion-controlled reactions. Convention:
pulse radiolysis cannot tell whether the mutual reaction proceeds
predominantly by dimerisation or disproportionation, but from
other work the two routes appear to be of comparable importance
(except for methyl).

     Hydrocarbon radicals react with oxygen at a rate which is
comparable to the rate of mutual reaction. It would be surprising
on chemical grounds if this were other than addition, and addition
has been verified in several ways including a) the absorption
spectrum of the species formed at pH 7 differs from that of O_2^-
(Figure 1 (16)) and b) the species formed cannot transfer an
electron to tetranitromethane whereas O_2^- can. The reaction may
therefore be written:-

$$R\cdot \; + \; O_2 \longrightarrow RO_2\cdot \qquad\qquad\qquad (25)$$

The radicals formed by addition of OH to olefines are

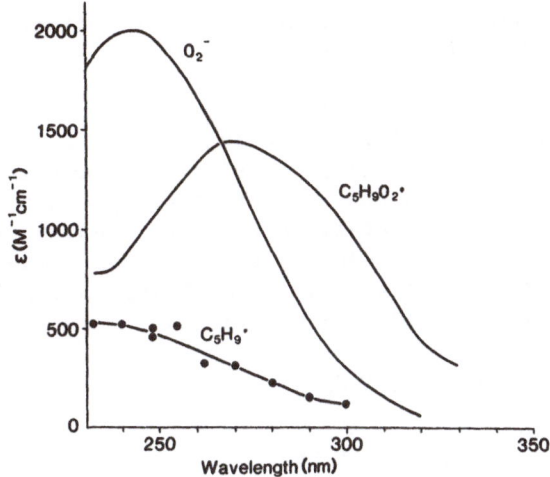

Figure 1. Absorption spectrum of $RO_2\cdot$ radicals ($C_5H_9O_2\cdot$) prepared by pulse radiolysis of aqueous cyclopentane containing N_2O and O_2 at pH 7. It can be seen that the spectrum is quite different from that of O_2^-. The spectrum of the $C_5H_9\cdot$ radical is also shown.

β-hydroxyalkyl radicals, e.g.

$$OH + CH_2 = CH_2 \longrightarrow \cdot CH_2\text{-}CH_2OH \qquad (26)$$

For the case of ethylene these react together with $2k \sim 5\text{-}6 \times 10^8 M^{-1}s^{-1}$ (17). They can also add to double bonds (the first propagation step in polymerisation) with a low rate constant, variously estimated as $3 \times 10^4 - 10^7 M^{-1}s^{-1}$, but the reaction cannot be observed directly by pulse radiolysis as the absorption spectrum is weak and would not be expected to vary significantly with chain length. In contrast to α-hydroxyalkyl radicals, β-hydroxyethyl radicals are not reducing agents. They will reduce Cu^{2+} to Cu^+, but this is probably through the intermediate formation of a compound containing a copper-carbon bond (17,18), e.g.:-

$$\cdot CH_2CH_2OH + Cu^{2+} \longrightarrow (Cu(III) - CH_2CH_2OH)^{\cdot 2+} \qquad (27)$$

$$(Cu(III) - CH_2CH_2OH)^{\cdot 2+} \longrightarrow Cu^+ + CH_2CH_2 + H^+ \qquad (28)$$

They also oxidise Cu^+ to Cu^{2+}, and this could also proceed through a compound with a metal-carbon bond, e.g.:-

$$\cdot CH_2CH_2OH + Cu^+ \longrightarrow (Cu(II) - CH_2CH_2OH)^+ \qquad (29)$$

$$(Cu(II) - CH_2CH_2OH)^+ \longrightarrow Cu^{2+} + C_2H_4 + OH^- \tag{30}$$

There are many other examples of organic free radicals reacting with metal ions through such routes (14).

Hydroxycyclohexadienyl radicals, formed by action of OH on benzene, react together at the diffusion-controlled rate and reduce Cu^{2+} and oxidise Cu^+ (19). A similar adduct is formed by action of OH on toluene as shown by spectroscopic and other evidence. However product analysis of irradiated N_2O-saturated toluene at pH 3 has shown bibenzyl to be a major product, implying that under acid conditions the hydroxymethylcyclohexadienyl radicals eliminate water to give benzyl which then dimerises:-

$$OH + C_6H_5CH_3 \longrightarrow HO\ C_6H_5\overset{\cdot}{C}H_3 \tag{31}$$

$$HO\ C_6H_5\overset{\cdot}{C}H_3 \xrightarrow{acid} C_6H_5\overset{\cdot}{C}H_2 + H_2O \tag{32}$$

$$2\ C_6H_5\overset{\cdot}{C}H_2 \longrightarrow (C_6H_5CH_2)_2 \tag{33}$$

This has been confirmed pulse radiolytically: almost immediately after a pulse, strongly acid solutions (e.g. 0.1M HClO$_4$) of toluene exhibit sharp peaks near 260, 305 and 320nm, characteristic of the benzyl radical, which have been shown to be due to OH attack by comparison of spectra with and without t-butanol (20). Similar peaks are seen after reaction of hydrated electrons with benzyl chloride or after reaction of O$^-$ with toluene. In even stronger acid (2-5M) there is evidence that the OH adduct does not change directly to the benzyl radical, but goes via a radical cation. Radical cations are seen more readily with other methylated benzenes (xylene etc.) and can also be formed in neutral solution by action of SO_4^- (from reaction of e_{aq}^- with persulphate). Spectra of the various species as seen for p-xylene are shown in Figure 2 (21). The sequence of the interconversion of the radicals formed by action of OH on methylated benzenes may now most simply be represented as follows (using toluene as the example):

$$OH + C_6H_5CH_3 \longrightarrow HO\ C_6H_5\overset{\cdot}{C}H_3 \underset{+OH^-}{\overset{-OH^-}{\rightleftharpoons}} C_6H_5\overset{\cdot+}{C}H_3 \atop \downarrow \atop C_6H_5\overset{\cdot}{C}H_2 + H^+ \tag{34}$$

Halides

Substitution of one of the hydrogen atoms in a saturated hydrocarbon by a halogen atom does not greatly affect the molecule' response to OH attack, since this still takes place by hydrogen atom abstraction. Like simple alkyl radicals, the halogenated alkyl radicals react with oxygen to form peroxy radicals. These

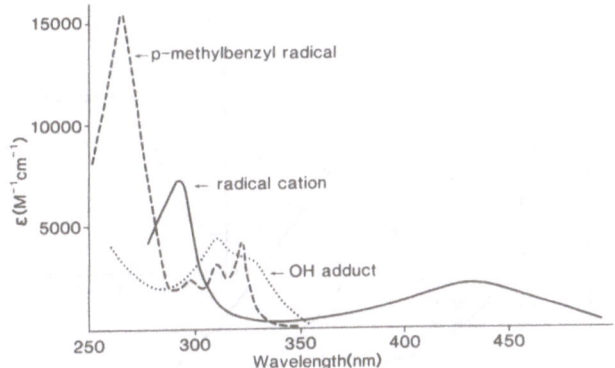

Figure 2.   Absorption spectra of species formed by action of OH on p-xylene.   The OH adduct is formed first.   Subsequently it changes to the radical cation.   Finally the radical cation changes to the p-methylbenzyl radical.

appear to be better oxidising agents than are alkyl peroxy radicals themselves, probably because of the electron-withdrawing properties of the halogen atoms (22).

Hydroxyl radicals react with unsaturated or aromatic halides by addition.   They add to the carbon atoms bearing the smallest number of halogen atoms.   Hydrogen halide can be eliminated within the duration of the pulse to give a more stable radical, e.g. (23):-

$$OH + CHCl = CHCl \longrightarrow (CH(OH)Cl - \dot{C}HCl) \longrightarrow \dot{O}CH = CHCl \qquad (35)$$

In simple cases, reaction of hydrated electrons with halides is followed by elimination of halide ion, although alternative reactions of electron adducts may be possible, as seen for instance in the reaction of hydrated electrons with p-fluorobenzonitrile, where fluorine-containing radicals can be formed, especially in acid solution where the electron adduct can protonate (24).   Where halide ions do become eliminated, the reactions provide a useful method of making a defined radical.   The formation of benzyl from benzyl chloride (see above) is one example where this has been useful in diagnosis.   Another example has been in the formation of the α and β radicals of aliphatic acids, e.g.:-

$$e_{aq}^- + CH_3CHClCO_2^- \longrightarrow Cl^- + CH_3\dot{C}HCO_2^- \qquad (36)$$

$$e_{aq}^- + CH_2ClCH_2CO_2^- \longrightarrow Cl^- + \cdot CH_2CH_2CO_2^- \qquad (37)$$

The radicals formed in reactions 36 and 37 differ greatly in absorption spectrum (Figure 3) (25). This is probably because in

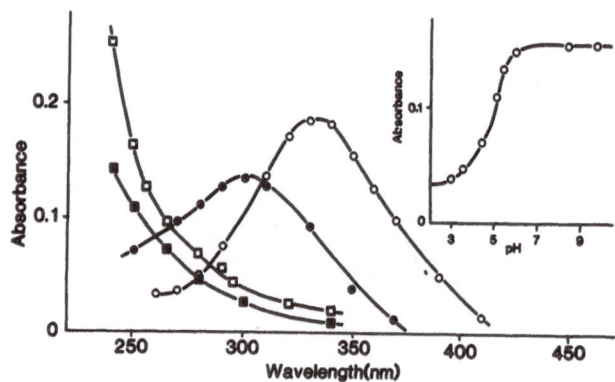

Figure 3. Absorption spectra of $CH_3CHCO_2^-$ (O), CH_3CHCO_2H (●), $\cdot CH_2CH_2CO_2^-$ (□) and $\cdot CH_2CH_2CO_2H$ (■). The inset shows the variation of absorbance of the α-radical with pH. From this curve the pK of the carboxyl group in the radical is 4.9, the same as in propionic acid itself.

the α radical the radical centre is conjugated with the carboxyl group whereas in the β radical it is not.

Alcohols and carbonyl compounds

The reducing radicals formed by OH attack on alcohols are of interest in view of their general utility in studying the pulse radiolysis of aqueous systems as well as for their own sake. Hydroxyl radicals are thought to abstract hydrogen atoms from α carbon-hydrogen, other carbon-hydrogen and oxygen-hydrogen bonds in that order of preference. The action of the α-radicals as reducing agents probably takes place through the donation of an electron to give a cation which stabilises by loss of a proton:-

$$RCHOH \longrightarrow [RCHOH]^+ \longrightarrow RCHO + H^+ \tag{38}$$

No such reaction takes place when the radical centre is on other carbon atoms, while the oxyl radical is an oxidising agent. The proportion of α-radicals have been found by measuring the amount of reduction of such added solutes as ferricyanide or tetranitromethane. The proportion of oxyl can be found by measuring the oxidation of iodide. Abstraction from other carbon-hydrogen positions is obtained by difference. Results of one set of determinations are given in Table 3 (26). It may be noted that

Table 3

Percentage of abstraction from various positions in alcohols

	α C-H	Other C-H	O-H
Methanol	93.0	-	7.0
Ethanol	84.3	13.2	2.5
1-Propanol	53.4	46.0	<0.5
2-Propanol	85.5	13.3	1.2
1-Butanol	41.0	58.5	<0.5
t-Butanol	-	95.7	4.3
Ethylene glycol	100	-	<0.1
1,2-Propanediol	79.2	20.7	<0.1
2,3-Butanediol	71.0	29.0	<0.1

except for ethylene glycol (which has four α- carbon hydrogen bonds and no other carbon hydrogen bond) the α-radical is always formed in a yield which is significantly less than 100%

α-Hydroxyalkyl radicals (in their neutral or anionic forms, according to pH) can usually also be prepared by allowing hydrated electrons to react with carbonyl compounds. This is not possible for the hydroxymethyl radical though, as formaldehyde exists as the hydrate in aqueous solution (methylene hydroxide, $CH_2(OH)_2$). β-Hydroxyalkyl radicals are produced when OH adds to olefins (see above). The β-hydroxyalkyl radicals are able to abstract hydrogen from the α- position of alcohols to give the α-hydroxy radical, but this is much too slow to be significant in pulse radiolysis ($k \sim 10^2 M^{-1}s^{-1}$).

The absorption spectrum of α-hydroxylalkyl radicals changes in alkaline solutions owing to dissociation of the hydroxyl hydrogen:-

$$R\dot{C}HOH \rightleftharpoons R\dot{C}HO^- + H^+ \qquad (39)$$

pK values obtained from absorption and conductivity measurements are in the region of 10.5-12.5, depending on the radical.

Hundreds of rate constants have been determined for reactions of α-hydroxyalkyl radicals with organic and inorganic compounds.

A few examples (all for the neutral form of the radicals) are given
in Table 4.   Although most reactions are electron transfers, the

Table 4

Rate constants for reactions of α-hydroxyalkyl radicals ($M^{-1}s^{-1}$)

	$\cdot CH_2OH$	$(CH_3)_2\dot{C}OH$
$Co(NH_3)_6^{3+}$	1.4×10^8	1.3×10^7
Cu^{2+}	1.1×10^8	4.5×10^7
$Fe(CN)_6^{3-}$	4.0×10^9	4.7×10^9
O_2	4×10^9	4×10^9
Zn^+	3×10^9	3×10^9
Ascorbic acid	$< 10^6$	1.2×10^6
Benzoquinone	5×10^9	5×10^9
Carbon tetrachloride	$< 10^6$	$10^9 - 10^9$
Cysteine	4.2×10^7	-
Methylene blue	3.4×10^7	4.4×10^9
2-Methyl-1,4-naphthoquinone	3.7×10^9	4.2×10^9
NAD^+	1.0×10^9	1.0×10^9

radicals are also able to react by addition, as for example in the
important case of reaction with oxygen:-

$$R\dot{C}HOH + O_2 \longrightarrow R\overset{O_2\cdot}{\underset{|}{C}}HOH \qquad\qquad (40)$$

- the evidence for which is similar to that given above for the
analogous reaction of alkyl radicals.   They are also able to add
to certain reduced forms of metal ions.

The radicals formed by action of OH on alcohols containing
two hydroxyl groups on adjacent carbon atoms (vic - glycols) are
able to undergo an acid-base catalysed water elimination giving a
more stable radical, e.g. (27):-

$$\cdot CHOH-CH_2OH \longrightarrow CHO-CH_2\cdot + H_2O \qquad\qquad (41)$$

Corresponding reactions of the general form:-

$$-COH - CHX - \longrightarrow -CO-CH- + H^+ + X^- \qquad (42)$$

can take place when X represents groups such as halogen, acetate or phosphate (cf. Reaction 35). Such reactions are important in the radiolysis of carbohydrates (28).

When hydroxyl radicals react with hydroxycarboxylic acids, they tend to abstract hydrogen from the carbon atoms to which the OH groups are attached. The radicals formed in this reaction would be expected to be the same as those formed by action of hydrated electrons on the corresponding carbonylcarboxylic acid. The simplest carbonylcarboxylic acid, glyoxylic acid, is in fact hydrated in aqueous solution, as $CH(OH)_2CO_2H$, but for higher homologues the two routes to the free radicals would be represented by equations such as those for lactic and pyruvic acids:-

$$OH + CH_3CH(OH)CO_2^- \longrightarrow CH_3\dot{C}(OH)CO_2^- + H_2O \qquad (43)$$

$$e_{aq}^- + CH_3COCO_2^- \xrightarrow{+H^+} CH_3\dot{C}(OH)CO_2^- \qquad (44)$$

The pK values of the OH groups are somewhat lower than those of α-hydroxyalkyl radicals (29). Using several different pairs (lactic/pyruvic, malic/oxalacetic, tartronic/ketomalonic, α-hydroxybutyric/α-ketobutyric, α-hydroxyglutaric/α-ketovaleric) it has been found that in every case radicals formed in the two ways react with duroquinone with the same rate constant (Table 5). The efficiencies are mostly close to 100% for the electron adducts to the carbonylcarboxylic acids, but less for the radicals formed by hydrogen abstraction from the hydroxycarboxylic acids. Low efficiencies are mainly attributable to the presence of alternative sites of attack by OH radicals (30).

Phenols

Hydroxyl radicals are strongly electrophilic, but not completely selective in their reactions. Accordingly they are able to add to phenols at any of the possible positions. The radicals of the dihydroxycyclohexadienyl type split off water in a reaction catalysed by acid or base to give resonance-stabilised phenoxyl radicals (31). The absorption spectra of the OH adduct to phenol and of the phenoxyl radical itself are shown in Figure 4.

Information about the position of addition can be gained from product analysis, and additional information has been obtained by pulse radiolysis of neutral N_2O-saturated solutions of phenol containing quinones and other oxidants (32). With a

Table 5

Rate constants and efficiencies for reaction with duroquinone of radicals formed by H- abstraction from hydroxycarboxylic acids or electron addition to carbonylcarboxylic acids (pH 7)

	$k, M^{-1}s^{-1}$	efficiency, %
$CH_3CH(OH)CO_2^-$	1.3×10^9	56
$CH_3COCO_2^-$	1.6×10^9	107
$^-O_2CCH_2CH(OH)CO_2^-$	7.7×10^8	48
$^-O_2CCH_2COCO_2^-$	8.2×10^8	107
$^-O_2CCH(OH)CO_2^-$	1.4×10^9	82
$^-O_2CCOCO_2^-$	1.2×10^9	98
$CH_3CH_2CH(OH)CO_2^-$	1.1×10^9	46
$CH_3CH_2COCO_2^-$	1.0×10^9	101
$^-O_2CCH_2CH_2CH(OH)CO_2^-$	1.1×10^9	50
$^-O_2CCH_2CH_2COCO_2^-$	1.1×10^9	·100

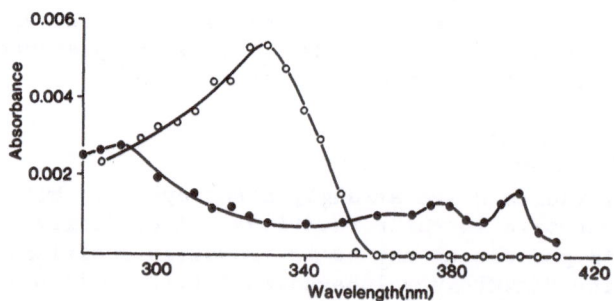

Figure 4. Absorption spectra in N_2O-saturated aqueous phenol, pH 6.2 O- after OH radicals have added to phenol, •- 400µs later by which time water has been eliminated to yield phenoxyl.

quinone with a one-electron reduction potential in the region +99 to -380mV, the semiquinone (in its anionic form) is produced in a yield about 80% of that of OH radicals. This is explicable if the o- and p-adducts to phenol are able to transfer electrons to quinone, e.g. according to:-

$$\text{(structures)} \qquad (45)$$

- but the adducts at the m- position and at the position occupied by the OH group cannot; or at least do so only with a low rate constant ($< \sim 10^7 \text{M}^{-1}\text{s}^{-1}$). With duroquinone and anthraquinone-2-sulphonate (one-electron reduction potentials at pH 7 of -240 and -380mV respectively) yields of semiquinone are only about 40% of that of OH radicals. This is consistent with the p-isomer reducing the quinone but the o-isomer not doing so with an appreciable rate constant. From product analysis and pulse radiolysis, the proportions of OH addition at the 1,2,3 and 4 positions are, respectively, 8%, 48%, 8% and 36%, showing that the hydroxyl substituent causes a distinct preference for attack at the ortho and para positions, consistent with its electron donating properties.

In acid or basic solution, phenoxyl radicals are formed at rates consistent with different isomeric adducts splitting off water at different rates. Combination of product determination with pulse radiolysis data enables the fastest reaction (rate-determining step reaction with H+, k $\sim 10^9 \text{M}^{-1}\text{s}^{-1}$) to be identified as elimination from the p- adduct:-

$$\text{(structures)} + H_2O \qquad (46)$$

Hydroxyl radicals add to hydroquinone as they do to phenol, and water loss gives the corresponding semiquinone radical (33). These will be discussed in the next contribution.

Phenoxyl radicals are not only formed through action of OH on phenols, but also by the action of other oxidising radicals such as Br_2^-. Rates are significantly higher in alkaline solution where the reaction is simple transfer of an electron from the phenolate anion. Phenoxyl radicals can also be formed when OH radicals attack benzene substituted with groups such as nitro or methoxy (34). In such cases the OH adduct at the position containing the substituent is able to eliminate nitrite, methanol etc. yielding phenoxyl. Another route to phenoxyl is provided by the rearrangement of anionic p-hydroxyphenyl radicals formed by action of hydrated electrons on p-bromophenol in alkaline solution (35):-

$$e_{aq}^- + \underset{Br}{\bigcirc}O^- \rightarrow \dot{\bigcirc}O^- + Br^- \qquad (47)$$

$$\bigcirc O^- + H_2O \rightarrow \bigcirc O\cdot + OH^- \qquad (48)$$

Phenoxyl radicals are powerful oxidising agents.

Sulphur Compounds

Pulse radiolysis of N_2O-saturated solutions of thiols, RSH, produces a species with an absorption maximum $\sim$ 400nm. It might have been thought that this would be the radical RS· since OH radicals are well known to abstract H from S-H groups. However the intensity of the absorption increases with both RSH concentration and pH over a wide range. For example with cysteamine ($NH_2CH_2CH_2SH$) at pH 7 the absorption produced in 3×10^{-2}M solutions is about ten times that in 10^{-4}M solutions. At a constan cysteamine concentration of 10^{-3}M, the absorption at pH 8 is about six times that at pH 6. The explanation is that RS· radicals formed in the first instance are able to come into equilibrium with the ionised form of the thiol (36):-

$$RS\cdot + RS^- \rightleftharpoons RSSR\cdot^- \qquad (49)$$

The same species can be formed by reaction of hydrated electrons with the corresponding disulphide, RSSR. Equilibrium constants for Reaction 49 are typically in the region of $1-3 \times 10^{-2}$M^{-1}, depending on the thiol.

Molecules which contain two -SH groups capable of being converted into an internal - S-S- group, e.g. lipoic acid:-

$$\underset{SH\ SH}{\overset{CH_2}{\underset{CH_2}{\bigwedge}}}CH-(CH_2)_nCO_2H \qquad \underset{S-S}{\overset{CH_2}{\underset{CH_2}{\bigwedge}}}CH-(CH_2)_nCO_2H \quad (50)$$

- can form intramolecular RSSR·$^-$ radicals on one-electron oxidation of the reduced form or one-electron reduction of the oxidised form (37). In acid solutions the RSSR·$^-$ of lipoic acid protonates, pK $\sim$ 5.5. Protonation of intermolecular $RSSR\cdot^-$ is followed immediately by dissociation into $RS\cdot$ and RSH.

Hydroxyl radicals add to organic sulphides, R_2S, and after a complex sequence of reactions give rise to relatively stable radical cations, $R_2S\cdot^+$, which are in equilibrium with the parent according to:-

$$R_2S\cdot^+ + R_2S \rightleftharpoons R_2SSR_2\cdot^+ \tag{51}$$

The $R_2S\cdot^+$ radicals have weak absorptions in the region of 300nm and the $R_2SSR_2\cdot^+$ have much stronger absorptions in the region of 500nm ($\varepsilon_{max} \sim 6000$ $M^{-1}cm^{-1}$). Similar species can be formed intra-molecularly from molecules containing two thioether groups. The radical cations $R_2SSR_2\cdot^+$ are formulated as containing three-electron bonds:-

$$\begin{array}{ccc} R & & R \\ \diagdown & + & \diagup \\ S & \therefore & S \\ \diagup & & \diagdown \\ R & & R \end{array}$$

- with two of the electrons being σ and one of them σ^*. Optical absorptions of a number of cations are explained in terms of $\sigma \rightarrow \sigma^*$ transitions (38), Cl_2^-, Br_2^- and I_2^- react with organic sulphides to form analogous species, e.g.:-

$$Br_2^- + R_2S \rightleftharpoons R_2S\text{-}Br\cdot + Br^- \tag{52}$$

- also considered to contain three-electron bonds (39). These are able to equilibrate with both the radical cations of Equation 51 as well as with halogen atoms, e.g.:-

$$R_2S\text{-}Br\cdot \rightleftharpoons R_2S + \cdot Br \tag{53}$$

$$Br\cdot + Br^- \rightleftharpoons Br_2^- \tag{54}$$

Hydroxyl radicals appear to both add and transfer electrons to disulphides, RSSR, but Br_2^-, Tl^{2+} and several other oxidising species react more simply, giving $RSSR\cdot^+$, which has a maximum absorbance at about 400nm. $RSSR\cdot^+$ does not react with R_2S but on the contrary the uncomplexed radical cation $R_2S\cdot^+$ appears to react according to (40):-

$$R_2S\cdot^+ + RSSR \longrightarrow R_2S + RSSR\cdot^+ \tag{55}$$

As well as reactions such as those mentioned here, numerous rate constants for protonations and deprotonations of the various sulphur radicals have been determined, as well as rate constants for radical-radical reactions. There is a substantial body of knowledge on the final stable products in irradiated aqueous

solutions of sulphur compounds, and the pulse radiolysis results
now enable increasingly more valid reaction mechanisms to be put
forward to account for these.

     Some of the work on which this article is based has been
supported by grants from the Cancer Research Campaign and the
Medical Research Council. The author is grateful to his colleagues
especially Drs. J. Butler and E.J. Land, for helpful discussions.

REFERENCES

1. Dorfman, L.M. and Adams, G.E.: 1973 "Reactivity of the hydroxyl
      radical in aqueous solutions" NSRDS-NBS 46, U.S. Department
      of Commerce, Washington DC.
2. Farhataziz and Ross, A.B.: 1977 "Selected specific rates of
      reactions of transients from water in aqueous solution
      III. Hydroxyl radical and perhydroxyl radical and their
      radical anions" NSRDS-NBS 59, U.S. Department of Commerce,
      Washington DC.
3. Ross, A.B. and Neta, P.: 1979 "Rate constants for reactions
      of inorganic radicals in aqueous solution" NSRDS-NBS 65,
      U.S. Department of Commerce, Washington DC.
4. Hart, E.J. and Anbar, M.: 1970 "The hydrated electron", Wiley-
      Interscience, New York, N.Y.
5. Anbar, M., Bambenek, M. and Ross, A.B.: 1973 "Selected
      specific rates of reactions of transients from water in
      aqueous solution. I. Hydrated electron" NSRDS-NBS 43,
      U.S. Department of Commerce, Washington DC.
6. Ross, A.B.: 1975 "Selected specific rates of reactions of
      transients from water in aqueous solution: Hydrated
      electron, supplemented data" NSRDS-NBS 43, Supplement,
      U.S. Department of Commerce, Washington DC.
7. Neta, P.: 1972, Chem.Rev. 72, pp. 533-543.
8. Anbar, M., Farhataziz and Ross, A.B.: 1975 "Selected specific
      rates of reactions of transients from water in aqueous
      solution. II. Hydrogen atom", NSRDS-NBS 51, U.S.
      Department of Commerce, Washington DC.
9. Capellos, C. and Bielski, B.H.J.: 1972 "Kinetic systems",
      Wiley-Interscience, New York, N.Y.
10. Hayon, E. and Simic, M.: 1974, Accounts Chem.Res., 7, pp. 114-
      121.
11. Neta, P.: 1976, Adv.Phys.Org.Chem., 12, pp. 223-297.
12. Boag, J.W.: 1968, Trans.Faraday Soc., 64, pp. 677-685.
13. Swallow, A.J.: 1973, MTP International Review of Science,
      Organic Chemistry, Series One, 10, pp. 263-291.
14. Swallow, A.J.: 1978, Progr.Reaction Kinetics, 9, pp. 195-365.
15. Ross, A.B.: 1981, private communication.

16. Rabani, J., Pick, M. and Simic, M.: 1974, J.Phys.Chem., 78, pp. 1049-1051.
17. Soylemez, T. and von Sonntag, C.: 1980, J.Chem.Soc., Perkin II, pp. 391-394.
18. Buxton, G.V. and Green, J.C.: 1978, J.Chem.Soc.Faraday 1, 74, pp. 697-714.
19. Bhatia, K. and Schuler, R.H.: 1974, J.Phys.Chem., 78, pp. 2335-2338.
20. Christensen, H.C., Sehested, K. and Hart, E.J.: 1973, 77, J.Phys.Chem., pp. 983-987.
21. Sehested, K., Holcman, J. and Hart, E.J.: 1977, 81, J.Phys. Chem., pp. 1363-1367.
22. Packer, J.E., Willson, R.L., Bahnemann, D. and Asmus, K.-D.: 1980, J.Chem.Soc. Perkin II, pp. 296-299.
23. Koster, R. and Asmus, K.-D.: 1977, Z.Naturf., 26B, pp. 1108-1116.
24. Klever, H. and Schulte-Frohlinde, D.: 1976, Ber.Bunsenges. Phys.Chem., 80, pp. 1259-1265.
25. Neta, P., Simic, M. and Hayon, E.: 1969, J.Phys.Chem., 73, pp. 4207-4213.
26. Asmus, K.;D. Mockel, H. and Henglein, A.: 1973, J.Phys.Chem., 77, pp. 1218-1221.
27. Bansal, K.M., Gratzel, M., Henglein, A. and Janata, E.: 1973, J.Phys.Chem., 77, pp. 16-19.
28. von Sonntag, C.: 1980, Adv.Carbohydrate Chem.Biochem. 37, pp. 7-77.
29. Simic, M., Neta, P. and Hayon, E.: 1969, J.Phys.Chem., 73, pp. 4214-4219.
30. Ahmad, M.S., Atherton, S.J. and Swallow, A.J.: unpublished.
31. Land, E.J. and Ebert, M.: 1967, Trans.Faraday Soc., 63, pp. 1181-1190.
32. Raghavan, N.V. and Steenken, S.: 1980, J.Amer.Chem.Soc., 102, pp. 3495-3499.
33. Adams, G.E. and Michael, B.D.: 1967, Trans.Faraday Soc., 63, pp. 1171-1180.
34. O'Neill, P., Schulte-Frohlinde, D. and Steenken, S.: 1978, Discuss.Faraday Soc., 63, pp. 141-148.
35. Schuler, R.H., Neta, P., Zemel, H. and Fessenden, R.W.: 1976, J.Amer.Chem.Soc., 98, pp. 3825-3831.
36. Adams, G.E., McNaughton, G.S. and Michael, B.D.: 1967 in "The Chemistry of ionization and excitation", ed. Johnson, G.R.A. and Scholes, G., Taylor and Francis, London, pp. 281-293.
37. Hoffman, M.Z. and Hayon, E.: 1972, J.Amer.Chem.Soc., 94, pp. 7950-7957.
38. Asmus, K.-D.: 1979, Accounts Chem.Res., 12, pp. 436-442.
39. Bonifacic, M. and Asmus, K.-D.: 1980, J.Chem.Soc. Perkin II, pp. 758-762.
40. Bonifacic, M. and Asmus, K.-D.: 1976, J.Phys.Chem., 80, pp. 2426-2430.

APPLICATION OF PULSE RADIOLYSIS TO THE STUDY OF MOLECULES OF
BIOLOGICAL IMPORTANCE

A.J. Swallow

Paterson Laboratories
Christie Hospital & Holt Radium Institute
Manchester, England

     Irradiation of suitable aqueous and non-aqueous systems
permits observation of numerous one-electron and other processes
which are of importance to the normal functioning of biological
systems as well as being fundamental to radiobiology. As well as
spectroscopic and kinetic data and values of pK, one-electron
reduction potentials can also be determined. Discussion is given
of quinones, carbohydrates (especially ascorbic acid), amino acids
and peptides, pyridine compounds, flavins, haem and haemoproteins
and vitamin B12. Pulse radiolysis can also be used to obtain
information about excited states, one especially important application being to measure triplet-triplet extinction coefficients.

INTRODUCTION

     The previous contribution has shown how pulse radiolysis can
be applied to aqueous solutions of organic compounds to provide
new information about the chemistry of organic free radicals.
Attention will now be focused on radicals which are specifically
of biological interest. It will also be shown how pulse radiolysis
can be applied to non-polar solutions to provide information about
the chemistry of biologically important molecules in a different
kind of unstable form, the electronically excited state. Recent
review articles (1,2) have dealt with aspects of these topics.

ONE-ELECTRON REDUCTION POTENTIALS

     Electron transfer is one of the most common reactions of
organic free radicals, and is of special interest in biology. If

J. H. Baxendale and F. Busi (eds.),
The Study of Fast Processes and Transient Species by Electron Pulse Radiolysis, 317–345.
Copyright © 1982 by D. Reidel Publishing Company.

the reducing power of the radical species donating the electron is
comparable to that of the species formed by acceptance of the
electron, an equilibrium will be set up, which for the simplest
case where protons are not involved could be expressed by:-

$$A^{\cdot -} + B \rightleftharpoons A + B^{\cdot -} \tag{1}$$

When the concentrations of the species in Reaction 1 reach equilibrium values, we have:-

$$K = \frac{[A][B^{\cdot -}]}{[A^{\cdot -}][B]} \tag{2}$$

Providing the rates of the forward and back reactions in Equation
1 are high, delivery of a pulse can often result in the achievement
of equilibrium before the radicals disappear by mutual reaction.
With species having appropriate absorption spectra, optical
measurements can then be used to measure K.

The equilibrium constant may be expressed in terms of the
standard one-electron reduction potentials of A and B:-

$$E(B/B^{\cdot -}) - E(A/A^{\cdot -}) = RTlogK \tag{3}$$

- where for potentials expressed in mV, the numerical value of RT
at 25°C is 59. Although two-electron reduction potentials are
known for numerous compounds, comparatively few accurate one-electron
reduction potentials are known. By setting up systems containing
reference compounds of known potential, measurements of equilibrium
by pulse radiolysis can be used to obtain values of one-electron
reduction potentials.

To measure the one-electron reduction of a compound such as
A or B by pulse radiolysis it is necessary to select, using
principles like those discussed in the previous contribution, the
highest concentration of A and B which will enable essentially all
OH radicals to be scavenged by 0.1-1M t-butanol, isopropanol or
formate additionally present in the solution. It helps to observe
equilibration if the relative concentrations of the two solutes
are such that in the first instance most of the reducing radicals
(e_{aq}^{-}, $CH_3\dot{C}HOH$, CO_2^{-}) react with one of them, say A. A typical
solution might contain 0.1M t-butanol, 10^{-4}M A and 10^{-5}M B.

In order to get an accurate measurement of equilibrium it is
desirable that the equilibrium value of $[A^{\cdot -}]$ should be comparable
to the equilibrium value of $[B^{\cdot -}]$. Achievement of this and the
preceding conditions will ensure that the equilibration follows a
course like that given in Figure 1 (3). If [A] were to be selected
as 10[B], it follows from Equations 2 and 3 that the differences

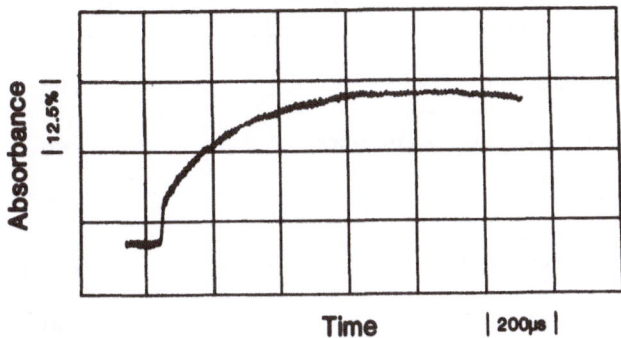

Figure 1. Oscilloscope trace showing the effect of delivering a
short pulse to a solution containing 0.1M t-butanol, 1.25×10^{-4}M
oxygen and 2×10^{-5}M 2,5-dimethylbenzoquinone (pH 7). Most of the
hydrated electrons react with oxygen, but a small fraction react
with the quinone causing a rapid increase in absorbance at 430nm.
The absorbance increases still further as the equilibrium
$O_2^- + Q \rightleftharpoons O_2 + Q^{\cdot-}$ becomes established.

between the one-electron potentials of the unknown and the
reference compound would have to be in the region of 59mV to get
$[A^{\cdot-}] \simeq [B^{\cdot-}]$ at equilibrium. If $[A]$ were to be selected as 100
$[B]$, the difference would have to be $\sim$ 118mV. Table 1 includes
some reference compounds which could be considered. It should be
noted that the radicals formed by OH attack on t-butanol or
isopropanol add to oxygen rather than transfer electrons to it,
(though CO_2^- transfers an electron to oxygen) so care should be
taken when using oxygen in solutions with t-butanol or isopropanol
as OH scavenger.

The pulse dose should introduce a concentration of $A^{\cdot-}$ which
is 10% or less of $[B]$. The largest change in absorbance will be
seen at the wavelength where there is the biggest difference
between $\varepsilon(A^{\cdot-}) + \varepsilon(B)$ and $\varepsilon(A) + \varepsilon(B^{\cdot-})$.

In practice, trial and error will normally be needed to
discover the best possible conditions. Concentration and dose
should be varied within the permissable limits to check the
validity of the measurement. Ideally more than one reference
compound should be used.

One-electron reduction potentials will vary with pH whenever
a hydrogen ion enters into the reaction. For example the existence
of the association:-

Table 1

One-electron reduction potentials at pH 7 (mV, vs. NHE)

Benzoquinone	+99
2,5-Dimethylbenzoquinone	-66
Oxygen	-155[*]
5-Nitropyridine	-191
Duroquinone	-240
p-Nitroacetophenone	-355
Anthraquinone-2-sulphonate	-380
Misonidazole	-383
p-Nitrobenzoate	-410
Paraquat	-450
Nitrobenzene	-486
5-Nitrouracil	-527
1,1'-Butylene-2,2'-bipyridylium	-635
NAD$^+$	-930

* - where $[O_2]$ is expressed in molar. If
$[O_2]$ is expressed in atmospheres the value
is -330mV.

$$A^{\cdot-} + H^+ \rightleftarrows AH^\cdot \tag{4}$$

will pull the Equilibrium 1 over to the left at pH values near the pK of $AH^\cdot$ and below. At pH 0 the one-electron reduction potential of A, $E_O(A,H^+/AH^\cdot)$ is related to $E(A/A^{\cdot-})$ by:-

$$E_O(A,H^+/AH^\cdot) = E(A/A^{\cdot-}) - RT\log K' \tag{5}$$

- where K' is the dissociation constant of $AH^\cdot$. The one-electron reduction potential $E(A/A^{\cdot-})$ is related to the two-electron reduction potential $E(A/A^{2-})$ by:-

$$E(A/A^{\cdot-}) + E(A^{\cdot-}/A^{2-}) = 2E(A/A^{2-}) \tag{6}$$

- where $E(A^{\cdot-}/A^{2-})$ is the one-electron reduction potential of the radical $A^{\cdot-}$, i.e. the second one-electron reduction potential of A. Such potentials can also be determined pulse radiolytically (4).

Quinones

"Semiquinones", intermediate between quinones and hydroquinones, have been classical free radicals for about fifty years. They can be prepared radiolytically by addition of OH to hydroquinones followed by acid-base catalysed elimination of water (5), a reaction which is analogous to the production of phenoxyl by action of OH on phenol as discussed in the previous contribution. Similarly OH adducts to methoxylated phenols or nitrophenols are able to eliminate methanol or nitrite respectively to yield semiquinones. Semiquinones are most simply prepared by reaction of e^-_{aq} or other reducing radicals with quinones. Some quinones are insoluble in water, and alcoholic solvents have been employed in which semiquinones are formed by an analogous reduction by solvated electrons and alcohol radicals. The physical chemistry of semiquinones as elucidated by pulse radiolysis and flash photolysis has recently been reviewed (6) and a bibliography is available (7).

Semiquinones have strong optical absorptions in easily monitored parts of the spectrum. In acid solutions where they are present in their uncharged forms, benzosemiquinone and its methyl substituted derivatives exhibit absorption peaks with maxima near 410nm, $\varepsilon_{max} = 4600$ $M^{-1}cm^{-1}$, independent of the number of methyl groups. In less acid, neutral or alkaline solutions, where they are anionic, values of λ_{max} shift to about 430nm and ε_{max} increases to about 6700 $M^{-1}cm^{-1}$. Similar shifts to increased wavelengths and higher extinction coefficients are seen for naphthoquinones in going to the anionic form, while with anthraquinones single peaks near 385nm, $\varepsilon_{max} \sim 1200$ change in the anionic form to split peaks at about 400 and 500nm with ε_{max} for both about 8000 $M^{-1}cm^{-1}$.

The differences in optical absorption have been used to determine the pK values of semiquinones. To a close approximation the pK values of benzosemiquinone and its methyl substituted derivatives increase with methyl substitution according to the formula:-

$$pK = 4.1 + 0.25n \tag{7}$$

- where n is the number of methyl groups. This is consistent with the electron donating properties of CH_3. The pK values of the naphtho- and anthra-semiquinones so far examined all lie between the extremes 3.2 to 5.4. Such values are lower than found for the one-electron reduced forms of simple aliphatic and aromatic ketones because of the high resonance stability of the semiquinone anion, which can be seen from the formula as represented:-

Ubisemiquinones and plastosemiquinones are of special biological importance. They are insoluble in water, but their pK values in methanol or aqueous isopropanol/acetone solution have been estimated to be in the region of 5-6 (6). It was in the reaction of O_2^- with quinones and the back reaction of semiquinones with oxygen that oxidation-reduction equilibria (Reaction 1) were first observed in the pulse radiolysis of aqueous solutions (8) although equilibria between aromatic anions in alcohol solutions had been seen previously (9). Some of the one-electron reduction potentials determined (or verified) by pulse radiolysis are included in Table 1. For benzoquinone and its methyl-substituted derivatives, the reduction potential $E(Q/Q^{\cdot-})$ varies with number of methyl groups, n, according to:-

$$E(Q/Q^{\cdot-}) = 100 - 85n \tag{8}$$

- presumably for the same reason as the systematic variation of pK (Equation 7). The one-electron reduction potentials of quinones can be combined with known two-electron reduction potentials to obtain one-electron reduction potentials for the reduction of the semiquinone to the hydroquinone. If the pKs of the species are known, reduction potentials can be calculated for any pH.

Rate constants for reactions of semiquinone free radicals vary up to a maximum of $\sim$ 2-3 x $10^9 M^{-1}s^{-1}$. When the semiquinone is reacting with a reactant whose one-electron reduction potential is identical to that of the quinone, the rate constants of both forward and back reactions are expected to be 1-5 x $10^7 M^{-1}s^{-1}$.

Interestingly, hydroxyl radicals seem to react with quinones to give semiquinones. For instance 2-methyl-1,4-naphthoquinone

forms a species having a similar absorption spectrum and pK to the
electron adduct, though the yield appears less than 100% (10).
This could be explained if addition occurs both at the benzene and
the quinone ring, addition to the quinone ring being followed by
keto-enol tautomerisation:-

(9)

Clearly the formation of a semiquinone from a quinone does not
necessarily mean that an electron has been transferred. The
reaction could be addition, as formulated here.

Carbohydrates: Ascorbic Acid

     In recent years radiation chemistry has produced striking
advances in the free radical chemistry of carbohydrates, but this
is one field where analysis of products after γ-radiation has been
more generally informative than pulse radiolysis (11). This is
because the products are diverse and can be separated and identified by skilful application of the techniques of carbohydrate
chemistry, whereas the different carbohydrate radicals do not in
general have characteristic absorption spectra. However pulse
radiolysis in conjunction with other techniques has made significant contributions to the free radical chemistry of ascorbic acid.
The parent ascorbic acid can exist in an oxidised form, dehydro-
ascorbic acid, as well as in the normal form (reduced) which has
pK = 4.17. The intermediate radical form has been known since
about 1960 from purely chemical experiments.

          ascorbic acid   dehydroascorbic acid

     The most straightforward studies have been made by forming
the radical from ascorbate in neutral or alkaline solutions. It
can be formed by action of OH or Br_2^- etc. Hydroxyl radicals may
add to the double bond after which H_2O is eliminated. Any such
process with Br_2^-, $(SCN)_2^-$ etc. is over much faster, so the
reaction can be regarded as electron transfer (12). Phenoxyl
radicals also oxidise ascorbate to the radical, also attributable
to simple electron transfer. A reaction of this type between the

phenoxyl radical of vitamin E and ascorbate has also been demonstrated in an air-saturated solution containing water, isopropanol
and carbon tetrachloride, and could be important in the antioxidant
role of the substances in biological systems (13). Perhaps
surprisingly, isopropanol radicals appear able to oxidise ascorbate
although they are normally regarded as powerful reducing agents
(14), but the rate constant is only $1.2 \times 10^6 M^{-1}s^{-1}$ (pH 6). The
reaction can be regarded as the abstraction of a hydrogen atom:-

$$(CH_3)_2\dot{C}OH + AH^- \longrightarrow (CH_3)_2CHOH + A^{\cdot -} \tag{10}$$

Whatever the mechanism, the net result in every case is the same,
the net removal of a hydrogen atom to form a highly resonance-
stabilised free radical:-

The same radical can also be formed by action of e_{aq}^- on dehydro-
ascorbic acid (15).

The ascorbate radical has an absorption peak at 360nm, well
separated from that of its parent. Although extinction coefficients
appear to vary, the shape of the spectrum as prepared by action of
OH does not change in going to more acid solutions until extremes
of acidity are reached. The most reliable value of the pK of the
radical has been determined by e.s.r. to be -0.45. The changes in
extinction are probably connected with complexities in the action
of OH on ascorbic acid in its undissociated form. The one-electron
reduction potential of the ascorbate radical has been determined
by equilibrating it with dihydroxybenzenes in alkaline solution.
Using literature pK values for ascorbic acid of 4.2 and 11.5 in
conjunction with the pK of the radical, -0.45, the one-electron
reduction potential of the ascorbate radical becomes +300mV at
pH7 and +990mV at pH 0 (4).

In the absence of any other species, ascorbate radicals
react with each other, presumably by disproportionation. At pH
values above 8 the rate constant is only $2k \sim 10^5 M^{-1}s^{-1}$ (15).

Amino Acids and Peptides

The importance of amino acids and peptides is that they are
the building blocks of proteins, the effect of radiation on which
is of interest in radiobiology. In contrast to many of the other
free radicals discussed here, amino acid and peptide radicals have
little to do with most of the normal biological functions of the

parent material but there are important exceptions as for instance
in the special cases of high potential copper proteins and systems
in which hydrogen peroxide is involved.

Hydroxyl radicals react with glycine, $NH_3^+CH_2COO^-$, to form
the radical $NH_2\dot{C}HCOOH$ which dissociates at the carboxyl group with
pK = 6.6 (16). Like the radicals formed from α-hydroxycarboxylic
acids, but unlike those from simple aliphatic carboxylic acids,
the glycine radical appears to be a good reducing agent. This is
probably because loss of an electron enables it to give a relatively
stable imino acid, NH = CHCOOH (cf. the formation of pyruvic acid,
$CH_3COCOOH$, when the lactic acid radical, $CH_3\dot{C}(OH)COOH$, acts as
reducing agent).

The cyclic peptides provide good models for the polypeptide
chain of proteins (17). Abstraction of H atoms by hydroxyl
radicals gives radicals which have the interesting property of
appearing to dissociate at the peptide hydrogen. For example
glycine anhydride gives:-

The reason for this unexpectedly low pK could be that in its
anionic form the radical is better represented by resonance-
stabilised structures such as:-

Hydroxyl radicals tend to attack peptides possessing terminal
amino groups in the uncharged form, $-NH_2$, by abstraction of H from
the amine group itself or from the neighbouring C-H. Protonation
of the group strongly deactivates these positions towards OH
attack, which then takes place at C-H groups further along the
peptide chain.

The high reactivity of aromatic groups such as benzene and
indole towards OH radicals means that for amino acids containing
such residues these groups are attacked as well as other parts.

For example OH adds to the benzene ring in phenylalanine and
tyrosine, in the latter case addition being followed by elimination
of H_2O to form the resonance stabilised phenoxyl radical as with
phenol (18):-

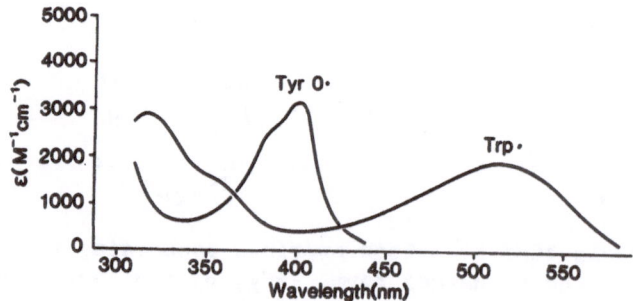

Likewise the action of OH on sulphur-containing amino acids is
dominated by the sulphur group.

The radical ions Br_2^- and $(SCN)_2^-$, the uncharged azide radical
$N_3^.$, and several other oxidising species are much more selective in
their action than is OH (19,20). In neutral solution, Br_2^-,
$(SCN)_2^-$ and $N_3^.$ all react rapidly with tryptophan ($k \sim 10^9 M^{-1}s^{-1}$),
and also, but much less rapidly, with tyrosine. They are
relatively unreactive to most other amino acids. In alkaline
solution they react with tyrosine ($TyrO^-$) at rates which are
comparable to those for reaction with tryptophan. The mechanism
can be regarded as simple electron transfer. The tryptophan
radical protonates with a pK of 4.3. The spectra of the radicals
formed from tryptophan and tyrosine at neutral pH is shown in
Figure 2 (20).

Figure 2.   Absorption spectrum of free radicals prepared by action
of azide radicals on tryptophan and tyrosine (neutral pH).

The radicals formed by one-electron oxidation of tryptophan

do not react rapidly with tyrosine and the radicals formed by one-
electron oxidation of tyrosine do not react rapidly with tryptophan.
Yet when the tryptophan residue is oxidised in a peptide which
also contains tyrosine, an intramolecular process takes place
resulting in a net transfer of the electron-deficient site (21):-

$$\text{Trp}^{\cdot} - \text{TyrOH} \longrightarrow \text{TrpH} - \text{TyrO}^{\cdot} \qquad (13)$$

The mechanism could be envisaged as electron or positive charge
transfer (hole transfer), in either case accompanied by proton
transfer, or as hydrogen atom transfer. The process can also take
place in proteins (22) and may have a bearing on attempts to use
selective free radical attack as a probe for the active site in
enzymes. It is curious that the reaction can be quite effective
when the two residues are held together in a peptide or protein,
and yet is ineffective in free solution. A contributing factor
must be that the electron or charge transfer process itself:-

$$\text{Trp}^{\cdot} + \text{TyrOH} \longrightarrow \text{Trp}^{-} + \text{TyrOH}^{\cdot +} \qquad (14)$$

- is strongly disfavoured (it would take place in the opposite
direction). It is also relevant that the process is greatly
speeded up in acid or alkaline solutions (23). However a fully
plausible mechanism has yet to be established.

Pyridine Compounds

     The pyridine nucleotides are essential components of many
biological oxidation-reduction processes. Their function is to
undergo a two-electron change, which for the oxidised form of
nicotinamide adenine dinucleotide (NAD^+) formally consists of the
addition of an electron and a hydrogen atom at the pyridine ring:-

$$(15)$$

In biological systems this process could take place by the direct
transfer of a hydride ion or by two separate one-equivalent steps.
Amongst other pyridine compounds of biological importance are the
bipyridylium salts, e.g. paraquat (methyl viologen):-

- whose mode of action is believed to involve simple one-electron
reduction to the radical form. The free radical forms of pyridine
compounds can conveniently be prepared by action of e_{aq}^-, CO_2^- or
$(CH_3)_2COH$ on the oxidised form of the compound, and in the case of
the pyridine nucleotides have also been made by action of Br_2^- and
other oxidising radicals on the reduced form.

Pyridinyl radicals possess characteristic absorptions in the
visible region of the spectrum. Using several simple pyridinium
compounds, additional absorption bands have also been found in the
infra-red, for example at 900nm for the 1-methyl-3-carbamidopyridir
radical. These are attributed to the promotion of the odd electror
to the next higher level as shown on the left hand diagram of
Figure 3 (24). Now pyridinium iodides are known to possess two

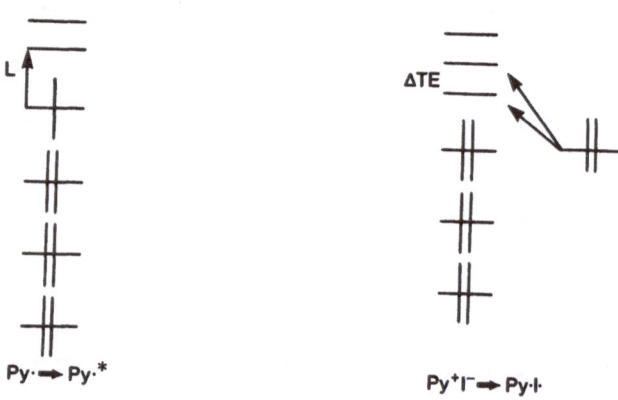

Figure 3. Excitation of the odd electron in pyridinyl radicals
(left) compared with transitions due to transfer of an electron
from iodide to pyridinium in pyridinium iodides (right).

charge-transfer bands attributable to promotion of the iodide
electron into two different levels. One of these levels is the
one occupied by the odd electron in the radical. The other is the
one to which the odd electron is promoted when the radical absorbs
infra-red light (right hand side of Figure 3). These assignments
demand that there should be a linear relationship between the
energies of the infra-red absorption bands of pyridinyl radicals
and the difference between the energies of the two charge transfer
bands in the corresponding iodide. This is indeed found (Figure
4).

The visible absorption bands of pyridinyl radicals have been
employed to measure pK values for numerous pyridinyl radicals.

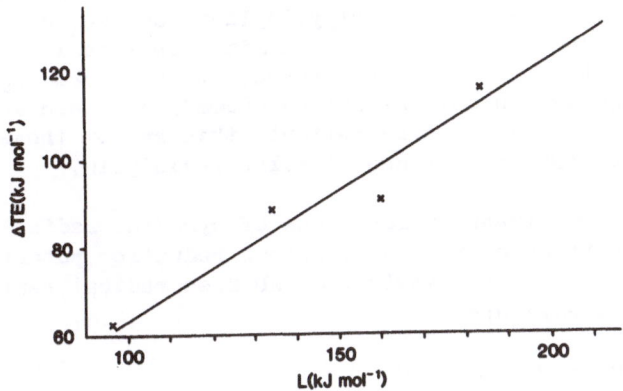

Figure 4. Linear relationship between transition energies in
pyridinyl radicals (L) and differences in transition energies in
pyridinium iodides (ΔTE).

Those for 1-substituted-nicotinamide and -isonicotinamide radicals
are of special interest, those from the former being significantly
more acid than those of the latter, e.g. the 1-cyclohexylnicotinamide
radical has pK = 1.3 and 1-cyclohexylisonicotinamide has pK = 2.1
(25). These pKs are attributable to protonation at the amide
oxygen:-

$$\text{(16)}$$

$$\text{(17)}$$

One-electron reduction potentials (pH 7) of numerous
bipyridylium salts are already known with certainty, and can be
employed as reference standards in determinations of one-electron

reduction potentials by pulse radiolysis. The most negative one
determined without pulse radiolysis is -640mV. A new one, for
1,1-butano-4,4'-dimethyl-2,2'-bipyridylium has been determined as
-735mV (26). The 1-methylisonicotinamide positive ion has -770mV
(27). Using these values as reference, an even more negative
value (average of two determinations -930mV) has been determined
for NAD^+. This value is more reliable than any of those previously
estimated for NAD^+ without use of pulse radiolysis.

The electron transfer reactions of pyridine radicals can be
understood in terms of the one-electron reduction potentials of the
parent compounds. For example the NAD free radical rapidly transfe
an electron to oxygen:-

$$NAD\cdot + O_2 \longrightarrow NAD^+ + O_2^- \tag{18}$$

In accordance with the parent one-electron reduction potential at
pH 7 (-450mV), the paraquat radical also rapidly transfers an
electron to oxygen. The one-electron reduction potential of O_2^-
(E O_2^-, $2H^+/H_2O_2$) is +865 at pH 7 and accordingly the paraquat
radical will also rapidly reduce O_2^-:-

$$PQ^{\cdot +} + O_2 \longrightarrow PQ^{2+} + O_2^- \tag{19}$$

$$PQ^{\cdot +} + O_2^- \xrightarrow{2H^+} PQ^{2+} + H_2O_2 \tag{20}$$

The formation of O_2^- from the paraquat radical is important as it
provides the key to explaining the phytotoxic action of this
compound (28).

Many years ago it was shown by enzymic and other tests that
NAD free radicals dimerise rather than disproportionate when they
react together:-

$$2NAD. \longrightarrow (NAD)_2 \tag{21}$$

1-methylnicotinamide radicals also dimerise as shown by the fact
that no protons are taken up during their mutual reaction, as
would have been expected if the oxidised and reduced forms were to
be formed (29):-

$$2Py\cdot \longrightarrow (Py)_2 \tag{22}$$

$$2Py\cdot + H^+ \nrightarrow Py^+ + PyH \tag{23}$$

In neutral solutions, 1-alkylisonicotinamide radicals are much
more long-lived than 1-alkylnicotinamide radicals, probably
because of the extra resonance possibilities afforded by their
structure (cf. left hand radical of equations 16 and 17). Their

mutual reaction proceeds increasingly more rapidly as the pH
decreases, owing to mutual reaction between the acidic and neutral
forms, which in this case is a disproportionation. Another
reaction of note is the addition of pyridinyl radicals to t-butanol
radicals which takes place in pulse radiolysis of pyridinium
compounds when t-butanol is used as OH scavenger:-

$$\text{(24)}$$

The product of the reaction resembles dihydropyridines and the
dimers in having an absorption spectrum with a peak near 340nm.
It resembles the dihydropyridines in exhibiting fluorescence
under UV light (39). The dimers do not fluoresce, probably
associated with the configuration being a folded one, with the two
rings close together.

The bipyridyl radicals are very stable in the absence of
oxidants, and this has permitted their reaction with cytochrome
P450 to be studied in an unusual way. Instead of using the pulse
to produce a small concentration of radicals in the presence of
a much higher concentration of cytochrome P450, as would normally
be done, a very large pulse dose was given, producing a concentration of radicals about ten times higher than that of the P450.
This still permitted the reaction to follow pseudo-unimolecular
kinetics, but with the solute rather than the radicals being
totally consumed. The radicals were found to reduce cytochrome
P450 with $k = 2.6 \times 10^7 M^{-1}s^{-1}$ and its camphor complex with
$k = 1.5 \times 10^6 M^{-1}s^{-1}$ (30).

Flavins

The 10- substituted -7,8-dimethylisoalloxazines, e.g.
riboflavin:-

- are known as flavins. Like the half-reduced quinones their half-
reduced forms are classical free radicals, fitting into the group
of resonance-stabilised radicals known as semiquinones. Flavin
semiquinones may be involved in the mode of action of flavin-
containing systems such as certain oxidising and dehydrogenating
enzymes as well as in certain photobiological processes. A great
deal is known about the chemistry of flavin semiquinones, much of
it obtained from purely chemical systems and from photochemistry,
but the unique potentialities of pulse radiolysis enable significan
contributions to be made.

The semiquinone forms of riboflavin have been produced from
riboflavin itself over an extremely wide range of acidities, i.e.
in the range $H_O = -3$ to pH 13, using formate to scavenge OH
radicals. Analysis of the changes in absorbance was consistent
with spectral changes in the parent material with pK values at
-8, 0.25 and 10.05 and in the semiquinone with pK at 2.3, 8.3 and
possibly somewhere else in the range 9-13. Measurements of
changes in absorption at $H_O - 1.1$ and pH 5.1 and 11.4 has enabled
the spectra of the various protonated forms to be established.
The decay of the semiquinone at pH 5.1 is independent of ionic
strength, so that the state of protonation of the various forms
could be specified (31). An interesting feature of the redox
chemistry of riboflavin is that in strongly acid solutions (e.g.
pH - 1.1) the equilibrium between the fully oxidised and fully
reduced forms and the semiquinone is in favour of the semiquinone
so that radiolysis leads to the formation of a completely stable
free radical. This had in fact been seen previously in experiments
with γ rays (32). The one-electron reduction potentials of riboflavin have been determined pulse radiolytically using duroquinone
and anthraquinone-2-sulphonate as reference compounds. The value
at pH 7 is $E_7^{\frac{1}{2}} = -292mV$ and the variation with pH has been found
to agree with that expected from the three pK values of the
parent riboflavin and two pKs of its semiquinone (33).

By using the separate molecules adenosine monophosphate and
riboflavin it has been found that any electrons taken up by
adenine are rapidly transferred to riboflavin. Accordingly the
free radical chemistry of flavin adenine dinucleotide (FAD) is
that of the flavin moiety (31). Using a number of bipyridylium
compounds as standards, the one-electron reduction potential of
FAD at pH 7 has been found to be $E_7^{\frac{1}{2}} = -236mV$ (34).

The enzyme lipoamide dehydrogenase in its oxidised form
contains both -S-S- and FAD residues. When CO_2^- reduces it, there
is no sign of the absorption in the region of 400nm characteristic
of -S-S-$^{\cdot-}$, but changes characteristic of the flavosemiquinone are
seen. This appears to be the first time that the formation of
this flavosemiquinone form of this enzyme has been demonstrated
(35).

Pulse radiolysis can provide no information about the site of
the protons in the various protonated states. Using lumazine,
there is evidence that the radical formed by immediate one-electron
reduction of the uncharged form gives a different semiquinone
from one-electron reduction of the deprotonated form followed by
protonation. This can be explained by a (partial) scheme such
as:-

(25)

- where the semiquinone may exist in various resonance forms. The
essence of the scheme is that radicals can exist in different
tautomeric forms, whose existence can be demonstrated by preparing
them by different routes. It is obviously not easy to establish
every detail of such reactions. 5-Deazaflavin is interesting as
the parent molecule combines pyridine-like character (in the central
ring) with flavin-like character:-

One-electron reduction by e_{aq}^{-} produces a radical which like NAD$\cdot$
and similar pyridinyl radicals is regarded as a strong reductant:-

- whereas the tautomer protonated at position 5 instead of position 1 is considered to be a mild oxidant (37).

Haem and Haemoproteins

Many proteins concerned with processes such as electron transfer and oxygen transport contain iron coordinated within a porphyrin ring as prosthetic group. The haemin molecule (chloroprotoferrihaem IX) models this group. It is insoluble in neutral aqueous solution, but can be dissolved in alkaline solution, where it exists as a dimer with the two iron atoms connected by an oxo bridge. Hydrated electrons reduce this form to the ferro form in 100% yield. The ferro form is thought to be monomeric, and various steps attributable to the necessary intermediate processes can be identified (38).

Haemin and deuterohaemin exist in the monomeric form at neutral pH when sufficient alcohol is present. Using deuterohaemin it has been found that this form can also be reduced to the ferro form, using either hydrated electrons or isopropanol radicals, but no intermediate steps are seen, consistent with the relaxation of the coordination sphere proceeding with $k > 5 \times 10^5 s^{-1}$. By reducing it in a solution containing a carefully chosen concentration of, for example, CCl_4, the reduced compound can be exposed to chlorinated methyl radicals ($\cdot CCl_3$) also formed by reduction. When this is done, spectroscopic changes are seen which are consistent with the addition of the chlorinated methyl radicals to the iron atom of the deuterohaem at approximately the diffusion-controlled rate. The formation of such adducts could be important in the metabolism of carbon tetrachloride (39).

Another model, one step closer to certain haemoproteins, consists of haemin solubilised in an SDS micelle. In this condition, monomeric haemin is believed to be located at one side of the micelle, with the haem edge projecting towards the water. In alkaline solution, where the haemin is thought to be coordinated to H_2O and OH^-, reduction by hydrated electrons or ethanol radicals is followed by a small change in spectrum consistent with protonation:-

$$[HO\text{-}Fe(III)(por)\text{-}H_2O]^0 + e_{aq}^- \longrightarrow [HO\text{-}Fe(II)(por)\text{-}H_2O]^-$$
$$\downarrow \qquad\qquad (26)$$
$$[H_2O\text{-}Fe(II)(por)\text{-}H_2O]^0 + OH^-$$

At pH 4, reduction is a simple one step process as both oxidised and reduced forms are thought to be bisaqua at this pH (40). Superoxide does not reduce micellised haemin in alkaline solution, but appears to bring it into an equilibrium at pH 4 consistent with the haemin having a one-electron reduction potential at this pH of about -30mV (41).

Cytochrome-c is a stable, well-characterised, readily available haemoprotein containing one haem group per molecule. It has the simple function in biological systems of transferring electrons through shuttling between ferric and ferrous forms. Hydrated electrons, CO_2^-, several organic free radicals and O_2^- are all capable of reducing ferricytochrome-c to the ferro form with a yield which is close to 100%. The rate constants tend to be faster in neutral solutions than in alkaline solution. With neutral solutions the reaction is a simple reduction, but in alkaline solution, fast reduction is followed by slow intramolecular changes, taking place over several hundred milliseconds. It is known that oxidised cytochrome-c contains a methionine residue coordinated to the central ion atom when the solution is neutral, and that this becomes displaced in alkaline solutions (pK $\sim$ 9). Reduced cytochrome-c has the methionine residue coordinated at all pH values. Thus the effect of pH on reduction is most simply expressed by:-

$$
\begin{array}{ccc}
Fe^{3+}met & \overset{pK\ 9}{\rightleftharpoons} & Fe^{3+} \\
\big\downarrow e^- & & \big\downarrow e^- \\
Fe^{2+}met & \overset{slow}{\longleftarrow} & Fe^{2+}
\end{array}
\qquad (27)
$$

- though there are various complications in the kinetics. It seems most likely that the slow rate of the intramolecular process in alkaline solution is basically due to a requirement for the reduced protein to undergo a conformational change to enable the methionine to enter the coordination sphere of the ferrous ion (42).

Although most radical reducing agents reduce cytochrome-c more slowly as the pH increases, the radical formed from pentaerythritol, $(CH_2OH)_3C\dot{C}HOH$, reduces cytochrome-c at least 160 times faster at pH 10 than at pH 7. This is consistent with the radical needing to approach the haem group to perform the reduction, and with the conformational change in alkaline solution opening up the molecule to enable the bulky pentaerythritol radical to get in (43).

When hydroxyl radicals react with ferricytochrome-c they produce absorption changes consistent with some of the ferro form being produced. Correspondingly, hydrated electrons are able to produce changes in ferrocytochrome-c resembling those caused by oxidation. Such changes are basically due to the reaction of OH and e_{aq}^- respectively with organic parts of the molecule to give radicals possessing reducing and oxidising powers respectively. Similar processes must be involved in the action of hydrogen atoms on cytochrome-c (44).

Numerous pulse radiolysis studies have also been made on haemoglobin as well as other proteins. Applications of pulse radiolysis to protein chemistry have been reviewed elsewhere (45).

Vitamin B12

This vitamin, also known as cyanocobalamin:-

- is a precursor of coenzyme B12, in which compound the cyanide group is replaced by adenosine covalently bound through its 5' carbon atom to the central cobalt atom. Coenzyme B12 is involved in several biochemical reactions including the dehydration of diols, e.g.:-

$$CH_2OH-CH_2OH \longrightarrow CH_3-CHO+H_2O \qquad (28)$$

The cobalt atom can exist in three different oxidation states, Co(III) (as in the formula above) Co(II) and Co(I). In the two latter forms the vitamin is known as B12r and B12s respectively. The three forms have very characteristic absorption spectra (Figure 5). There is also a form of the vitamin in which the

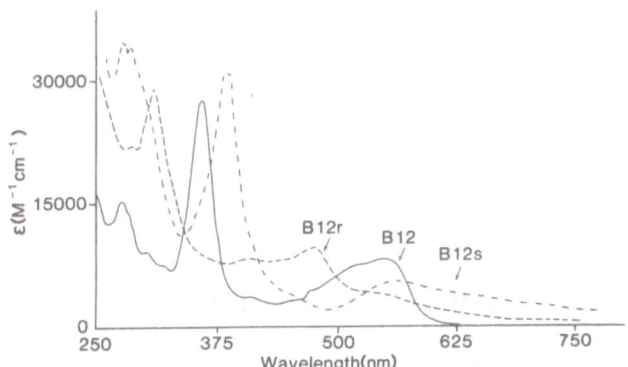

Figure 5. Absorption spectra of vitamin B12 in three different oxidation states: B12 = Co(III), B12r = Co(II), B12s = Co(I).

cyanide ligand is replaced by water or hydroxide (pK 8). This form, which is often called vitamin B12a, has sharp peaks at 351 or 355nm instead of at 361nm for B12 itself.

Hydrated electrons reduce vitamin B12 to vitamin B12r, but with only about two thirds efficiency (46,47). The cyanide ligand must be lost during the reduction, but there is no sign of this happening in a separate step, so ligand loss probably occurs with a first order rate constant $\geqslant 10^6 s^{-1}$. Reaction of hydrated electrons with coenzyme B12 is also fairly inefficient ($\sim$ 80%) and no step is seen attributable to slow loss of the adenosine (48). The comparatively low efficiencies are surprising, as any reaction of hydrated electrons with the benzimidazole or the corrin rings might have been expected to lead to reduction of the central cobalt atom to which these groups are coordinated. A partial explanation is provided by the presence of six amide and one peptide group, with which hydrated electrons are moderately reactive, and which are linked to the corrin ring only by saturated bonds. Analogous inefficiencies are seen with nitrobenzoatopenta-mminecobalt(III) complexes where, for the para ligand for example, protonation of the electron adduct to the ligand prevents observable transfer to the cobalt, and even without protonation, internal electron transfer occurs only with $k = 2.6 \times 10^3 s^{-1}$ (49).

CO_2^- reduces B12a at the usual diffusion-controlled rate (two thirds efficiency) but appears to reduce B12 itself only with a very low rate, too low to be measured by the usual pulse radiolysis method (47). O_2^- does not induce B12 but B12r reacts with oxygen, consistent with vitamin B12 having a strongly negative reduction potential.

CO_2^- reduces B12r to B12s with 100% efficiency. This is
consistent with B12r having a reduction potential of about −600mV
(47). Hydroxyisopropanol radicals react similarly. The most
interesting observation however is that in acid solution B12r can
be oxidised to the cobalt(III) state by the t-butanol or by $\cdot CH_2CH$
radicals (formed by OH abstraction from ethylene glycol followed
by water elimination, $H\dot{C}(OH)CH_2OH \longrightarrow \cdot CH_2CHO$ (48). In alkaline
solution both t-butanol radicals and $\cdot CH_2CHO$ add on, the latter
reaction being:-

$$Co(III) + \cdot CH_2CHO \longrightarrow Co(III)-CH_2CHO \qquad (29$$

This reaction is relevant to the action of the diol dehydrase.
Similar additions of radicals have been seen with the water-
soluble model compound, cobalt(II)sulphophthalocyanine (50,51).
This compound exists as the dimer in aqueous solution, and
addition is followed by dissociation, yielding the monomeric compound containing a cobalt-carbon bond as previously seen with the
vitamin.

Another reaction about which pulse radiolysis has provided
information is the known reaction of B12s with N_2O, which follows
the stoichiometry:-

$$2Co(I) + N_2O \longrightarrow 2Co(II) + N_2 \qquad (30$$

Since this reaction is stoichiometric, it cannot involve the inter
mediate formation of free hydroxyl radicals, as these are unspecif
in their action on corrinoids. Vitamin B12s has been prepared in
the absence of N_2O by action of CO_2^- on B12r. The rate of reactio
with N_2O increases with acidity, from k $\sim$ 200M^{-1}s^{-1} at pH 8 to
$\sim$ 1200M^{-1}s^{-1} at pH 3.5. On the alkaline side the reaction could
consist of entry of N_2O into the coordination sphere of the cobalt
followed by a sequence of reactions such as:-

$$(31)$$

- followed by reaction between the vitamin B12a and vitamin B12s.
On the acid side the reaction could be a hydride ion transfer:-

$$Co(III)H^- + N_2O \longrightarrow Co(III) + N_2 + OH^- \qquad (32)$$

- also followed by reactions between N12a and B12s. The reaction
between Co(III) and Co(I) proceeds with a rate constant of 3.2×10^7
(52) or $1.5 \times 10^7 M^{-1}s^{-1}$ (53) in acid solution, where the ligand on
the Co(III) is H_2O, but with a much lower rate constant in alkaline
solution, consistent with the ligand changing to OH. Vitamin B12
itself does not react with B12s at a measurable rate.

NON-AQUEOUS SOLUTIONS: EXCITED STATES

Free radicals are not the only unstable forms of molecules
which are of biological interest. Excited states are also important, especially for photobiology. They cannot conveniently be
formed by pulse radiolysis of aqueous or alcoholic solutions but
can usefully be prepared by pulse radiolysis in non-polar solvents
such as cyclohexane, hexane or benzene. Pulse radiolysis is
especially useful in the study of the triplet state since the
unique method of excitation permits the formation of triplets from
molecules which have only a low efficiency for crossover from
singlet to triplet without a triplet donor having to be present.

One of the most important applications of pulse radiolysis to
the study of triplet states is the determination of triplet-triplet
extinction coefficients (54). The method relies upon transfer of
energy from one solute to another:-

$$A^T + B \longrightarrow A + B^T \qquad (33)$$

To measure an unknown triplet-triplet extinction coefficient by
pulse radiolysis it is necessary to prepare the solute in a solution which also contains a reference solute whose triplet is well
characterised. The concentration of one of the solutes should be
one or two orders of magnitude higher than the other so that in
the first instance radiolysis will give rise predominantly to only
one kind of triplet. The nature of the radiolytic processes is
not important in the present context but it may include such
processes as neutralisation of solvent and solute positive and
negative ions, energy transfer from solvent to solute, and singlet-
triplet intersystem crossing. After completion of Reaction 33,
the solution will contain triplets of the second solute. Ideally
this reaction should be fast compared with decay of A^T or B^T, but
correction for decay can be made if necessary.

In the original work (55) cyclohexane solutions were employed
with benzophenone at a high concentration ($\sim 10^{-1}M$) as a triplet

donor. In the absence of a triplet acceptor, triplet benzophenone abstracts a hydrogen atom from the solvent:-

$$(C_6H_5)_2CO^T + C_6H_{12} \longrightarrow (C_6H_5)_2\dot{C}OH + C_6H_{11}\cdot \qquad (34)$$

- to give the benzophenone ketyl radical whose extinction coefficient in cyclohexane has been measured to be $3700M^{-1}cm^{-1}$ at its absorption maximum of 545nm. In the presence of increasing high concentrations of an acceptor whose triplet energy level is more than about $12kJmol^{-1}$ less than that of benzophenone, Reaction 33 increases the yield of acceptor triplet at the expense of benzophenone ketyl until, at high acceptor concentrations ($\sim 10^{-2}M$) essentially all donor triplets are scavenged. At such concentration there will be a residual yield of benzophenone ketyl which has been formed by processes such as protonation of the benzophenone radical anion or H atom scavenging. Figure 6 (55) shows a typical

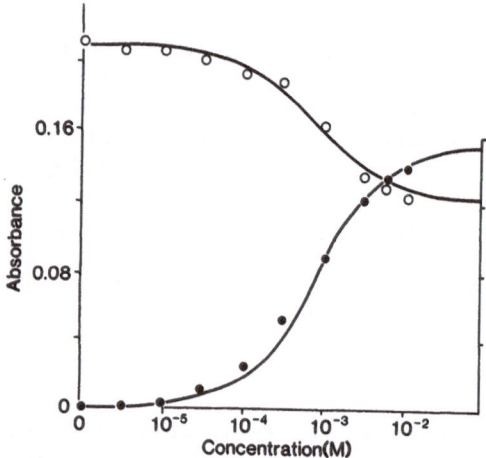

Figure 6. Absorbance produced by delivery of a pulse to cyclohexan solutions containing $10^{-1}M$ benzophenone and anthracene at various concentrations (abscissa): O absorbance to 542.5nm due to benzophenone ketyl, ● absorbance at 420nm due to anthracene triplet

dependence of triplet yield on concentration. Since Reaction 33 produces one triplet for every benzophenone ketyl lost, the unknown extinction coefficient can be derived.

Benzophenone is not a suitable reference compound for use in cyclohexane or hexane solutions if the benzophenone ketyl radicals react with the solute of interest, as has been observed for example with duroquinone. It is also unsuitable if the acceptor triplet

cannot be accurately measured optically in the solution. In such
cases other reference compounds, whose extinction coefficients
have in turn been determined against benzophenone have to be
considered, or benzene could be employed as solvent. The determination will then consist of an intercomparison of triplets rather
than of triplet with a ketyl. Biphenyl has proved to be generally
useful as a triplet donor. Its triplet level is nearly as high as
that of benzophenone, and its extinction coefficient is higher
by an order of magnitude than that of the benzophenone ketyl
radical. There is little absorption above that 400nm so that a
clear region of the spectrum is available for the straightforward
observation of the acceptor triplet. In cyclohexane or hexane its
triplet is longer lived by two orders of magnitude than the
benzophenone triplet, so that quite low concentrations of acceptor
can be used, which is especially convenient if the acceptor is
difficult or expensive to obtain in large amounts.

   Some triplet-triplet extinction coefficients which have been
obtained by pulse radiolysis are given in Table 2. The full
absorption spectrum of the β-carotene triplet it shown in Figure 7

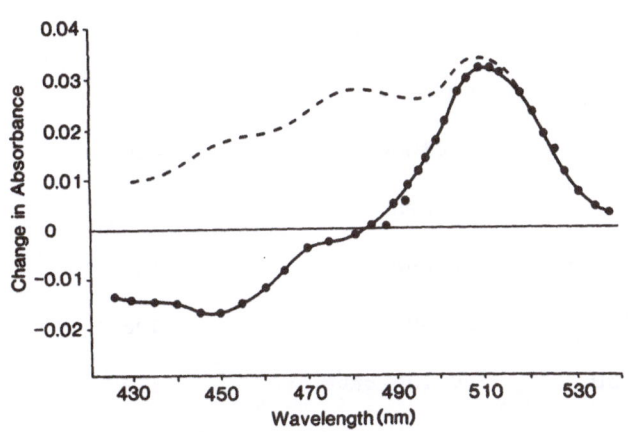

Figure 7. Changes in absorbance, ΔA, produced by delivering a pulse
to a hexane solution containing β-carotene (—●—). β-carotene
absorbs light strongly so in the experiment a sensitiser was present
so that a low concentration of β-carotene could be used. The
pulse results in excitation of singlet β-carotene to the triplet
state, so that one singlet is lost for every triplet formed. The
triplet spectrum (---) is obtained by adding ΔA to the absorbance
corresponding to the singlets depleted.

Table 2

Triplet-triplet extinction coefficients

	Solvent	Wavelength maximum (nm)	Extinction Coefficien ($M^{-1}cm^{-1}$)
Angelicin	benzene	450	4,700
Anthracene	benzene	429	45,550
	cyclohexane	422	64,700
Bergapten	benzene	450	10,200
Bilirubin	benzene	500	8,800
Biphenyl	benzene	367	27,100
	cyclohexane	361	42,800
β-Carotene (all t-)	hexane	515	242,000
Duroquinone	benzene	490	6,950
	cyclohexane	490	5,330
Naphthalene	benzene	423	13,200
	cyclohexane	414	24,500
Psoralen	benzene	450	8,100
Retinal (all -t)	hexane	445	114,000
Retinol (all -t)	hexane	405	74,000
Ubiquinone-30	cyclohexane	430	6,550
Xanthotoxin	benzene	480	10,000

(56). Knowledge of extinction coefficients enables important further information to be obtained. For example use of triplet extinction coefficients in conjunction with data from laser flash photolysis enables singlet-triplet crossover efficiencies to be measured. For β-carotene the crossover efficiency is exceptionally small, < 0.001, in fact which is relevant to the mode of action of β-carotene in biological systems.

A triplet will not efficiently donate energy to an acceptor unless its energy level is sufficiently above that of the acceptor triplet. Hence pulse radiolysis can be used to establish the energy of a triplet of unknown level. Bilirubin, a substance which accumulates in excess in neonatal jaundice, is one example whose energy level has been determined in this way. In benzene solution the triplet of anthracene (176 kJ mol^{-1}) was found to donate energy efficiently to bilirubin, but triplet naphthacene (123 kJ mol^{-1}) did not. Slow donation of energy was found with perylene (151 kJ mol^{-1}). Hence the energy of the bilirubin triplet must be $\sim$ 150 kJ mol^{-1} (57). Knowledge of the energy level has enabled aspects of the photooxidation of bilirubin to be clarified.

Some of the work on which this article is based has been supported by grants from the Cancer Research Campaign and the Medical Research Council. The author is grateful to his colleagues, especially Drs. J. Butler and E.J. Land, for helpful discussions.

REFERENCES

1. Wardman, P.: 1978, Rep.Progr.Phys. 41, pp. 259-302.
2. Adams, G.E. and Wardman, P.: 1977 in Pryor, W.A. (ed.) "Free radicals in biology", 3, pp. 53-95.
3. Ilan, Y.A., Czapski, G. and Meisel, D.: 1976, Biochim.Biophys. Acta, 430, pp. 209-224.
4. Steenken, S. and Neta, P.: 1979, J.Phys.Chem., 83, pp. 1134-1137.
5. Adams, G.E. and Michael, B.D.: 1967, Trans.Faraday Soc., 63, pp. 1171-1180.
6. Swallow, A.J.: 1980 in "Function of quinones in energy conserving systems", to be published by Academic Press N.Y.
7. Swallow, A.J., Ross, A.B. and Helman, W.P.: 1981, Radiat.Phys. Chem., 17, pp. 127-140.
8. Patel, K.B. and Willson, R.L.: 1973, J.Chem.Soc.Faraday, I, 69, pp. 814-125.
9. Arai, S. and Dorfman, L.M.: 1968, Adv.Chem., 82, pp. 378-386.
10. Rao, P.S. and Hayon, E.: 1973, Biochim.Biophys.Acta, 292, pp. 516-533.
11. von Sonntag, C.: 1980, Adv.Carbohydrate Chem.Biochem., 37,

pp. 7-77.

12. Schuler, R.H.: 1977, Radiat.Res. 69, pp. 417-433.

13. Packer, J.E., Slater, T.F. and Willson, R.L.: 1979, Nature, 278, pp. 737-738.

14. Redpath, J.L. and Willson, R.L.: 1973, Int.J.Radiat.Biol., 23, pp. 51-65.

15. Bielski, B.H.J., Comstock, D.A. and Bowen, R.A.: 1971, J.Amer. Chem.Soc., 93, pp. 5624-5629.

16. Neta, P., Simic, M. and Hayon, E.: 1972, J.Phys.Chem., 76, pp. 3507-3508.

17. Hayon, E. and Simic, M.: 1971 Intra-Science Rep., 5, pp. 357-369.

18. Land, E.J. and Ebert, M.: 1967 Trans.Faraday Soc., 63, pp. 1181-1190.

19. Adams, G.E., Aldrich, J.E., Bisby, R.H., Cundall, R.B., Redpath, J.L. and Willson, R.L.: 1972, Radiat.Res., 49, pp. 278-289.

20. Land, E.J. and Prutz, W.A.: 1979, Int.J.Radiat.Biol., 36, pp. 75-83.

21. Prutz, W.A. and Land, E.J.: 1979, Int.J.Radiat.Biol., 36, pp. 513-520.

22. Prutz, W.A., Butler, J., Land, E.J. and Swallow, A.J.: 1980, Biochem.Biophys.Res.Commun., 96, pp. 408-414.

23. Prutz, W.A., Land, E.J. and Sloper, R.W.: 1981, J.Chem.Soc., Faraday I, 77, pp. 281-292.

24. Kosower, E.M., Land, E.J. and Swallow, A.J.: 1972, J.Amer. Chem.Soc., 94, pp. 986-987.

25. Neta, P. and Patterson, L.K.: 1974, J.Phys.Chem., 78, pp. 2211-2217.

26. Anderson, R.F.: 1980, Biochim.Biophys.Acta, 590, pp. 277-281.

27. Farrington, J.A., Land, E.J. and Swallow, A.J.: 1980, Biochim.Biophys.Acta, 590, pp. 273-276.

28. Farrington, J.A., Ebert, M., Land, E.J. and Fletcher, K.: 1973, Biochim.Biophys.Acta, 314, pp. 372-381.

29. Kosower, E.M., Teuerstein, A., Burrows, H.D. and Swallow, A.J.: 1978, J.Amer.Chem.Soc., 100, pp. 5185-5190.

30. Debey, P., Land, E.J., Santus, R. and Swallow, A.J.: 1979, Biochem.Biophys.Res.Commun., 86, pp. 953-960.

31. Land, E.J. and Swallow, A.J.: 1969, Biochemistry, 8, pp. 2117-2125.

32. Swallow, A.J.: 1955, Nature, 176, pp. 793-794.

33. Meisel, D. and Neta, P.: 1975, J.Phys.Chem., 79, pp. 2459-2461.

34. Anderson, R.F.: 1976, Ber.Bunsen-Ges.Phys.Chem., 80, pp. 969-972.

35. Elliot, A.J., Munk, P.L., Stevenson, K.J. and Armstrong, D.A.: 1980, Biochemistry, 19, 4945-4950.

36. Moorthy, P.N. and Hayon, E.: 1975, J.Phys.Chem., 79, pp. 1059-1062.

37. Goldberg, M., Pecht, I., Kramer, H.E.A., Traber, R. and

Hemmerich, P.: 1981, Biochim.Biophys.Acta, 673, pp. 570-593.

38. Butler, J., Jayson, G.G. and Swallow, A.J.: 1976, J.Chem.Soc., Faraday I, 72, pp. 1391-1402.

39. Brault, D., Bizet, C., Morliere, P., Rougee, M., Land, E.J., Santus, R. and Swallow, A.J.: 1980, J.Amer.Chem.Soc., 102, pp. 1015-1020.

40. Evers, E.L., Jayson, G.G. and Swallow, A.J.: 1978, J.Chem. Soc., Faraday I, 74, pp. 418-426.

41. Butler, J., Jayson, G.G., Robb, I.D., Saunders, M.B. and Swallow, A.J.: unpublished.

42. Land, E.J. and Swallow, A.J.: 1974, Biochim.Biophys.Acta, 368, pp. 86-96.

43. Simic, M.G. and Taub.: 1978, Biophys.J., 24, pp. 285-292.

44. Shafferman, A. and Stein, G.: 1975, Biochim.Biophys.Acta, 416, pp. 287-317.

45. Klapper, M.H. and Farragi, M.: 1979, Quart.Rev.Biophys., 12, pp. 465-519.

46. Faraggi, M. and Leopold, J.G.: 1973, Biochem.Biophys.Res. Commun., 50, pp. 413-420.

47. Blackburn, R., Erkol, A.Y., Phillips, G.O. and Swallow, A.J.: 1974, J.Chem.Soc.Faraday I, 70, pp. 1693-1701.

48. Blackburn, R., Kyaw, M., Phillips, G.O. and Swallow, A.J.: 1975, J.Chem.Soc.Faraday I, 71, pp. 2277-2287.

49. Simic, M.G., Hoffman, M.Z. and Brezniak, N.V.: 1977, J.Amer. Chem.Soc., 99, pp. 2166-2172.

50. Ahmad, K., Blackburn, R. and Swallow, A.J.: 1981, in preparation.

51. Ferraudi, G. and Patterson, L.K.: 1980, J.Chem.Soc.Dalton, pp. 476-480.

52. Blackburn, R., Kyaw, M. and Swallow, A.J.: 1977, J.Chem.Soc. Faraday I, 73, pp. 250-255.

53. Ryan, D.A., Espenson, J.H., Meyerstein, D. and Mulac, W.A.: 1978, Inorg.Chem., 17, pp. 3725-3726.

54. Bensasson, R. and Land, E.J.: 1978, Photochem.Photobiol.Rev., 3, pp. 163-191.

55. Land, E.J.: 1968, Proc.Roy.Soc.A, 305, pp.457-471.

56. Land, E.J.: 1980, Biochimie, 62, pp. 207-221.

57. Land, E.J.: 1976, Photochem.Photobiol., 24, pp. 475-477.

STRUCTURE AND DYNAMICS OF PARAMAGNETIC TRANSIENTS BY PULSED EPR AND NMR DETECTION OF NUCLEAR RESONANCE

Alexander D. Trifunac

Chemistry Division
Argonne National Laboratory
Argonne, Illinois 60439
U.S.A.

ABSTRACT

Structure and dynamics of transient radicals in pulse radiolysis can be studied by time resolved EPR and NMR techniques. EPR study of kinetics and relaxation is illustrated. The NMR detection of nuclear resonance in transient radicals is a new method which allows the study of hyperfine coupling, population dynamics, radical kinetics, and reaction mechanism.

INTRODUCTION

The study of transient radicals produced by pulsed radiolysis, as carried out using magnetic resonance tools, involves the determination of radical hyperfine coupling constants, kinetics, and spin population dynamics. The coupling constants not only provide a definite spectral assignment but also provide insight into spin delocalization on the radical and thus its structure. The determination of radical kinetics is experimentally useful when radicals without distinct optical absorption bands are investigated. In addition, the magnetic resonance methods allow us to study various manifestations of the spin system, i.e., non-equilibrium population dynamics, that is, CIDEP and CIDNP, and spin relaxation phenomena. The study of Chemically Induced Magnetic Polarization (CIMP) provides useful insights into radical reaction mechanisms.

J. H. Baxendale and F. Busi (eds.),
The Study of Fast Processes and Transient Species by Electron Pulse Radiolysis, 347–362.

TIME RESOLVED EPR

The hyperfine coupling, i.e., the electron spin density at various nuclei, tells us about the structure of the radical species. The majority of EPR studies deal with the observation and analysis of hyperfine couplings of a great variety of radicals.

Time resolved EPR suffers from a decrease of sensitivity, and from line broadening at the very short times (<1 μsec) so one often cannot resolve the very closely spaced lines. Depending on the experimental conditions, hyperfine couplings that are smaller than 1-2 gauss cannot be resolved. One can still analyze the width of the whole spectrum and get an idea of the size of the hyperfine coupling, but there are better methods for that. Time resolved EPR is best suited for the study of transient radicals that have well resolved lines.

I want to emphasize again that all this refers to the study of very short-lived radicals (μsec lifetimes). When radicals live hundreds of μsec, one can do conventional high resolution EPR.

In my view, most of the usefulness of time-resolved EPR studies has been in the area of dynamics, or in observing transient radicals not seen otherwise. The structural studies of various radical hyperfine couplings, while important, have been secondary.

Kinetics and Relaxation

Both kinetic and relaxation data are obtainable by studying the time dependence of transient (z) magnetization (1,2).

The transient magnetization develops and decays in time. In the reactions where a single radical is present, e.g., $\cdot CH_2COO^-$, the creation and destruction of magnetization is describable in terms of a kinetic equation (1):

$$\dot{M}_z = -T_R(t)^{-1}M_z - T_1^{-1}(M_z - M_z^0) + \dot{E}(t)M_z^0 \tag{1}$$

The loss in magnetization M_z occurs because of the chemical decay of radicals where the lifetime $T_R(t)$ is defined as:

$$T_R(t)^{-1} = (\frac{\dot{R}}{R}) = -(\frac{d\ln R}{dt}) \tag{2}$$

where R is the concentration of radicals.

This definition of $T_R(t)$ is general enough as it allows the quantity to be either positive or negative (radical creation or decay). The second term in eq. (1) describes spin relaxation of M_z toward its equilibrium value $M_z^0(t)$ with time constant T_1. M_z^0 is proportional to the radical concentration at time (t) so from eq. (2):

$$\dot{M}_z^0 = T_R(t)^{-1} M_z^0 \qquad (3)$$

The last term in eq. (1) represents the creation of the non-equilibrium population by chemical reaction. This is expressed through a time dependent enhancement factor E(t). When we have, as in the acetate radical, a dominant bimolecular decay, we define a second order lifetime T_c where k_d is the bimolecular rate constant:

$$T_c^{-1} = 2k_d R(0) \qquad (4)$$

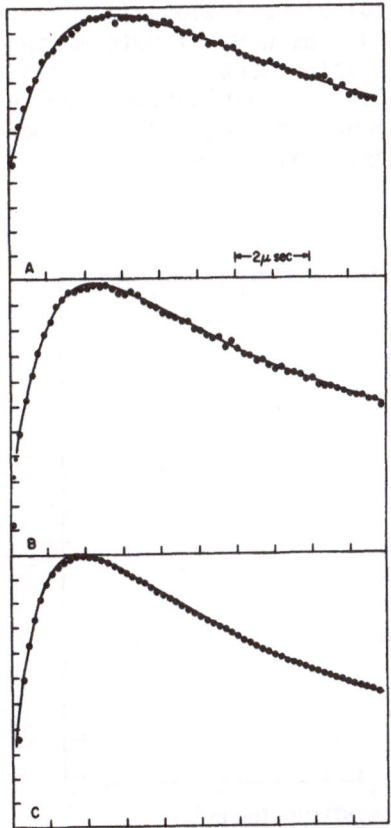

Figure 1. High field line of $\cdot CH_2CO_2^-$ radical at three radical concentrations. Data (circles) and computer fit (solid line) are illustrated.

The equations illustrated above can be numerically integrated by computer and used to analyze the experimental data. M_z is computed for various values of T, T_c, and E. Additional baseline offsets, first order rate constants, and initial enhancement can be considered as well. An example of such curve fitting is illustrated in Figure 1. By doing such data analysis, one can determine T_c, i.e., the second order reaction rate, or by using a radical with known kinetics, one can determine its reactions with another substrate by determining the psuedo first order rate constants.

This analysis using pulsed EPR represents a considerable simplification over the type of analysis used for CW-EPR. Still, it is not as straightforward as analysis of optical data. The polarization and relaxation are the complicating factors. However, the study of these complicating factors provides additional insights into the radical reaction dynamics. For example, if we examine how radical relaxation varies (actually decreases) as radical concentration increases (as we increase the dose) we can learn about the details of radical collisions (3). Figure 2 illustrates the results. The various ways of determining T_1 all tell us that there is a remarkable speeding up of spin relaxation at high radical concentrations. The explanation for this is that spin-spin interaction (Heisenberg exchange) becomes more prevalent at high radical concentrations. This exchange interaction is

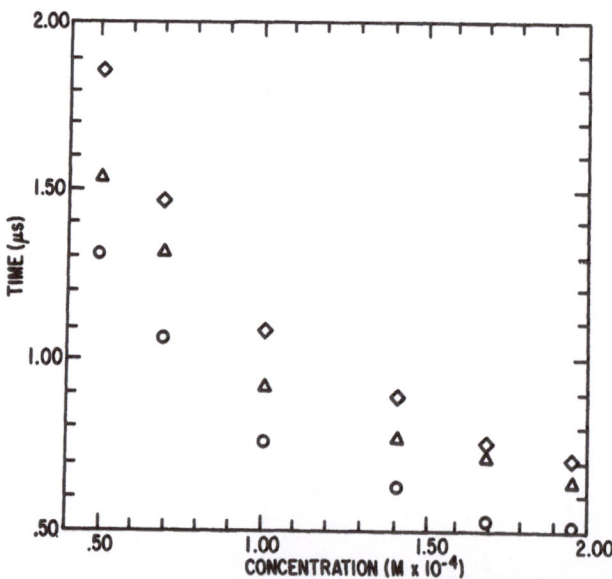

Figure 2. T_1 measurements at various $\cdot CH_2CO_2^-$ radical concentrations: $\bigcirc$ - asymptotic analysis; $\triangle$ - inversion recovery; $\diamondsuit$ - least squares computer fitting.

important during radical encounters only, and thus we have a measure of the non-reactive encounters by measuring the decrease of
T_1 with radical concentration. On the other hand, the measurement
of radical kinetics gives information on the reactive encounters.
From this we conclude that there are 2 to 5 times more non-reactive
than reactive encounters. Of course, the analysis is considerably more complex than I have indicated, and there are several
complicating factors. The problems arise at extremely high radical
concentration where substantial second order decay occurs during
the time (τ) between the microwave pulses. This causes the observed signal (M_z) to appear smaller than it actually is. At lower
radical concentrations, or if the radical decay is first order,
the problem does not arise.

In closing, pulsed EPR is the method that has superior time
resolution, comparable sensitivity, and simplified kinetic analysis
compared to the continuous wave time-resolved EPR. The problems
of spectral resolution at very short times are similar in the two
approaches, and there are problems peculiar to the pulsed EPR
method. But, as I will illustrate later, inherent advantages of
the pulsed EPR approach make this, in the opinion of the author,
a method of choice. Remember that the pulsed EPR spectrometer
components are just added to the CW EPR bridge, so without much
difficulty, one can do either.

NMR DETECTION OF NUCLEAR RESONANCE

The previously illustrated NMR experiment is a steady-state
experiment. A straightforward modification of the experiment
opens a way to a new spectroscopic method (4).

When a radio frequency (rf) irradiation is supplied to the
reacting sample after irradiation with the electron beam pulse,
the nuclear spin populations in the diamagnetic products show
substantial perturbations (Figure 3). The frequency applied during the radical lifetime is the ENDOR frequency (5-50 MHz) perturbing the electron nuclear spin population in the transient radical. Our experiment, as applied in pulse radiolysis, is based
on a flow system utilizing two magnets. The ability to use a variable magnetic field for irradiation and the pulsed rf irradiation
makes our experiment a time resolved one and opens a wide range of
possibilities. The technique of NMR-detected nuclear resonance
(NMR-NR) provides information about the mechanism and kinetics of
radical reactions and about populations of the magnetic energy
levels of transient radicals in all magnetic fields, and is a very
accurate method for determining hyperfine coupling constants of
transient radicals.

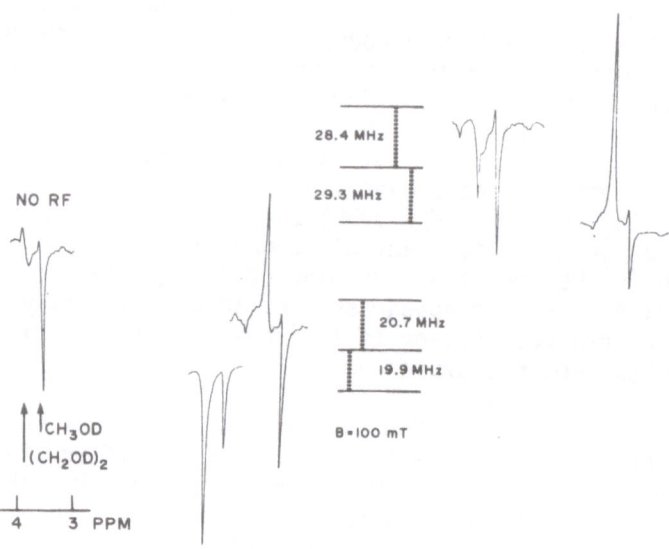

Figure 3.   The NMR-NR effect in pulse radiolysis of 0.1
M. Methanol in D$_2$O at 100 mT (1000 gauss). (A) NMR
spectrum observed when electron irradiation is carried
out with appropriate rf frequency (spectra on the right)
and without rf (spectrum on the left).

For the purpose of illustration, we consider the methanol
radiolysis. Pulse radiolysis of aqueous (D$_2$O) methanol initially
produces the radicals ·OD, D·, and e$_{aq}^-$ as well as other species.
The dominant reaction and polarization pathways in the subsequent
methanol reactions are listed below in Eqs. 5-8.

$$CH_3OD \xrightarrow{\ \cdot OD\ } \cdot CH_2OD \tag{5}$$

$$2(\cdot CH_2OD) \longrightarrow (CH_2OD)_2 \tag{6}$$

$$\cdot CH_2OD + D\cdot \longrightarrow CH_2DOD \tag{7}$$

$$e_{aq}^- + \cdot CH_2OD \longrightarrow CH_2OD^- \xrightarrow{\ D^+\ } CH_2DOD \tag{8}$$

The NMR-NR effect considered herein is shown in Figure 3.
The intensity of the NMR signal due to ethylene glycol is plotted
as a function of the frequency of the rf irradiation pulse supplied during the lifetime of the ·CH$_2$OD radical to the electron

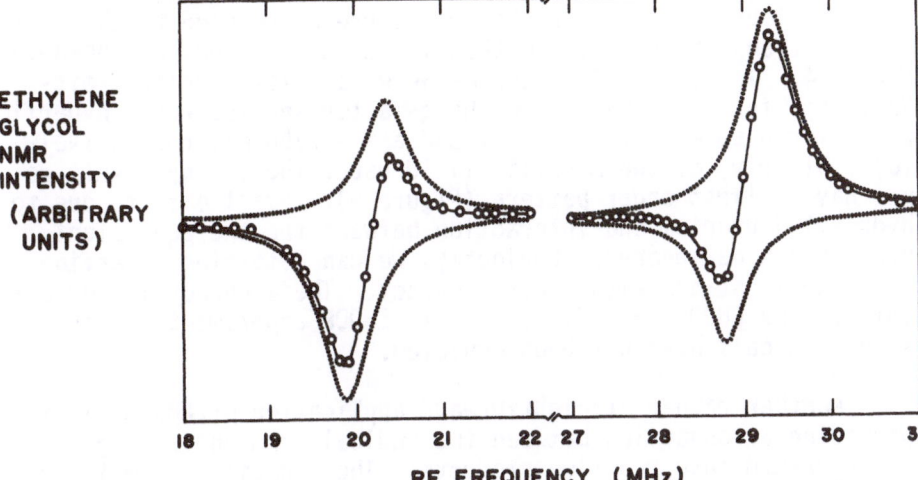

ETHYLENE
GLYCOL
NMR
INTENSITY
(ARBITRARY
UNITS)

RF FREQUENCY (MHz)

Figure 4. NMR-NR effect as in Figure 3. The experimental points (circles) were fitted with Lorentzian
components (dotted curves) of the solid curve.

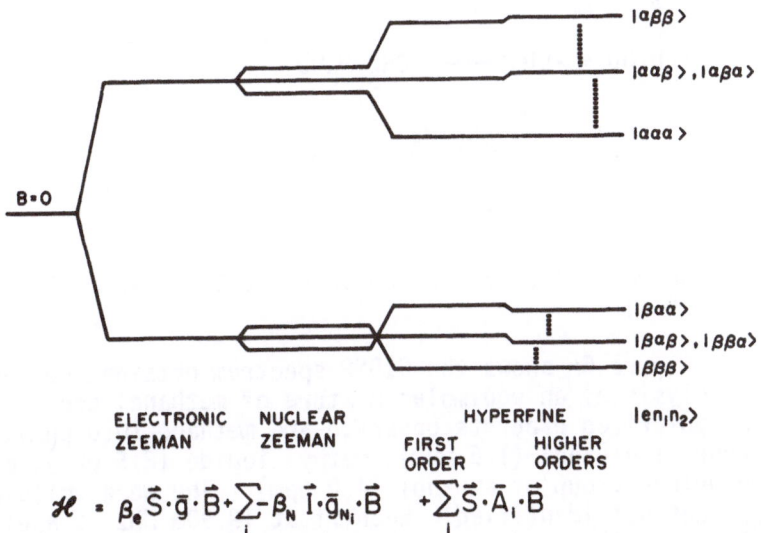

$$\mathcal{H} = \beta_e \vec{S} \cdot \vec{g} \cdot \vec{B} + \sum_i -\beta_N \vec{I} \cdot \vec{g}_{N_i} \cdot \vec{B} + \sum_i \vec{S} \cdot \vec{A}_i \cdot \vec{B}$$

Figure 5. Energy levels and the spin Hamiltonian for
spin systems like ·CH$_2$OH radical in intermediate magnetic fields (∿100 mT).

irradiation cell, which is in a 100 mT (1000 gauss) magnetic field
(Figure 4).

The four peaks in the observed curve correspond to the four nuclear transitions of the methanol radical, $\cdot CH_2OD$, as shown in Figure 4. Transition frequencies were calculated using Breit-Rabi type formulas (5), where the g-factor and isotropic hyperfine parameter used were g = 2.00317 and a = -1.750 mT, respectively (6). Clearly, at the magnetic field used, the energy levels do not have a first order pattern (Figure 5). Small effects due to hydroxyl deuterium and interaction between the two equivalent protons can be ignored. Obviously, we can determine hyperfine coupling constants with great accuracy. The accuracy is comparable to the ENDOR method, but so far ENDOR experiments on transient radicals have not been achieved.

Another rather straightforward application of NMR-NR is in providing a connection between the radicals and their fragments incorporated into reaction products. The radical system in the radiolysis of methanol-methyl iodide mixture in D_2O is used to illustrate this point (Eqs. 9-13, Figure 6),

$$CH_3 \xrightarrow{e_{aq}^-} \cdot CH_3 \tag{9}$$

$$CH_3OD \xrightarrow{\cdot OD} \cdot CH_2OD \tag{10}$$

$$\cdot CH_2OD + \cdot CH_3 \longrightarrow CH_3CH_2OD \tag{11}$$

$$2(\cdot CH_3) \longrightarrow CH_3CH_3 \tag{12}$$

$$\cdot CH_3 + \cdot D \longrightarrow CH_3D \tag{13}$$

and the reactions of the $\cdot CH_2OD$ radical, as in methanol radiolysis (Eqs. 5-8).

Figure 6A shows the CIDNP spectrum obtained during the pulse radiolysis of an equimolar mixture of methanol and methyl iodide. The polarized products observed are methane (0.6 ppm), the methyl group of ethanol (1.6 ppm), methyl iodide (2.5 ppm), and the methylene group of ethanol (4.0 ppm). The weak emission at 3.1 ppm was not identified. When rf at 19.900 MHz is applied (Figure 6B), only the products from the $\cdot CH_2OD$ fragment are affected. These signals are observed at 4.0 ppm, which illustrate the strong effect on the quartet from ethanol (ethylene glycol would also be within this intense signal), and at 3.6 ppm, which shows an emission for CH_2DOD not observed in Figure 6A. For a determination of which polarizations arise from the $\cdot CH_3$ radical, a radio frequency of 36.500 MHz is applied. The results which are shown in Figure 6C indicate the signals affected are methane at 0.6 ppm, which

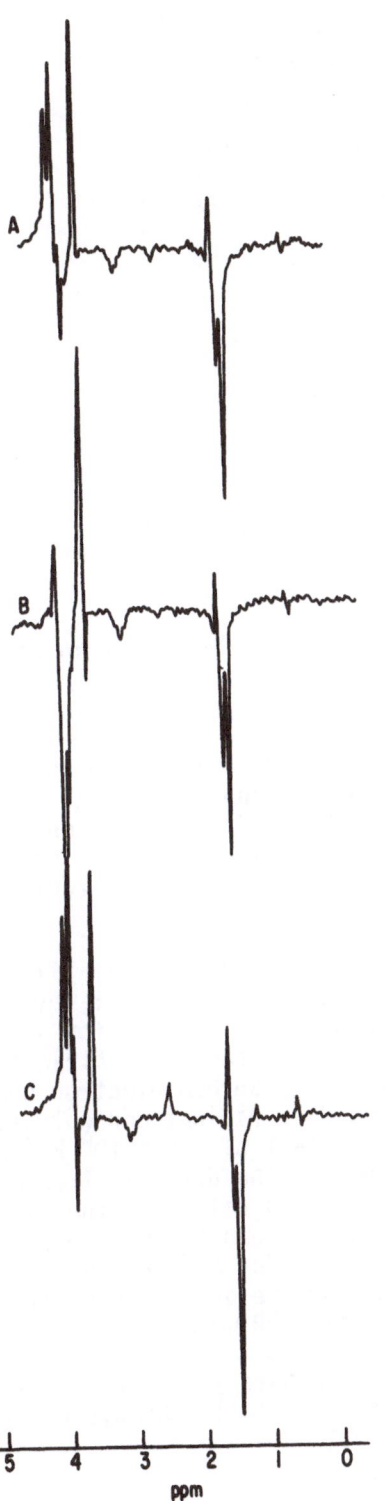

Figure 6. Pulse radiolysis in methanol-methyl iodide in D_2O at 100 mT: (A) with electron beam; (B) A with rf at 19.9 MHz; (C) A with rf at 36.5 MHz.

exhibits a slight increase in intensity, methyl iodide at 2.5 ppm, which shows complete inversion of its signal as compared to Figure 6A, and the triplet of methanol at 1.6 ppm, showing substantial increase in intensity. Interestingly, ethane is even observed at 1.2 ppm (weak enhanced absorption) which would have gone undetected in Figure 6A. In a more complex radical reaction system, where it may be less obvious what part of the radical product comes from which radical, the NMR-NR approach should provide a beneficial adjunct to the CIDNP study.

In the course of our study, it became apparent that lower rf power was needed to effect slower reacting (longer lived) radicals. For example, in the methanol radiolysis glycol polarization can be influenced at very low rf power while even at the highest rf power available, the effect on methanol is always smaller. Thus, different radical lifetimes of the CIDNP cage product (methanol) vs. escape product (glycol) are differentiated.

Short rf pulses (0.5-2 μ-sec) are used to probe the time dependence of the NMR-NR signal intensity. The plots thus obtained are shown in Figure 7. The signal intensity vs. time

NMR-NR OF METHANOL RADIOLYSIS
(D$_2$O, 2 μs RF PULSE, B=100 mT)

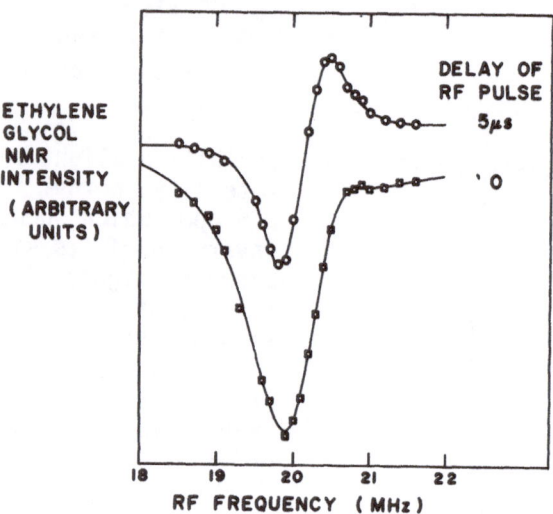

Figure 7.   Time dependence of NMR-NR effect in methanol
radiolysis.

of the rf pulse after the electron beam (at time zero) should
represent radical kinetics as long as the nuclear T_1 in the radicals is appreciably longer than the chemical decay (second-order
half-life) of the radicals. This is the case with the ·CH$_2$OD radical in the concentration range utilized (10^{-3}-10^{-4} M radical con-
centration/pulse).

From the second-order plot, the second-order half-life of the
·CH$_2$OH radicals is ∿6 μs. The radical concentration is evaluated
to be 6 × 10^{-4} M since the bimolecular rate constant for ·CH$_2$OH
is known (3 ± 1) × 10^8 M^{-1} s^{-1}) (10). One can also obtain the
radical concentration directly from the absorbed electron beam
intensity. This latter approach yields a radical concentration
of 10^{-3} M and a second-order rate constant of 1.6 × 10^8 M^{-1} s^{-1}.
However, given the variables involved in the determination of the
radical concentration, it is preferable to calibrate the experimental system by using a radical with a known second-order rate
constant, as we have illustrated above. The accuracy of rate
constants obtained by NMR-NR should compare quite favorably to
those obtained by usual optical or fast EPR methods.

The kinetic applications of NMR-NR have yet to be fully understood and developed as we have to analyze fully the dynamic problems encountered.

CIDNP and CIDEP Contributions

We present a model for the observed intensities based on the radical pair theory of CIDEP and CIDNP (7). Again, we consider the NMR-NR spectrum of $\cdot CH_2OD$ radical in methanol radiolysis, observed through ethylene glycol product. The Boltzman distribution is overshadowed by these polarizations, as determined by the presence of only a barely detectable ethylene glycol NMR signal in the irradiation solution after the nuclear spins have had time to relax. Populations will be predicted for the $\cdot CH_2OD$ energy levels using the radical pair theory of CIMP (8), for various possible pairs involving $\cdot CH_2OD$. An adequate spin Hamiltonian for a radical pair is

$$\mathcal{H} = \beta H \cdot (g_1 S_1 + g_2 S_2) - J(1/2 + 2S_1 \cdot S_2) +$$

$$\sum_i a_{1i} S_1 \cdot I_i + \sum_k a_{2k} S_2 \cdot I_k \tag{14}$$

where the first term represents the electronic Zeeman energy for the two unpaired electrons. S_1 and S_2 are the electronic spin operators for the two components. The next term represents the scalar exchange interaction of magnitude 2J and the remaining two terms represent the isotropic components of the hyperfine interactions in the two members of the pair. The nuclear Zeeman interaction, although significant in the calculation of energies, is not required for calculating polarizations. In each case considered, two freely diffusing radicals form the pair (i.e., so-called F-pair). During the initial encounter, those pairs with singlet electronic spin phasing tend to form product (with no CIMP), leaving behind predominantly triplet radical pairs. Combination products, in which the members of the pair react during a subsequent encounter, will be favored by a rapid singlet-triplet crossing rate, whereas scavenging products, with reaction between a member of the radical pair and a freely diffusing radical, will be favored by slow singlet-triplet crossing. Nuclear spin polarization (CIDNP) results from this selection process. Electron spin polarization (CIDEP) also develops after there is an excess of triplet radical pair character, through intersystem crossing and exchange during radical encounters. The singlet-triplet crossing rate is given by

$$a_n = (2h)^{-1}[(g_1 - g_2)\beta H + \sum_i a_{1i} m_{1i} - \sum_j a_{2j} m_{2j}] \tag{15}$$

where g_1 is the g-factor for radical 1 and a_{1i} and m_{1i} are the hyperfine constant and nuclear spin quantum number, respectively, for nucleus k of radical 1. The relationship in Eq. 15 applies to the high applied magnetic field case, in which non-secular terms in the electron spin can be ignored and only $S-T_0$ mixing (not $S-T_{\pm 1}$) is considered.

Since we can do experiments using various magnetic fields, we have chosen for this illustration H = 100 mT, because the four nuclear transitions of $\cdot CH_2OD$ are then easily resolvable (Figure 5). At higher fields, the transition frequencies approach a first order pattern, reducing the observed effect due to overlap of positive and negative peaks. The signal-to-noise ratio also deteriorates because the NMR-NR effect is smaller, relative to the increased CIDNP net effect which arises principally from the e_{aq}^- + $\cdot CH_2OD$ pair which has a large g-value difference. At lower applied fields, however, deviations from the high field approximation may become significant and the intense H_1 (rf) fields cannot be used if one wishes to observe well resolved NMR-NR transitions. Of course, the larger the nuclear hyperfine coupling, the higher is the magnetic field one can utilize.

It will be useful to consider separately the possible contributions of the net and multiplet effects, which arise from the first term and the last two terms, respectively, of the right-hand side of Eq. 15. In usual CIMP spectra, the net effect gives rise to an equal polarization for each of the lines in an EPR or NMR multiplet, whereas the multiplet effect has zero as the sum of the polarizations in each multiplet. Irradiating $\cdot CH_2OD$ with rf will affect these spin level populations. A 13 μs rf pulse beginning 0.5 μs before the electron beam pulse was used. With the maximum available H_1 in the 10-20 mT range, a $\pi/2$ pulse would be 2-4 μs. However, the reactive transient radicals have a distribution of lifetimes, creating a range of effective pulse widths (tip angles). It is not feasible at this time to treat quantitatively the effects on the populations caused by a range of rf pulse tip angles. Thus, we will qualitatively consider the rf-induced population shifts by assuming saturation (i.e., equalization of populations), even though saturation may not be occurring. This assumption leads to predicted relative NMR-NR intensities which agree with the experimental spectrum when the ratio of the population shifts due to the nuclear and electronic multiplet effects (n_M/e_M) is 1.7. The value of n_N/e_M, the ratio between the magnitudes of the nuclear net and electronic multiplet effects obtained from the observed peak areas, is small relative to the estimated uncertainties in the areas. NMR-NR spectra measured as a function of time using 2 μs rf pulses (Figure 7) are qualitatively similar to Figure 4, but vary in the ratio n_M/e_M. Not surprisingly, this ratio is also different for other chemical systems. Even though the present data are integrated over a range of radical lifetimes during which CIMP develops and relaxes and radical decay takes place, it has been possible, using NMR-NR, to obtain information about relative energy level populations in $\cdot CH_2OD$ that is not available from CIDEP alone or otherwise. As predicted for high fields, the CIDEP

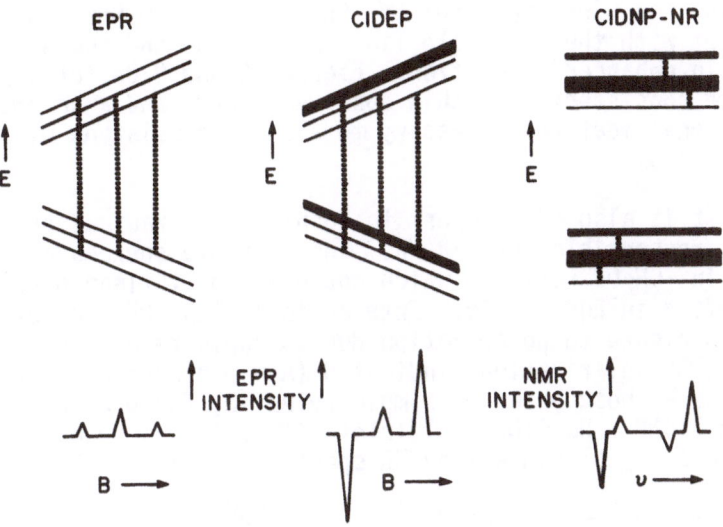

Figure 8. Comparison of spin population as observed in EPR experiments on ·CH$_2$OH radical with Boltzmann population (left), with CIDEP (middle), and as observed by NMR-NR (right).

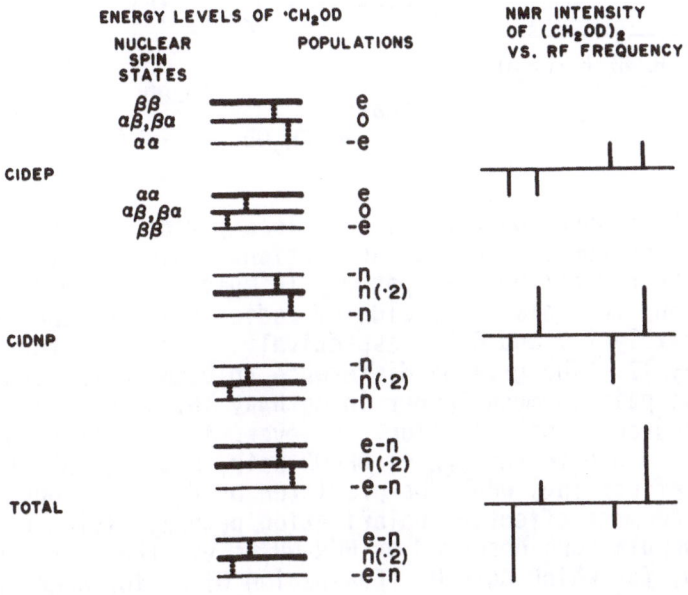

Figure 9. Population schemes as observed in NMR-NR showing how CIDEP and CIDNP contributions are combined.

fields, the CIDEP spectrum of $\cdot CH_2OD$ at ca. 10 GHz consists of a triplet with the low field line in emission and the high field line in enhanced absorption. Figures 8 and 9 depict a population scheme consistent with this observed CIDEP, and a scheme with the additional features necessary to also be consistent with the NMR-NR spectrum.

It is also of interest to identify the radical pair interactions responsible for NMR-NR. The possible sources of polarization in $\cdot CH_2OD$ radicals which could yield ethylene glycol are specified in Eqs. 16-18. Interactions with $\cdot OH$ are not expected to contribute to polarization due to rapid relaxation in $\cdot OH$ (9). The CIDNP polarizations indicated (A = enhanced absorption, 0 = none) are those based on simple rules (8) for predicting high field CIDNP. No CIDNP multiplet effect is observable by NMR in ethylene glycol since the NMR spectrum consists of a single line.

$$\overline{e_{aq}^- + \cdot CH_2OD} \xrightarrow{\text{escape}} \cdot CH_2OD \xrightarrow{\cdot CH_2OD} (CH_2OD)_2 \text{ (A)} \qquad (16)$$

$$\overline{D\cdot + \cdot CH_2OD} \xrightarrow{\text{escape}} \cdot CH_2OD \xrightarrow{\cdot CH_2OD} (CH_2OD)_2 \text{ (A)} \qquad (17)$$

$$\overline{\cdot CH_2OD + \cdot CH_2OD} \begin{cases} \xrightarrow{\text{cage}} (CH_2OD)_2 \text{ (0)} & (18a) \\ \xrightarrow{\text{escape}} \cdot CH_2OD \xrightarrow{\cdot CH_2OD} (\dot{C}H_2OD)_2 \text{ (0)} & (18b) \end{cases}$$

At fields above about 75 mT, ethylene glycol is found to have a CIDNP spectrum in enhanced absorption. This observation shows that the radical pair in pathway 16 must be involved. In neutral solution, the relative yields of radicals $\cdot CH_2OD$ and $\cdot D$ are approximately 2.7 and 0.6, respectively, favoring pathway 16 over pathway 17. The g-value difference Δg between the members of the radical pair is much larger in pathway 16, also favoring this route as a source of polarization. However, if the NMR-NR effect were primarily due to the $e_{aq}^- + \cdot CH_2OD$ pair, a detectable NMR-NR net CIDNP effect (n_N) would be predicted by Eq. 15. Thus, it is likely that the most effective polarization pathway giving rise to the spin populations observed by NMR-NR is via the pair $\cdot CH_2OD + \cdot CH_2OD$, for which $\Delta g = 0$. Evaluation of a_n for each spin state of the radical pair (Eq. 15) shows that even though no CIDNP multiplet effect is observable in ethylene glycol, the nuclear states of the pair still differ in their reaction rates and that the nuclear levels of $\cdot CH_2OD$ are depopulated at different rates. The perturbing rf pulses permit NMR detection of this polarization, as well as the CIDEP polarization. The CIDNP multiplet effect predicted for the escaping radicals has the same sign of n_M as that observed (opposite to that predicted for the cage reaction). This is reasonable since the time of rf irradiation (effective pulse length) can be much longer for the longer-lived escaping radicals, allowing partial saturation to occur. This conclusion is also consistent with the time dependence of the NMR-NR effect, which appears to follow second order kinetics (with a half life of about 6 μs at the radical concentrations used).

In summary, the spin populations in a transient radical ob-served using NMR-NR can be explained using the radical pair theory of CIMP. The multiplet effects in both CIDEP and CIDNP have been observed for a symmetric radical pair using NMR-NR. As illus-trated, the CIDNP and CIDEP contributions to the spin populations of transient radicals observed by NMR-NR are separable. Experi-ments are in progress to measure other types of spin systems, to extract the kinetics of the CIDEP and CIDNP processes, and to find out how the radical chemistry of the given system is revealed through the variation in the electron/nuclear population ratios. We believe that the NMR-NR technique will develop into a powerful new technique for the study of transient radicals in solution in both radiation and photochemistry.

ACKNOWLEDGMENT

Work supported by the Office of Basic Energy Sciences, Division of Chemical Sciences, U. S. Department of Energy, under Contract W-31-109-Eng-38.

REFERENCES

(1) Trifunac, A.D. and Thurnauer, M.C.: 1979, *Time Domain Elec-tron Spin Resonance*, ed. L. Kevan and R. N. Schwartz, John Wiley & Sons, Inc., New York.
(2) Trifunac, A.D., Norris, J.R., and Lawler, R.G.: 1979, J. Chem. Phys. 71, p. 7380.
(3) Syage, J.A., Lawler, R.G., and Trifunac, A.D., to be pub-lished.
(4) Trifunac, A.D. and Evanochko, W.T.: 1980, J. Am. Chem. Soc. 102, p. 4598.
(5) (a) Breit, G. and Rabi, I.I.: 1931, Phys. Rev. 38, p. 2082. (b) Poole, C.P. and Farach, H.A.: 1972, *The Theory of Mag-netic Resonance*, Wiley-Interscience, New York, p. 203.
(6) Laroff, G.P. and Fessenden, R.W.: 1973, J. Phys. Chem. 77, p. 1283.

(7)   Nuttall, R.H.D. and Trifunac, A.D.: 1981, Chem. Phys. Lett.
      81, 151 (1981).
(8)   *Chemically Induced Magnetic Polarization*: 1977, L. T. Muus
      et al., eds. (NATO ASI), D. Reidel, Publishers, Dordrecht.
(9)   (a) Verma, N.C. and Fessenden, R.W.: 1976, J. Chem. Phys.
      65, p. 2139.
      (b) Trifunac, A.D., Johnson, K.W., Clifft, B.E., and Lowers,
      R.H.: 1975, Chem. Phys. Lett. 35, p. 566.
(10) Ivan, Y., Rabani, J., and Henglein, A.: 1976, J. Phys. Chem.
      80, p. 1558.

TRANSIENTS IN LOW TEMPERATURE AQUEOUS GLASSES

Fernand Kieffer

Laboratoire de Physicochimie des Rayonnements,
associé au C.N.R.S.
Université de Paris-Sud, 91405 Orsay, France.

Abstract. Low temperature glasses differ from liquids essentially
by the suppression of molecular diffusion. Their study permits, on
the time scale of minutes and hours, the observation of processes
which take place in nanoseconds and microseconds in room temperature liquids. Some luminescence techniques which are not generally
applicable in the study of liquids are briefly described. Four
types of low temperature matrices are examined in some detail,
mainly as regards presolvated and solvated electrons. Finally the
subsequent fate of solvated electrons is discussed. Pulse techniques applied to glasses at 77 or even 4 K enable one to approach
initial events more closely.

1. THE GLASSY STATE.

Fast processes can be studied in two ways :
1. by rapid techniques
2. by slowing down the processes.
It is of course possible to combine (1) and (2) and this has
been done frequently in recent years in order to obtain information about earlier stages of radiation effects.

The slowing down of reactions is achieved by cooling. Reactions which proceed in liquids with an activation energy have their
rate constants reduced very roughly by 1/2 for a decrease in
temperature of 10°C. This is only a first step. At a sufficiently
low temperature the liquid solidifies. This means that translational motion of molecules ceases and reactive species can no longer diffuse towards reaction partners : they are frozen in, and

363

J. H. Baxendale and F. Busi (eds.),
The Study of Fast Processes and Transient Species by Electron Pulse Radiolysis, 363–397.

only reactions with neighbouring molecules such as hydrogen abstraction, are possible. Thus unstable intermediates, radicals or ions which of course are not the primary products, can be observed at leisure, either seemingly stable or still undergoing some change. The transformation of these species can also be accelerated ad libitum by warming-up.

The solids obtained by cooling can either be crystalline or glassy. While in a crystal the constituent atoms, ions or molecules are stacked in a regular array, in glasses the absence of long-range order is shown by X-ray or electron diffraction studies. Glasses are a particular type of amorphous materials, transparent, isotropic and rigid. A distinctive feature is the existence of a "glass transition". Familiar examples of room temperature glasses are window glass, pyrex, silica ; a lollipop is an organic glass made of sugar and containing some solutes such as flavouring and colouring matters.

Radiation chemists have studied crystals, and much work still goes on with single crystals, but low temperature glasses became very popular when it appeared that they stabilise not only radicals and radical-ions, but also electrons which are hardly observable in crystals. Their transparency makes possible optical absorption studies and their isotropicity results in random orientation of additives or intermediates ; this may be an advantage because any measurements are free from orientational complications. Thus γ-radiolysis of low temperature glasses became the "poor man's pulse radiolysis".

Glasses are generally prepared by rapid cooling of liquids. This is accompanied by a gradual increase in viscosity. As is shown in Fig. 1, heat capacity changes gradually through a temperature region where the liquid becomes very viscous and is called a supercooled liquid, and then suddenly changes at a temperature called the glass transition temperature, T_g.

The glass transition has been defined(1) as the "phenomenon in which a solid amorphous phase exhibits with changing temperature a more or less sudden change in the derivative thermodynamic properties, such as heat capacity and expansion coefficient, from crystal-like to liquid-like values".

The temperature at which this change is observed varies with the time scale over which the mesurement is performed. This is so because the glass transition is not a thermodynamic transition (which would be independent of cooling or heating rate, like a melting point), but has a kinetic origin : it depends on structural changes which are slow near T_g. The phenomenon is observed when this structural relaxation time is of the order of the time scale of the experiment. Since viscosity is also determined by this relaxation time, T_g is usually associated with a viscosity of 10^{13} poise, but a lower value (e.g. 10^{11} P) could be adopted from measurements with a faster technique.

Table I. Glass transition temperatures and viscosities at 77 K of some glasses.

Glass	T_g (K)	$\eta_{77\ K}$ (P)
Silica	1600	
Pyrex	790	
Polystyrene	360	
Sucrose	340	
Glycerol	178	2.6×10^{57}
Decalin	134	
Methanol	99	1.1×10^{25}
Ethanol	93	2.1×10^{19}
Methylcyclohexane	86	1.2×10^{18}
2-Methylpentane	78	2.4×10^{13}
3-Methylpentane	77	2.2×10^{12}
Isopentane	65	

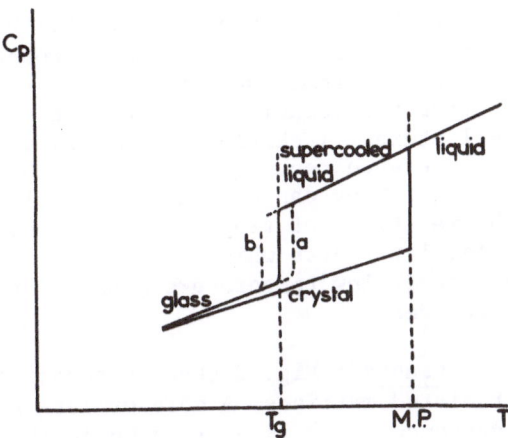

Figure 1. Heat capacity diagram of glass transition, compared to crystal melting.

What liquids are liable to form glasses on cooling ? There is no generale rule. The problem is a kinetic one, involving the rate of crystal nucleation and growth on the one hand, and the rate of cooling on the other. Some organic or aqueous glasses can be obtained easily (e.g. aqueous NaOH, ethanol), others require special precautions to ensure rapid cooling (e.g. methylcyclohexane). According to the thermal treatment, some of these substances can be obtained at will in a crystalline or a glassy state. Thus MCH and ethanol crystallise upon simple warming.

Summing up, upon cooling, below the glass transition the glass is - on a time scale of days or years - in an unstable state, but this is a state of mechanical stability which, contrary to the metastable supercooled liquid state, exhibits the thermodynamic properties of a solid.

2. TECHNIQUES USED IN MATRIX ISOLATION STUDIES.

From the point of view of radiation chemistry, the initial stagescan be considered to be identical with those described by my colleagues for the liquid state. But instead of reacting rapidly the unstable species formed (except the H˙ atom), are trapped by the rigid matrix and may be studied over quite long periods of time by techniques of EPR or optical absorption spectroscopy, or their transformation or disappearance may be followed during a temperature rise or upon optical exitation or the application of electric fields, by these techniques or by luminescence or conductivity measurements. Fast techniques may be applied to the study in the glassy state, much work has been done with microsecond and nanosecond pulse techniques, and some even with subnanosecond techniques. I shall not describe the techniques which have already been considered in preceding lectures. I should like merely to give a short description of luminescence techniques which are highly sensitive and can be very rapid. They can of course be applied only in those cases where the reactions of trapped species lead to the formation of excited molecular states which deactivate radiatively. Recombination luminescences of low temperature glasses have been studied for nearly two decades.

Recombination luminescences : Fig. 2 shows schematically what is observed in many irradiated matrices. A more or less slowly decaying isothermal luminescence (ITL) is emitted while the sample is kept at the temperature of irradiation. When the sample is heated, luminescence intensity increases and passes through one or more thermoluminescence maxima ("glow peaks") before falling off to zero when the reactive species have disappeared. Photoluminescence is the luminescence increase above ITL observed when the sample is illuminated in the absorption bands of trapped negative species. "Stimulation spectra" are based on the measurement of photoluminescence intensities after short stimulations with monochromatic light of different wavelengths.

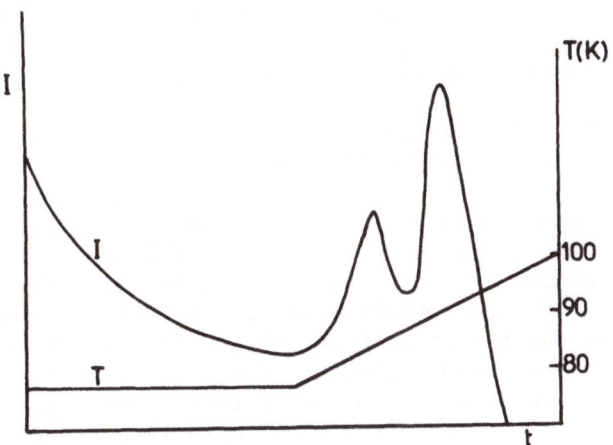

Figure 2. Post-irradiation luminescence intensity as a function of time at constant temperature (ITL) and during linear heating (thermoluminescence).

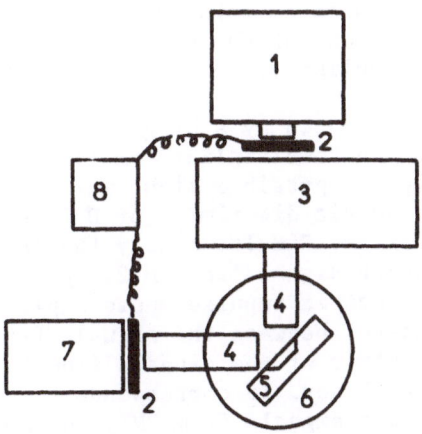

Figure 3. Block diagram of set-up for photostimulated luminescence studies. 1 : Xenon lamp, 2 : shutters ope-rated by synchroniser 8, 3 : monochromator, 4 : light guides, 5 : sample holder, 6 : cryostat, 7 : photomul-tiplier.

Fig. 3 shows a block diagram of a simple set-up for the determination of stimulation spectra. These spectra have been used in a qualitative way to check the presence of suspected negative ions, to identify trapped positive species from the lifetime of stimulated phosphorescence, and also to distinguish between bound-bound and bound-free transitions of electrons trapped in a given matrix.

Electroluminescence is observed when an electric field is applied to a sample containing trapped electrons.

ITL can be studied on a short time scale with pulse radiolysis or flash photolysis techniques. A very refined technique is coincidence counting in which single photoelectrons are counted in coincidence with the ionizing particles. This technique consists in viewing the sample with two photomultiplier tubes, one of which works under "single photoelectron" conditions, i.e. the down-stream electronics are adjusted to reject any signal corresponding to the simultaneous ejection of more than one photoelectron from the cathode. Under these conditions, the probability of a single photoelectron leaving the photocathode at any instant is proportional to the number of photons impinging on it. Thus if a light flux varying with time falls on the cathode, the probability of a single photoelectron being ejected from it varies with time according to the same law.

The set-up is shown diagramatically in Fig. 4. The sample (a dilute solution of a scintillator in the matrix to be studied) contains a weak α- or β-source emitting only about 10^4 particles per second, so that piling up of two or more ionisation events within a microsecond is unlikely.

The scintillation produced by a single ionisation event is detected by the monitor photomultiplier (PM 1) which is placed so as to receive the highest possible light intensity from the scintillation, and is entered via discriminator D 1 as zero time in a time to amplitude converter. Simultaneously the TAC is gated to admit the first subsequent signal from PM 2. Upon reception of this signal it produces an electric impulse whose amplitude is proportional to the time interval between the signals from PM 1 and PM 2. The amplitude spectrum of the signals from the TAC is analysed by a multichannel analyser. It corresponds to the histogram of the time intervals between signals from PM 1 and PM 2 and repre - sents the evolution of luminescence intensity with time.

The fact that all signals from PM 2 corresponding to more than one photoelectron, or arriving outside a preselected time interval with respect to the signals from PM 1, are rejected, reduces noise tremendously and makes this method one of the most sensitive for the study of weak luminescences. It was developed in a nuclear physics environment where coincidence counting of particles had

been a familiar technique and where the study of scintillators
required precisely such a technique. When improved photomultipliers became available in the 1960's, it was therefore quickly
developed and utilised for scintillation and other excitation
transfer studies and made possible some precise determinations
of short fluorescence lifetimes(2). The adaptation to work at
liquid nitrogen temperature is fairly simple.

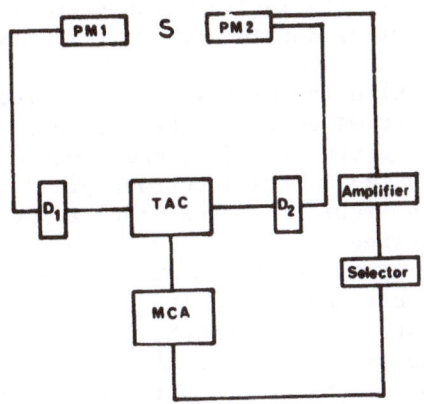

Figure 4. Simplified block diagram of set-up for coincidence photon counting. S : sample, PM 1 : monitor photomultiplier, PM 2 : photomultiplier working under "single
photoelectron " conditions, D_1 and D_2 : discriminators,
TAC : time to amplitude converter, MCA : multichannel
analyser.

To my knowledge this method has not been widely employed in
radiation physics and chemistry. This seems a pity because, apart
from relative simplicity, it has another advantage : the dose rate
delivered to the sample is very low. For particles with an energy
of 1 MeV it is of the order of 10^{10} $eVg^{-1}s^{-1}$, i.e. less than 1 $radh^{-1}$;
since convenient storage times are of the order of 2 to 24 hrs, the
total dose remains well below 100 rad, so that radiolytic alterations of the sample can be neglected.

3. RADIATION CHEMISTRY IN GLASSY MATRICES.

The simultaneous development of research in pulse radiolysis
at room temperature and in γ-radiolysis of low temperature glasses

arose from a renewed interest in the role of ionic intermediates.

Up to 1936 it was usually considered that <u>ionic</u> intermediates alone played an important role in the effect of <u>ionising</u> radiation on chemical systems. Then, Eyring et al.(3) showed that
atomic intermediates could explain the kinetics of various reactions, among them radiolysis. Many reaction mechanisms in radiation chemistry were found to involve atoms or free radicals. The
use of radical scavengers and the development of EPR techniques
for the detection, identification and localisation of free radicals led to considerable progress in radiation chemistry, and
ionic intermediates went out of fashion for some twenty years.

In the late 1950' s, with the discovery of low temperature
ionic polymerisation, it became apparent that ionic species must
play a role in a number of cases. In 1962, the study of the radiolysis of aqueous solutions led to the spectroscopic detection of
the "solvated" electron in liquid solutions (4a) and of the
"trapped" electron in alkaline aqueous glasses (4b), and very
soon in a whole series of matrices of the alcohol, sugar, amine,
ether and finally hydrocarbon families. The two methods of pulse
radiolysis at room temperature and of matrix isolation at 77 K
gave quite comparable results, one on the time scale of microsecond
and very soon of nanoseconds, the other on the scale of minutes
or hours.

A combination of the pulse radiolysis and matrix isolation
techniques was first used in a spectroscopic study in 1968(5).
And in 1970, Richards and Thomas (6) published absorption spectra
obtained by ns pulse radiolysis of glasses at 77 K showing that
the end-of-pulse spectra, extending much further into the infra-
red than the known e_t spectra, transformed into the latter within
a few microseconds (Fig. 5).

Simultaneously, Higashimura et al.(7) obtained the first absorption spectra in glassy matrices irradiated with γ-rays at 4 K,
showing similar, but stable, absorptions in the infra-red. Baxendale and Wardman(8) observed analogous changes in cooled liquid
alcohols subjected to ns pulse radiolysis and found that the time
scale of these changes varied with viscosity. And Hunt et al.(9)
extended these observations to room temperature liquids with picosecond pulse radiolysis.

The study of fast processes in radiation chemistry finally
boils down essentially to the study of the fate of electrons in
different states and of that of the positive holes which arise
when an electron is ejected from its orbit and travels through the
matrix. Because free or quasi-free electrons are extremely short-
lived, the appearance of initially localised electrons is as yet
inaccessible to experiment, and experimental work concerns electron
which are already in some "trapped" state, "presolvated" or "sol-

vated". Such electrons can be prepared by the action of "ionising" radiation or by photodetachment from solute ions such as $Fe(CN)_6^{3-}$ or by photoionisation of solutes with a low ionisation potential, such as TMPD (tetramethyl-p-phenylene diamine) and other aromatic amines, indole and derivatives (tryptophan), phenolate..., which are chosen principally according to their solubilities in the matrix studied.

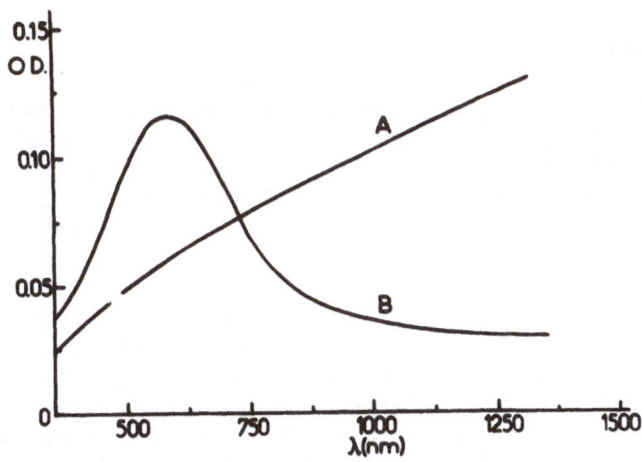

Figure 5. Absorption spectra observed in pulse radiolysis of ethanol glass at 77 K. A :at the end of a 10 ns pulse, B : after 4 μs (Richards and Thomas(6)).

The division of my lectures into one on aqueous glasses and one on organic glasses is of course rather arbitrary, and I shall draw on results from both when necessary. Van Leeuwen(10) recently proposed to subdivide glassy matrices into 4 groups, from the point of view of the absorption spectra of trapped electrons observed at very short times. Aqueous glasses belong to two of these groups :

<u>alkaline glasses,</u> in which at all temperatures between 4 and 300 K, only one absorption band is observed around 600 nm. Some strongly polar organic glasses such as methanol glass also show this behaviour.

some "exceptional" glasses : ice, strong saline solutions (LiCl, $MgCl_2$, BeF_2 and a 1:1 mixture of ethylene glycol and water). In these glasses both deeply trapped electrons ($\lambda_{max} \approx 600$ nm) and shallowly trapped ones ($\lambda_{max} \approx 3$ μm) are observed, the yield of the latter increasing with decreasing irradiation temperature.

The other two groups distinguished by van Leeuwen are :

less polar glasses, such as ethanol and higher alcohols, in which at least two kinds of trapped electrons are observed, with absorption maxima around 1700 and 600 nm, the early distribution being temperature dependent.

non-polar glasses, such as hydrocarbons, with only one slightly temperature dependent absorption maximum around 1800 nm.

4. SOME EXAMPLES OF AQUEOUS GLASSES.

4.1. Pure ice

Pure ice is only obtained in crystalline form, and is considered here only for the sake of completeness and in order to show important differences with respect to aqueous glasses. At 77 K, the most prominent unstable species detected by EPR is the OH˙ radical, (G(OH) = 0.8). Detection of e_t^- was impossible because it is formed in much lower yield (G (e_t^-) = 5 x 10^{-4}), so that its EPR signal was masked by that of OH˙ ; Kawabata(11) interpreted his results as showing that electron traps do not pre-exist in crystalline ice, but are created by the irradiation. He succeeded in detecting the e_t^- signal in D_2O containing 10^{-2} M NH_4F (which increases the yield, but does not alter the absorption spectrum) by the following procedure : irradiation at 77 K with a dose of 7·Mrad, annealing to 125 K for 6 min under white light, and reirradiation at 77 K to only 0.1 Mrad. EPR spectra were then measured before and after illumination and the difference was taken. A line width of 2.7 G was obtained (cf. 6.5 G in alkaline ice and 3.2 G in deposited vapour of D_2O and Na, where D_2O is probably glassy). The traps in this case where self-trapping is excluded seem to be vacancies in the ice lattice. In H_2O ice Kawabata could not detect the EPR signal of e_t^-.

Optical absorption studies after 30 ns pulse radiolysis(12) at 76 K show the presence of a visible band (λ_{max} = 650 nm) and an IR band which was measured up to 2350 nm in pure D_2O ice and fits a Lorentzian function which corresponds to λ_{max} = 2950 nm. The low energy side of the visible band decays much faster than the band as a whole.

Both bands are observed when the sample is pulsed at higher temperatures, as Fig.6 shows : the absorption measured (100 ns after the start of a 40 ns pulse) at 2350 nm increases with temperature through a maximum near 150 K and falls off to zero at 220 K the absorption measured at 650 nm, on the contrary, decreases from 77 to 230 K and increases steeply from 250 to 270 K. The visible and IR bands decay at different rates and are clearly due to two different species. The very fast decay ($\tau_{1/2} \approx 1$ μs) explains

discrepancies on yields between different authors. Buxton et al.
(12) conclude from their results that e^-_{vis} arise from in-spur
formation of vacancies by the neutralisation reaction of D_2O^+ with
a quasi-free electron giving rise to D + OD + vacancy, followed
by trapping of another quasi-free electron within the spur. They
suggest that e^-_{IR} are trapped in cavities natural to the crystal
structure, around which little orientational polarisation takes
place . The observed geminate decay kinetics are most probably due
to the reaction e^-_t + D_2O^+ $\longrightarrow$ OD + D.

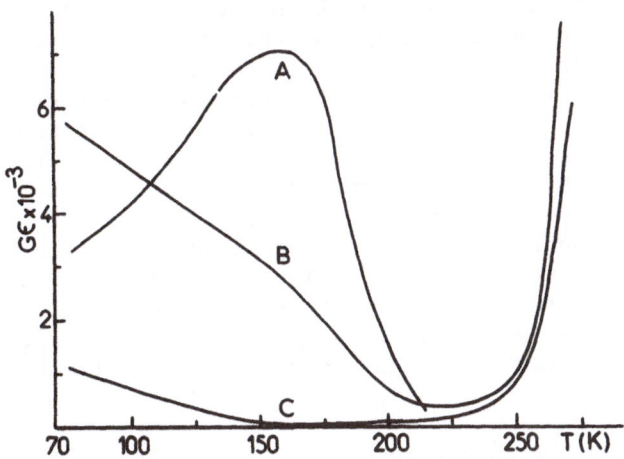

Figure 6. Effect of temperature on absorption in pure
crystalline D_2O(40 ns pulse radiolysis). A : measured at
2350 nm 100 ns after start of pulse, B : measured at 650
nm 100 ns after start of pulse, C : measured at 650 nm
3 μs after start of pulse (Buxton et al.(12)).

A luminescence spectrum with two bands around 370 and 530 nm
was obtained by the same authors, in good agreement with results
observed by Steen and Holteng(13) during X-irradiation of H_2O and
D_2O ice at 77K.

Conductivity measurements on crystalline ice were made by
Warman et al.(14) with ns pulse radiolysis at temperatures from
-10 to -190°C, with the microwave absorption method. They determined the decay kinetics of highly mobile electrons and distinguished
4 temperature regions with different kinetic behaviour. In the upper one (-40 to 0°C) it seems that (assuming that the species formed by trapping of the mobile electron, T^-, rapidly converts to a

fully "solvated" form, e^-_s), the yield of e^-_s is controlled by
competition between the highly thermally activated trapping process and the weakly temperature-dependent ion recombination process. This explains the increase in $G(e^-_s)$ with increasing temperature, already shown before.

Close to the melting point the localisation of "dry" conduction electrons occurred in less than 1 ns. Since the rate of localisation should represent an upper limit to the rate of appearance
of e^-_s, it should be possible to observe a growth of the visible
absorption band over a few hundred ps. Warman and Jonah(15) tes-
ted this with picosecond pulse radiolysis. An absorption with
λ_{max} = 660 nm was found within about 30 ps after the beginning of
a 25 ps pulse. At -5°C this absorption grows to about twice its
initially observed value in 2 ns and then decays slightly over the
following 2 ns. As the temperature is lowered the post-pulse growth
is less and less pronounced, and at -28.5°C only a decay of the
absorption from the initial value is observed. The authors explain
these observations by the formation of two species : Species A is
responsible for the prompt absorption ; its yield is not strongly
temperature dependent and it decays over a few hundred picoseconds.
Species B is responsible for the "long-lived" absorption observed
over nanoseconds and microseconds. The yield of B decreases with
decreasing temperature, and at -5°C it is formed on a time scale of
several hundred picoseconds. The authors identify it with an electron which has become "solvated" by dipolar relaxation of the medium.

There is no definite evidence of a blue shift, since there
is a difference of only $\approx$15 nm between the end-of-pulse and the
500 ps spectra ; a "presolvated" state would be expected to absorb at much lower energies. Only speculative interpretations can
be given of the nature of the prompt spectrum.

A half-time of electron solvation in the range 200-500 ps
can be deduced from a rough subtraction of growth and decay kinetics. This is longer than the half-time of localisation of highly mobile electrons estimated to be 80 ps(14). The difference may
represent the time required for trap deepening by dipolar relaxation. These times are about four orders of magnitude shorter than
the dielectric relaxation time of ice $(10^{-5}s)$ and can be explained
only if trapping occurs at the lattice vacancies present at concentrations of 10^{-4} to 10^{-3} M.

The case of electron solvation in ice is particularly interesting in that it occurs on a time-scale which is completely incompatible with the dielectric relaxation time of the medium.

Quite recently, Gillis and coworkers(16) pulsed H_2O and D_2O
ice at 73 K and at 6 K. At 73 K they found results similar to tho-

se obtained earlier(12) with D_2O which had been chosen for its
better IR transmission. The 650 nm absorption decays much more rapidly in H_2O than in D_2O ice (respective half lives : 25 and 350
ns) and the decays give fairly linear plots of OD versus log t,
suggesting a tunnelling mechanism for recombination. In the vacancy formation mechanism by an in-spur neutralisation reaction(12),
involving formation of D + OD, these fragments should have sufficient kinetic energy to be expelled some distance from the formed
vacancy. Because of momentum conservation, OH should have 1/18 of
the available energy in H_2O and OD 1/10 in D_2O, so that OD could
be expected to be further away from the vacancy in which the electron is trapped than OH. If e^-_{vis} decays by reaction with OH by
tunnelling, a small difference in distance can explain the difference in half-lives : it is sufficient that half the e^-_{vis} be trapped within respectively 20.2 Å and 22.6 Å of the hydroxyl radical
in H_2O and D_2O.

At 6 K, in H_2O ice, the ratio of infrared to visible absorption is much larger than at 73 K. A notable feature is the rapid
decay of the visible band ($\tau_{1/2}$ = 75 ns). In fact it decays faster than the IR band.

4.2. Acid glasses.

Concentrated solutions ($\approx$ 5 M) of H_2SO_4, $HClO_4$, H_3PO_4 freeze
to clear glasses. After γ-irradiation at 77 K, no trapped electrons are found in these glasses, probably because electrons are
scavenged by H^+ to form H· atoms ($G(H·) \approx 1$-2). These are detected
by EPR as a characteristic doublet and their concentration has
been shown to increase linearly with dose up to some 50 Mrad when
a stationary concentration is reached.

It seems that in ns pulse radiolysis of 8 M HCl/H_2O and
DCl/D_2O glasses both visible and infrared absorbing trapped electrons, decaying within 10 μs, are formed, but the authors do not
comment on these results(16).

4.3. Alkaline glasses.

On cooling, strong solutions, up to 10 M or more, of NaOH or
KOH in water, form glasses in which paramagnetic species (trapped
electrons and O^- ions) can be detected by EPR. Both are formed with
yields of about 2.7/100 eV by γ-radiolysis at 77 K. A word must
be said in favour of such concentrated solutions of alkali hydro-
xides or of various salts which form glasses. The concentration of
additive is often higher than 10 M and it may seem at first thought
that some of these solutions contain more additive than water, but
this is not so, simply because of the low molecular weight of water. A litre of water contains about 50 moles (cf. $\approx$ 10 moles for
1l of 3-methylpentane) and water molecules are well in excess of
solute. Even so, it is certainly an oversimplification to consider

these matrices as water when 25% or so of the radiation energy is
absorbed by solute molecules and when these could be expected to
participate in the solvation shell. But, as early as 1964, Moorthy
and Weiss(17) concluded from a comparison of line widths of the
EPR singlets in H_2O and D_2O glasses containing LiOH, NaOH or KOH,
that the nearest neighbours of the unpaired electron must be water
molecules : for the three alkali metal ions, line widths were found
to be 16.1 ± 1.5 G in H_2O and 5.8 ± 0.5 G in D_2O. Moreover, from
the absence of hyperfine splitting they concluded that the unpaired electron cannot be localised on any single water molecule. This
was an early hint to the e^-_t structure which we shall see later.

The reactions leading to the formation of e^-_t and O^- can be
written as follows :

$$H_2O \rightsquigarrow H_2O^+ + e^- \rightarrow e^-_t$$

$$\quad + H_2O \rightarrow H_3O^+ + \dot{O}H$$

$$\dot{O}H + OH^- \rightarrow O^- + H_2O$$

O^- is detected by EPR as a broad singlet, with linewidth
about 50G. Since O^- has a half-filled p-orbital, it can react with
an electron to form the O^{2-} ion, although electrostatic repulsion
does not favour this reaction. Thus it behaves as a trapped hole,
and Ershov et al.(18) observed the simultaneous decay of O^- and
e^-_t at 163 K, suggesting mutual reaction. Trapped electrons are
formed in 10 M NaOH glass with a yield about 10000 times higher
than in crystalline ice. Their absorption spectrum at 77 K lies
in the visible, with λ_{max}= 585 nm. Illumination in any part of
the absorption band renders e^-_t mobile, as was shown by the good
coincidence of the curves representing photocurrent(19) or stimulated luminescence(20) versus wavelength of exciting light with
the absorption spectra(Fig.7,A). This is quite different from the
relationship observed in pure crystalline ice between photobleaching quantum efficiency and exciting wavelength(19)(Fig.7,B),
which shows that e^-_t are significantly bleached only by light
corresponding to the high energy tail of the absorption spectrum.
Thus it seems that in alkaline ice electrons can be optically excited only into the conduction band whereas in crystalline ice
there is a threshold of photon energy below which excitation occurs to a bound excited state. In this respect alkaline glasses
are an exception among polar glasses which generally exhibit a
threshold below which the electron does not escape from its trap.

Absorption spectra in NaOH glass were determined also at
4.2 and 1.6 K(21,22). As can be seen in Fig.8, for 10 M NaOH,

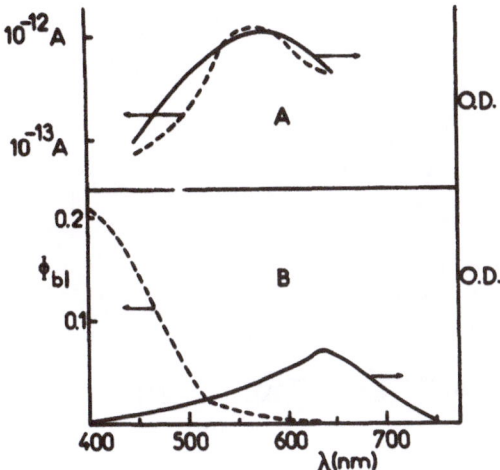

Figure 7. A. Photoconductivity spectrum (---) and absorption spectrum (——) of γ-irradiated 10 M NaOH at 77 K (Kevan(19)).
B. Photobleaching efficiency(---) and absorption spectrum (——) of γ-irradiated single crystal of ice at 77 K (Kevan(19)).

a slight red-shift of the maximum is observed with lowering of temperature, as well as considerable broadening of the red side of the spectrum. Table II summarises the data on λ_{max} and band width at half-height, $W_{1/2}$:

Table II. Spectral characteristics of electrons trapped in X-irradiated NaOH glasses at 77,4 and 1.6 K(Dolivo and Kevan (22)).

Matrix	λ_{max} (nm)			$W_{1/2}$ (eV)		
	77K	4K	1.6K	77K	4K	1.6K
10 M NaOH	585	620	650	0.80	1.50	1.27
18 M NaOH	610	675	675	1.20	1.39	1.30

378 F. KIEFFER

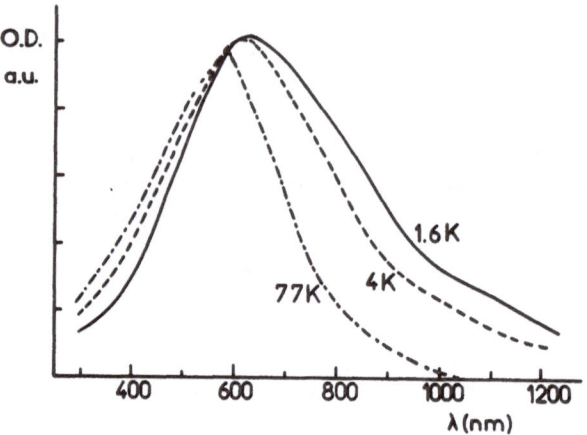

Figure 8. Absorption spectra of e_t^- in 10 M NaOH glass at 77, 4 and 1.6 K (Dolivo and Kevan(21)).

From this increasing red shift with decreasing temperature it seemed reasonable to attribute the red part of the spectrum to "presolvated" electrons. This attribution is supported by photo bleaching at 850 nm which allows a decomposition of the absorption band into two sub-bands (see Fig. 9), with quite comparable results for the 10 M and 18 M glasses, indicating that a tentative attribution of the red shift observed at 77 K between 10 M and 18 M NaOH to a particular species, (Na^+ - e^-) pairs, is probably not justified.

The notion of presolvated electrons arose when pulse radiolysis was applied to low temperature glasses and when "steady state" radiolysis was extended to liquid helium temperatures. In room temperature liquids and in glasses at 77 K, optical absorptior bands are similar with moderate displacements of λ_{max}, and can be taken to correspond to solvated electrons with an equilibrium distribution of solvent molecules around them. "Presolvated" electrons, with λ_{max} shifted, sometimes considerably, towards the infra-red, correspond to a non-equilibrium distribution of solvent molecules. The attainment of equilibrium, through reorientation of surrounding molecules or bonds, or through tunnelling of the electron towards a "better" trap, appears very nicely in Fig. 10 which shows the evolution of the absorption spectrum of 1-propanol

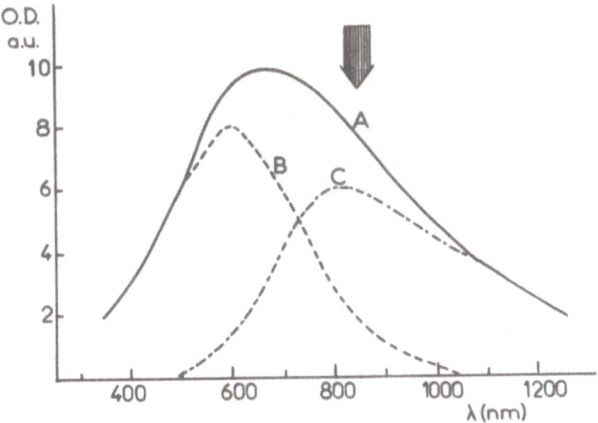

Figure 9. Absorption spectra of 18 M NaOH irradiated at
4 K. A : before photobleaching, B : after bleaching at
850 $\pm$ 25 nm, C : bleached spectrum obtained by subtraction (Dolivo and Kevan(21)).

at 77 K after a 20 ns pulse(23).

    Kevan et al.(24) compared absorption spectra and photoconduc-
tivity spectra in a number of glassy matrices at 4 K, in order to
gain an insight into the nature of the transitions involved. In
the case of 10 M NaOD/D_2O glass, they found a good agreement between the two types of spectra, with λ_{max} = 620 nm, indicating that
the transitions are predominantly from the bound state to the conduction state. But with respect to 77 K data, the red edge of the
optical absorption band is shifted more to the red than the red
edge of the photoconductivity band, and hence a small contribution
from bound-bound transitions seems to exist at 4 K. Such bound-
bound transitions merely mean that the electron does not escape
from its trap. They may or may not mean that the "presolvated" electron becomes solvated as it relaxes from the excited state.

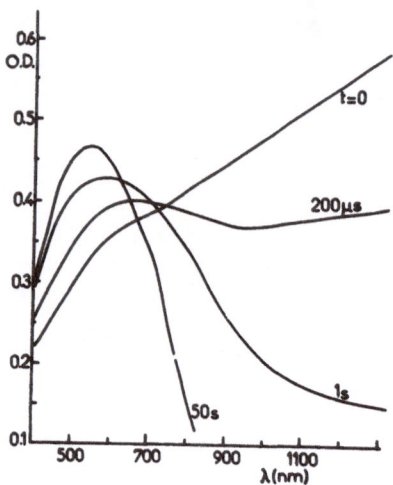

Figure 10. Spectral changes in 1-propanol at 77 K after a 20 ns pulse. (Baxendale and Sharpe (23)).

Like cooling to 4 K, pulse radiolysis of NaOH glass at 77 K shows a slight red-shift of the absorption in comparison with the γ-radiolysis results : the spectrum at 100 ns of 6 M NaOH glass published by Miller(25) shows λ_{max} = 620 nm (cf. γ-irradiation: 585 nm).

4.4. Salt glasses.

Many salts in concentrated aqueous solutions form glasses at low temperature. Relatively little EPR work was done on these glasses, but optical absorption studies are very abundant.

Among the different salt glasses studied, two types can be distinguished:

Those whose optical absorption shifts but slightly between 77 K and 4.2 K, e.g. 5 M K_2CO_3 or 7 M $NaClO_4$; their width at half-height does not increase markedly, but their tail extends into the infra-red up to about 1300 nm at 4 K and to 1700 nm at 1.6 K (22).

Those in which a strong infra-red band peaking above 2000 nm appears at 4 K on the minutes scale(20,26), or on the ns or μs scale in pulse radiolysis(27,28) : e.g. 9.5 M LiCl, 2.5 M $MgCl_2$, 7.5 M BeF_2. The second type is particularly interesting because the IR-absorbing electrons are here a well distinguishable species in shallow traps. For the LiCl glass(Fig. 11), λ_{max} at 77 K is 590 nm ; at 4 K it is shifted to 695 nm and the IR band peaks beyond 2400 nm (it has been estimated to be around 3600 nm). The

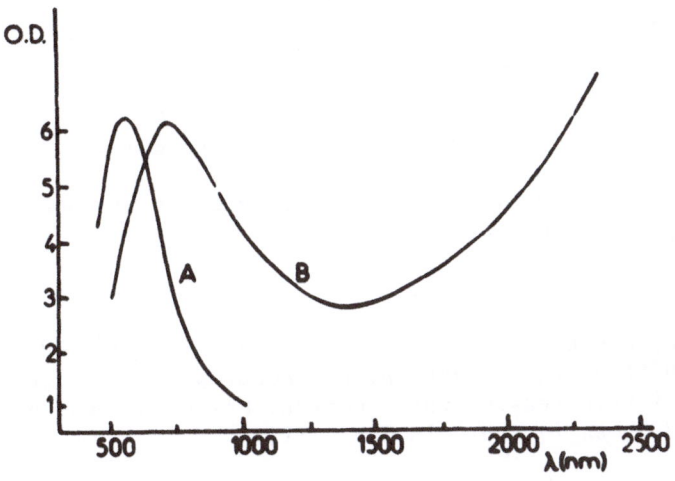

Figure 11. Absorption spectra of e_t^- in 10 M LiCl/D_2O glass.A : 77 K, B : 4 K. (Dolivo and Kevan(26)).

photoconductivity spectrum(29) shows a broad maximum near 1400 nm, i.e. at greater energies than the IR absorption maximum. No photo conductivity peak seems associated with the visible absorption peak, but from bleaching experiments it appears that some contribution is present at these wavelenghts, but is normally hidden by the more intense spectrum due to the IR band. A photoconductivity spectrum taken at 85 K peaks at 450 nm, i.e. also at energies

greater than the visible absorption peak (590 nm at 77 K). Hence
both the "visible electron " at 85 K and the "infra-red electron"
at 4 K show bound-bound and bound-continuum transitions. A similar conclusion was reached by the study of stimulation spectra
in 12 M LiCl/D$_2$O glass(20): at 77 K a stimulation maximum of
475 nm corresponds to the absorption maximum of 590 nm, and at
4 K the respective figures are 625 and 695 nm for the visible
band. In the IR, a stimulation spectrum could not be obtained because of solvent absorption.

In the dark, the IR absorption decays perceptibly in 10-20
min at 4 K, but is stable for at least 1 hour at 1.6 K (26). At
76 K, the IR absorption appears only in pulse work ; it decays on
a μs time scale, and it seemed at first that there was no concomitant increase of the visible absorption(27). It was concluded
that there was no or little conversion of the shallowly trapped
electrons into deeply trapped ones. The decay of the former may
take place predominantly through recombination with cations. We
shall see later that the same research team succeeded four years
later in evaluating the contribution of this conversion in the
disappearance of presolvated electrons.

Photobleaching at 4 K, with light of 550 nm, of a LiCl glass
irradiated at 77 K diminishes the visible band and produces an
IR band(26). The mechanism is obvious : one of the reactions which
electrons ejected from visible traps may undergo is retrapping in
shallow traps to form e_{IR}.

4.5.Ethylene glycol—water glass.

Glasses can be obtained from different EG/H$_2$O mixtures, and
a gradual shift of the e_t^- absorption spectrum and a narrowing of its
band width with increasing water content has been found by Ershov
and Pikaev(30), as is shown in Table III.

Table III. Absorption of e_t^- in glassy EG/H$_2$O mixtures
at 77 K (30).

EG/H$_2$O	λ_{max} (nm)	$W_{1/2}$ (eV)
100/0	502	1.63
90/10	507	1.47
80/20	520	1.41
60/40	540	1.25
40/60	554	0.98
20/80	585	0.87

The most popular mixture is the 50/50 one. This is easily
cooled to a hard glass which has some tendency to cracking. It
was one of the first glasses to be studied at 4 K, and with respect to e_t^- spectra it can be considered as analogous to the LiCl
type glasses.

Higashimura et al. studied its EPR(7) and optical absorption
spectra(21). They found an EPR singlet with $\triangle H$ = 14 G at 77 K
and 3 G at 4 K. The latter is bleached by light of $\lambda >$1000 nm,
whereas the former is not. Warming the sample irradiated at 4 K
to 77 K changes the 3 G signal into the 14 G one. This change is
irreversible.

They also found an optical absorption band shifted from
585 nm at 77 K to 714 nm at 4 K, and the appearance of an IR band
extending above 2000n m at 4 K. Warming to 77 K caused the IR
band to disappear while the intensity of the visible band increased. The IR band could also be bleached with light of $\lambda >$1000 nm,
but without increase of the visible band. If the glass is warmed
slowly, the IR band decreases gradually from the longer wavelength
side and disappears completely at about 40 K. So in this case,
contrary to the LiCl/H_2O case, the disappearance of e_{IR}^- occurs
much more evidently to some extent by transformation to the fully
solvated state.

In pulse radiolysis of EG/D_2O at 76 K, an end-of-pulse spectrum was found(27) with a visible band peaking at $\approx$ 650 nm and
an IR band extending beyond 1900 nm. The decay of the IR band is
accompanied by a growth of the visible band lasting about 10 μs.
But this growth only amounts to 20% of the IR decay, so that trap
deepening (by dipole relaxation, or by tunnelling or hopping to
deeper traps) is far from accounting alone for the disappearance
of e_{IR}^-. Fig. 12 shows that, on the nanosecond time scale at 6 K,
the IR band is far more important in EG/D_2O 50/50 glass than in
pure EG glass. This suggests that e_{IR}^-, rather than being a "presolvated" electron, is in an environment of water molecules mainly.
The same conclusion has been drawn by Gillis at al.(40) for
LiCl/D_2O glasses in which the relative intensity of the IR band
decreases considerably as LiCl content increases from 6 to 8 to
15 M.

The photoconductivity spectrum obtained at 4 K(24) (Fig. 13)
shows a threshold near 1600 nm and a maximum near 1000 nm, and a
high energy tail down to about 400 nm. There is no evidence for
an ultra-violet band as was observed at 85 K (λ_{max} = 350 nm).
Thus both at 85 K and 4 K conductivity maxima are considerably
blue-shifted relative to the absorption maxima : 585 $\rightarrow$ 350 nm,
$>$ 2000 $\rightarrow$ 1000 nm. Both for solvated and presolvated electrons,
bound-bound transitions and bound-continuum transitions occur in
the absorption spectrum.

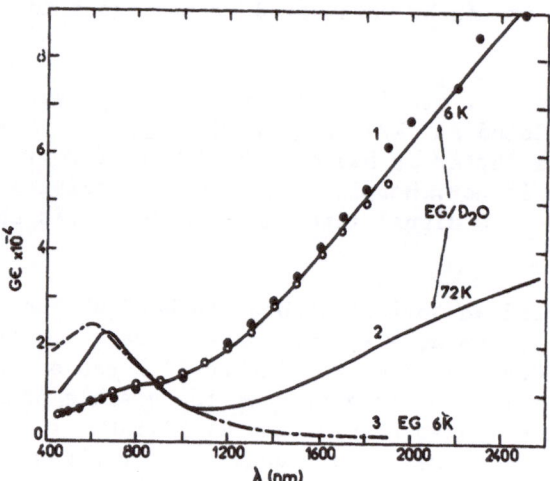

Figure 12. Absorption spectra observed 100 ns after start of a 40 ns pulse. Curve 1 : EG-d$_6$/D$_2$O (•) and EG-h$_6$/D$_2$O (°) at 6 K ; curve 2 : the same at 72 K ; curve 3 : pure EG-h$_6$ at 6 K. (Cygler, Klassen et al. (31)).

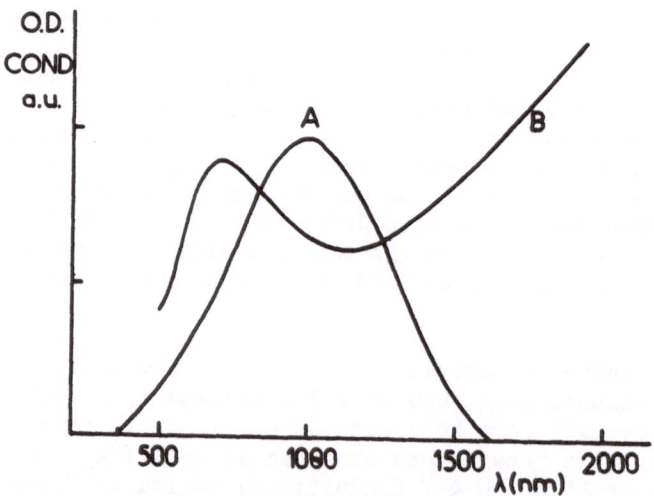

Figure 13. Photoconductivity spectrum (A) and absorption spectrum (B) of EG/D$_2$O glass irradiated at 4 K (Rice et al.(24)).

In this glass, the stimulation spectra were in disagreement (20), at least at 4 K where no maximum appeared around 1000 nm ; from a threshold near 1900 nm the photostimulation band rose gradually to pass its maximum at about 630 nm. At 77 K, the threshold appeared around 800 nm and no maximum was reached yet at 450 nm, the lower wavelength limit because in these experiments e_t^- were produced by photoionisation of tryptophan , and below 450 nm, direct excitation of this solute was possible (great band width of stimulation wavelength). From the appearance of the low energy wing of the spectrum, its maximum could well lie around 350 nm, as in the photoconductivity spectrum.

The disagreement at 4 K may mean that the 630 nm stimulation maximum corresponds to the 714 nm absorption band, i.e. concerns solvated electrons only. According to photoconductivity results, at 630 nm transitions are essentially bound-continuum in the high energy tail of the IR absorption band. In photostimulation bound-bound transitions in the visible band near 630nm could lead to tunnelling from the excited state to the cation, thus giving rise to luminescence. This is impossible in photoconductivity where the electron must be excited to the conduction band, and therefore photoconductivity is a more reliable technique. Apparently no such effect occurs with the presolvated electrons.

5. THE FATE OF TRAPPED ELECTRONS.

Fully relaxed ("solvated") trapped electrons are a stable species only in so far as the matrix isolates them from possible reaction partners. Thus a temperature rise leads to their disappearance, generally near T_g. At lower temperatures, they may be kept for hours or days without any considerable decay. But unrelaxed ("presolvated") electrons disappear on a time scale of nanoseconds or minutes, according to the temperature. When reactive molecules ("scavengers") are added to the matrix, it is found that even "solvated" electrons disappear.

The decay kinetics of e_t^- have been studied in various aqueous glasses, and it is generally observed that e_t^- decays linearly with the logarithm of time. The concentration of e_t^- is measured either by EPR or optical absorption spectroscopy.

In irradiated 10 M NaOH glass both e_t^- and an efficient acceptor, O^-, are present. Upon heating, both species disappear close to T_g (> 150 K). This suggests(18) the occurrence of the reaction

$$e_t^- + O^- \longrightarrow O^{2-}$$

at a diffusion-controlled rate (like the corresponding liquid-phase reaction of e_{aq}^-). At lower temperatures, a slow simultaneous decay of e_t^- and O^- is observed. Zamaraev(33) studied the decay of O^- by EPR and that of e_t by both EPR and optical spectroscopy. Fig. 14 shows the decay of both species. It is based

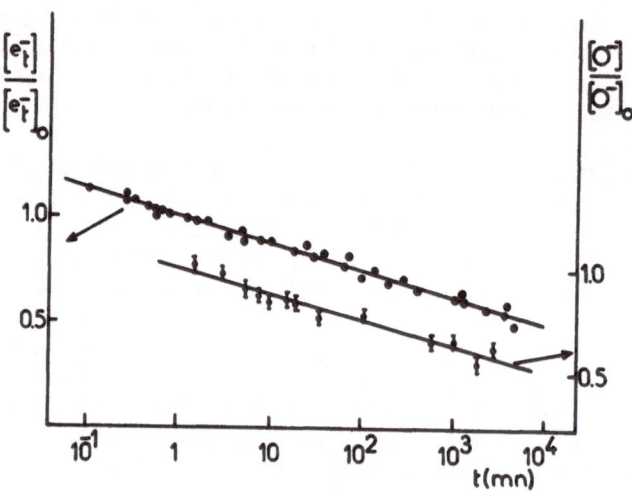

Figure 14. Decay curves of e_t^- and O^- in 10 M NaOH glass, from EPR measurements after β-and γ-irradiations to different doses, at 77 K and 120 K. (Zamaraev and Khairutdinov(33)).

on data obtained by EPR with β- and γ-irradiation, doses of about 5 to 50 Mrad, and temperatures of 77 and 120 K ; it includes data of a sample stored at 4 K between EPR measurements at 77 K. All these experimental points lie on the same straight lines.

The decay can be expressed by the following logarithmic law:

$$\frac{\left[e_t^-\right]}{\left[e_t^-\right]_0} = 1 - B \ln \left(\frac{t}{t_0}\right)$$

where $\left[e_t^-\right]$ is the concentration of e_t^- at time t_0, i.e. immediately after irradiation, and B is a constant.

The average distance between e_t^- and O^- is known from EPR data to be about 30-35Å, the true distance for the majority of pairs is > 20 Å.

All these observations agree with a long range electron transfer by a tunnelling mechanism : temperature-independent reaction rate, exclusion of diffusion between reactants too far apart to react, and logarithmic kinetics.

Similar decay kinetics were observed in different glasses, in the presence of scavengers. The earliest application of a tunnelling calculation to the decay of e_t^- seems to be a short note by Tsujikawa et al.(34), calculating the decay rate constant in MTHF as a function of ion-pair separation. Both rate constant and ion-pair separation had been determined experimentally from EPR measurements by Smith and Pieroni(35), and from their rate constant, Tsujikawa's graph gave a separation value in excellent agreement with the experimental value.

When the e_t^- spectrum does not decay uniformly, the decay observed at different wavelengths is found to follow a linear log t relationship, with a steeper slope at the longer wavelengths. An example of this is shown in Fig.15 in which relative concentra-

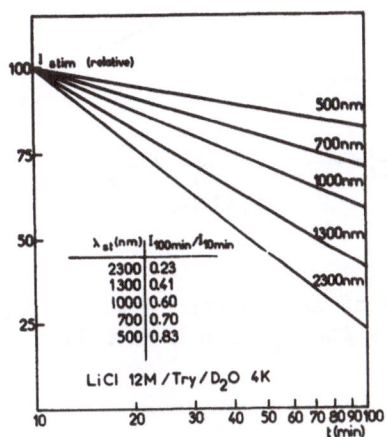

Figure 15. Decay of e_t^- in 12 M LiCl/D_2O, prepared by photoionisation of tryptophan, observed by stimulated luminescence response (proportional to $\left[e_t^-\right]$) in different parts of the spectrum (Kieffer and Klassen, unpublished results).

tions of e_t^- absorbing at different wavelengths were measured by the stimulated luminescence intensity at these wavelengths, at

4 K, in a LiCl/D_2O glass containing tryptophan. The e_t^- were prepared by the photoionisation of tryptophan.

In pulse radiolysis, the same decay law is valid on a much
shorter time scale, and the same wavelength dependence is apparent. A blue shift of the spectrum with time may even show up as
an increase in absorption at short wavelengths. This can be seen
in Fig.16 which shows that in 6 M NaOH at 77 K absorption decays
faster at 875 nm than at 750 nm between 10^{-6} and 10^2 seconds and
that at 550 and 400 nm, absorption actually increases up to
10^{-2}s. This indicates trap relaxation and deepening.

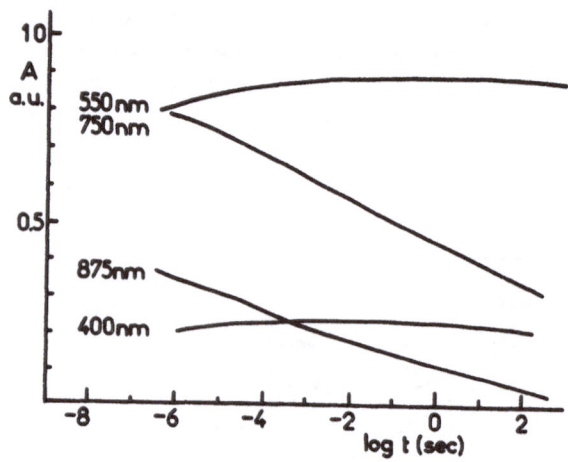

Figure 16. Change of absorbance A with time at different wavelengths for e_t^- in 6 M NaOH at 77 K (Miller
(25)).

The effect of temperature is shown in Fig. 17. Practically
no difference appears between 77 and 143 K in the evolution of
absorption at 550 nm. But at higher temperatures, the growth of
absorption at 550 nm lasts for shorter and shorter times, and
is followed by a fast decay. The contribution of diffusion to the
reaction of e_t^- with O^- becomes apparent above 143 K, accelerating
the decay of e_t^-.

Scavengers which react with e_t^- can have very different efficiencies. Thus Fig.18 shows the decay of e_t^- in 8 M NaOH in the
presence of two efficient scavengers (curve B) and of two inef-

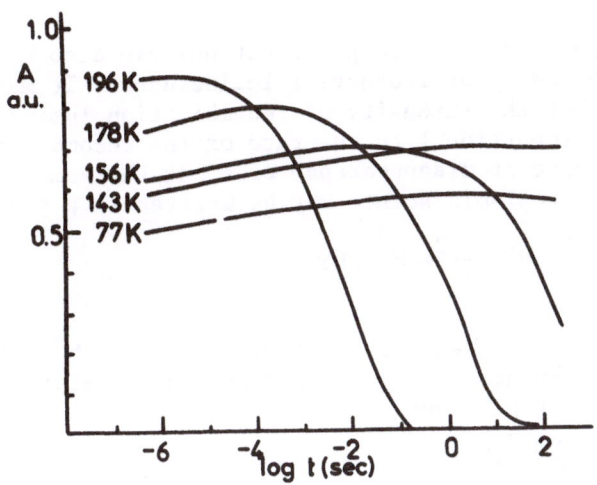

Figure 17. Effect of temperature on change of absorbance A at 550 nm for e_t^- in 6 M NaOH (Miller (25)).

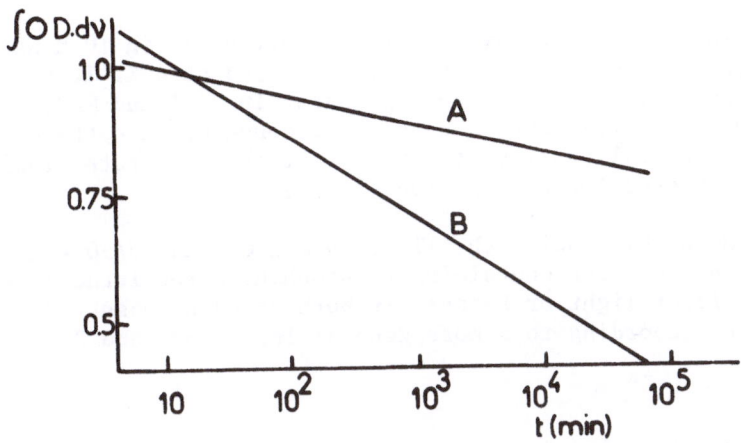

Figure 18. Decay of total trapped electron concentration (integrated over the absorption spectrum) in 8 M NaOH containing : curve A : inefficient scavengers (phenolate, chloroacetate), curve B : efficient scavengers (NO_2^-, NO_3^-). (Kroh and Stradowski(36)).

ficient scavengers, as the optical density integrated over the e_t^- spectrum versus log time. A much higher reaction rate is achieved with nitrate and nitrite ion than with phenolate and chloroacetate.

The disappearance of trapped electrons can also be followed by observing the decay of isothermal luminescence. It must simply be remembered that the intensity of recombination luminescence at any time is proportional to the rate of the recombination reaction, i.e. the rate of disappearance of e_t^- at the same time. The reaction leading to luminescence may be written in a general form:

$$e_t^- + M^+ \xrightarrow{\ 1\ } M^* \xrightarrow{\ 2\ } M + h\nu$$

Step 2 is a fast radiative deexcitation (and M^* can be expected to decay exponentially). Step 1 should therefore be the rate-determining step. Hence

$$I = k \left(- \frac{d\left[e_t^-\right]}{dt} \right)$$

and the decay of ITL should follow a kinetic law which is a differentiated form of that describing the decay of e_t^- , i.e. the linear relationship with log time. Such a law is

$$I = kt^{-1}$$

This form of decay has been observed(37) in 10 M NaOH glass, both pure and containing phenol, at 77 and at 4 K. It has also been observed in many organic glasses. In LiCl and $MgCl_2$ glasses, Buxton et al.(27) found a good coïncidence between the change with time in optical density at 2350 nm and the integrated luminescence intensity over the same period of time.

Moan(38) studied the ITL of 9 M LiCl and 50/50 ethylene glycol/H_2O glasses containing tryptophan, after irradiation with ultra-violet light or X-rays. In both cases he observed decay kinetics according to a more general law (Debye and Edwards law):

$$I = kt^{-m}$$

Plotting log I versus t, he obtained at 77 K linear plots (Fig. 19) with slopes m close to 1 in the absence of scavengers. He found that m was independent of tryptophan concentration and considered this as an argument in favour of geminate recombination and against the diffusion model on which Debye and Edwards(39) had based their development of the relation $I = kt^{-m}$, with values of m changing with solute concentration.

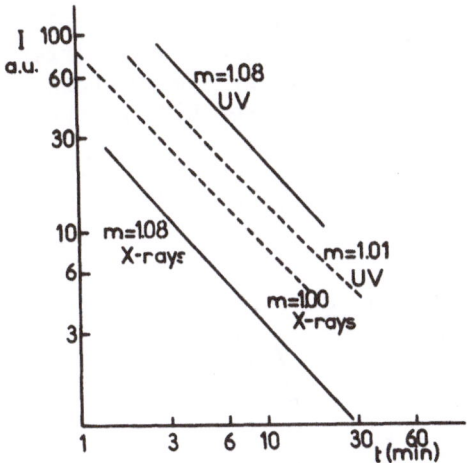

Figure 19. Decay of ITL at 77 K in 1/1 EG/H$_2$O (———)
and 9 M LiCl glasses (---) containing tryptophan, after
exposure to X-rays or to 254 nm UV light (Moan(38)).

Using the same graphical representation, Santus et al.(40)
found linear log I vs. log t relationships, with values of m fairly close to 1 in 8 M KOH and EG/H$_2$O glasses containing tryptophan
or 1-methyltryptophan. In very elegant experiments Gillis and
Walker(41) produced e$_{IR}$ in γ-irradiated LiCl and MgCl$_2$ glasses
(containing only e$_{vis}$) by 694 nm light from a Q-switched ruby
laser. They were then able to follow the decay of e$_{IR}$ by absorption in the wavelength region 1340–1780 nm and found the usual
logarithmic decay between 2 μs and 1 ms. They also studied the
decay of the recombination luminescence over 8 decades of time
from 100 ns to 10 s and found a linear log I vs. log t plot with
a slope m = 1 $\pm$ 0.05 at times longer than about 10 μs. The initial part of their curve is a hump which looks very much like an
exponential decay in these coordinates, and corresponds to some
superimposed fluorescence process. Their conclusions are :
 - the transitions involved in the 694 nm absorption are
   bound-free transitions ;
 - the quantum yield for this electron release is at least
   0.12 ;
 - luminescence is observed only in those aqueous glasses
   which form e$_{IR}$ (no luminescence in hydroxide and K$_2$CO$_3$
   glasses under these conditions);

- e$_{IR}^{-}$ do not significantly convert to e$_{vis}^{-}$, but decay in a manner consistent with tunnelling to geminate reaction partners. We shall see that this last conclusion is now seriously challenged.

Klassen et al.(42) have recently tried to correlate decays of e$_{IR}^{-}$, e$_{vis}^{-}$ and Cl$_2^{-}$ in LiCl/D$_2$O glasses at three different LiCl concentrations, 6, 9.5 and 12 M, at 72 K. Fig.20 shows the absorption spectra of these species in the three LiCl glasses. The

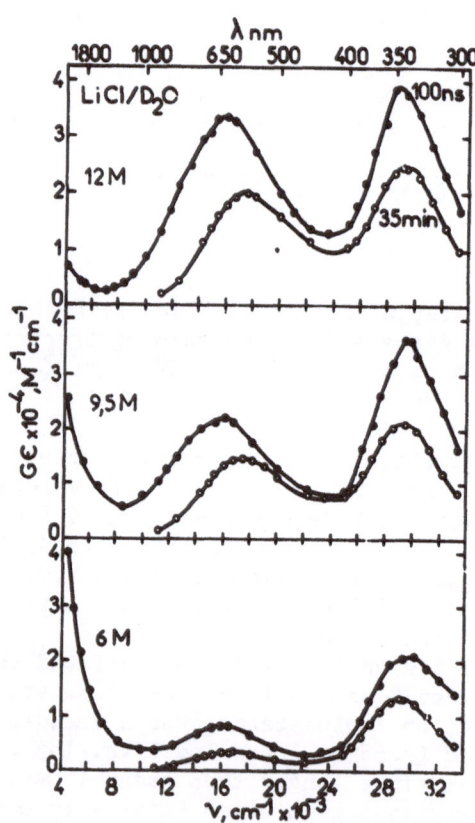

Figure 20. Absorption spectra of irradiated LiCl/D$_2$O glasses at 72 K. ● measured 100 ns after start of a 40 ns pulse, o measured 35 min after 1-6 40 ns pulses. (Klassen et al.(42)).

different reactions considered probable in irradiated $LiCl/D_2O$ glasses are

$$D_2O \xrightarrow{} D_2O^+ + e^-$$

$$Cl^- \xrightarrow{} Cl + e^-$$

$$e^- \longrightarrow e^-_{vis}, e^-_{IR}$$

$$D_2O^+ + Cl^- \longrightarrow D_2O + Cl \qquad (1)$$

$$D_2O^+ + D_2O \longrightarrow D_3O^+ + OD \qquad (2)$$

$$OD + Cl^- \longrightarrow OD^- + Cl \qquad (3)$$

$$Cl + Cl^- \longrightarrow Cl_2^- \qquad (4)$$

$$(e^-, e^-_{vis}, e^-_{IR}) + D_2O^+ \longrightarrow D_2O \qquad (5)$$

$$(e^-, e^-_{vis}, e^-_{IR}) + D_3O^+ \longrightarrow D_2O + D \qquad (6)$$

$$(e^-, e^-_{vis}, e^-_{IR}) + Cl_2^- \longrightarrow 2\ Cl^- \qquad (7)$$

$$(e^-, e^-_{vis}, e^-_{IR}) + OD \longrightarrow OD^- \qquad (8)$$

The authors are mainly interested in reaction (7). As can be seen in Fig. 21, when they bleached e^-_{vis} by light of $\lambda > 450$ nm (i.e. not absorbed by Cl_2^-), they found that e^-_{vis} and Cl_2^- decayed to practically the same degree ; this indicates that photobleached e^-_{vis} decays almost totally by reaction (7). From the dark decays, in which a decrease in e^-_{IR} is accompanied by some increase in the e^-_{vis} band, at different times in 9.5 and 6 M $LiCl/D_2O$, they concluded that there is at least a 6% conversion of e^-_{IR} to e^-_{vis} between 10^{-7} and 10^{-5}s in 9.5 M glass and between 10^{-4} and 10^{-1}s in 6 M glass. Similarly it appeared that, between 10^{-7} and 10^{-4}s, the dark decay of the e^-_{vis} takes place almost entirely by reaction (7), whereas practically no e^-_{IR} takes part in this reaction. Only e^-_{IR} decay can be correlated with light emission(27).

In the time interval 10^{-4} - 2100 s, it appears that Cl^- must also take part in another reaction than (7), but nothing is known about this reaction.

As to the environment of e^-_{vis} and e^-_{IR}, these authors think that probably e^-_{vis} is trapped in the vicinity of Cl_2^-, hence in water it is initially associated with Cl^-, i.e. in regions of solvated Li^+ and Cl^- ions. On the contrary, e^-_{IR} seems trapped away from Cl_2^-, i.e. probably in a region of free water.

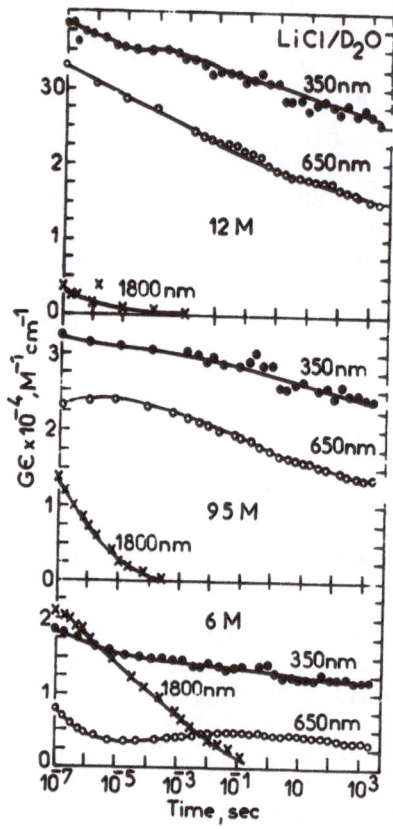

Figure 21. Changes in the three absorption bands of
irradiated LiCl/D_2O glasses at 72 K (Klassen et al.(42)).

6. CONCLUSION

From the foregoing it is obvious that the problems of "presolvated" and "solvated" electrons are far from solved. In some
aqueous glasses, such as NaOH, only a slight red shift of the
e_t^- absorption spectrum is observed at low temperatures or at
short times. The red tail is attributed to presolvated electrons,
and photoconductivity experiments show that for both solvated and
presolvated electrons, the transitions involved are predominantly
bound-continuum transitions. In other glasses; such as LiCl,

and also ethylene glycol-water mixtures, the presolvated electron band is displaced much further into the infra-red. In these matrices there is an important contribution of bound-bound transitions in both the presolvated and the solvated state. It can be a matter of faith to decide whether there is or not a conversion of presolvated to solvated electrons. Thus whilst in EG/H_2O such a conversion has always seemed obvious from experimental data, in LiCl glass it had first seemed that presolvated electrons disappeared essentially by recombination with geminate partners, but it appeared later that a conversion to the solvated state was not negligible.

The geometrical structure of the solvated electron with an equilibrium configuration of solvent molecules is now well known, thanks to various types of very refined electron magnetic resonance studies such as electron spin echo modulation. Kevan(43) has determined the distances and orientations of the surrounding molecules of the first solvation shell in several glasses. For aqueous glasses he proposes the structure shown in Fig. 22. The electron is surrounded by six approximately equivalent water molecules, with one OH bond of each oriented towards the electron.

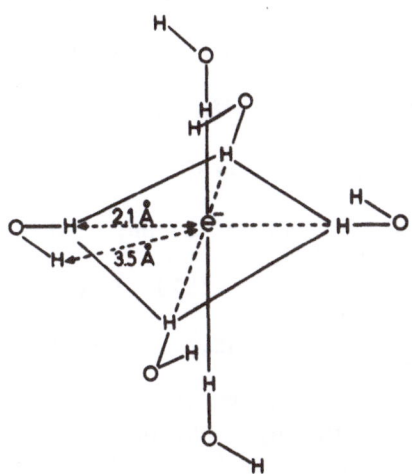

Figure 22. Geometrical structure of the solvated electron in aqueous glasses, from electron spin echo modulation analysis. (Kevan(43)).

No direct information of this type is as yet available on presolvated electrons in their non-equilibrium configuration of solvent molecules. Their solvation shell consists of disoriented

molecular dipoles around the electron, and by comparison with
the solvated state it has been shown that the average difference
in electron-proton distance between the average structures of
presolvated and solvated states is only 0.2-0.3 Å, so that we do
have an approximate knowledge of the geometrical structure of
presolvated electrons too.

On the contrary, except for crystalline ice, nothing is
known about the formation of the presolvated electron which, in
all nanosecond work, even at 4 K, was found to be present by the
end of the pulse, and it may well require picosecond techniques,
combined with cooling to 4 K or less, to obtain some results on
the initial localisation of the electron.

REFERENCES

1. Materials Advisory Board, Nat. Ac. Res. Council 1968, p. 243.
2. e.g. Voltz, R., Klein, J., Heisel, F., Lami, H., Laustriat, G.,
   and Coche, A. : 1966, J. Chim. phys. 63, pp. 1259-1264.
3. Eyring, H., Hirschfelder, J. O., and Taylor, H. : 1936,
   J. Chem. Phys. 4, pp. 479-491 ; 570-575.
4a. Hart, E.J., and Boag, J.W. : 1962, J. Am. Chem. Soc. 84, pp.
    4090-4095.
4b. Schulte-Frohlinde, D., and Eiben, K. : 1962, A. Naturf. 17a,
    pp. 445-446.
5. Land, E.J., and Swallow, A.J. : 1968, J. Chem. Phys. 49,
   pp. 5552-5553.
6. Richards, J. T., and Thomas, J. K. :1970, J. Chem. Phys. 53,
   pp. 218-224.
7. Higashimura, T., Noda, M., Warashina, T. and Yoshida, H. : 1970,
   J. Chem. Phys. 53, pp. 1152-1155.
8. Baxendale, J. H., and Wardman, P. : 1973, J. Chem. Soc. Faraday Trans. I 69, pp. 584-594.
9. Gilles, L., Aldrich, J. E., and Hunt, J. W. : 1973, Nature
   Phys. Sc. 243, pp. 70-72.
10. van Leeuwen, J. W. : 1981, Ph. D. Thesis, Utrecht.
11. Kawabata, K. : 1976, J. Chem. Phys. 65, pp. 2235-2242.
12. Buxton, G. V., Gillis, H. A., and Klassen, N. V. : 1977,
    Can. J. Chem. 55, pp. 2385-2395.
13. Steen, H. B., and Holteng, J.A. : 1975, J. Chem. Phys. 63,
    pp. 2690-2697.
14. Warman, J. M., de Haas, M. P., and Verberne, J. B. : 1980,
    J. Phys. Chem. 84, pp. 1240-1248.
15. Warman, J. M., and Jonah, C. : 1981, Chem. Phys. Letters 79,
    pp. 43-46.
16. Trudel, G. J., Gillis, H. A., Klassen, N. V., and Teather, G. G. :
    1981, Can. J. Chem. 59, pp. 1235-1240.
17. Moorthy, P. N., and Weiss, J. J. : 1964, Phil. Mag. 10,
    pp. 659-674.

18. Ershov, B. G., Pikaev, A. K., Glazunov, P. I., and Spitsyn, V. I. : 1963, Dokl. Akad. Nauk SSSR 149, pp. 363-366.
19. Kevan , L. : 1972, J. Phys. Chem. 76, pp.3830-3838.
20. Klassen, N. V., Teather, G. G., and Kieffer, F. : 1979, Can. J. Chem. 57, pp. 1488-1499.
21. Hase, H., Noda, M., and Higashimura, T. : 1871, J. Chem. Phys. 54, pp. 2975-2978.
22. Dolivo, G.,and Kevan, L. : 1979, J. Chem. Phys. 70, pp. 2599-2604.
23. Baxendale, J. H., and Sharpe, P. H. G. : 1976, Int. J. Rad. Phys. Chem. 8, pp. 621-624.
24. Rice, S. A., Dolivo, G., and Kevan, L. : 1979, J. Chem. Phys. 70, pp. 18-25.
25. Miller, J. R. : 1975, J. Phys. Chem. 79, pp. 1070-1078.
26. Dolivo, G., and Kevan, L. :1979, J. Chem. Phys. 70, pp. 5489-5493.
27. Buxton, G. V., Gillis, H. A., and Klassen, N.V. : 1976, Can. J. Chem. 54, pp. 367-381.
28. Nguyen, T. Q., Walker, D. C., and Gillis, H. A. : 1978, J. Chem. Phys. 69, pp. 1038-1043.
29. Rice, S. A., Dolivo, G., and Kevan, L. : 1978, J. Chem. Phys. 68, pp. 4864-4869.
30. Ershov, B. G., Makarov, I. E., and Pikaev, A. K. : 1967, Khim. Vysok. Energ. 1, pp. 472-479.
31. Cygler, J., Gillis, H. A. , Klassen, N. V., and Teather, C. G. : 1981, Rad. Phys. Chem. 17, pp. 379-382.
32. Gillis, H. A., Teather, G. G., and Buxton, G. V. : 1978, Can. J. Chem. 56, pp. 1889-1897.
33. Zamaraev, K. I., and Khairutdinov, R. F. : 1974, Chem. Phys. 4, pp. 181-195.
34. Tsujikawa, H., Fueki, K., and Kuri, Z. : 1965, Bull. Chem. Soc. Jap. 38, p.2210.
35. Smith, D. R., and Pieroni, J. J. : 1965, Can. J. Chem. 43, pp. 876-887.
36. Kroh, J., and Stradowski, C. : 1975, Int. J. Rad. Phys. Chem 7, pp. 23-38.
37. Ershov, B. G., and Kieffer, F. : 1974, Chem. Phys. Letters 25, pp. 576-578.
38. Moan, J. : 1974, J. Chem. Phys. 60, pp. 3859-3865.
39. Debye, P., and Edwards, J.O. : 1953, J. Chem. Phys. 20, pp. 236-239.
40. Aubailly, M., Bazin, M., and Santus, R. : 1975, Chem. Phys. Letters 31, pp. 340-343.
41. Gillis, H. A., and Walker, D. C. : 1976, J. Chem. Phys. 65, pp. 4590-4595.
42. Klassen N. V., Adams, R. J., Teather, G. G., and Ross, C. K. : 1980, J. Phys. Chem. 84, pp. 3609-3613.
43. Kevan, L. : 1980, J. Phys. Chem. 84, pp. 1232-1240 ; 1981, Acc. Chem. Res. 14, pp. 138-145.

LABILE SPECIES AND FAST PROCESSES IN LIQUID ALCOHOL RADIOLYSIS

Gordon R. Freeman

Chemistry Department, University of Alberta Edmonton, Canada, T6G 2G2

ABSTRACT

The radiolysis of liquid ethanol is used to illustrate the types of labile species and fast processes that occur during the radiolysis of alcohols. Reactive solutes alter product yields. Kinetic analysis of product yields as functions of solute concentration, temperature and pressure illuminate the processes. The analysis involves known physical properties of the solvent, such as dielectric constant and dielectric relaxation times, and their temperature dependences. More detailed information about electron solvation and reaction rates has been obtained by various pulse techniques.

J. H. Baxendale and F. Busi (eds.),
The Study of Fast Processes and Transient Species by Electron Pulse Radiolysis, 399–416.

INTRODUCTION

The main products from the radiolysis of a liquid C_n alcohol are hydrogen, a C_n carbonyl compound and a C_{2n} diol [1]. The search for details of the reaction mechanism necessarily uncovered properties of the reaction intermediates. Fascination with rapidly improving electronic devices during the past decade has focussed attention almost exclusively on the reaction intermediates, and the overall mechanism has faded into the background. For example, kinetic analysis of electron scavenging yields in alcohols as a function of temperature and solute concentration included the effect of the time dependence of the dielectric constant as the solvent relaxed around each newly created charge [2-4]. Information about the relaxation process has been extended by time resolved optical absorption spectrum measurements of localized electrons at low temperatures [5-7] and short times down to ~ 10^{-11} s [8-13].

One hopes that in the not too distant future there will be new experimental and theoretical attacks on the broad front of the overall reaction mechanisms of radiolysis. Much remains to be done. A large problem that needs attention is the stochastic treatment of the kinetics of reactions in spurs that contain more than one ion-electron pair, and mixtures of ions, electrons and neutral radicals. A subtopic in this problem is the mechanisms of electron and positive charge migration in polar and nonpolar liquids.

The present discussion of liquid alcohols will begin with the overall mechanism, using ethanol as an example. Some of the processes will then be examined in more detail.

OVERALL REACTION

Major Products

The stoichiometric equation for the formation of the major products in ethanol are [14]:

$$C_2H_5OH \longrightarrow\hspace{-1.2em}\rotatebox{0}{\sim\!\!\sim\!\!\sim}\hspace{0.3em} H_2 + CH_3CHO$$

$$2C_2H_5OH \longrightarrow\hspace{-1.2em}\rotatebox{0}{\sim\!\!\sim\!\!\sim}\hspace{0.3em} H_2 + (CH_3CHOH)_2$$

The yields of radiolysis products are conventionally listed as G values, which are the number of molecules of the product formed per 100 eV of energy absorbed by the system. The G values of hydrogen, acetaldehyde and 2,3-butanediol are listed in Table 1.

TABLE 1

γ Radiolysis of liquid ethanol: major product yields[a]

Product	G		
	0.01^{b}	1^{b}	100^{b}
H_2	5.8 ± 0.1	5.1 ± 0.2	4.15 ± 0.15
CH_3CHO	3.7 ± 0.2	3.0 ± 0.2	2.0 ± 0.2
$(CH_3CHOH)_2$	2.1 ± 0.4	2.1 ± 0.4	2.1 ± 0.4

a. T = 295 ± 4K, dose rate = $(0.1 - 100)10^{15}$ eV/gs. Ref. 14.
b. Dose in units of 10^{18} eV/g.

     The yields of hydrogen and acetaldehyde decrease with
increasing irradiation dose (Table 1). This means that one or
both of the products react with radiolysis intermediates, such
as free radicals, ions or electrons. The subsequent reaction of
a product is called a secondary reaction; it destroys the product
and alters the intermediate. The yield of 2,3-butanediol is independent of dose (Table 1), so the secondary reactions do not alter
the yield of the precursor of the diol. A mechanism that is consistent with these observations is as follows.

$$C_2H_5OH \xrightarrow{\quad\quad} C_2H_5OH^+ + e^- \tag{1}$$

$$e^- + nC_2H_5OH \rightarrow e^-_s \tag{2}$$

$$C_2H_5OH^+ + C_2H_5OH \rightarrow C_2H_5O\cdot + C_2H_5OH_2^+ \tag{3}$$

$$e^-_s + C_2H_5OH_2^+ \rightarrow H + C_2H_5OH \tag{4}$$

$$e^-_s \rightarrow C_2H_5O^-_s + H \tag{5}$$

$$C_2H_5O^-_s + C_2H_5OH_2^+ \rightarrow 2C_2H_5OH \tag{6}$$

$$C_2H_5O\cdot + C_2H_5OH \rightarrow C_2H_5OH + CH_3\dot{C}HOH \tag{7}$$

$$H + C_2H_5OH \rightarrow H_2 + CH_3\dot{C}HOH \tag{8}$$

$$2CH_3\dot{C}HOH \rightarrow CH_3CHO + C_2H_5OH \tag{9}$$

$$2CH_3\dot{C}HOH \rightarrow (CH_3CHOH)_2 \tag{10}$$

where e^-_s is a solvated electron. We will examine details of some of these reactions later, but for the moment we need only note that the ions and electrons ultimately generate H atoms and alcohol radicals. The H atoms react with alcohol to form H_2 and $CH_3\dot{C}HOH$ radicals; reaction (8) is exothermic by ~ 45 kJ/mol [15]. Reaction (7) is exothermic by ~ 35 kJ/mol [15], so it probably occurs to a significant extent. If any $C_2H_5O\cdot$ radicals survive long enough to react with other radicals, the most plentiful in the system being $CH_3\dot{C}HOH$, the disproportionation products would be acetaldehyde and ethanol, the same as those of (9); the combination product would be ethyl hemiacetal, $CH_3CH(OH)OC_2H_5$, which decomposes easily to acetaldehyde and ethanol and would probably have been analyzed as such.

The more reactive radicals H and $C_2H_5O\cdot$ produce the less reactive $CH_3\dot{C}HOH$ radicals, which can only wander around in the solution until they meet each other, then disproportionate to acetaldehyde and ethanol or combine to form 2,3-butanediol. We are neglecting exchange reactions with the solvent, because they do not change the composition of the solution: $CH_3\dot{C}HOH$ + $C_2H_5OH \longrightarrow C_2H_5OH + CH_3\dot{C}HOH$.

The concentrations of products in the solution increase with dose. The acetaldehyde can react with solvated electrons.

$$e^-_s + CH_3CHO \longrightarrow CH_3CHO^-_s \tag{11}$$

$$CH_3CHO^-_s + C_2H_5OH^+_2 \longrightarrow CH_3\dot{C}HOH + C_2H_5OH \tag{12}$$

The loss of a solvated electron to (11) prevents the formation of an H_2 through (4) or (5) plus (8). Thus the net yields of hydrogen and acetaldehyde are reduced by (11) at high doses. However, the $CH_3\dot{C}HOH$ radical that would have been formed by (8) is replaced by that produced in (12), so there is no change in the butanediol yield or in the gross rate of formation of acetaldehyde (Table 1).

The dose effect on the hydrogen and acetaldehyde yields can be removed by the addition of a mineral acid, such as hydrochloric or sulfuric. Then (11) cannot compete with (4).

Minor Products

A small amount of methane was produced by the radiolysis of ethanol (Table 2). This means that a small amount of C–C bond cleavage occurred.

$$CH_3CH_2OH \xrightarrow{\wedge\wedge\wedge} \cdot CH_3 + \cdot CH_2OH \tag{13}$$

The methyl radicals would abstract hydrogen from ethanol, the reaction being exothermic by ~ 42 kJ/mol.

$$\cdot CH_3 + C_2H_5OH \longrightarrow CH_4 + CH_3\dot{C}HOH \qquad (14)$$

The $\cdot CH_2OH$ radicals are less reactive than $\cdot CH_3$, which is generally true when comparing alkyl radicals and their α-hydroxy analogs. Thus $\cdot CH_2OH$ mainly undergoes disproportionation and combination with other radicals, the most abundant in solution being $CH_3\dot{C}HOH$.

$$\cdot CH_2OH + CH_3\dot{C}HOH \longrightarrow CH_2O + CH_3CH_2OH \qquad (15)$$

$$\cdot CH_2OH + CH_3\dot{C}HOH \longrightarrow CH_3OH + CH_3CHO \qquad (16)$$

$$\cdot CH_2OH + CH_3\dot{C}HOH \longrightarrow CH_3CH(OH)CH_2OH \qquad (17)$$

The measured yields of methane, formaldehyde, 1,2-propanediol and carbon monoxide, the last resulting from decomposition of $\cdot CH_2OH$ fragments that were formed in highly excited states, indicate that $G(CH_3OH) \approx 0.1 - 0.2$. This small amount of methanol would not have been detected, due to interference by ethanol.

The product water (Table 2) means that C-O bond cleavage

TABLE 2

γ Radiolysis of liquid ethanol: minor product yields[a]

Product	G		
CH_4	0.6		
H_2O	0.5		
C_2H_6	0.2		
C_2H_4	0.2		
H_2CO	0.2		
$CH_3CH(OH)CH_2OH$	0.13	R_1R_3[b]	
CO	0.1		
$CH_3CH(OH)C_2H_5$	0.08	R_1R_4[b]	
$CH_3CH(OH)CH_2CH_2OH$	0.05	R_1R_2[b]	

a. T = 298K, dose $\approx$ (10 - 100)10^{18} eV/g, dose rate $\approx$ (1 - 10)10^{15} eV/g s. Ref. 14.

b. $CH_3\dot{C}HOH = R_1$, $\dot{C}H_2CH_2OH = R_2$, $\dot{C}H_2OH = R_3$, $CH_3\dot{C}H_2 = R_4$.

also occurred.

$$C_2H_5OH \xrightarrow{\wedge\!\wedge\!\wedge} \cdot C_2H_5 + \cdot OH \qquad (18)$$

The hydroxyl radicals abstract from ethanol, while the ethyl radicals react mainly with other radicals.

$$\cdot OH + C_2H_5OH \longrightarrow H_2O + CH_3\overset{\bullet}{C}HOH \qquad (19)$$

$$\cdot C_2H_5 + CH_3\overset{\bullet}{C}HOH \longrightarrow C_2H_4 + CH_3CH_2OH \qquad (20)$$

$$\cdot C_2H_5 + CH_3\overset{\bullet}{C}HOH \longrightarrow C_2H_6 + CH_3CHO \qquad (21)$$

$$\cdot C_2H_5 + CH_3\overset{\bullet}{C}HOH \longrightarrow CH_3CH(OH)C_2H_5 \qquad (22)$$

The H abstraction reactivity of $\cdot C_2H_5$ (actually $\cdot CH_2CH_3$) is similar to that of $\cdot CH_2OH$ and much smaller than that of $\cdot CH_3$ [15]. The reactivity decreases with increasing electron donating strength of groups α to the radical position.

For similar reasons the hydrogen atom abstracted from ethanol is mainly from the carbon attached to the hydroxyl group; reactions (7), (8), (14), (19). The yield of product that incorporates $\cdot CH_2CH_2OH$ radicals, that is 1,3-butanediol, is very small (Table 2). Judging from the yields of radical combination products, the average value of the ratio of rate constants k_{23}/k_{24} is $\geqslant 70$ for the radicals R· in the liquid at 298K.

$$R\cdot + CH_3CH_2OH \longrightarrow RH + CH_3\overset{\bullet}{C}HOH \qquad (23)$$

$$R\cdot + CH_3CH_2OH \longrightarrow RH + \cdot CH_2CH_2OH \qquad (24)$$

This concludes the brief discussion of minor products. We return now to the main mechanism.

NONHOMOGENEOUS KINETICS AND FAST PROCESSES

Product yields may be altered by the addition of a solute to the liquid prior to irradiation. For example, addition of nitrous oxide reduces the hydrogen yield and produces nitrogen. Reaction (25) competes with (4) and (5).

$$e^-_s + N_2O \longrightarrow N_2 + O^-_s \qquad (25)$$

The yields of nitrogen from solutions of nitrous oxide in ethanol at 161K and 363K are shown in Figure 1 [16]. Kinetic analysis [17] of these and other results, such as the effect of the addition of a strong acid to increase the contribution of reaction (4), requires refinement of the mechanism represented by reactions (1) –

(10). For example, the cross-over of the curves in Figure 1 indicates that the reaction kinetics are not the same at $[N_2O] < 10^{-3}M$ as at $[N_2O] > 10^{-3}M$. The shapes of the curves at the lower concentrations are governed by the homogenous kinetics of reactions in normal solutions. The curve shapes at the higher concentrations are determined by the nonhomogeneous kinetics of reactions in spurs.

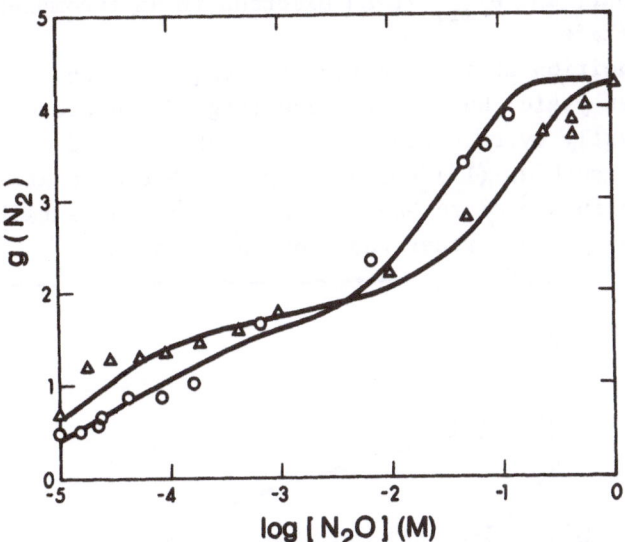

Fig. 1 – Nitrogen yields from nitrous oxide solutions in ethanol at 161K ($\triangle$) and 363K ($\bigcirc$). The yields were corrected for a small amount of nitrogen formed by the direct radiolysis of nitrous oxide at $[N_2O] > 0.1$ M, so are represented by $g(N_2)$. The experimental points are from ref. 16. The curves were calculated from a kinetics model that included the time dependent dielectric constant, ref. 17.

The preceding reactions (1) - (4) are therefore subdivided and extended as follows [17].

$$C_2H_5OH \xrightarrow{\rotatebox{0}{\wedge}} [C_2H_5\dot{O}H^+ + e^-] \qquad (1')$$

$$\xrightarrow{\rotatebox{0}{\wedge}} [C_2H_5\dot{O}H^{+*} + e^-] \rightarrow [CH_3CHOH^+ + H + e^-] \qquad (1'')$$

$$[e^- + nC_2H_5OH] \rightarrow [e^-_{irl}] \rightarrow [e^-_s] \qquad (2')$$

$$[C_2H_5\dot{O}H^+ + C_2H_5OH] \rightarrow [C_2H_5O\cdot + C_2H_5OH_2^+] \qquad (3')$$

$$[CH_3CHOH^+ + C_2H_5OH] \rightarrow [CH_3CHO + C_2H_5OH_2^+] \qquad (3'')$$

$$[e^-,\ e^-_{irl},\ e^-_s + C_2H_5\overset{\bullet+}{O}H^+,\ CH_3\overset{+}{C}HOH^+,$$
$$C_2H_5\overset{+}{OH_2}] \longrightarrow [\text{geminate neutralization}](4')$$
$$\longrightarrow e^-_s + C_2H_5\overset{+}{OH_2} \ (\text{free ions}) \ (4'')$$

where the square brackets indicate that the enclosed species are within a spur, and e^-_{irl} is an electron in an incompletely relaxed localized state.

Addition of the electron scavenger nitrous oxide reduces the hydrogen yield, but not to zero (Fig. 2). Hydrogen is therefore produced partly by reactions that are not affected by electron scavenging, such as (1") followed by (8). Hydrogen atoms are proposed as an intermediate because the olefin 1,3-pentadiene reduces the hydrogen yield farther than does nitrous oxide [17,18].

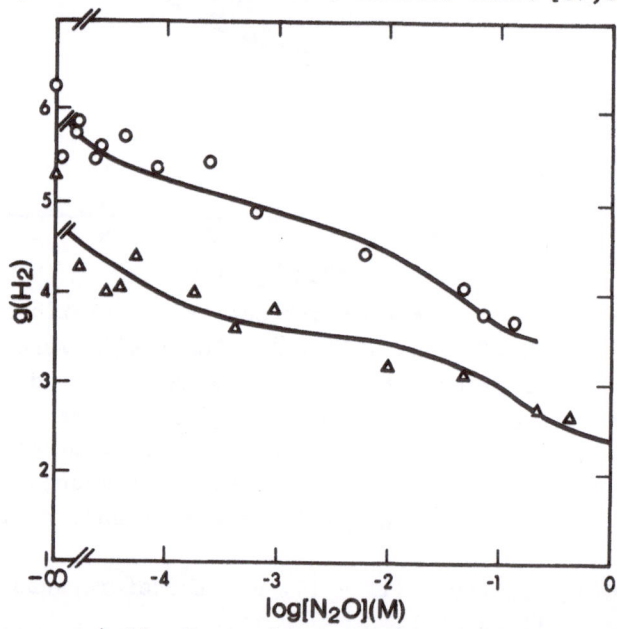

Fig. 2 - Hydrogen yields from nitrous oxide solutions in ethanol at 161K (△) and 363K (○). The curves were calculated from a kinetics model that included the time dependent dielectric constant. Ref. 17.

Reactions (3') and (3") appear to be rapid [18]. In the liquid phase they correspond to the shift of a proton along a hydrogen bond so that the normal O-H and hydrogen bonds switch places

$$-\overset{\bullet+}{O}-H--\alpha \quad \longrightarrow \quad -\overset{\bullet}{O}--H\overset{+}{-\alpha}$$
$$=\overset{+}{O}-H--\alpha \quad \longrightarrow \quad =O--H\overset{+}{-\alpha}$$

The reactions are accompanied by rearrangement of solvent dipoles in the field of the ion, so the final $C_2H_5OH_2^+$ would be more precisely represented by $C_2H_5OH_2^+{}_{,s}$ to designate the solvated state.

The electrons generated by the radiation in reaction (1) initially possess excess kinetic energy. They lose the energy through collisions with the liquid. Most of them become thermalized before they have escaped the coulomb field of their parent ions, so some of them are drawn back to undergo geminate neutralization. Thermal energy tends to cause the species to diffuse at random, and some of the ions and electrons wander far enough from each other that their mutual coulombic attraction becomes small compared to thermal energy kT. They are then "free ions", not correlated in pairs. The fraction $\emptyset_{fi}$ of ions and electrons that become free ions is related to properties of the solvent through equation (26) [19].

$$\emptyset_{fi} = \exp(-E_c/kT) = \exp(-\xi^2/\varepsilon ykT) \qquad (26)$$

where E_c is the coulombic interaction energy of the ion and electron at the instant of thermalization, y is the distance between them at that time, ξ is the electronic charge, ε is the dielectric constant of the medium between the ion and electron, k is Boltzmann's constant and T is the Absolute temperature.

In a given system all of the electrons do not have the same thermalization range, that is, there is a distribution of y values, so equation (26) should be written:

$$\emptyset_{fi} = \int_0^\infty N(y)\exp(-\xi^2/\varepsilon ykT)dy \qquad (26')$$

where N(y)dy is the fraction of thermalized electron-ion pairs that have initial separation distances between y and y + dy.

Reactions (2') and (4') are written in a manner that indicates that they occur over time periods that may overlap. The electron-ion pairs that have the smallest y values undergo geminate neutralization before the electrons and ions are fully solvated, and perhaps even before the electrons form a localized state. The dielectric constant of the liquid between the ion and electron increases with time (Fig. 3) over a period that coincides with the occurrence of (2'), (3') and (3"). In fact, essentially the same solvent rearrangement results in the increased ε and the settling of the electron and ion into the equilibrium solvated state. The time required is ~ 10^{-11} s in ethanol at 300K [2,12]. The value of ε used is therefore an average over the time required

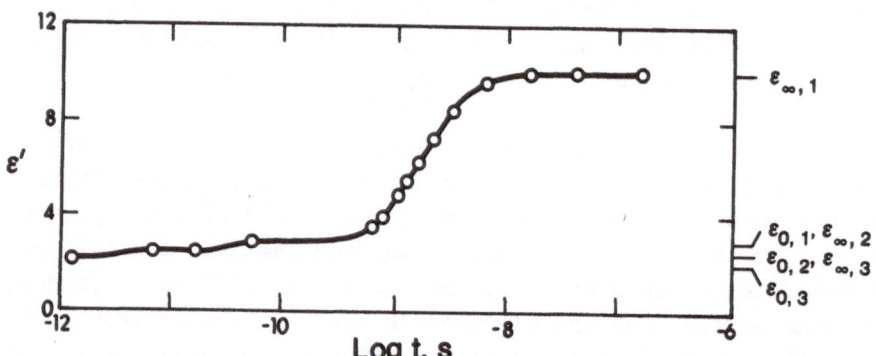

Fig. 3 – Dielectric constant of 1-octanol at 293K as a function of
time after the instantaneous application of a constant
electric field. Data from ref. 23, with the substitution
$t = (2\pi f)^{-1}$, where f = oscillating field frequency.
$\varepsilon_{0,3} = 2.2$, $n_D^2 = 2.04$.

for reactions (4') and (4") to occur, which is dependent on y.
Equation (26') becomes

$$\phi_{fi} = \int_0^\infty N(y)\exp(-\xi^2/\varepsilon_{av}ykT)dy \tag{26"}$$

The average dielectric constant may be approximated by [20]:

$$\varepsilon_{av} = \varepsilon_\infty + \frac{\varepsilon_0 \tau}{t_{gn}} \ln\left(\frac{1 + [(\varepsilon_\infty - \varepsilon_0)/\varepsilon_0]\exp(-\varepsilon_\infty t_{gn}/\varepsilon_0 \tau)}{1 + [(\varepsilon_\infty - \varepsilon_0)/\varepsilon_0]}\right) \tag{27}$$

where ε_∞ is the value of the dielectric constant at $t = \infty$ (often
designated ε_s), ε_0 is the value at $t = 0$, τ is the dielectric relaxation time of the liquid, and the geminate neutralization time
t_{gn} varies approximately as y^3 [2,21],

$$t_{gn} \approx 2.3 \times 10^6 \varepsilon_{av}y^3/(\mu_+ + \mu_-). \tag{28}$$

where μ_+ and μ_- are the mobilities of the charges. Equations (27)
and (28) are solved by successive approximations.

The value of τ to use in equation (27) is not one of the
conventionally listed values and can only be estimated approximately. The dielectric relaxation time of a liquid is commonly determined from its microwave absorption spectrum. The energy of a
microwave photon in the region of the absorption maximum, which
is usually at greater than 1 cm wavelength, is less than 1×10^{-4}

eV. This is much smaller than kT, which is 2.6×10^{-2} eV at 25°C. The microwaves therefore do not appreciably perturb the normal thermal agitation of the system during the measurement. However, the sudden creation of a charged species in a liquid would significantly alter the normal thermal motion of the neighboring molecules. The ion-dipole interactions would increase the speed with which the nearest-neighbor dipoles would line up with the field of the ion, thereby decreasing the dielectric relaxation time in the immediate vicinity of the ion.

From an Arrhenius plot of conventional τ values for ethanol at different temperatures [22,23] one may determine that the main dipole reorienation in liquid ethanol has an activation energy of 18 kJ/mol and that the unperturbed relaxation time τ_{up} is given by equation (29).

$$\begin{aligned} \tau_{up} &= 8 \times 10^{-14} \, e^{18,000/RT} \\ &= 8 \times 10^{-14} \, e^{2200/T} \quad , \text{ s} \end{aligned} \tag{29}$$

The ion-dipole interaction exerts a torque on the dipole, which reduces the activation energy of the rotation. For randomly oriented molecules, the average torque U exerted on a molecule with dipole moment μ by a singly charged ion at a distance r through a medium of dielectric constant ε is roughly

$$U = \int_0^{\pi/2} \frac{\xi\mu}{\varepsilon r^2} \sin\theta \, d\theta \Big/ \int_0^{\pi/2} d\theta$$

$$= 0.64 \, \xi\mu/\varepsilon r^2 \tag{30}$$

Putting $\xi = 1.60 \times 10^{-19}$ C, μ in units of Cm, r in m, converting c^2 to 9×10^9 Jm, and multiplying U by Avogadro's number N_0, equation (30) becomes

$$UN_0 = 5.5 \times 10^{14} \, \mu/\varepsilon r^2 \, , \text{ J/mol.} \tag{30'}$$

For ethanol the gas phase value of μ is 5.7×10^{-30} Cm, which we apply to the liquid phase for lack of better information.

The average torque on the molecules in the first solvation layer about an ion, taking $r = 4 \times 10^{-10}$ m and $\varepsilon \approx 4$, would correspond to 5 kJ/mol. If this were applied simply to reduce the activation energy from 18 to 13 kJ/ml it would decrease the reorientation time 7-fold. In addition, the somewhat diffuse charge

distribution of the localized electron would tend to neutralize
hydrogen bonds in its vicinity, which would remove a major hindrance
to dipole reorientation and increase the rate by an order of magnitude. These perturbing effects decrease rapidly with increasing r.
In the second solvent layer, taking $r \approx 8 \times 10^{-10}$ m and $\varepsilon \approx 8$,
the 0.6 kJ/mol torque would tend to reduce the reorientation time
only 1.3-fold. Molecules cannot move independently of each other
in the liquid phase, so the motions in the first layer would speed
those in the second and concomitantly be slowed by them. The net
effect of the perturbations would be reorientation in the first
layer around a localized electron in a time $\tau_p \approx (0.01 - 0.1)\tau_{up}$.
The acceleration due to the torque and hydrogen bond neutralization
would decrease rapidly with distance, and the rate of relaxation
of the third or fourth layer would be expected to be about the
same as in the bulk liquid. The acceleration might be somewhat
smaller around a cation; there would be no neutralization of the
hydrogen bonds, but the torque would be greater due to the more
tightly localized charge.

Dielectric relaxation measured by microwaves in alcohols
does not follow the simple Debye equation, which represents the
real part of the complex dielectric constant $\varepsilon'(\nu)$ as follows:

$$\varepsilon'(\nu) = \varepsilon_o + \frac{\varepsilon_\infty - \varepsilon_o}{1 + (2\pi\nu)^2\tau^2} \tag{31}$$

where ν is the frequency of the oscillating electric field. The
frequency (or time) dependence of the dielectric constant of an
alcohol has been somewhat arbitrarily subdivided into three Debye
regions whose relaxation times τ_1, τ_2 and τ_3 have relative magnitudes $\tau_1/\tau_2 \approx 10$-50 and $\tau_2/\tau_3 \approx 10$ [23]. The authors found that the
larger values of τ_1/τ_2 occurred when $\tau_1 > 1$ ns, that is, in long
chain alcohols at room temperature and in smaller alcohols at
lower temperatures. The long relaxation time τ_1 has been attributed to the hydrogen bonded structure of the liquid and the time
required to break a hydrogen bond prior to reorientation, while
τ_2 and τ_3 have been more speculatively attributed to reorientation
of "free monomer" or R-O- and -O-H groups [23-26]. The τ_1 process
accounts for ~ 80% of ε in many alcohols, but there is not general
agreement about whether the ratio τ_1/τ_2 increases of decreases
with increasing temperature, or whether it goes one way in some
alcohols and the opposite in others [23,25,27]. The lowering of
the energy of the localized electron to form the solvated state
is attributable mainly to the charge-accelerated τ_1 process, which

accounts for most of the equilibrium dielectric constant. The
process is much faster than that observed by microwave dispersion
for three reasons: (a) the charge of the electron neutralizes the
hydrogen bonds of the –O–H groups that it encompasses, thereby
facilitating rotation; (b) the strong electric field near the
electron exerts a torque greater than kT on adjacent dipoles; (c)
the field beyond the first solvation layer becomes attenuated as
the dipoles in the first layer orient in the field, which reduces
the time constant for the net relaxation process to $\tau' = \varepsilon_0 \tau / \varepsilon_\infty$
[28,29]. The τ' process is referred to as relaxation in the pre-
sence of a constant charge, while the conventionally measured values of τ refer to relaxation in a constant field. The value of
ε_0 for the τ_1 process is ε_∞ for the τ_2 process, and ε_0 for the
τ_2 process is ε_∞ for that of τ_3 (Fig. 3). The value of $\varepsilon_\infty / \varepsilon_0$ is
5-6 for τ_1 in ethanol temperatures near 300K [22]. Although τ_2 and
τ_3 have so far been too short to measure in methanol [30] and
ethanol [23], one may take $\varepsilon_{0,3} \approx n_D^2$ which is 1.85 for ethanol
near 300K, whence $(\varepsilon_\infty / \varepsilon_0)_2 (\varepsilon_\infty / \varepsilon_0)_3 \approx 2.3$ [31].
          Electrons in incompletely relaxed localized states absorb
light at longer wavelengths than do those in the equilibrium solvated state (Fig. 4). As $e^-_{irl} \rightarrow e^-_s$ takes place the absorbance
at long wavelengths decays and that at short wavelengths grows.

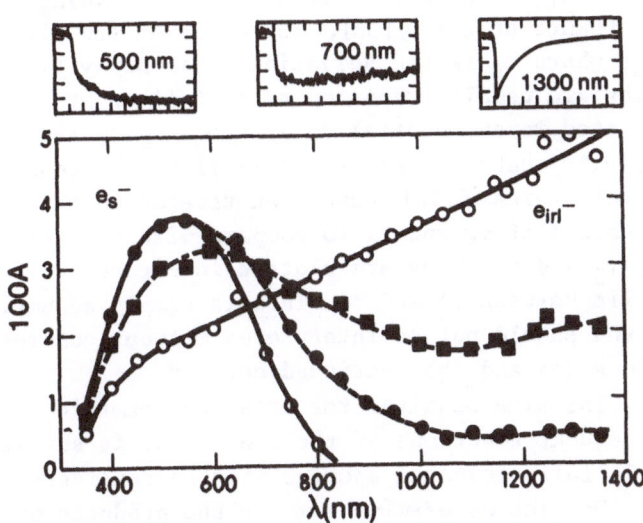

Fig. 4 – Optical absorption spectra of localized electrons in 1-
     propanol at 152K following a 5 ns pulse of 11 MeV elec-
     trons. Dose $\approx 6 \times 10^{16}$ eV/g. O , at end of pulse,■ , at
     65 ns; ● , at 200 ns; ◐ , at 1 µs. Top, oscilloscope
     traces at 50 ns/div. Ref. 6.

By converting plots such as those in fig. 4 to absorbance against photon energy, one may compare the loss of area under the low energy end of the curve to the growth of area under the high ener̄gy end. There is more decay at low energies than growth at high energies [32,33]. Assuming that the oscillator strength [34] of the electrons in this absorption region is unaffected by the band shift, one finds that the decay at low energies, for example at $\lambda \gtrsim 800$ nm in Fig. 4, is due to both reactions (2') and (4'). The latter reaction must involve the migration of charges. Since charge migration occurs during this time period it is possible that some of the growth of absorbance at the shorter wavelengths is due to electrons migrating from shallower to deeper traps [6,33]. The relative contributions of the two processes, dipole relaxation about a given site and electron migration to deeper traps, to the spectral shifts have not yet been decided. Migration would probably become less important as the traps get deeper, so its main contribution would occur at very short times and from very shallow traps ($\lambda > 1000$ nm).

The rate of decay of absorbance at $\lambda \approx 1300$ nm is not greatly different from the rate of growth at 514 nm [11]. The growth is not simple first order, but approximate relative time constants can be obtained by plotting $\ln [(A_{max} - A_t)_{514}/ (A_{max} - A_o)_{514}]$ against t for $A_t \approx (0.8 - 1.0)A_{max}$, where A_o is the absorbance at t = 0 [33]. The slope of the plot may be taken as $-\tau_s$, where τ_s is the solvation time. Decay times at long wavelengths are usually reported as half lives, $t_{\frac{1}{2}}$. Half lives can be converted to exponential decay times τ_d by the relation $\tau_d = t_{\frac{1}{2}}/0.69$. Data for electrons in liquid 1-propanol at different temperatures [6,12] have been treated in this manner to obtain values of τ_s and τ_d to compare with the relaxation times τ_1', τ_2', τ_3' and τ_2. They are plotted in Figure 5. The values of τ_s and τ_d lie between τ_2' and τ_1'. This is simply an empirical correlation and should not be interpreted without considering the perturbations (a) and (b) mentioned earlier.

The time required for free ion formation by reaction (4"), and subsequent reactions of the free ions, is sufficiently long that only fully solvated species need be considered as free ions.

Now let us examine some of the products of the geminate neutralization reaction (4'). Addition of 1 mM nitrous oxide scavenges the free ion solvated electrons. Higher concentrations attack also electrons in spurs. On going from 1 mM to 100 mM nitrous oxide in ethanol at 161K the nitrogen yield increases by 1.4 units (Fig. 1), while the hydrogen yield decreases by only 0.7 units

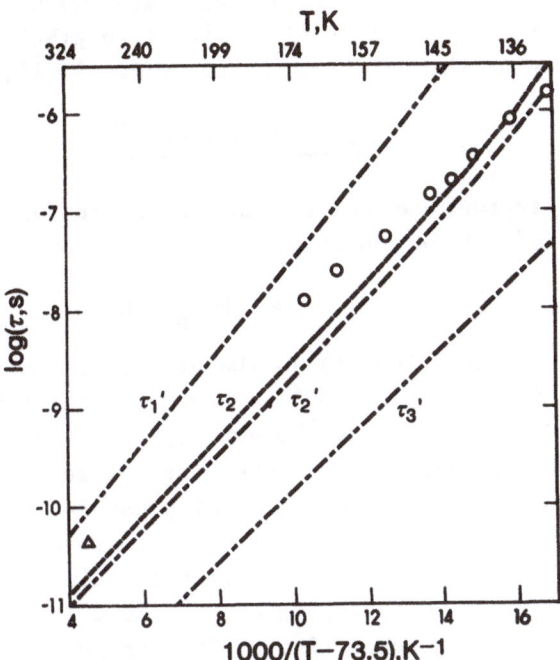

Fig. 5 - Cole-Davidson plot of τ_d at 1300 nm (O) and τ_s at 514 nm (△) for the optical absorbance of electrons in liquid 1-propanol. Data for τ_d were taken from ref. 6; τ_s was estimated by comparing data for 1-propanol and 1-butanol in ref. 12, using $A_o \approx 0.5\ A_{max}$; see text. The dashed lines represent the unperturbed relaxation times τ_1', τ_2, τ_2' and τ_3', obtained from data in refs. 23 and 25a.

(Fig. 2). This means that electrons that undergo geminate neutralization are involved in two kinds of reaction, only one of which produces hydrogen. The ethoxy radical or acetaldehyde formed in reaction (3') or (3") might intercept the electron before it reaches the cation. For example:

$$[e^-_s + C_2H_5OH_2^+] \longrightarrow [C_2H_5OH + H] \qquad (4a')$$

$$[e^-_s + C_2H_5O\cdot] \rightarrow [C_2H_5O^-_s] \xrightarrow{[C_2H_5OH_2^+]} 2C_2H_5OH \qquad (4b')$$

Reaction of e^-_s with acetaldehyde would ultimately produce a $CH_3\dot{C}HOH$ radical as in (12). Approximately half of the geminate neutralizations produce hydrogen, so (4a') represents half of the total (4').

It should be mentioned that later work raised the upper plateau of the nitrogen yield in Fig. 1 to $g(N_2) = 4.9$ [35], which may be taken as the total ionization yield in liquid ethanol.

DECOMPOSITION OF e^-_s

Electrons can react with hydroxylic solvents to form the oxy anion and a hydrogen atom.

$$e^-_s \longrightarrow RO^-_s + H \tag{32}$$

An example was reaction (5) in the ethanol radiolysis mechanism. The activation energy of (32) in alcohols and water is in the vicinity of 20 kJ/mol, while the entropy of activation is in the vicinity of -90 J/mol K [1,36,37]. The volume of activation, obtained from kinetics measurements at pressures up to several thousands of atmospheres, is -21 cm^3/mol in methanol and -22 cm^3/mol in ethanol at 295K [38]. The reaction is driven by the difference between the solvation energies of RO$^-$ and e$^-$, which are roughly 450 and 150 kJ/mol, respectively. The gas phase reaction between a thermal electron and an alcohol or water molecule,

$$ROH + e^- \longrightarrow RO^- + H \tag{33}$$

would be endothermic by about 300 kJ/mol (estimated from the RO-H bond dissociation energies and the electron affinities of RO·). Thus (33) does not occur to an appreciable extent.

The rate controlling process in reaction (32) is the rearrangement of the solvent molecules about the reaction site. The activation energies of dielectric relaxation in water and in C_1 to C_3 alcohols are in the region 16-26 kJ/mol, nearly the same as the activation energies of the corresponding reaction (32). The large negative entropies and volumes of activation of (32) result from the more ordered structure of the solvent about the RO$^-$ than about the e$^-$, and the greater electrostriction about the more confined charge on the -O$^-$. The structure of the transition state must be close to that of the product state [1].

REFERENCES

1. G.R. Freeman, in "Actions chimiques et biologiques des radiations", M. Haissinsky (ed.), 14, 73 (1970), Masson et Cie,
   Paris; references therein.

2. G.R. Freeman, Adv. Chem. Ser. 82, 339 (1968).

3. K.N. Jha and G.R. Freeman, J. Chem. Phys. 48, 5480 (1968).

4. W.H. Hamill, J. Phys. Chem. 73, 1341 (1969).

5. J.T. Richards and J.K. Thomas, J. Chem. Phys. 53, 218 (1970).

6. J.H. Baxendale and P. Wardman, J. Chem. Soc., Faraday Trans-
   actions I, 69, 584 (1973); Nature, 230, 449 (1971).

7. H. Hase, T. Warashina, M. Noda, A. Namiki and T. Higashimura,
   J. Chem. Phys. 57, 1039 (1972).

8. M.J. Bronskill, R.K. Wolff and J.W. Hunt, J. Chem. Phys. 53,
   4201 (1970).

9. G. Beck and J.K. Thomas, J. Phys. Chem. 76, 3856 (1972).

10. P.M. Rentepis, R.P. Jones and J. Jortner, J. Chem. Pys. 59,
    766 (1973).

11. W.J. Chase and J.W. Hunt, J. Phys. Chem. 79, 2835 (1975).

12. G.A. Kenney-Wallace and C.D. Jonah, Chem. Phys. Lett. 39, 596
    (1976).

13. Y. Wang, M.K. Crawford, M.J. McAuliffe and K.B. Eisenthal,
    Chem. Phys. Lett. 74, 160 (1980).

14. G.R. Freeman, "Radiation Chemistry of Ethanol", NSRDS-NBS 48,
    U.S. Dept. of Commerce, Washington, D.C., 1974.

15. S.W. Benson, J. Chem. Ed. 42, 502 (1965).

16. J.C. Russell and G.R. Freeman, J. Phys. Chem. 72, 816 (1968).

17. K.N. Jha and G.R. Freeman, J. Chem. Phys. 51, 2846 (1969).

18. J.J.J. Myron and G.R. Freeman, Can. J. Chem. 43, 381 (1965),
    and references therein.

19. L. Onsager, Phys. Rev. 54, 554 (1938).

20. K.N. Jha and G.R. Freeman, J. Chem. Phys. 51, 2839 (1969).

21. G.R. Freeman, second article in the present series, "Basics

of radiation chemistry".

22. F. Buckley and A.A. Maryott, "Tables of Dielectric Dispersion Data", NBS Circular 589, Washington, D.C., 1958.

23. S.K. Garg and C.P. Smyth, J. Phys. Chem. $\underline{69}$, 1294 (1965).

24. E. Bauer and D. Massignon, Trans. Faraday Soc. $\underline{42A}$, 12 (1946).

25. (a) R.H. Cole and D.W. Davidson, J. Chem. Phys. $\underline{20}$, 1389 (1952)
    (b) F.X. Hassion and R.H. Cole, J. Chem. Phys. $\underline{23}$, 1756 (1955).

26. L. Glasser, J. Crosley and C.P. Smyth, J. Chem. Phys. $\underline{57}$, 3977 (1972).

27. W. Dannhauser and R.H. Cole, J. Chem. Phys. $\underline{23}$, 1762 (1955).

28. R. Schiller, Chem. Phys. Letters $\underline{5}$, 176 (1970).

29. H. Fröhlich, "Theory of Dielectrics", 2nd edn., Clarendon Press, Oxford, 1958, p. 72.

30. B.P. Jordan, R.J. Sheppard and S. Szwarnowski, J. Phys. D: Appl. Phys. $\underline{11}$, 695 (1978).

31. The factor 2.3 includes all processes faster than that represented by τ_1, and probably contains a contribution from infrared frequencies to which one could ascribe a $\tau_4 \approx 10^{-13}$ s [26]. See the legend of Fig. 3.

32. J.H. Baxendale and P.H.G. Sharp, Int. J. Radiat. Phys. Chem. $\underline{8}$, 707 (1976).

33. K. Okazaki and G.R. Freeman, Can. J. Chem. $\underline{56}$, 2305 (1978).

34. Oscillator strength, $f = 3.5 \times 10^{-5} \int_0^\infty \varepsilon(E)dE$, where $\varepsilon(E)$ is the decadic molar absorbancy at photon energy E(eV).

35. T.E.M. Sambrook and G.R. Freeman, Can. J. Chem. $\underline{53}$, 1521 (1975)

36. G.L. Bolton, M.G. Robinson and G.R. Freeman, Can. J. Chem. $\underline{54}$, 1177 (1976).

37. J.H. Baxendale and P. Wardman, Chem. Comm. 429 (1971).

38. G.L. Bolton, K.N. Jha and G.R. Freeman, Can. J. Chem. $\underline{54}$, 1497 (1976).

LABILE SPECIES AND FAST PROCESSES IN LIQUID ALKANES

F. Busi

Istituto di Fotochimica e Radiazioni
di Alta Energia, C.N.R.
via de Castagnoli 1, 40126 BOLOGNA
Italy

Abstract: The initial reactions and yields of the
ions, free radicals and excited states which are generated by
the absorption of ionising radiation in liquid alkanes are
discussed.

INTRODUCTION

The primary processes induced by transfer of energy from
ionising radiation to molecules of liquid hydrocarbons are simply
excitation and ionization:

1) $RH \rightarrow RH^*$

2) $RH \rightarrow RH^+ + e^-$

In principle various excited states may be formed and also
a variety of ions which originate from the loss of fragments by
excited parent ions. Beginning with these, further new ions may
be formed by reaction of the primary ones with the parent
molecules often with the simultaneous formation of free radicals.
The excited states may also produce free radicals and atoms by
decomposition.

With molecules of the complexity of hydrocarbons the variety
of species which may be produced in these early reactions is so
large that to obtain a detailed mechanism of how the final radiolysis products are arrived at is almost impossible except for a
few simple cases.

J. H. Baxendale and F. Busi (eds.),
The Study of Fast Processes and Transient Species by Electron Pulse Radiolysis, 417–431.

However, the irradiation of alkanes and alkane solutions provides a very useful method for the study of organic cations and anions formed by reaction of the alkane ions and electrons with solutes, of excited states both of the alkanes and of solutes, and of free radicals. To make use of these systems requires a knowledge of the basic radiation chemistry in terms of yields and reactions of the ions, excited states and free radicals produced in the pure liquid systems.

IONIC SPECIES

The alkane cations formed may have excess energy and as a result dissociate. Fragmentation is common in the gas phase but in liquids the process is expected to be less important since excited cations may be deactivated.

In pure liquid alkanes these initial ions are formed in "spurs" i.e. groups of pairs of cations and electrons along the track of the fast electrons produced by the radiation. There can be several ion pairs in each spur and the spurs will be separated by some hundreds of nanometers.

When the ions have lost their excess energy (thermalised) they begin to diffuse and as a result an ion or electron may escape the Coulombic attraction of its geminate partner and become homogeneously distributed in the liquid. Alternatively, it may be so close to its partner that escape is not possible and neutralisation occurs. In the first process the ions are said to become "free" and in the second, "geminate recombination" is said to occur. In alkanes most of the ions undergo geminate recombination and this is the predominant source of the radiolysis products. The "free" ions also ultimately undergo neutralization according to homogenous diffusion controlled kinetics unless they react with added solutes.

Free Ion Yields

The fraction of ions which become free, Φ_{fi}, depends on the system and is given by:

$$\Phi_{fi} = \exp(-z^2/\varepsilon kTy)$$

where y is the separation distance of the electron and its parent ion at the thermalization distance, z is the charge, ε the dielectric constant, T the temperature and k the Boltzman constant. This equation is an expression of the competition between electrostatic interaction and thermal motion of the ions.

The yields of free ions (usually expressed as G_{fi} = number of ions produced per 100 eV absorbed) is measured experimentally by three methods: ion scavenging, conductivity and ion collection using the clearing field technique (1). The scavenging technique involves the quantitative determination of a chemical product formed by reaction of a solute with the electrons or positive ions. The conductivity technique has been described earlier in these proceedings. The clearing field technique consists of pulse irradiating the liquid in a two-electrode cell, and applying an electric field to collect all the free ions at the electrodes. The concentration, n, of the ions after the pulse would normally be given by:

$$dn/dt = I\ G_{fi}/100 - \alpha n^2$$

where I is the dose rate (eV/cm^{3}s), α the ion recombination rate coefficient. Hence, here the recombination term, αn^2, must be kept very small so that effectively all the charge is collected at the electrodes, which means that very low doses must be used. The value of G_{fi} is calculated from the dose and the amount collected. G_{fi} values, obtained by the different techniques, are given for various alkanes in Table 1.

TABLE 1:   Free ion yields in liquid alkanes (2)

Alkane	G_{fi}	Alkane	G_{fi}
Methane	1.13	n-Decane	0.117
Ethane	0.158	Cyclopropane	0.049
Propane	0.166	Cyclopentane	0.155
Butane	0.193	Cyclohexane	0.148
n-Pentane	0.17	2-Methylpentane	0.148
n-Hexane	0.12	2,3-Dimethylbutane	0.192
n-Heptane	0.1313	2,3,4-Trimethylpentane	0.174
n-Octane	0.124	Methylcyclohexane	0.122
n-Nonane	0.117	Neopentane	0.86

Ions in Spurs

Yields. In alkanes these form the bulk of the ions - 95% or more. Attempts have been made to determine the total yield of ions, G_{ti}, i.e. those in spurs, G_{gi}, together with the free ions, G_{fi}, by using scavengers which give a measurable product on reaction with alkane cations or electrons - more usually the latter. The reactions then involved are:

$$RH^+ + e_s^- \rightarrow RH \text{ or } RH^*$$

$$e_s^- + S \rightarrow P - \text{measurable product}$$

If the system were simple, clearly at high enough concentrations of scavenger S all e_s^- react with S and the product yield $G(P)$ will be a constant and a measure of G_{ti}. However, it is found that very high concentrations of S are required and because of secondary reactions, e.g. with radicals, it is rarely possible to obtain conditions which $G(P)$ is constant.

Nitrous oxide is perhaps the least complicated scavenger since it reacts stoichiometrically with e_s^- to give unreactive N_2, and is otherwise unreactive except to excited alkanes with which it also gives N_2. Thus the limiting value of $G(N_2)$ which appears to be attained in Fig.1 with c-hexane would give $G_{ti} + G(RH^*)$.

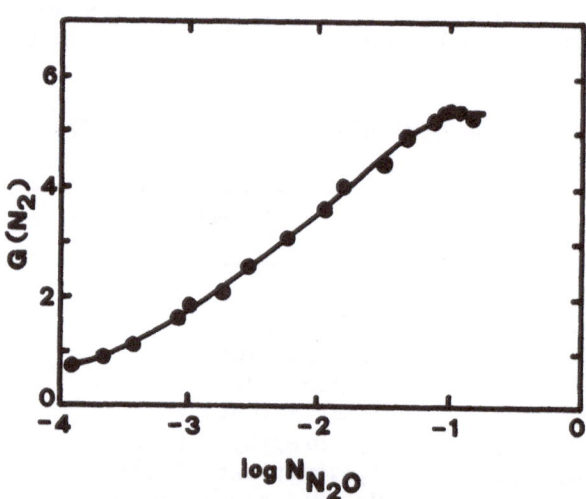

Figure 1.   Nitrogen yields from irradiated c-hexane
               containing nitrous oxide (36).

Experiments of this kind with ethane, propane, c-propane, neopentane, methylcyclohexane and cyclohexane all give the same value for $G(N_2) = G_{ti} + G(RH^*) = 5.3 \pm 0.3$.

Another approach (9) uses an empirical equation for $G(P)$ which has been found to have wide application viz:

$$G(P) = G_{fi} + G_{gi} \cdot \frac{(\alpha[S])^{\frac{1}{2}}}{1+(\alpha[S])^{\frac{1}{2}}}$$

where α is an empirical parameter which derives from the kinetics of the above competing reactions and varies with the scavenger. Reasonable agreement with experiment is obtained for a variety of scavengers (10) using values of G_{fi} and G_{gi} which are consistent to better than 10%. For cyclohexane $G_{gi} = 4.0$ is a reasonable mean value which is 97% of the total ion yield.

A theoretical treatment (11) uses a non-homogeneous kinetic model starting from an initial ion distribution function and deriving the probability that the ions of a pair, at a given separation distance, react with a solute before recombination.

Kinetics

In the absence of solutes to react with, the free ions combine in a bimolecular reaction. The recombination can be followed using spectroscopic and conductimetric kinetic measurements following a radiation pulse.

The pulsed liquids show optical absorption in the visible and infra-red due to alkane cations and electrons (3,4,5,6). Fig.2 shows the absorption spectrum attributed to cyclohexane cation and Fig.3 that to electrons in methylcyclohexane.

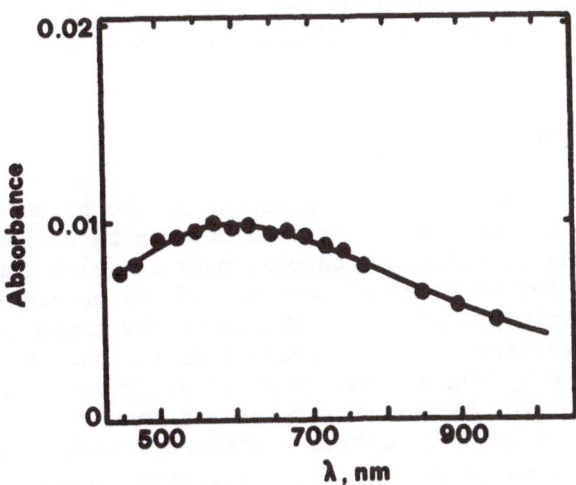

Figure 2.   Spectrum of cyclohexane cations $C_6H_{12}^+$ (3).

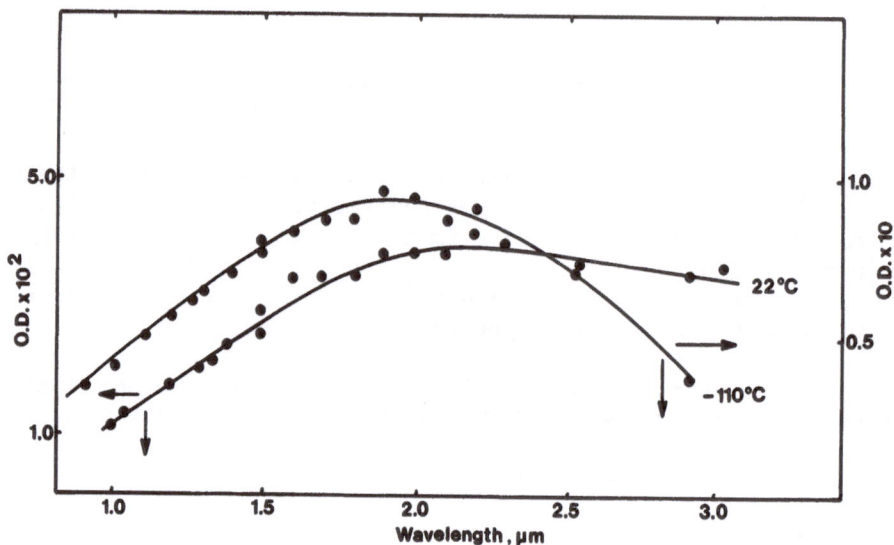

Figure 3.　　　　Spectrum of e_s^- in methyl cyclohexane　(6).

Using these absorptions, the neutralisation reactions have been followed (4,5) and, for example, rate constants k = 1.0 x 10^{14} and 7.1 x 10^{13} $m^{-1}.s^{-1}$ obtained for ion recombination in methylcyclohexane and n-hexane at room temperature. These values agree with those obtained using the better conductivity technique (8).

These very high rates arise from the fact that ions are combining in a low dielectric constant medium and that the electrons have abnormally high diffusion coefficients in hydrocarbons.

FREE RADICALS (21,24,37)

The presence of free radicals produced by the breakdown of the parent alkane molecule under irradiation can be deduced indirectly from the radiolysis products, many of which can be readily accounted for in terms of radical-radical combinations. Thus a characteristic product from alkanes is the compound produced by dimerisation of the radicals from the parent molecule which have lost one hydrogen atom. More directly radicals can be observed and identified before they react and lose their radical character by the EPR and NMR methods described elsewhere in this volume. They may also be observed by absorption spectroscopy and their dimerisation kinetics followed.

Yields (21). Estimates of free radical yields have been made using their reactions with appropriate scavengers. The commonest of these is iodine which reacts:

$$R\cdot + I_2 \rightarrow RI + I$$

to give the alkyl iodide (or hydrogen iodide with H atoms). Assuming the I atoms produced all gave I_2 (probably since they are relatively unreactive) then a measure of the iodine removed during radiation gives the total amount of free radicals formed. Again it must be established that there is a sufficient concentration of scavenger to compete with radical-radical reactions, that no other reactions occur, e.g. with excited states or ions, to remove scavenger, and also that secondary reactions do not interfere (12). Other scavengers used in this way are anthracene, oxygen, nitric oxide and galvinoxyl (12,13,14). The general results are not completely satisfactory in that they do not agree better than ± 20% when differing scavengers are used, but they do provide a rough indication for a variety of liquids. In a few cases which have been the subject of very careful study the values are more reliable.

Yields of individual radicals have been obtained (15-20) for certain liquids using radioactive scavengers which label the products of their reactions with the radicals. Thus, with radioactive iodine, the methyl radical will give radioactive methyl iodide which can be detected and assayed using isotope dilution techniques. Some of the results obtained are given in Table 2.

Table 2.   Radical yields in some alkanes (21).

Alkane	Radical
Ethane	Me, 0.6; Et, 3.8
n-Hexane	Me, 0.7; Et, 0.3; n.Pr, 0.3; Bu, 0.27; Hexyl, 4.1.
2,2-Dimethylbutane	Me, 1.0; Et, 1.2; Bu, 0.6; Hexyl, 1.2.
c-Hexane	c-Hexyl, 4.

There have been attempts to summarise the observations on yields and types of radicals formed from alkanes (22,23) in terms of empirical relations and Fig.4 shows an example of this.

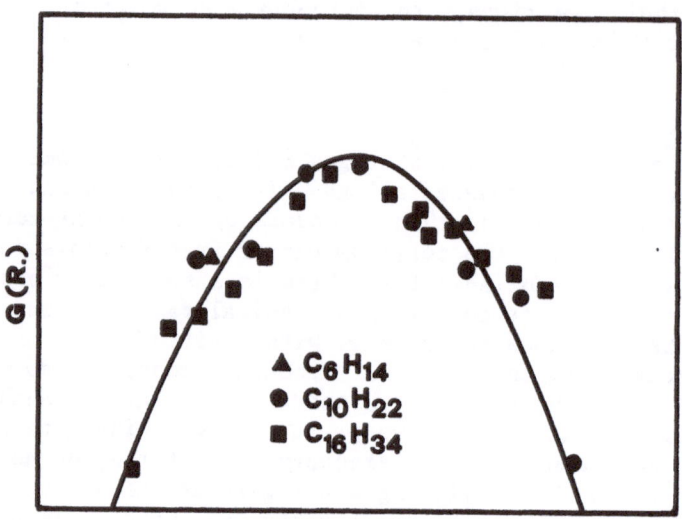

Figure 4.   Yields of fragment radicals as a function of their
carbon number.   (37).

In general, the number of fragment radicals increases the
more the carbon chain is branched but the yield of hydrogen atoms
(of which the hydrogen yield is a partial measure) decreases.
Table 3 shows the variation of C-H and C-C cleavage for a range
of alkanes and how it parallels the hydrogen yield.

Table 3. $G(H_2)$ and C-H, C-C bond cleavage in alkanes (24).

Alkanes	$R = \dfrac{G(C-H)}{G(C-C)}$	$G(H_2)$
2,2-dimethylbutane	0.40	2.0
2,3-dimethylbutane	0.70	2.9
3-methylpentane	1.40	3.4
n-hexane	3.30	5.0
cyclohexane	6.25	5.8
methylcyclohexane	3.45	4.8
1,1-dimethylcyclohexane	1.33	3.3

Roughly speaking it can be said that the total yield of
radicals in alkanes is about 4-6/100 eV absorbed and that of these
hydrogen atoms are always a substantial amount.

Reactions

Because of the variety of radicals formed it is not normally possible to study individual radicals produced in alkane systems. The radical C_6H_{11} predominant in c-hexane radiolysis, is an exception, and its dimerisation and reactions with oxygen and other reagents have been characterised using pulse radiolysis.

The radicals can, however, be useful in producing other radicals by addition to unsaturated compounds, by hydrogen abstraction, or by addition of, for example, oxygen. Hydrogen atoms are particularly useful in this respect in these systems.

EXCITED STATES

Formation of excited molecules of the medium occurs through Coulombic interaction with the fast electrons generated in the ionization processes. Energy transfer from slow electrons to molecules does not obey the optical selection rules and therefore singlet, triplet and other optically forbidden excited states in the whole spectrum may be produced. Nevertheless, in a condensed medium the upper excited states convert very rapidly to the lowest levels which are usually the only excited states experimentally observable.

Another process which may yield molecules with excess energy is ion recombination:

$$RH^+ + e^-_{solv} \rightarrow RH^*$$

The energy involved in the neutralization reaction is given (25) by:

$$E = I_p - (z^2/\varepsilon r_o + Q)$$

where I_p is the ionization potential of RH (ca. 10 eV); r_o is the distance between the positive ion and the electron before the final diffusion jump, z is the electronic charge, ε is the dielectric constant of the medium and Q the solvation energy of the neutralizing ions. The latter, which arises from polarisation of the hydrocarbon can be calculated to be about 2 eV. Assuming that the electron tunnels back to the positive ion from a distance r_o = 10 Å the Coulombic term $(-z^2/\varepsilon r_o)$ is only about 0.7 eV. Thus the available energy for excitation of the neutral molecule is likely to be several eV higher than the first excited states of alkanes, which are found from the u.v. absorption to be ca. 6 eV.

The neutralization may give singlet or triplet states, depending on the spin states of the neutralizing ions. In liquid alkanes about 96% of the neutralization takes place in the spurs in a few nanoseconds, while spin relaxation generally requires microseconds. However, it has been pointed out recently that the loss of spin correlation may take place in nanoseconds as a consequence of a process related to CINDP and due to isotropic hyperfine coupling but quantitative evaluation of the process is not available (26). Recombination between unrelaxed geminate ions will produce singlets, whereas that between ions of different origins may produce other states – in particular triplets. Therefore the ratio of singlet to triplet excited states depends on the ion pair population in the spurs. Using ion pair population distribution derived from the gas phase, equal probabilities for singlet and triplet formation have been calculated, although there is as yet no firm experimental evidence for triplets in pure irradiated liquid alkanes.

The excited singlet states of certain alkanes make their presence shown by emission of fluorescence during and after irradiation. Lipsky and Hirayama (27) have shown that the same short-lived excited state fluorescing in the u.v. is produced both by ultra-violet light and ionizing radiation. No agreement exists on the mechanisms of formation or yields of the fluorescence produced by electrons or γ-rays. Picosecond electron pulse radiolysis gives results which do not allow a distinction between direct excitation and ion recombination as the origin of the fluorescence in any of the hydrocarbons studies. The results of emission studies of irradiated alkane solutions have been interpreted by Lipsky *et al.* (27) in favour of ion recombination. Similar studies have been interpreted by Busi *et al.* (28,29) in favour of formation predominantly by direct excitation.

Lipsky *et al.* analysed the fluorescence intensity changes of bicyclohexane in the presence of perfluorodecalin or perfluoromethylcyclohexane. The solutes can quench the excited singlet of the solvent and scavenge the solvated electrons. The authors used the following equation:

$$G(S) = \left[G^+(1-p^+) + G^* \right](1-p^*)$$

where G^+ and G^* are the yields of S from ion neutralization and direct excitation respectively, p^+ and p^* are the probability of geminate neutralization scavenging and of quenching respectively. The value of p^+ has been assumed to be given by the empirical equation known to apply to ion scavenging:

$$p^+ = \frac{\sqrt{\alpha\,[S]}}{1+\sqrt{\alpha\,[S]}}$$

where $|S|$ is the solute concentration and α is an empirical parameter determined only by the solvent-solute kinetics. p^* is related to the fluorescence yield of the solution, Φ, and the pure solvent, Φ^0 by:

$$\Phi = \Phi^0(1 - p^*)$$

It then follows:

$$\frac{1}{R} = \frac{1}{\delta} + \left[1 + \frac{1}{\sqrt{\alpha\,[S]}}\right]$$

where $R \equiv (1 - G_s\Phi^0/G_s^0\Phi) = \delta^+ p^+$, and $\delta^+ = G^+/G_s^0$. Using the above equation δ^+ can be determined from the intercept of a plot of R^{-1} vs $C^{-\frac{1}{2}}$. For bicyclohexyl $\delta^+ = 0.9 \pm 0.1$ which means that the predominant mechanism of excited state formation is via the ions. However, the low sensitivity of the method in the calculation of δ^+ and the use of perfluorocycloalkanes as solutes limit the reliability of the conclusions.

Busi and Casalbore (29) have analysed the fluorescence from alkanes solutions irradiated in the presence of nitrous oxide. The solute reacts with alkane excited singlet and with solvated electrons and we have the following reactions occurring:

$$RH^* \longrightarrow RH + h\nu$$

$$RH^+ + e_s^- \longrightarrow RH^*$$

$$RH^* + N_2O \longrightarrow RH + N_2 + O$$

$$N_2O + e_s^- \longrightarrow N_2 + O^-$$

Nitrogen is produced, with unit efficiency, in both N_2O reactions.

Thus, the quenching of RH^* directly by N_2O is a simple competition between quenching and emission in homogeneous solution and should lead to Stern-Volmer kinetics. However, quenching by reaction of N_2O with e_s^- before RH^* is formed is not so simple since recombination and e_s^- reactions are spur reactions which do not follow homogeneous reaction kinetics. In this situation the Stern-Volmer equation would not be expected to apply.

Figure 5 shows a Stern-Volmer plot of results obtained for methylcyclohexane. The best fit is linear and the intercept is 1.

Hence, the authors have concluded that the contribution of ion recombination to the fluorescence yield is negligible in methylcyclohexane. Thus, although perhaps different hydrocarbons may behave differently in these circumstances, it is clear that this problem of the origin of the excited states is still unresolved.

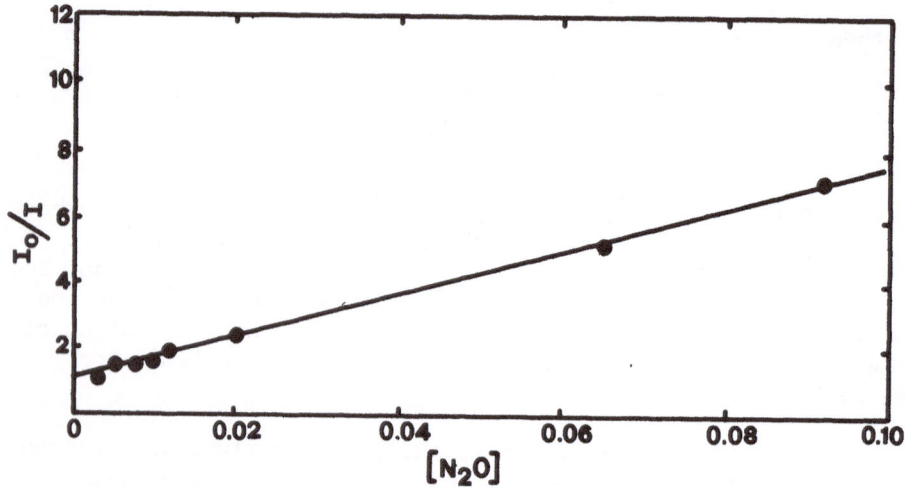

Figure 5.   Stern-Volmer plot for quenching of pulsed electron excited methylcyclohexane fluorescence at 220 nm by nitrous oxide (29).

Yields.   Two methods have been used to determine the yields of alkane excited states produced by ionising radiation.

In one,   (31,32) the absolute value of the emission is measured and, using the quantum yield of emission determined by u.v. excitation, the yields of excited states are calculated (Table 4). The doubtful factor here is the quantum yield value, since this is known to be a function of the exciting energy and although this is known in u.v. excitation, there is no information as to the energy level from which the emission occurs in molecules excited by ionising radiation. However, recent work (32) claims to have resolved this difficulty and concluded that for c-hexane $G(RH^*) = 1.45 \pm 0.15$.

Table 4: Yields of fluorescence of irradiated liquid alkanes

propane	1.8 ± 0.5^a
n-butane	1.9 ± 0.7^a
n-pentane	1.7 ± 0.6^a
n-hexane	1.6 ± 0.5^a ; 1.4 ± 0.3^c
n-heptane	1.5 ± 0.5^a ; 1.1 ± 0.3^c
n-octane	1.5 ± 0.6^a
cyclohexane	1.5 ± 0.4^a ; 0.51 ± 0.15^f ; 1.45 ± 0.15^e
methylcyclohexane	$1.9 - 2.2^d$; 0.95 ± 0.2^c
cyclo-octane	2.2 ± 0.5^a ; 1.46 ± 0.3^c
dodecane	$3.3 - 3.9^b$
hexadecane	$3.3 - 3.9^b$
cyclopentane	1.8 ± 0.5^a
cycloheptane	1.6 ± 0.6^a
cyclodecane	2.3 ± 0.5^a
cis-decalin	3.4^b
bicyclohexyl	3.5^b

a - ref.(30) ; b - ref.(27) ; c - ref.(38) ; d - ref.(31);
e - ref.(32) ; f - ref.(28).

The second method (33) is based on the transfer of energy from RH* to an acceptor e.g. benzene, toluene, or xylene, which itself fluoresces. This fluorescence can be used to calculate the number of excited acceptors produced which, if all RH* transfers its energy, is a measure of the amount of RH* present initially. The problem here is that the acceptor excited states may also be produced *via* ionic processes e.g. formation of acceptor cation and neutralization by e_s^-. Calculations (34) have indicated that the error due to this may amount to 30%. In the absence of such complications the variation of fluorescence intensity with acceptor concentration should follow Stern-Volmer kinetics and the extrapolation of intensity to infinite acceptor concentration will give the acceptor emission when all RH* have transferred energy. Measurements on the emission in identical conditions from the pure acceptor whose yield is known can then be used to calculate RH* yields. In fact the emission from the acceptor does not exactly follow the Stern-Volmer relation, presumably because of the other ways in which the acceptor excited state can be produced. However, the alkane fluorescence quenching which is not subject to this interference should also follow Stern-Volmer kinetics and in fact is found to do so (33,35). Using this method with p-xylene in c-hexane as the acceptor, $G(RH^*)$ = 0. 51 has been obtained

Clearly the big discrepancy between the two methods is very unsatisfactory and further work is required. The results obtained to date are summarised in Table 4.

REFERENCES

1.   Schmidt, W.F. and Allen, A.O.; 1968, J.Phys.Chem., 72, 3730.
2.   Allen, A.O.: 1976, NSRDS-NBS57
3.   Bös, J., Brede, O., Mehuert, R., Nilsson, G., Samskog, P.-O. and Reitberger, T.: 1979, Radiochem.Radioanal.Lett., 39, 353.
4.   Baxendale, J.H., Bell, C. and Wardman, P.: 1971, Chem.Phys. Lett., 12, 347.
5.   Baxendale, J.H., Bell, C. and Wardman, P.: 1973, J.C.S. Faraday I, 69, 776.
6.   Baxendale, J.H. and Busi, F.: Unpublished results.
7.   Baxendale, J.H. and Rasburn, E.J.: 1974. J.C.S. Faraday I, 70,705.
8.   Baxendale, J.H., Keene, J.P. and Rasburn, E.J.: 1974, J.C.S. Faraday I, 70, 718.
9.   Warman, J.H., Asmus, K.D. and Schuler, P.H.: 1968, Advan. Chem.Ser., 82, 25.
10. Schuler, R.H., and Infelta, Pierre P.: 1972, 76, 3812.
11. Freeman, G.R.: 1967, J.Chem.Phys., 46, 2822.
12. Schuler, R.H.: 1958, J.Phys.Chem., 62, 37.
13. Schuler, R.H.: 1964, J.Phys.Chem., 68, 3873.
14. Laplam-Masanet, J. and Ivanoff, N.: 1962, Proceedings of the Tihany Symposium, 75.
15. Holroyd, R.A. and Klein, G.W.: 1962, Int.J.Appl.Radiat. Isotopes,: 13, 493.
16. Holroyd, R.A. and Klein, G.W.: 1965, J.Am.Chem.Soc., 87, 4983
17. Holroyd, R.A.: 1966, J.Am.Chem.Soc., 88, 5381.
18. Holroyd, R.A.: 1966, J.Phys.Chem., 70, 1341.
19. Holroyd, R.A. and Klein, G.W.: 1965, J.Phys.Chem., 69, 194.
20. Holroyd, R.A. and Klein, G.W.: 1964, Int. J. Appl. Radiat. Isotopes, 15, 633-641.
21. Holroyd, R.A.: "Fundamental processes in radiation chemistry" pp.413-514, Interscience, 1968, and "Aspects of hydrocarbon radiolysis", pp.1-32, Academic Press, 1968.
22. Isildar, M.: 1976, Thesis, Carnegie-Mellon University, Pittsburg.
23. Schuler, R.H. and Kuntz, R.R.: 1963, J.Phys.Chem., 67, 1004.
24. Földiak, G.: 1978, Radiat.Phys.Chem., 11, 267.
25. Freeman, G.R.: 1968, Radiation Res.Rev., 1, 1.

26. Haberkorn, R.: 1977, Chem.Phys., 19, 165.
27. Walter, L., Hirayama, F. and Lipsky, S.: 1976, Int.J.Radiat. Phys.Chem., 8, 237.
28. Busi, F., Flamigni, L. and Orlandi, G.: 1979, Radiat.Phys. Chem., 13, 165.
29. Busi, F. and Casalbore, G.: 1981, Gazzetta Chim.Italiana, 11, 12.
30. L. Wojnàrovits and G. Földiak, 1980, Acta Chim. Acad. Sci. Hung. 105, 27.
31. Walter, J. and Lipsky, S.: 1975, Int.J.Radiat.Phys.Chem., 7, 175.
32. Tak Choi, H., Askew, D. and Lipsky, S.: Work submitted to Radiat.Phys.Chem.
33. Baxendale, J.H. and Mayer, J.: 1972, Chem.Phys.Lett., 17,458.
34. Busi, F.: 1980, Radiat.Phys.Chem., 16, 101.
35. see Ref (28).
36. Busi, F., Flamigni, L. and Roda, A.: 1975, Int.J.Radiat. Phys.Chem., 7, 589.
37. Földiak, G.: 1980, Radiat.Phys.Chem., 16, 451.
38. Busi, F. and Casalbore, G.: Work submitted to Gazzetta Chim. Italiana.

THE DYNAMICS OF ELECTRONS AND IONS IN NON-POLAR LIQUIDS

John M. Warman

Interuniversity Reactor Institute
Mekelweg 15, 2629 JB Delft, The Netherlands

ABSTRACT

The present state of knowledge of the parameters affecting
ionisation and recombination, and the reaction kinetics and
mobilities of electrons and molecular ions in non-polar liquids is
reviewed. A historical perspective of the development of this
knowledge is given and where possible, pictorial representations
of the processes involved are attempted.

J. H. Baxendale and F. Busi (eds.),
The Study of Fast Processes and Transient Species by Electron Pulse Radiolysis, 433–533.
Copyright © 1982 by D. Reidel Publishing Company.

INTRODUCTION

The present paper will be concerned with the nature of ions and
ionic processes in liquid media for which the principal ion—medium
interaction is very rapid (ca 10^{-15} s) electronic polarisation of
the molecules, resulting in a relative dielectric constant of
approximately 2. Particular attention will be given to those
aspects which are conceptually far removed from the normal experience of those more familiar with the behaviour of ions in
either the aqueous or the gaseous phases. This will include the
ionisation process itself and subsequent geminate recombination
and escape. Also, the behaviour of excess electrons and solvent
radical cations and radical anions will be discussed in some detail.
In non solvating media these species can display orders of magnitude higher diffusion coefficients and reaction rate constants
than would be expected on the basis of the viscosity of the
medium alone.

In polar media, full dielectric relaxation requires nuclear
rearrangements and hence takes on the order of picoseconds or
more. It is probable therefore that the view of ionic processes
gained from studies in dielectric liquids does in fact give a
glimpse of the processes occurring and the species present in the
early stages of ionisation in more polar systems. The data obtained
give an insight into the electronics of amorphous, molecular
systems which could be of considerable biological relevance since
it would seem most unlikely that nature would have missed the
tremendous potential of charge and energy transport over large
distances via electronic mechanisms in highly structured and
compartmentalised biological systems.

Almost by definition, non—polar dielectric liquids do not
support the stable existence of isolated ionic species. The main
reason for this is not, as is sometimes thought, the weak
"solvating power" of the medium since this is in fact quite strong.
For example, using the Born expression [1] for the solvation

energy ΔG_{sol} in eV for an ion of radius R in Å

$$\Delta G_{sol} = \frac{7.2}{R} \frac{\varepsilon-1}{\varepsilon} \qquad (1)$$

the total solvation energy associated with the sodium (0.97 Å) and chloride (1.80 Å) ions in a medium of dielectric constant 2 is found to be almost 6 eV (140 kcal/mol). This is more than suffi- cient to overcome the 1.5 eV difference between the gas phase values of the ionisation potential of sodium and the electron affinity of the chlorine atom. The fact that sodium chloride does not undergo ionic dissociation in for example hydrocarbon liquids despite this apparent stability, is of course due to the large coulomb binding energy (1.5 eV for a 5 Å separation) which pertains in the low dielectric constant system and which prevents the ions from separating. For water this binding energy is reduced to the order of the mean thermal energy of the molecules of the medium even for distances of only a few angstroms and diffusional separation of the ions can take place.

In view of the above, the question arises as to how one can then study the properties of isolated ions in media for which even stable ion pairs cannot dissociate. The answer lies in the formation of pairs of ions with a distance separating them sufficiently large to ensure at least a reasonable, if small, probability of escape, by diffusional motion, from their mutual coulombic attraction. The probability, P, that a pair of ions initially separated by a distance r_o undergo complete separation is given by the Onsager expression [2]

$$P = \exp(-r_c/r_o) \qquad (2)$$

where r_c is the distance from an ion at which the coulombic potential energy is equal to $k_B T$ with k_B the Boltzmann constant i.e.

$$\frac{e^2}{4\pi\epsilon_o\epsilon_r r_c} = k_B T \tag{3}$$

At room temperature in a dielectric 2 liquid the escape distance, r_c, is 290 Å which means that to have even a 1% probability of escape a pair of ions must be separated by at least 63 Å. It must perhaps be considered a fortunate accident of nature that on exposure of dielectric liquids to high energy (X, γ or particle) radiation, a significant fraction of the electrons produced on ionisation of individual molecules does in fact manage to reach such large distances before becoming thermalised.

In what immediately follows an outline will be given of the important steps (in the authors' opinion) which have led to the present understanding of the role of ionic processes in irradiated non-polar liquids. As such therefore this represents the foundation which was necessary in order to be able to sensibly apply radiation techniques to the study of ions and their reactions in these media.

Historical Perspective

It had been known since the turn of the century that normally insulating liquids became conducting when exposed to high energy radiation. This phenomenon had attracted considerable experimental and theoretical attention over the years as is well documented in the review of Adamczewski which appeared in 1965 (initially in Polish) [3]. The majority of early work was however concerned with the degree of ionisation in the presence of relatively high electric fields which undoubtedly perturbed the ionisation process itself. It was in fact only at about the time of writing of Adamczewski's review, that reliable, accurate measurements of the yield of separated ions in the absence of an applied electric field were being attempted [4,5].

For n-hexane, which was at that time considered to be a

characteristic hydrocarbon liquid, these measurements gave a yield of approximately 0.1 for the number of escaped or "free" ion pairs per 100 eV of absorbed energy ($G_{fi} \simeq 0.1$). Since the average energy required for ionisation of a molecule in the liquid phase was not expected to be much different from the ca 25 eV found in the gas phase, the yield of 0.1 $(100 \text{ eV})^{-1}$ indicated an overall two to three percent probability of escape for the ion pairs initially formed. While the percentage was small, it still corresponded to the effective formation of all ion pairs with a separation of approximately 70 Å, equal to many molecular diameters.

Partially on the basis of the very small yields of separated ions it was thought by many at the time that ionic processes must play a negligible role in the chemistry of irradiated non-polar liquids and their solutions. This opinion was to a large extent based on the estimate of Samuel and Magee [6] that those electrons which did not undergo escape would return to the positive ion within a time of less than a picosecond. Taking a rate constant of $10^{10} \text{ M}^{-1}\text{s}^{-1}$ as maximum for a diffusion controlled ion scavenging reaction in a liquid of viscosity 1 cP, it could readily be deduced that, even for a concentration of scavenger as high as 0.1 M, much less than 1% of the geminately recombining ion pairs would be available for reaction. The expected G value of product from this source taking a value of 4 $(100 \text{ eV})^{-1}$ for the total ion yield would then be only 0.04 $(100 \text{ eV})^{-1}$ and still appreciably less than the yield of escaped ions.

This general opinion was held despite the (in hindsight) convincing arguments of Hamill [7,8], Willard [9] and Schuler [10] that ionic reactions must be important based mainly on the large yields ($\gg 0.1$ $(100 \text{ eV})^{-1}$) of alkyl radicals formed in irradiated hydrocarbon solutions of alkyl halides. The large alkyl radical yields were attributed by the authors to dissociative electron attachment reactions which had been observed to occur in electron beam experiments in the gas phase.

$$e^- + RX \rightarrow R^\cdot + X^- \qquad (4)$$

The implications of these early studies were largely ignored, most probably because of the fascination at the time of the majority of radiation chemists with the solvated electron in water and the newly developed tool of pulse radiolysis.

It took in fact the combined application of nitrons oxide, a pet electron scavenging solute of the aqueous radiation chemists, and, pulse radiolysis techniques to the non-polar liquids to revive a more general interest in these systems. The actual experiments involved the measurement by Scholes and Simic [11], of nitrogen yields from irradiated solutions of nitrous oxide in cyclohexane and the measurement of the yield of biphenyl anions in a solution of biphenyl in cyclohexane on a microsecond timescale by Scholes, Simic, Adams, Boag and Michael [12]. The former experiment showed large yields of nitrogen to be formed, up to a G value of approximately 4 at 0.1 Molar. This could only reasonably be explained by occurrence of the electron attachment reaction

$$e^- + N_2O \rightarrow N_2 + O^- \qquad (5)$$

The yield of biphenyl anion found in the microsecond pulse radiolysis experiments was however determined to be only about 0.1 $(100 \text{ eV})^{-1}$, or of the same magnitude as the free ion yields found in conductivity experiments. The enormous difference in yields using steady state and pulse radiolysis techniques was seen by many at the time as an almost irresolvable conflict.

Free ion yields were reexamined using more refined conductivity methods [13-15] but no drastic change was found and a value of ca 0.1 $(100 \text{ eV})^{-1}$ became almost accepted as the yield of escaped ions in dielectric liquids. The agreement between the low values of G_{fi} determined in conductivity experiments and the low yield of ions observed on a microsecond timescale using optical pulse radiolysis was further substantiated by Keene, Land and Swallow [16] in a study of the formation of both negative and positive ions from several aromatic solutes dissolved in cyclo-hexane.

Possible doubts that the conditions of pulse radiolysis were in some way responsible for the low ion yields were removed by the experiments of Hagemann and Schwarz [17]. They showed that when the neutral product of electron attachment to benzyl chloride (the benzyl radical)

$$e^- + C_6H_5CH_2Cl \rightarrow C_6H_5CH_2^{\cdot} + Cl^- \tag{6}$$

was monitored under pulse radiolysis conditions via its well known absorption, large yields were found similar to those found in the steady-state γ radiolysis experiments.

Additional evidence for the participation of much larger yields of ions than G_{fi} in scavenging studies accumulated, also from the side of positive ion reactions. Thus Williams [18-20] demonstrated the formation of significant yields of HD from solutions of deuterated ammonia and alcohol in irradiated hydro-carbon liquids. This could only be readily explained in terms of proton transfer reactions such as

$$RH^+ + ND_3 \rightarrow ND_3H^+ + R^{\cdot} \tag{7}$$

followed by formation of deuterium atoms on neutralisation which produced HD by abstraction from the hydrocarbon

$$ND_3H^+ + e^- \rightarrow ND_2H + D \tag{8}$$

$$D + RH \rightarrow HD \tag{9}$$

Among other reactions which could only be attributed to a positive

ion precursor, Ausloos, Scala and Lias [21,22] demonstrated the
occurrence of the H_2 transfer ion-molecule reaction between a
saturated hydrocarbon and cyclopropane in the liquid phase.

$$C_nH_{2n+2}^+ + c-C_3H_6 \rightarrow C_nH_{2n}^+ + n-C_3H_8 \tag{10}$$

This reaction, because of its specificity, was to prove important
in the further development of the understanding of ionic processes.
An important additional effect observed by Ausloos et al. was the
increase in yield of positive ion scavenging on addition of an
electron scavenger. This was ascribed to a lengthening of the
recombination time on conversion of the electron to an anion!

It was on the basis of his positive ion scavenging experiments
that Williams first noted the now almost axiomatic linear dependence of the product yield due to ion scavenging on the square
root of the solute concentration. Extrapolation of this plot for
ND_3 [19] indicated a limiting, "zero concentration" yield of the
same magnitude as the value of G_{fi} from the conductivity experiments. Further confirmation of the square root dependence for
low concentrations of both positive ion and electron scavengers
was demonstrated by Sherman [23], for nitrous oxide in cyclohexane,
by Rzad and Schuler [24] for cyclopropane in normal hexane and
cyclohexane, and for alkyl halides in n-hexane, cyclohexane and
iso-octane by Rzad, Warman, Asmus and Schuler [25-27].

It was on the basis of the extensive alkyl halide data that
equation (11), relating the yield of product from ion scavenging
$G_i(S)$ to the concentration [S], was found to describe ion
scavenging data over a very wide concentration range [26]

$$G_i(S) = G_{fi} + G_{gi} \sqrt{\alpha_s[S]}/(1 + \sqrt{\alpha_s[S]}) \tag{11}$$

The parameters α_s in (11) was referred to as the reactivity of the
solute towards the ion or electron and it was thought that the

empirical parameter G_{gi}, which was initially found to have a
value of approximately 4 could possibly be associated with the
yield of geminately recombining ions.

Equation (11) helped to bring a degree of order to the flood
of scavenging data which had followed the report of Scholes and
Simic. The situation was such that in a review of 1968 [28],
only 4 years later, 12 different groups could be named who had
published data on nitrogen yields from irradiated hydrocarbon
solutions of nitrous oxide. Of particular note in the mountain of
data, apart from those already mentioned, were the results of
Sagert [19,30] and of Rajbenbach [31,32] who demonstrated the
use of saturated perfluorocarbons as electron scavengers in
liquid hydrocarbons.

The square root concentration dependence of ion scavenging
was provided with a theoretical basis by the work of Hummel [33]
who, using Monchick's modification of the Smoluchowski diffusion
equation, derived a limiting low concentration scavenging expression given by

$$G_i(S) = G_{fi} + K \sqrt{k_s [S]/D(+,-)} \tag{12}$$

where K was a constant, k_s was the rate constant for scavenging
by the solute and $D(+,-)$ was the sum of the diffusion coefficients
of the primary positive and negative ions. This equation was to
receive further development later but served the important role
at the time of giving substance to the reactivity parameter α_s
and indicating, not unexpectedly, its direct relationship to the
rate constant for reaction and, more important, its inverse
dependence on the sum of the primary ion diffusion coefficients.
From the increase in positive ion scavenging efficiency observed
on addition of an electron scavenger to cyclohexane [34] it was
therefore possible to conclude that the diffusion coefficient of
the electron was 16 to 33 times larger than that of the primary

positive ion. This automatically indicated an abnormally high diffusion coefficient for the electron, a conclusion which had previously been reached [35] on the basis of the considerably smaller concentration of electron scavengers, compared with positive ion scavengers, required to give the same product yield.

At the time of publication of Hummel's paper [33] other, somewhat less tractable, theoretical approaches were already in existence [13,19,36-40]. A fundamental difference in the basic mechanism of ionisation used to explain the experimental observations was also to be found between different workers. Thus Freeman had proposed [13,36] that a spectrum of ranges of electrons was produced which was a reflection of the energy spectrum of secondary electrons. The most energetic electrons, having the greatest range, were therefore mainly responsible for the yield of separated ions. Alternatively Mozumder and Magee [40], argued that the large separation distances were due mainly to outward motion of subexcitation electrons with energies of only a few electron volts or less which was relatively slowly degraded by excitation of vibrational and librational modes of the molecules. Experimental support for the latter proposal was to be forthcoming.

While the theoretical explanations remained somewhat diverse, it was apparent to all concerned that the lifetime of geminately recombining ion pairs must be much longer than the initial calculations of Samuel and Magee [6] had suggested if the scavenging data were to be explained. In their early microsecond optical pulse radiolysis study Keene, Land and Swallow [16] had in fact looked for an initial rapid decay of geminately recombining solute ions by investigating the absorption transient within their 2 µs long pulse. No indication was found for a significantly higher yield compared with that present on a timescale of tens of microseconds.

As had been suggested by the last named authors, it was left to the development and application of nanosecond time resolution

pulse radiolysis techniques to provide conclusive evidence for the
relatively slow decay of geminate ion pairs. This Thomas, Johnson,
Klippert and Lowers did in 1968 [41] when they published their
classic oscilloscope trace showing the initial rapid decay
("geminate spike") of the biphenyl radical anion in liquid cyclohexane over tens to hundreds of nanoseconds. This result provided
a certain degree of closure as far as the general aspects of
ionic processes in non-polar liquids were concerned tying together as it did the results of the conductivity measurements on
free ion yields and the steady state scavenging data. This closure
was perhaps best illustrated by the good fit to the biphenyl
anion decay kinetics [42] obtained using the expression derived
for the concentration dependence of scavenging, equation (11),
by applying the Laplace transform method [43].

Any complacency felt was however quickly dispelled by a
succession of surprises which were to follow. The first of these
was in fact reported in the same year as the nanosecond pulse
radiolysis data of Thomas et al. It proved to be of extreme
importance since it provided evidence on the basis of which a
choice could be made between the then current models of ionisation.
Also, it opened up the minds of radiation chemists to the fact
that even liquids with very similar chemical and physical properties could differ considerably when it came to the process of
ionisation. The observation, which was reported by Tewari and
Freeman [44] using a conductivity method, Schmidt and Allen [45]
using the clearing field technique, Capellos and Allen [46] using
optical pulse radiolysis and Rzad and Warman [47] using scavenger
product yields, was that the yield of free ions could in fact vary
substantially even within the family of saturated hydrocarbon
liquids. A value as high as 0.8 $(100 \text{ eV})^{-1}$ was in fact observed
[44,45] in liquid neopentane. The importance of the degree of
sphericity of the individual molecules could be clearly seen for
the first time!

These observations lent support to the Mozumder and Magee model [40] of ionisation since the ranges of highly energetic secondary electrons were not expected to be sensitive to molecular structure whereas a dependence on structure of the more intimate interaction between subexcitation electrons and molecules could be imagined.

The seeds of the next surprise were already contained in the free ion yield measurements of Tewari and Freeman [44] in the form of large conductivity overshoots in neohexane and neopentane which decayed faster than the time resolution of their equipment. Experiments by Minday, Schmidt(L.D.) and Davis [48] and by Schmidt(W.F.) and Allen [49] with more highly purified liquids and better time resolution confirmed that the previous overshoots were due to the formation of very high mobility electrons. The mobility of 93 $cm^2 V^{-1} s^{-1}$ reported for tetramethylsilane (TMS) [49] was approximately five order of magnitude larger than any mobility value previously found in a molecular liquid and was only a factor of approximately 5 lower than the mobility which had been found for electrons in the atomic liquid, argon [50,51]. Equally as interesting was the three orders of magnitude difference in mobility found between TMS [49] and the straight chain saturated hydrocarbon n-hexane [48]. This illustrated even more clearly the dramatic influence of sphericity on the electronic properties of molecular liquids.

A mobility measurement which held other consequences, was that of the electron in liquid cyclohexane [49] which was found to be 0.35 $cm^2 V^{-1} s^{-1}$. This was not a factor of about 30 greater than molecular ion mobilities as had been concluded from the effect of electron scavengers on positive ion scavenging in steady state studies but closer to a factor of 1000. This could only be explained by an excessively high mobility for the primary positive ion in liquid cyclohexane. This conclusion was to be verified later by kinetic [52-54] and direct conductivity [55] measurements.

For the large number of chemists who had been involved in the basically physical discoveries of the previous years it was particularly rewarding to see the large electron mobility values translated into rate constants for reaction with solutes several orders of magnitude larger than those for normal diffusion controlled reactions. These initial kinetic measurements on electron reactions were carried out by Baxendale, Bell and Wardman [56] and Richards and Thomas [57] who monitored the recently found infra red absorption of electrons, and by Bakale, Gregg and McCreary [58] and Beck and Thomas [59] using fast, relaxation conductivity methods. While the absolute values of the first electron rate constants were not particularly unexpected, in view of the known high mobility values, their determination was still of considerable significance since it was the final key needed to be able to quantitatively relate chemical change via scavenging to the physical processes of ionisation and decay.

The intervening decade has certainly seen no diminution of effort nor of surprises. For example the great importance of electron energy level measurements; the complex behaviour of ion yields and mobilities at low densities; the intricacies of field dependences; the complexity of kinetics in high mobility liquids and many more aspects of the continuing saga of the life of an ion in a dielectric liquid were completely unexpected 10 years ago. The subject itself is still dynamic and has by no means become textbook material. What follows therefore is the author's own, possibly biased view of the present state of "generally accepted" knowledge in the field.

IONISATION

Discussions about ionisation in condensed media are frequently fraught with semantic difficulties. This problem is particularly

pronounced in the case of non-polar liquids because of the fact
that even when two oppositely charged ions are separated by a
distance of as much as 400 Å, they still have only a 50%
probability of completely escaping their mutual coulombic attraction. Strictly speaking therefore, the two ions are still coupled
together and in this sense can be considered to be an ion pair.
On the other hand, the diffusional motion of the individual ions
separated by many molecular diameters is not expected to be
strongly perturbed by the coulomb field and they can therefore be
treated as separate entities with the same physico-chemical
properties they would have if completely isolated. The use of the
term ion pair in what follows includes such weakly correlated
pairs of ions and should not be confused with the more specific
usage of ion pair in certain other fields where it is usually
associated with more tightly bound, "contact" or "solvent
separated", pairs of ions. It is also perhaps worth mentioning
that the concept of ionisation as used in dilute gases, involving
as it does the separation of ions to infinity is completely
inapplicable to condensed systems.

The starting point for recent theoretical treatments of
ionisation in non polar liquids has been the single ion pair
model. In this model correlated pairs of oppositely charged ions
are taken to be formed at time zero in thermal equilibrium with
the medium and sufficiently far from other pairs that cross
coulombic interaction can be neglected. The distance between the
ions of a pair can be taken to be the same for all pairs (delta
function) or, more realistically, to be given by a probability
distribution function $F(r)$ which gives for the fraction of ion
pairs with a separation between r and $r + dr$

$$dN(r)/N = F(r)dr \qquad (13)$$

Subsequent development with time, including neutralisation of
correlated ions (geminate recombination) and complete separation

or escape, is then a problem involving the diffusional motion of the two particles in the presence of the coulomb field for which the Smoluchowski equation is taken as the basic starting point [60].

It is apparent that the single ion pair model is a realistic representation of the situation when ionisation is a result of the absorption of low energy photons, providing of course that the photon flux is not too great. For high energy radiation however, even when the linear energy transfer is small (high energy electrons, γ and X-rays), it is known that ionisation events can occur quite close to each other [61]. Effects due to multiple ion pairs would of course be reflected in the requirement of a distribution function somehwat different for the high energy radiation case than for photoionisation when results of a single ion pair model were compared with experiment.

In practice it is argued that even for high energy, low LET radiation, the single ion pair picture should be a reasonable representation of reality if the timescale of interest is not too short [60]. The argument is based on the expectation that for high energy electrons a large fraction of energy-loss events will in fact result in the formation of only a single ion pair, and the reflection that where multiple pairs are involved one pair of ions is likely to considerably outlive the others.

Differences in the ion pair distribution functions resulting from ionisation by high energy radiation and photon absorption might be expected to result from the, on average, much larger kinetic energies of the electrons ejected in the former case. This would have been so had the range of electrons been determined by energy loss processes in the range of a few to tens of electron volts. In fact, as mentioned in the introduction, the evidence bears out that the distance an electron travels from its sibling positive ion before thermalisation is mainly controlled by energy loss processes in the 1 eV and lower range as was suggested

by Mozumder and Magee [40]. The ion pair separation distribution
functions resulting from "ionising radiation" and from photoionisation are therefore expected to be quite similar for photon
energies sufficient to result in an initial kinetic energy of
one eV or so for the ejected electron.

A distinct problem in the application of theory to ionisation
in non-polar liquids is definition of time zero or in other words,
the time at which the ion pair can be considered to be in thermal
equilibrium with the medium. An automatic consequence of inefficient loss of electron energy to internal nuclear motions of the
molecules and collective motions of the medium, which is used to
explain the long range, is that the thermalisation process can in
fact take an appreciable time. In table I are listed the times

TABLE I

Thermalisation times estimated, τ_{th}^{e}, and measured, τ_{th}^{m}, for the
liquids shown

Compound	T($^{\circ}$K)	τ_{th}^{e} (ps)[a]	τ_{th}^{m} (ps)[b]
Xe	161	5500	6500
Kr	121	2600	4400
Ar	87	1000	900
CH_4	111	22	–
$C(CH_3)_4$	296	1.8	–
TMS	296	1.1	–
iso-octane	296	<0.8	–

a) estimated from data in ref. (113) using equation (10) of ref.
(115) with $|n| = 0.5$ and $K = 1$ eV; b) ref. (116)

estimated to be required for initially hot electrons to cool to
within 10% of thermal for some atomic and molecular liquids. The

very long times of nanoseconds required for electron thermalisation
in the rare gas liquids have been confirmed recently by experiment
as shown in the table. The thermalisation time however decreases
substantially as the complexity of the molecules increases but
is still seen to be a comparatively long picosecond or so for
liquid neopentane.

In liquids composed of anisotropic molecules the thermalisation time is most probably considerably less than a picosecond
and it is possible that in these systems the formation of a
localised state actually precedes complete degradation of the
electron kinetic energy as was originally suggested by Freeman [13,
37]. In electron attaching liquids such as CCl_4 localisation by
attachment will almost certainly occur at electron kinetic
energies in excess of thermal.

It is apparent that the assumption, implicit in the single
ion pair model, that the time required for thermalisation is
negligibly short, is unrealistic for the rare gas systems and is
indeed in some doubt even for the high mobility molecular liquids.
It is in fact known that considerable reductions in the escape
probability occur in the latter liquids for solute concentrations
of 10^{-2} M and above [44,62-64] which is quite probably due to
scavenging of electrons while still hot. Since the geminate recombination of a large fraction of electrons is known to occur
on a timescale of picoseconds even in low mobility ($\mu < 1$ $cm^2V^{-1}s^{-1}$)
liquids [65,66], see table VIII and IX, it is apparent that for
the high mobility liquids a large fraction of electrons could
undergo recombination without in fact ever becoming thermalised.
This is a problem which has been previously recognised [60].

It is perhaps worth pointing out that the problem of electron
thermalisation is one which is common to all sources of the
ionising impulse whether this be megavolt electrons or, single or
multiple photons and whether or not the electron is ejected from
a solvent molecule, a dissolved solute or a solid surface in

contact with the liquid. To a large extent therefore the charge separation process and the resultant further processes of escape, recombination and scavenging might be expected to be similar in all of these cases. An interesting and developing field is the intercomparison of different modes of ionisation.

While the majority of investigations of the properties of electrons and ions in dielectric liquids have used high energy electrons as the primary source of ionisation, there are several examples of the use of photons. For example the double photon pulsed laser ionisation of pyrene used by Beck and Thomas [59] for electron rate constant measurements; the electron energy level measurements of Holroyd and others [50,67-72] using single photon ionisation of a metallic surface or a dissolved solute, and recently the energy level measurements carried out by Schmidt using synchroton radiation to directly photoionise the solvent [73].

One of the main aims of recent theoretical and experimental work has been the elucidation of the initial "separation" distribution function $F(r)$. The approach has been to assume a given distribution function and compare the predictions of theory using that function with the results of experiment. The trial functions most frequently chosen, apart from the delta function, are the three-dimensional Gaussian for which

$$F(r) = \frac{4r^2}{\pi b^3} \exp(-r^2/b^2) \tag{14}$$

and the exponential with a short range cut-off or neutralisation radius, R_a, of 10 to 20 Å

$$F(r) = \frac{1}{b} \exp([R_a - r]/b) \tag{15}$$

The different distribution functions are compared in figure 1, with the characteristic distances, b, being chosen to give the

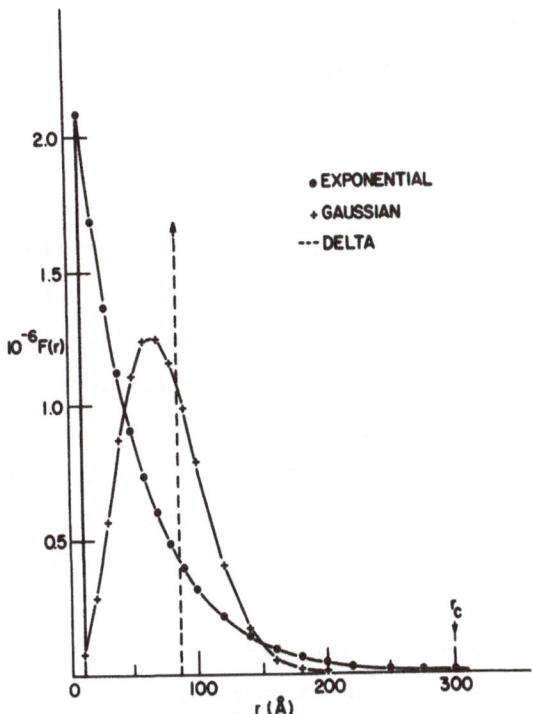

Figure 1. For a liquid of dielectric constant 2.0 at room temperature the initial ion pair separation distributions shown would
all result in a total 3% escape probability. This corresponds to
the situation for ionisation of cyclohexane if the total initial
yield of ion pairs is 5 per 100 eV absorbed energy. The mathematical form of the distribution functions is given in the text.

same probability of escape (0.03) in a liquid of dielectric

constant 2.0 at room temperature.

To date, comparisons of theory with experiment, in the form

of scavenging curves and effects of electric field on the escape

probability, have tended to favour the exponential form of the

distribution function [74-76]. Possible reasons why this particular

distribution might be expected rather than a statistical gaussian

range distribution have been put forward [74]. As will be seen

below the predictions of the single ion pair model are definitely

in good agreement with experiment at long times and for low

solute concentrations. Under these conditions however the predicted results are insensitive to the particular initial distribution of ion pair separation distances.

RECOMBINATION AND ESCAPE

As mentioned in the previous section, it is now generally accepted
that treatment of the further development in time of the "initial"
ion pair separation distribution must involve solution of the
Smoluchowski equation in the presence of a central force field.
This takes into account the combined effects of the tendency to
recombine under the influence of coulomb attraction and a net
tendency for outward movement due to the random diffusional motion
of the ions. After several attempts to solve the problem, using a
variety of approximations, it would appear that the nut can now
be considered to have been succesfully cracked albeit by somewhat
different approaches but which lead to consistent results. For
a deeper insight into the mathematical complexities involved, the
reader is referred to the last review of ionisation in non-polar
liquids [60] and to publications dealing with the application of
the theory [74-84].

   The assumptions involved in the treatment are those already
mentioned above, i.e. that the initial (time zero) situation is
that of isolated pairs of ions in thermal equilibrium with the
medium and with a distribution of separation distances within the
pairs. An additional assumption in further calculation is that the
ions remain in thermal equilibrium. This implies that the rate of
transfer of any excess energy of the particles to the medium is
sufficiently high to ensure that no potential energy loss due to
drift in the field direction is converted into thermal energy of
the ions. This condition could well break down at short distances
for high electron mobility (low thermalisation rate) liquids. Some
consequences of this have been discussed by Mozumder [85].

A problem can also arise even without hot electron effects coming into play due simply to the very rapidly varying field experienced by an ion moving in a coulomb potential and the finite time required to reach an equilibrium drift velocity corresponding to the field strength applicable at a given time. A simple calculation shows that a non-localised electron of mobility μ $cm^2 V^{-1} s^{-1}$ would begin to have difficulty following the rapidly changing field at a distance r given by

$$r^3 \simeq 20m\mu^2/4\pi\epsilon_o\epsilon_r \qquad (16)$$

In a medium of dielectric constant 2 this corresponds to distances of 9.4, 44 and 202 Å for free electron mobilities of 1, 10 and 100 $cm^2 V^{-1}s^{-1}$ respectively.

As for the other conditions mentioned here and in the previous section, it is apparent that problems with the assumptions involved in the theoretical treatment of the single ion pair model would be expected to be greatest for systems in which the excess electrons remain highly mobile, i.e. $\mu \geq 10$ $cm^2 V^{-1}s^{-1}$. Or, in positive terms, the present theoretical approach would be expected to predict the behaviour of ions quite well for non-polar liquids in which the electron mobilities are on the order of 1 $cm^2 V^{-1}s^{-1}$ or less.

We now proceed to a consideration of the results of the calculations and a comparison of these results with experimental data.

Correlated Ion-Pair Kinetics

It is particularly helpful for the less mathematically inclined to have a physical picture of the processes occuring. This has to a certain extent been provided by the calculations of De Zeeuw, Infelta and Hummel [60,78,83] in the form of a numerical solution of the problem, beginning with all pairs of ions initially separated by the same distance r_o. They then proceeded to show the development of the distribution of distances as a function of time taking neutralisation to occur at a distance R_a. The neutralisation rate at R_a was taken to be a variable parameter.

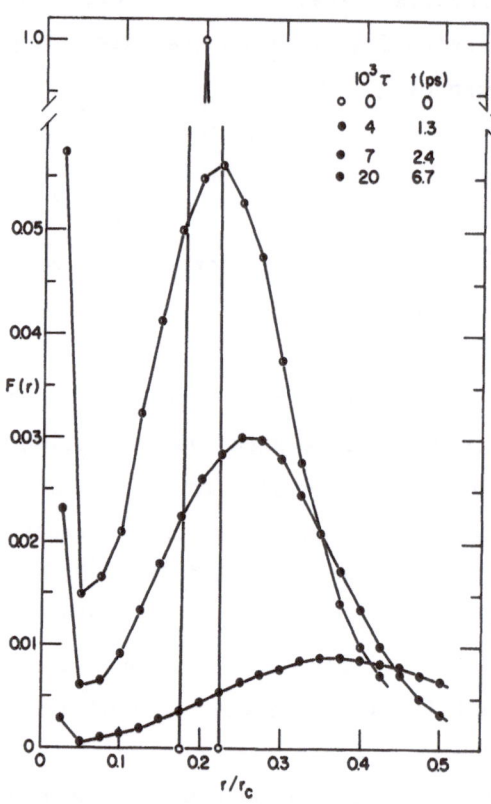

Figure 2. The change with time of the ion pair separation distribution function after beginning with all ion pairs separated by $r/r_c = 0.2$. Data from reference (60). The radius at which neutralisation occurs is $r/r_c = 0.025$ or ca 7.5 Å for a dielectric constant of 2. The temporary build-up at the reaction radius is due to a finite reaction velocity. The absolute time given in picoseconds is determined from the generalised time $\tau = Dt/r_c^2$ taking $r_c = 300$ Å and a total ion mobility of 1 $cm^2 V^{-1}s^{-1}$.

The results shown in figure 2 were obtained using a non-infinite neutralisation rate and illustrate very well the inward flow

of electrons to the positive ion with temporary build-up at the
reaction radius and at the same time the gradual broadening of
the remaining distribution and displacement of the maximum to
longer distances eventually resulting in the escaped ion fraction
$W(\infty)$.

An interesting point is that the distribution does not split
into two parts (separate maxima) with one moving inwards and the
other outwards, but rather leaks at both ends. This is conceptually
very important since it is usual to differentiate between geminate
ions and free ions as if they were completely separate entities
at all times. For example at a time when the number of surviving
ion pairs is twice that of the number which will escape (denoted
γ and called the "equivalent geminate ion lifetime"), it is
customary to think that one is observing a yield of geminately
recombining ion pairs equal to the yield of ion pairs which are
already free or at least destined to become free. This, as
figure 2 shows, is not the case.

It is apparent, on the basis of the foregoing, that the actual
time taken to become a free ion is a rather difficult parameter
to define. This will be discussed further in the section on free
ion yields.

The total fraction of ions still present at time t (the
survival probability $W(t)$) is obtained by integration, from the
reaction radius to infinity, of distributions such as those shown
in figure 2 and of course integration over the initial distribution of separation distances where a distribution other than a
delta function is considered. The calculations are usually made
in terms of the generalised parameter τ which can be related to
real time t by the expression

$$\tau = Dt/r_c^2 \tag{17}$$

where D is the sum of the diffusion coefficients of the ions and, as before, r_c is the Onsager escape distance.

A reciprocal square root dependence of $W(t)$ on time was found in the limit of infinite time even in early treatments [79], viz.

$$W(t)_{t \to \infty} = W(\infty) \left[1 + (r_c^2/\pi Dt)^{\frac{1}{2}} \right] \tag{18}$$

or in terms of τ

$$W(\tau)_{\tau \to \infty} = W(\infty) \left[1 + 1/\sqrt{\pi\tau} \right] \tag{19}$$

Because of the limiting square root dependence expected, the results of full calculations are plotted as $W(\tau)/W(\infty)$ against $1/\sqrt{\tau}$ in figure 3.

The three lines shown in figure 3A were obtained using the delta, gaussian and exponential distributions shown in figure 1. These were all normalised to give the same value of $W(\infty)$, (=0.03). In figure 3B the full lines were obtained using a delta function with different values of r_0/r_c such that the values of $W(\infty)$ were 0.018, 0.048 and 0.135. The lower dashed line in figure 3B is the linear behaviour predicted on the basis of the limiting relationship given above (equation (19)).

Two important conclusions can be drawn from figure 3, as pointed out recently by Van den Ende et al. [86]. Firstly, the limiting square root dependence is in fact seen to be only an extreme limiting case which is tangential to the results of the full calculations at infinite time. Secondly, the value of $W(\tau)/W(\infty)$ is in fact independent of the distribution function and of the absolute value of $W(\infty)$ up to a survival probability at least 4 times the escape probability if the latter is approximately 0.2 or less. This corresponds to a yield of 0.5 $(100 \text{ eV})^{-1}$ for molecular liquids with $G_{fi} \approx 0.12$.

As a corollary of this latter observation it is apparent that in order to obtain experimental information which could lead to a

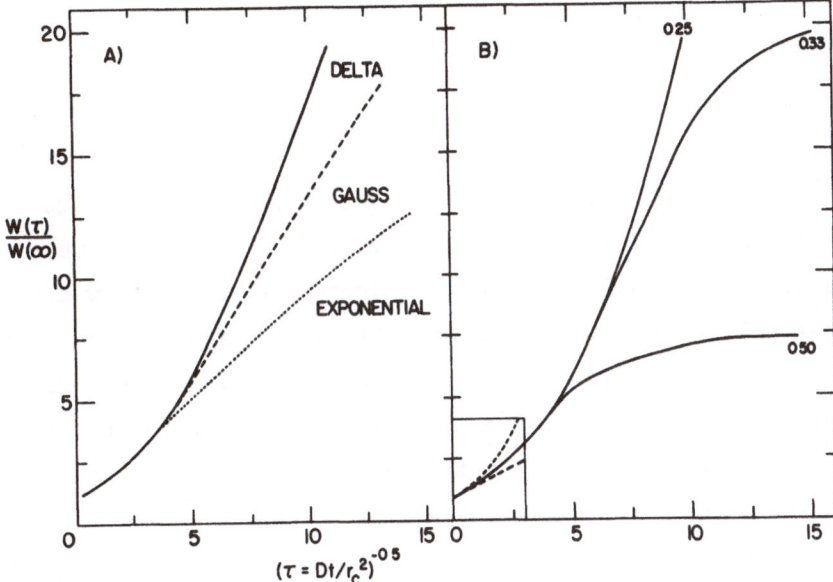

Figure 3. The dependence of the correlated ion pair survival probability divided by the probability for complete escape, $W(\infty)$, on the reciprocal square root of the generalised time $\tau = Dt/r_c^2$: A) for different distribution functions (see figure 1) all of which give $W(\infty) = 0.03$; B) for delta functions with values of r_0/r_c of 0.25, 0.33 and 0.50 giving $W(\infty) = 0.018$, 0.048 and 0.135 respectively. The dashed lines in the left bottom corner of 3B are the dependence as predicted by the prescribed diffusion treatment of Mozumder [79] (upper) and the limiting reciprocal square root dependence predicted by general theory (lower). The data are taken from reference [86].

differentiation between the different possible distribution functions, the decay kinetics of ions would have to be followed on a timescale considerably shorter than that given by $1/\sqrt{\tau} = 5$ and preferably on the order of $1/\sqrt{\tau} = 10$ or more. Using equation (17) this is found to correspond to a real time given by

$$t \lesssim 0.01 \ r_c^2/D \tag{20}$$

or in terms of the mobility sum at room temperature with a liquid of $\epsilon = 2$

$$t \leq 3.4 \times 10^{-12}/\mu \tag{21}$$

for μ in $cm^2 V^{-1} s^{-1}$.

For molecular ions in a liquid of viscosity 1 cP ($\mu \simeq 10^{-2}$ $cm^2 V^{-1} s^{-1}$), condition (21) gives times of a few nanoseconds or less. It is in fact on this kind of timescale that relatively large yields $[W(\tau)/W(\infty) \sim 10]$ of geminate ions are observed in optical pulse radiolysis studies and these might be worthy of a fuller kinetic analysis than at present has been carried out. A disadvantage of the optical studies is that almost invariably a solute is added to the system in order to result in the formation of an ion with a reasonably large extinction coefficient. It is obvious that to obtain relevant kinetic information, the timescale of scavenging by the solute must be an order of magnitude shorter than the timescale of the observations or in other words a few hundred picoseconds for the present purposes. For scavenging rate constants on the order of $10^{12} M^{-1} s^{-1}$, (see below) approximately 10^{-2} M solute is therefore necessary. This has as a disadvantage that, from effects on free ion yields, it is known [44,62-64] that concentrations on this order can in fact perturb the initial distribution thus negating the aim of the experiment. An approach would be preferred in which no species (e.g. the red absorption of the electron in some low mobility liquid) is studied under the combined conditions of timescale and mobility required to obtain information relevant to the form of the distribution function.

The use of optical absorption data to check the long time distribution-independent region of the theoretical calculations involves other difficulties of which the main one is the relatively high doses (several kilorads) required to observe the small yields present. As a result a considerable amount of homogeneous recombination occurs on the timescale of interest. In the case of the commonly used solute biphenyl, indications of other complications in the decay kinetics at long times have been found in recent experiments [86,87].

It has been shown that the time dependence of the ion-pair yield can be studied using the microwave conductivity method with total doses considerably less than in optical experiments [86,88]. This results in corrections for homogeneous recombination of only a few percent on timescales of several microseconds. The results of these experiments have been found to be in good agreement with theoretical calculations. This is illustrated in figure 4A for a

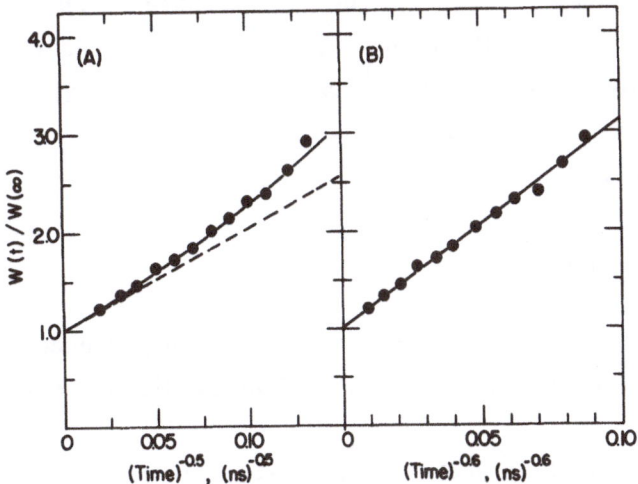

Figure 4. A comparison of theory, lines, with experiment, points, for the distribution independent long-time decay of ion pairs as shown in the lower corner rectangle in figure 3B. The points are for a solution of CO_2 (electron scavenger) and C_6H_6 (positive ion scavenger) in cyclohexane. The absolute timescale for the calculated curves was determined using the measured mobility of 8.8×10^{-4} $cm^2V^{-1}s^{-1}$. The dashed line in 4A is the limiting reciprocal square root dependence expected from theory. In 4B is illustrated the linearisation of the full calculation and experimental points when plotted against $t^{-0.6}$. The data are from references (86), (88) and (208).

solution of benzene as positive ion scavenger and CO_2 as electron scavenger in liquid cyclohexane. In these experiments the value of $G\mu$ was measured as a function of time and by extrapolation to infinite time $G_{fi}\mu$ was obtained. On dividing by the value of G_{fi} = 0.15 for cyclohexane the sum of the mobilities of the ions present $(8.8 \times 10^{-4} \, cm^2 v^{-1} s^{-1})$ was found. This could then be used to obtain the diffusion coefficient and hence calculate the relevant real time axis for the calculated curve using equation (17).

In figure 4A the deviation of the experimental points and the calculated full line from the limiting square root dependence is further illustrated. It has recently been found empirically [88] that the theoretical curve, over the region where it is insensitive to the form of the distribution, can be linearised if $W(\tau)/W(\infty)$ is plotted against $\tau^{-0.6}$ rather than $\tau^{-0.5}$. This is illustrated in figure 4B. From the slope of such a plot the survival probability is found to be given by

$$W(\tau)/W(\infty) = 1 + 0.6/\tau^{0.6} \tag{22}$$

The consequences of this for the concentration dependence of ion scavenging for low scavenger concentrations is discussed in a later section.

From equations (22) and (17) it is possible to derive a value for the equivalent geminate ion lifetime γ, i.e. the time at which the separation distribution is such that the overall probability for escape is equal to the probability for geminate ion recombination.

$$\gamma = 0.43 \, r_c^2/D \tag{23}$$

For the primary ions in cyclohexane γ is found to be 0.55 ns whereas for molecular ions of total mobility $8 \times 10^{-4} \, cm^2 v^{-1} s^{-1}$, γ = 180 ns.

To date, direct experimental studies of correlated ion pairs on a timescale short enough to be able to definitely distinguish between an exponential and a gaussian initial distribution function

have not been reported. The calculated real time dependences of
the survival probability for the three distributions shown in
figure 1 are plotted in figure 5. The parameters used to obtain

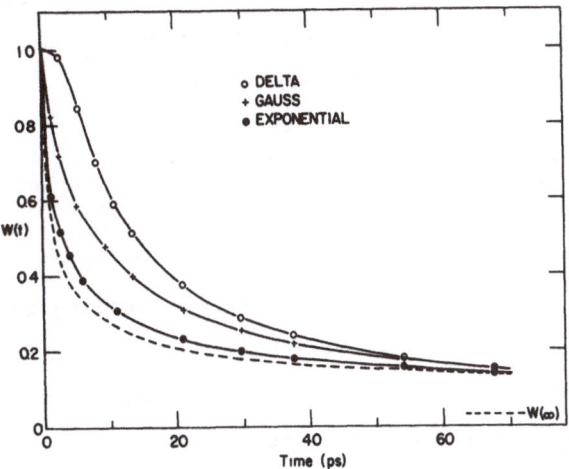

Figure 5. The predicted real time dependence of the ion pair
survival probability in liquid cyclohexane for the distributions
shown in figure 1 taking 0.24 $cm^2V^{-1}s^{-1}$ for the sum of the
electron and radical cation mobilities, and a total initial ion-
pair yield of $5(100 \ eV)^{-1}$. The dashed line is that obtained using
the expression derived by Laplace transformation of the empirical
concentration dependence of scavenging in cyclohexane, equation
(24), using a measured reactivity to rate constant ratio of 2.3
picoseconds (see table VIII).

the absolute timescale were r_c = 290 Å and μ = 0.24 $cm^2V^{-1}s^{-1}$ with
an escape probability of 0.03. The curves therefore represent the
predicted decay kinetics of the primary ions in cyclohexane, if
the total initial yield of ion pairs is taken to be $5(100 \ eV)^{-1}$.

The possibility of flexibility in the choice of the total
ion yield, at least on the low side, can make decisions as to the
"correct" distribution more difficult than might appear from the
significant differences illustrated in figure 5. For example if
a gaussian distribution is taken with a total yield substantially
lower than 5 and a correspondingly higher value of $W(\infty)$, in order
to fit the free ion yield, then the resulting curve of yield

against time becomes difficult to distinguish from that using an
exponential with G_i = 5 except at very short times [83,88].

Also shown in figure 5 is the time dependence suggested on
the basis of the kinetic equation for the surviving yield derived
from the empirical scavenging equation of Warman, Asmus and
Schuler [42] (discussed further below)

$$G(t) = G_{fi} + G_{gi} \exp(\lambda t) \; \mathrm{erfc}(\lambda t)^{\frac{1}{2}} \qquad \qquad (24)$$

The absolute timescale was obtained by using a value of
$\lambda = 4.3 \times 10^{11} \; s^{-1}$ determined from combined scavenging and absolute
rate constant studies (see table VIII below). The form of the
decay given by (24) is seen to resemble most closely that calculated using the exponential distribution.

Because of the complexity of the kinetics, it is somewhat
meaningless to talk of an average lifetime for geminately recombining ions. In the view of the author, however, it would seem
necessary to have some parameter for conversational purposes which
gives an idea of the timescale over which a significant degree of
decay of ion pairs has occurred. A reasonable phenomenological
choice for the "geminate recombination time", τ_{gi}, would seem to
be the mean time for scavenging pertaining to the scavenger concentration for which half of ion pairs which undergo geminate
recombination are scavenged by a solute S. In terms of the
empirical scavenging equation, (11), τ_{gi} is then given by α_s/k_s
which is equal to the reciprocal of λ in equation (24). At τ_{gi},
57% of the initial ion-pairs which would undergo geminate recombination have already done so. For cyclohexane, the geminate lifetime defined in this way is 2.5 ± 0.3 picoseconds as is shown in
tables VIII and IX. For n-hexane and iso-octane the geminate
lifetimes are found to be 15 ps and 0.7 ps respectively. The
latter value is seen to be of the same order of magnitude as the
estimated upper limit for the electron thermalisation time in

this liquid given in table I.

The Free Ion Yield

As pointed out in the previous section, it is in the strictest sense mistaken to consider the ion pair population at any particular time to be made up of geminate ion pairs and free ions. Rather it can be said that at time t, the probability of escape for those ion pairs remaining, $P_{fi}(t)$, will be $W(\infty)/W(t)$ which, using expressions (22) and (17) (for $W(\infty)/W(t) \gtrsim 0.2$), gives

$$P_{fi}(t) = \left[1 + 0.6\left(\frac{r_c^2}{Dt}\right)^{0.6}\right]^{-1} \tag{25}$$

For certain experimental applications and theoretical discussions it is useful to have a measure of the "time required for escape", τ_{fi}. If we arbitrarily define this as the time at which the ion-pairs remaining have an overall 90% probability of becoming completely separated then τ_{fi} will be given by

$$\tau_{fi} = 16.7 \, r_c^2/D \tag{26}$$

Taking again as an example cyclohexane, the escape times for the primary ions ($\mu = 0.24 \text{ cm}^2\text{V}^{-1}\text{s}^{-1}$) and for molecular ions ($\mu = 8 \times 10^{-4} \text{ cm}^2\text{V}^{-1}\text{s}^{-1}$) are found to be 23 ns and 6.8 μs respectively. It is interesting to note that the escape time is four orders of magnitude longer than the geminate recombination time τ_{gi} given in the previous section.

Accurate determination of the yield of free ions is of considerable importance for a quantitative theoretical understanding since, for a given initial distribution function $F(r/b,0)$ where b is the characteristic distance of the distribution, G_{fi} is given by

$$G_{fi} = G_i \int_0^\infty F(r/b,0)\exp(-r_c/r)d(r/b) \qquad (27)$$

The characteristic distance is therefore defined if G_{fi} is known and a given value of the total initial yield of ion pairs is chosen.

Since the early measurements of Freeman [4], and Allen and Hummel [5] in n-hexane, an enormous number of free ion yields have been determined in a variety of non-polar liquids and a compilation has been published [89]. More recent work in which the effect of fluid density has been studied is to be found in reference [90]. Values for some readily available organic solvents are listed in table III below. The general tendency for more spherically symmetrical molecules to display larger G_{fi} values which was apparent in the first reports of Tewari and Freeman [44] and Schmidt and Allen [45] has been substantiated in work which followed.

Because the Onsager escape distance appears in the exponential term in (27), the free ion yield is very sensitive to variations in dielectric constant and temperature of the medium. A better comparison between different media and at the same time a better physical insight is gained when G_{fi} is converted into the characteristic range parameter, b, for a given distribution function and assumed total yield of ion pairs. It has become almost standard to use the method of obtaining b introduced by Schmidt and Allen [45,91], who took a gaussian distribution function for F(r) and a total yield of 4.3 $(100 \text{ eV})^{-1}$. This common usage is of course to be preferred over the proliferation of b values based on individual, in-house distribution functions facilitating as it does comparison of reported data. However, this should not be taken as an acceptance of the gaussian distribution as giving the best description of ionic processes in general. As mentioned in the previous section, the exponential form would in fact seem to be preferred. It would seem worthwhile signifying the fact that the b value is obtained using a gaussian distribution by use of for example the symbol b_{gp} as is done by Freeman. Since the total yield used also determines the b value derived, a convention in which this is also indicated might be preferred. For example $b(4.3)_g$ for the Schmidt-Allen usage, or $b(5)_e$ for a total yield of 5 and an exponential distribution.

An additional interest in the b parameter stems from its obvious relationship to the electron mobility [92] via the effective mean free path of electrons which is discussed in a later section.

It is of interest in the context of the present paper to investigate briefly the consequences that a change in b value has for the kinetics of the initially formed correlated ion pairs. It can be said immediately and possibly surprisingly that the magnitude of this parameter has no influence on the long time decay kinetics in the form of $W(t)/W(\infty)$ for values of b less than $r_c/2$. Thus the value of the equivalent geminate ion lifetime γ and the free ion escape time τ_{fi} are both dependent only on r_c and the sum of the mobilities of the ions composing the correlated pair and are not influenced by the average initial distance of separation of the ions. This has been previously emphasized by Mozumder [93].

The lifetime for shorter lived ion pairs does however follow expectations and increase as the average initial separation distance increases if the sum of the diffusion coefficients is the same. For the simplest case of a delta distribution, the value of the generalised time τ at which half of the ion pairs have recombined, $W(\tau) = 0.5$, is found from calculations to be 0.011, 0.019 and 0.082 for values of r_o/r_c of 0.29, 0.33 and 0.50. In addition, from the results of scavenging studies given in table IX, it can be seen that the larger value of $b(4.3)_g$ in iso-octane (95 Å [63]) compared with that in cyclohexane (66.1 Å [63]) is reflected in a considerably larger value of the product of the geminate ion lifetime and the mobility, $\mu\tau_{gi}$, for the former compound

$(3.7 \times 10^{-12} \ cm^2V^{-1})$ than for the latter $(0.7 \times 10^{-12} \ cm^2V^{-1})$.

Homogeneous ion recombination

Having once escaped their mutual coulombic attraction, ions may freely diffuse in the bulk medium and are available to the investigator for a deeper study of their physico-chemical properties. In the absence of reactive impurities, disappearance will occur by homogeneous recombination with other free ions. The reaction is second order in the ion concentration N_i which results in a time dependence given by

$$N_i(t)^{-1} = N_i(o)^{-1} + \alpha t \qquad (28)$$

where α is the ion recombination rate constant.

For two ions which undergo neutralisation at a separation distance R_a with a velocity of approach v, the rate constant for recombination is given by, [60]

$$\alpha = \mu e/\epsilon_o\epsilon_r \ [1 - \exp(-r_c/R_a) + r_c D \exp(-r_c/R_a)/R_a^2 v] \qquad (29)$$

In the derivation of equation (29), it is implicitly assumed that the ions remain at all times in thermal equilibrium with the medium. This condition might be expected to break down for very high electron mobility liquids where energy exchange with the medium is inefficient. This possibility will be discussed later.

For low dielectric constant media the ratio r_c/R_a is so large (R_a being expected to be on the order of 10 Å) that equation (29) reduces to

$$\alpha_D = \mu e/\epsilon_o\epsilon_r \qquad (30)$$

This expression is most frequently referred to as the Debye

equation [94] in liquid circles although it was also derived for the dense gas case by Langevin [95] and by Harper [96] and is referred to by their combined names by those working in gas phase recombination studies.

By substituting the diffusion coefficient for the mobility using the Einstein relation

$$D = \mu k_B T / e \qquad (31)$$

and the Onsager radius in place of ε_r using

$$r_c = e^2 / 4 \pi \varepsilon_o \varepsilon_r k_B T \qquad (3)$$

one finds that equation (30) is in fact equivalent simply to a diffusion controlled reaction with a reaction radius equal to the Onsager radius e.g.

$$\alpha_D = 4 \pi r_c D \qquad (32)$$

Several measurements have been made of the recombination coefficient not only for molecular ions but also for electrons in a variety of hydrocarbon liquids [44,97-102] and in a hydrocarbon mixture of n-hexane and neopentane [102]. The latter, mixture measurements span a range of electron mobilities from 0.08 to 60 $cm^2 V^{-1} s^{-1}$. The values of the parameter $\alpha \varepsilon_r / \mu$ have been found to be within experimental error of the value of 1.81×10^{-6} cmV expected from equation (30) apart from a measurement in liquid argon [97] which was substantially lower and a note by Allen and Holroyd [98] that their value for tetramethylsilane seemed to show a slight (10 to 20%) negative deviation.

It is sometimes thought that, in order to check whether the Debye equation is valid for a system, firstly a value of the sum of the mobilities of the ions must be available and secondly the absolute value of the concentration in the kinetic measurements

must be known. In a conductivity experiment however, neither of
these requirements are necessary since the decay kinetics of the
conductivity itself can be used to check "Debye". This can be seen
by substituting in equation (28) for N_i using

$$\sigma = eN_i\mu \tag{33}$$

and for α using equation (30). This results in

$$\sigma(t)^{-1} = \sigma(o)^{-1} + t/\varepsilon_o\varepsilon_r \tag{34}$$

A check of the Debye relation is therefore simply that the first
half life of the conductivity be given by

$$t_{\frac{1}{2}} = \varepsilon_o\varepsilon_r/\sigma(o) \tag{35}$$

or more correctly that the slope of a reciprocal conductivity
versus time plot is equal to $1/\varepsilon_o\varepsilon_r$.

It is perhaps worth pointing out that if the lifetime of a
conductivity transient is required to be a microsecond or longer,
then, from equation (35), the initial conductivity must be less
than 1.8×10^{-5} ohm^{-1}m^{-1}.

A knowledge of the mobility or the absolute concentration is
of course required if an absolute value of α is of interest. The
absolute values of the recombination rate constant tend to be
bewilderingly high for chemists used to normal diffusion controlled
processes. Thus in neopentane [97,102] and TMS [98] α is almost
10^{17} M^{-1}s^{-1} which outstrips by six order of magnitude the values
customary in ionic studies in polar liquids. Electron concentrations considerably less than nanomolar are therefore necessary in
high mobility liquids if ion lifetimes are to be much longer than
a few tens of nanoseconds.

While no significant deviations from the Debye equation have
been found for molecular liquids to date, recombination rate

constants lower than α_D are to be expected for liquids in which the effective electron mean free path becomes comparable to the Onsager radius. The possibility then arises that the electron does not undergo a collision while in the coulomb field of the positive ion. Since then it can lose non of the kinetic energy gained due to its motion in the field, it will escape via an open orbit unless it happens to pass within the relatively small reaction radius, R_a, of the positive ion. The rate controlling step in recombination under these circumstances becomes the rate at which electrons can dissipate energy in the medium while within r_c rather than the rate of diffusion into the Onsager sphere.

This is a problem which has received considerable experimental [103,104] and theoretical [105,106] attention in gaseous systems recently. As a rough criterion, diffusion would be expected to play the predominant role in recombination if the electron thermalisation time, τ_{th}, is less than the characteristic time, r_c^2/D, for recombination and escape

$$\tau_{th} < r_c^2/D \tag{36}$$

For the liquids TMS and neopentane at room temperature the condition is fulfilled if the thermalisation time is less than about 4 ps. Recent estimates of τ_{th} are in fact approximately 1 ps for these liquids (see table I). It is possible that the slightly lower value than expected of α for TMS found by Allen and Holroyd [98] is an indication of a slight effect due to inefficient energy exchange in this liquid.

For liquid methane ($\mu = 400$ cm^{2}V^{-1}s^{-1} at 111°K) the thermalisation time is required to be less than 15 ps which is in fact of the same magnitude (table I) as estimated for τ_{th}. A recombination rate constant less than α_D would therefore be expected for this liquid. The only literature value available [97] however suggests no significant difference from α_D. The low values of α found for the liquid rare gases [97,107] are in agreement with the long thermalisation times for these liquids (table I). This potential problem is obviously one which should receive attention in the future.

A parameter of somewhat more than academic interest, is the concentration of free ions for which the timescale of escape is comparable with the timescale calculated for homogeneous recombination. Under these conditions the kinetics will obviously be complex and equations (30) or (32) may no longer be applied to the recombination process. In order to be able to apply the Debye equation therefore, the inequality

$$4\pi r_c DN_i \ll D/r_c^2 \tag{37}$$

corresponding to

$$N_i \ll 4\pi r_c^3 \tag{38}$$

must be satisfied. For a liquid of dielectric constant 2 at room temperature, this results in the condition that the concentration of ions formed must be considerably less than 3×10^{15} per cubic centimetre or 5×10^{-6} moles per litre. Using high energy radiation, taking a G value of 1 for ion formation, this concentration is obtained for a total dose of 5 krad. Doses of kilorads are frequently used in optical absorption pulse radiolysis measurements. This could possibly lead to kinetic complications in these studies. Consequences of high ion concentrations have been considered by Mozumder and Magee [108].

MOBILITIES

Knowledge of the mobility of an ion is of utmost interest since, not only does it provide a measure of the diffusion coefficient via the Einstein relation which is essential for a quantitative understanding of the kinetics of ionic processes but, in addition, it can provide information about the microscopic interactions between the ion and its surroundings. Because of this, an enormous amount of experimental effort has been expended on mobility measurements, particularly of electrons, in recent years. Many of the important aspects of this work such as field strength dependences and intermediate density studies will receive only brief mention in what follows since it is the intention to limit the discussion as much as possible to "normal" systems. Before going into the specifics of the mobilities of different ion types there follows a short and simple discussion of the concept of mobility.

General

In the absence of an electric field an ion undergoes Brownian type, diffusional motion as for a neutral entity. However, on applying an electric field of strength E V cm^{-1} the normally isotropic thermal velocity distribution of the ion begins to develop a net velocity component, ω, in the direction of the field. For ions of mass M the equation of motion in the direction of the field can be written

$$M \, d\omega/dt = Ee - M\omega G \tag{39}$$

where Ee is the accelerating force due to the applied field and $M\omega G$ is a decelerating or frictional force due to interaction of the ion with the medium. Integration of equation (39) shows that the velocity ω increases with time after application of the field according to

$$\omega(t) = \frac{Ee}{MG} \left[1 - \exp(-Gt) \right] \tag{40}$$

and eventually reaches a constant, equilibrium value $\omega(\infty)$ given by

$$\omega(\infty) = \frac{Ee}{MG} \tag{41}$$

The friction coefficient G has the units of frequency and, as can be seen from equation (40), its reciprocal is the mean time required for momentum relaxation of the ion, τ_m, i.e. the mean time required to randomise any net momentum component in a given direction. In the majority of "dc" (constant voltage) experiments, ion motion is studied on a timescale much longer than τ_m. Equilibrium conditions therefore apply and the steady state drift velocity $\omega(\infty)$ is measured. Since for low field strengths the drift velocity is in fact proportional to the field strength it is more convenient to use the field strength independent parameter $\omega(\infty)/E$ which is the mobility, μ.

From the above, the mobility is seen to be given by

$$\mu = \frac{e}{M} \tau_m \tag{42}$$

In situations where the ion motion can be considered to be controlled by momentary isotropic scattering events such as for ions in dilute gas systems, τ_m is an average time between such collisions and its reciprocal ν_m is an average collision frequency.

For molecular ions in condensed systems the concept of momentary collisions with scattering centres is obviously inapplicable. However for electrons in certain media, in which high electron mobilities are found, scattering can be described in terms of momentary collisions with density fluctuations in the medium (phonons). The collision frequency given by

$$\nu_m = e/m\mu \tag{43}$$

therefore regains some physical significance and the related parameter, the effective mean free path for momentum transfer, λ_m,

$$\lambda_m = \bar{m v}\mu/e \qquad (44)$$

can provide useful physical insights which are particularly valuable for the chemist. A relationship which is quite useful to remember is that the mean free path in angstroms, as given by (44), is equal to 0.64 times the mobility in $cm^2V^{-1}s^{-1}$ at room temperature ($\bar{v} = 1.1 \times 10^7$ cm s^{-1}). In equations (43) and (44) the electron rest mass, m, is given rather than the effective mass m^*. This can be larger or smaller than m depending on the nature of the interaction between the electron and the molecular lattice but is usually not significantly different from m which is often used as a first approximation if m^* is unknown.

For highly localised ions in the condensed phase a degree of physical insight into the microscopic processes controlling motion can be obtained via the Einstein equation relating the diffusion coefficient to the residence time of a particle between jumps τ_j and the mean distance travelled per jump, σ_j

$$D = \sigma_j^2/6\tau_j \qquad (45)$$

The average residence time of an ion can then be determined from the mobility using

$$\tau_j = e\sigma_j^2/6\mu k_B T \qquad (46)$$

As an example of the times involved for molecular ions at room temperature we take $\sigma_j = 5$ Å and $\mu = 5 \times 10^{-4}$ $cm^2V^{-1}s^{-1}$ which gives a value of 33 picoseconds for τ_j. This time is quite long compared to the rotation time of molecules (e.g. ca 1 ps at room temperature for cyclohexane [109]) or the actual time of approximately 2 ps taken to jump a distance of 5 Å with the mean thermal

velocity of 2.5×10^4 cm s^{-1} applicable to a molecule of molecular
weight 80. The net motion of a highly localized ion in the direction of the applied field can be seen as resulting from a slight
statistical preference for jumps to occur to a lower energy state
in the field direction.

Electrons

As pointed out in the introduction, the electron mobilities first
reported in molecular liquids immediately indicated the great
diversity of values which was to be found in later work. For
liquid hydrocarbons at room temperature for example, mobility
values have been found to range from 70 for neopentane [110] to
as low as 0.013 for trans decalin [111]. Large differences have
been found even between geometrical isomers. For example the
values of 2.2 and 0.03 for cis- and trans-butene-2 respectively
[92] and a value for the cis isomer of decalin a factor of 6 higher
[112] than that given above for the trans isomer.

Rather than being related to any particular bulk property of
the liquids studied, the electron mobility has been found to reflect more or less the shape of the constituent molecules [44,110,
113,114]. Thus, in the case of the geometrical isomers mentioned
above, the cis forms are in both cases more compact structures
than the trans. The large influence of molecular shape on the
properties of electrons had been inferred by Tewari and Freeman
[44] and by Schmidt and Allen [45] even before mobility measurements were made, on the basis of their measured free ion yield
values.

The direct relationship between electron mobility, $\mu(-)$, and
the degree of sphericity of the constituent molecules, is illustrated in table II by the changes in $\mu(-)$ which occur on successive substitution of the hydrogen atoms of methane with methyl
groups. A drop in $\mu(-)$ by many orders of magnitude occurs even
on replacing only one hydrogen atom to form ethane which in the

TABLE II

Illustration of the dependence of the electron mobility on degree of anisotropy of the molecules of the medium.

Compound	μ^a $(cm^2\ V^{-1}s^{-1})$	T $(^{\circ}K)$	$E_{\mu}^{\ b}$ (kcal)
methane	430	140	$-0.2^{c,d}$
ethane	0.8	200	1.6^a
propane	0.12	200	3.4^a
iso-butane	5	294	1.8^e
neo-pentane	70	294	0.32^a

(a) ref. (110), (b) E_{μ} is $-k_B d\ln\mu/d(1/T)$ at the temperature shown, (c) ref. (117), (d) ref. (118), (e) ref. (119)

process destroys the isotropy of the methane structure. This decrease continues on formation of propane but is reversed on substitution of the third and fourth hydrogens which results in a return to pseudo-spherical geometry in the form of neopentane.

In table III are given the electron mobilities found in some readily available and commonly used laboratory solvents listed in order of the magnitude of $\mu(-)$. Also listed in tables II and III, where available, are values of the temperature dependence of the mobility in the form of activation energies, E_{μ}, obtained from the slopes of Arrhenius plots of the mobility i.e. assuming the relationship

$$\mu = A\ exp(-E_{\mu}/k_B T) \qquad (47)$$

TABLE III

Some electron properties in common solvents at room temperature.

Compound	μ_o (cm^2 v^{-1}s^{-1})	E_μ (kcal mole^{-1})	v_o^k (eV)	G_{fi}	10^{12}k(e+SF$_6$) (M^{-1}s^{-1})	10^{12}k(e+CH$_3$Br) (M^{-1}s^{-1})
tetramethylsilane	100[a]	0	-0.55	0.74[l]	210[a]	-
2,2,4 trimethylpentane (iso-octane)	5.3[a]	1.2[a]	-0.17	0.33[l]	58[a]	28[f]
cyclopentane	1.1[b]	3.0[i]	-0.19	0.16[l]	-	-
cyclohexane	0.23[a,c]	3.0[i]	0	0.15[l]	4.0[a]	4.1[f]
n-pentane	0.15[c,d]	4.2[j]	0	0.15[l]	-	-
benzene	0.13[e]	7.6[e]	-0.14	0.081[e]	-	-
cis-decalin	0.10[f]	-	-	-	1.6[f]	1.1[f]
n-hexane	0.08[a,c,g]	4.5[a]	+0.05	0.13[l]	2.0[a]	1.2[f]
toluene	0.07[e]	3.4[e]	-0.22	0.093[e]	-	-
trans-decalin	0.013[h]	4.5[h]	-	0.13[h]	0.12[f]	0.21[f]

(a) ref. (65), (b) ref. (49), (c) ref. 98, (d) ref. (92), (e) ref. (120), (f) ref. (112), (g) ref. (121), (h) ref. (111), (i) ref. (122), (j) ref. (119), (k) ref. (110), (l) ref. (63).

to be obeyed. Equation (47) does in fact describe mobility values in the majority of nonpolar liquids quite well in the temperature range between the melting point and the normal boiling point. Very complex dependences of the electron mobility on temperature (density) are found for densities substantially lower than that of the normal liquid.

A general tendency is found towards higher activation energies as the absolute magnitude of the mobility decreases although, with the exception of benzene, a limit of approximately 4.5 kcal for

E_μ seems to be reached for low mobility liquids. The pseudo-
spherical compounds display almost no activation energy and in
fact for argon and methane E_μ is slightly negative for temperatures just above their triple points. Apart from this qualitative
difference in the effect of temperature between the (pseudo)
spherical systems and "low mobility" liquids, other significant
differences have also been found. Thus, at high applied fields
$\mu(-)$ decreases for the spherical systems but increases for
liquids with mobilities on the order of or less than $1 \ cm^2V^{-1}s^{-1}$
[113,133-135]. Also a pronounced long wavelength optical absorption has been found to be associated with electrons in low
mobility liquids [56,57,136] but is absent in high electron
mobility media.

In general the properties of electrons in the pseudo-spherical
molecular liquids resemble very closely those of electrons in the
liquid and solid rare gases and in the crystalline semiconductor
materials silicon and germanium [137-141]. As in these latter
cases therefore, electrons are considered to occupy a completely
extended, quasi-free state within the lattice of atoms or
molecules. Introduction of anisotropy into the molecular polarisability apparently results in a degree of localisation of the
ground state of the electron. This is thought to be possibly due
to the presence of local regions of lower potential energy in the
medium due to preferrential orientation of the molecules. This
will be discussed further after first attempting to present a
brief account of the present state of understanding of the
electron in high mobility liquids.

Without going into detail, two types of experimental result
should be mentioned as having determined the course of thinking
(if not complete understanding) of the electron-in-non-polar-liq-
uids problem. The first was the extension of mobility measurements to densities considerably lower than that of the normal
liquid. This demonstrated the existence of a pronounced maximum

at a density somewhat greater than the critical density [114, 123-132]. The second was the determination of the energy level of the electron, V_o, by photoemission experiments under the same conditions of density. This showed that at the density of maximum mobility the energy of the electron in the liquid with respect to vacuum passed through a minimum [142,143]. From this, the conclusion was drawn that the mobility was being controlled to a large extent by fluctuations in the ground state energy level of the electron resulting from fluctuations in the density of the medium. The impedance to electron motion due to such fluctuations would then pass through a minimum for $dV_o/dN = 0$.

This conclusion, initially based on experimental observation, was in fact a rediscovery of the deformation potential scattering theory developed for the solid state by Shockley [144-146] and embodied in the equation

$$\mu = \frac{\pi}{\sqrt{3}} \frac{\hbar^4 e \rho u_1^2}{E_1^2 m^{5/2} k_B T (k_B T_e)^{\frac{1}{2}}} \tag{48a}$$

In equation (48a), ρ is the specific gravity of the medium, u_1 the longitudinal sound velocity, T_e the electron temperature and E_1 is the change in level of the bottom of the conduction band per unit dilation of the medium and is related to V_o by

$$E_1 = -N \frac{dV_o}{dN} \tag{48b}$$

Substituting in (48a) for E_1 from (48b) and for $(\rho u_1^2)^{-1}$ the isothermal compressibility χ_T gives then

$$\mu = \frac{\pi}{\sqrt{3}} \frac{\hbar^4 e}{(V_o')^2 N^2 m^{5/2} k_B T (k_B T_e)^{\frac{1}{2}}} \tag{48c}$$

Almost identical equations to (48c) have been derived more recent-

ly with slight differences in the numerical factors [147,148]. It
can now be stated with some certainty that mobility equations
containing explicitly the energy level fluctuation parameter V_o'
have superceded the expression initially developed by Lekner [149]
for liquid argon which held sway for several years.

A relatively simple pictorial representation of the basic
theory can be obtained by considering the interaction between the
electron and the molecules of the medium. For isolated atoms or
molecules this interaction is controlled by a long range charge-
induced dipole attractive force due to polarisation and a short
range repulsion due to electrostatic and exchange forces. This
is effectively the situation in the dilute gas which is illustrated
by the upper diagrams in figures 6 and 7 (see over). For the
majority of the time the electron moves freely in three dimensions, in a state of uniform potential (V = 0), between the
molecules and only briefly undergoes deflections on passing close
to a molecule by "rolling in and out" of the surrounding potential
well. The energy level of the electron in the free state is
insensitive to the number density of the medium and the magnitude
of the mobility is controlled by the isolated scattering events.

When the mean distance between molecules becomes less than
that at which the potential energy of interaction is on the order
of kT ($|V(r_{kT})| = k_B T$) the situation becomes somewhat complicated.
The potential energy of the electron with respect to a particular
molecule is still governed by the attractive and repulsive forces
operative in the isolated atom case but now modified (screened)
by similar interactions with surrounding molecules. The situation
is illustrated schematically in figures 6 and 7 using a perfect
lattice representation. In the particular case shown the molecular
centres have been taken to be closer together than the distance
of the minimum in potential energy for the isolated atom-electron
case since the Van der Waals attractive interaction between
non-polar molecules is of shorter range (r^{-6} as opposed to r^{-4})
than the charge interaction. That is, for the normal liquid or

Low Density (d≫r_{kT})

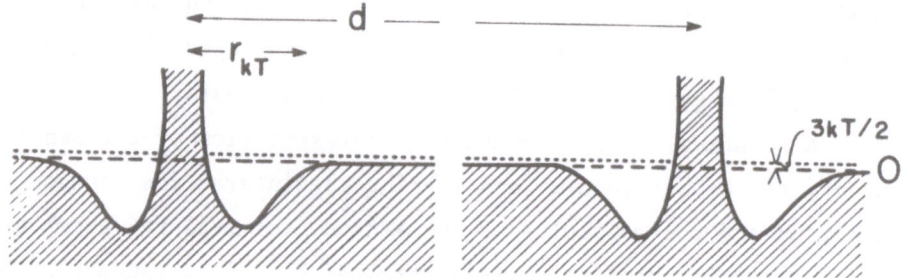

High Density (d<r_{kT})

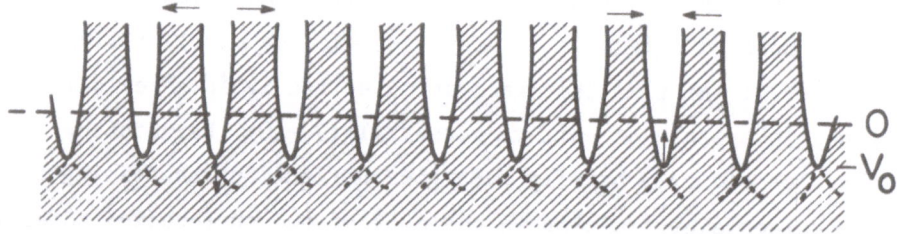

Figure 6. Potential energy diagrams for an electron interacting with polarisable atoms. In the upper case, the atoms are sufficiently far apart that the interaction can be considered to be simply the combined long range attractive polarisation interaction (equal to −kT at r_{kT}) and a hard core repulsive interaction with the isolated atom. In the lower case the interaction with an individual atom is basically the same but is now "screened" by interactions with other atoms and is of a range which is limited by the distance between atoms. The lowest energy level, denoted V_0, is localised in one dimension but completely delocalised, for a perfect lattice, in higher dimensions as is shown in the lower part of figure 7.

A) Low Density

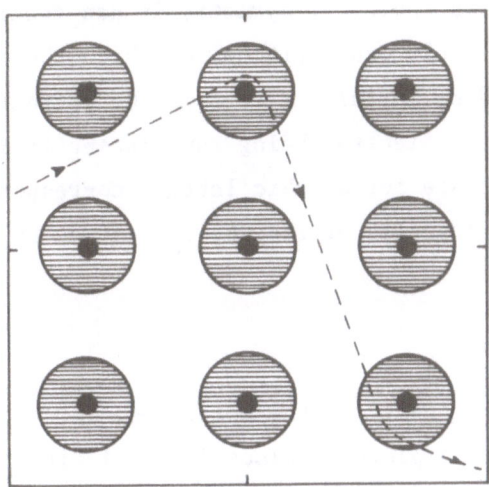

B) High Density

<u>Figure 7</u>. Two dimensional square lattice representations of the electron energy diagrams in figure 6. The black areas are for potential energies greater than zero (net repulsion), the shaded areas for a net attractive potential greater than kT. The white areas are regions of constant potential energy and "infinite" extent. The electron state is represented by the dashed lines.

solid situation the molecules tend to be somewhat closer together
than an electron would like. Irrespective of the exact form of
the potential, from its spherical symmetry it can readily be
deduced that there is only one position for which the potential
energy is uniform and of infinite extent in the perfect lattice.
This corresponds to the electron being equidistant from nearest
neighbour atoms and hence for a cubic lattice corresponds to
close-packed spherical shells of radius, r_{cp}, given by

$$r_{cp}^3 = \frac{3 \times 0.74}{4 \pi N} \qquad\qquad (49)$$

where N is the number density of the molecules.

The state r_{cp} is completely delocalised allowing the electron
to adopt a Maxwellian kinetic energy distribution with a characteristic temperature equal to that of the medium. In the lower
diagram in figure 6 is shown by the arrows how fluctuations in
the molecular distances due to deviations from the average lattice
positions result in fluctuations in the energy level of the
delocalised state. It is this coupling between thermal density
fluctuations and the state of the electron which is termed the
deformation potential and which ensures that the mobility of the
electron is non-infinite. It is considered [144] that only those
density fluctuations of an extent on the order of the de Broglie
wavelength of the electron will have a net effect on the electron
momentum and hence result in scattering.

In terms of this much simplified picture it is possible to
see the difference between electrons in the dilute gas and in the
condensed phase as being similar to the difference between man in
space where three dimensional motion is unrestricted (apart from
momentary collisions with other material objects) and that of man
on earth where the combined gravitational and intermolecular
forces restrict free motion to a two dimensional spherical shell.

In terms of figure 6, a changeover from one kind of motion to

the other would be expected to occur in the density region for
which the interaction energy at the average close-packed spherical
shell radius, r_{cp}, is equal to $k_B T$, i.e., taking repulsive interactions to be negligible

$$\frac{\alpha e^2}{8\pi\epsilon_o r_{kT}^4} = k_B T \qquad (50)$$

where α is the polarisability of the atom or molecule. For α in
cubic angstroms, the density in molecules per cm^3 corresponding
to condition (50) is, using (49),

$$N_{kT} = 3.5 \times 10^{19} \, (T/\alpha)^{0.75} \qquad (51)$$

For polarisabilities of 2.6 and 4.0 $\overset{o}{A}{}^3$, corresponding to methane
and xenon respectively, equation (51) gives values of N_{kT} of
9.1 x 10^{20} and 6.6 x 10^{20} cm^{-3} for a temperature of 200 ^oK.
Deviations from dilute gas behaviour have in fact been found in
this density region for these compounds [125,132]. The ideal-gas
pressure corresponding to the above densities is on the order of
30 atmospheres. For comparison the pressure at which the mean
distance between molecules is equal to the de Broglie wavelength
of an electron ($<\lambda>$ = 87 $\overset{o}{A}$ at 296 ^oK) is only approximately one
tenth of an atmosphere.

The picture of electrons discussed above is strictly only
valid for crystalline atomic solids. It has been found however
for both atomic and pseudo-spherical molecular systems that the
electron mobility decreases only by a factor of approximately
two on melting [51,151,152] and that in addition this factor is
what would be expected from the change in compressibility based
on solid state theory [51,151,152]. As far as electrons are
concerned therefore, the concepts developed for the solid state
appear to be equally applicable to the liquid phase at least at

close to the freezing point, despite the loss of long range order.

As pointed out previously, the optical absorptions and thermally activated mobilities of electrons in anisotropic systems have been taken to indicate that in these media the lowest energy excess electron state is localised. Whether the electron is localised in a cavity or bubble state as in helium [150], or is restricted to constantly changing amoeboid like regions of low total energy involving preferrential orientations of several molecules is as yet uncertain although the latter would seem to be preferred at present. One thing which is apparent is that completely extended, zero kinetic energy states such as those in isotropic systems are no longer self evident for anisotropic molecules. This is illustrated in figure 8 for a simulation of

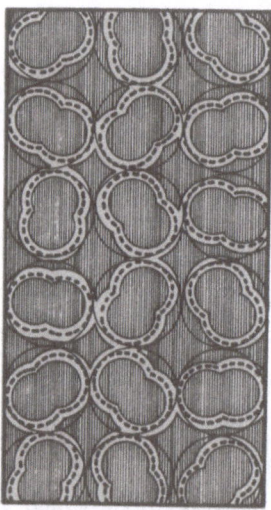

Figure 8. A square lattice representation of the electron medium interaction for the case of a non-isotropically polarisable molecule allowed to freely rotate at lattice positions. Positions of equi-potential are shown by the peanut-shaped full and dashed lines. This figure should be compared with figure 7 for a high density, isotropically polarisable system.

ethane in which the molecules are held at lattice positions but

allowed to freely rotate. Complete contact between the equi-
potential, now peanut shaped, shell around all molecules no longer
exists for any given potential energy level. Increasing the
thickness of the shells which corresponds to increasing the
kinetic energy does however allow an increasing amount of overlap
between states of equal total energy.

Whether or not it is physically realistic, a two state model
of electron mobilities in low mobility liquids has been used by
several workers and can prove quite useful. In this model two
distinct states of the electron are considered to exist in the
medium, a high mobility, delocalised state of mobility μ_d and a
very low mobility, highly localised state of lower energy and
mobility μ_1. The populations of these states are taken to be in
thermal equilibrium with an activation energy being required to
move from the lower to the upper state. Accordingly the overal
mobility is given by

$$\mu(-) = \mu_1 + (\mu_d - \mu_1) \exp(-\Delta H_{d1}/k_B T) \qquad (52)$$

Frequently it is then taken that μ_1 is on the order of molecular
ion mobilities and so much smaller than μ_d that equation (52) can
be approximated by

$$\mu(-) = \mu_d \exp(-\Delta H_{d1}/k_B T) \qquad (53)$$

The activation energy determined experimentally, E_μ, is then seen
to be equal to the enthalpy difference between the two mobility
states. The value of the preexponential factor A in equation (47)
has been found [119,122,147,153-155] to be of the same magnitude
as the mobilities in the spherical molecular systems in agreement
with expectations based on this model.

Interesting experiments have been carried out on liquid
mixtures of hydrocarbons in which both changes in the electron
mobility and the value of V_o [156] have been measured. For

n-hexane in neopentane [154] the mobility was found to decrease
exponentially with increasing mole fraction of hexane, x_h,
according to

$$\mu(-) = 70 \exp[-6.8x_h] \qquad (54)$$

This effect provides a convenient but, as yet, little used method
of concocting specific mobility or V_o values when required, as
for example in kinetic studies. Mixture experiments on methane
and ethane [157] have shown a more complex dependence of the zero
field mobility on mole fraction than suggested by equation (54).

Also of considerable interest are results obtained on the
effect of low concentrations of polar molecules, such as alcohols,
on the electron mobility in non-polar liquids [158,159]. These
indicate the interaction of electrons with molecular clusters and
are obivously a first step in closing the gap between the polar
and non-polar systems.

Solvent Radical Cations (Holes)

From the increase in the positive ion scavenging efficiency of
cyclopropane brought about by the addition of electron scavengers,
the conclusion was reached that the mobility of the primary
positive ion, $\mu(+)$, in cyclohexane was a factor of approximately
20 less than that of the electron [34]. This was thought at the
time to be due, not unexpectedly, to an excessive mobility of the
electron. When the electron mobility in cyclohexane was later
determined to be not 20 times but almost 1000 times larger than
the mobility of a molecular ion [49], the necessary consequences
for the mobility of the positive ion became apparent.

Confirmation of the high mobility of the solvent radical
cation of cyclohexane came initially in the form of optical pulse
radiolysis results on the formation kinetics of the radical cations
of the solutes pyrene, biphenyl and TMPD [52-54]. The formation

rates observed required rate constants for reaction with the
solvent positive ion more than an order of magnitude larger than
for normal diffusion controlled reactions. The direct observation
of a highly mobile positive ion in pulse irradiated cyclohexane
using a conductivity technique [55] provided the final confirmation that positive charge transport could take place in a non-
polar liquid without the necessity of molecular displacement.

An interesting aspect of the optical pulse radiolysis studies
was the finding that, of the six hydrocarbon liquids investigated
(cyclopentane, n-hexane, cyclohexane, methylcyclohexane, iso-
octane and cyclooctane), only the two compounds containing a six
membered carbon ring system gave indications of the formation of
an excessively reactive positive ion [54]. This specificity was
confirmed by later conductivity studies in which particular
attention was given to attempting to observe highly mobile solvent
cations in different liquids. The results are given in table IV
(see over) with in column 3 the mobility of the electron $\mu(-)$, in
column 5 the average mobility of the molecular ions, $\mu(S^+,S^-)/2$
in the fully scavenged system, and in column 4 the mobility of the
solvent radical cation, $\mu(+)$, if there was an indication that this
was significantly larger than the average molecular ion mobility.
From the absence of a value of $\mu(+)$ it may be concluded that
either the solvent radical cation has a close to normal mobility
or possibly has a large mobility but a lifetime less than a few
nanoseconds.

The factors influencing the transfer of charge subsequent to
ionisation of a solvent molecule are illustrated in figure 9 (see
over) for a hypothetical system composed of diatomic molecules.
There is little doubt that distortion in the nuclear coordinates
of the ion as compared to those of the neutral molecule will play
a primary role in determining whether or not charge transfer can
readily take place. It would appear that the rigid six-membered
ring system ensures a close matching of nuclear coordinates

TABLE IV

The room temperature mobilities in $cm^2 V^{-1}s^{-1}$ of electrons, $\mu(-)$, radical cations, $\mu(+)$, and the mean mobility for molecular ions, $\mu(S^+,S^-)/2$, determined using the microwave conductivity method.

Compound	G_{fi} [a]	$\mu(-)$	$10^3 \mu(+)$ [c]	$10^3 \mu(S^+,S^-)/2$
cyclopentane	0.16	0.92	d	0.58^e
cyclohexane	0.15	0.23	9.5	0.40^f
n-hexane	0.13	0.071	d	1.1^e
benzene	0.08	0.12	d	0.60^g
methylcyclohexane	0.12	0.044	2.6	0.45^f
cyclooctane	0.17	0.076	d	0.17^e
iso-octane	0.33	5.3	d	0.84^e
trans-decalin	0.13^h	0.013	9.0	0.26^e
cis-decalin	0.13^h	0.10	2.0	0.17^e
tetramethylsilane	0.74	100^b	d	1.2^e

(a) from ref. (89) except decalins, (b) ref. (110), (c) ca 10^{-2} M SF_6 present as electron scavenger, (d) no significant change on addition of ca 10^{-2} M positive ion scavenger S_+, (e) S_+ benzene, (f) S_+ ammonia, (g) S_+ toluene, (h) from CH_3Br scavenging. The mobility data are from references 55, 111, 112 and 160.

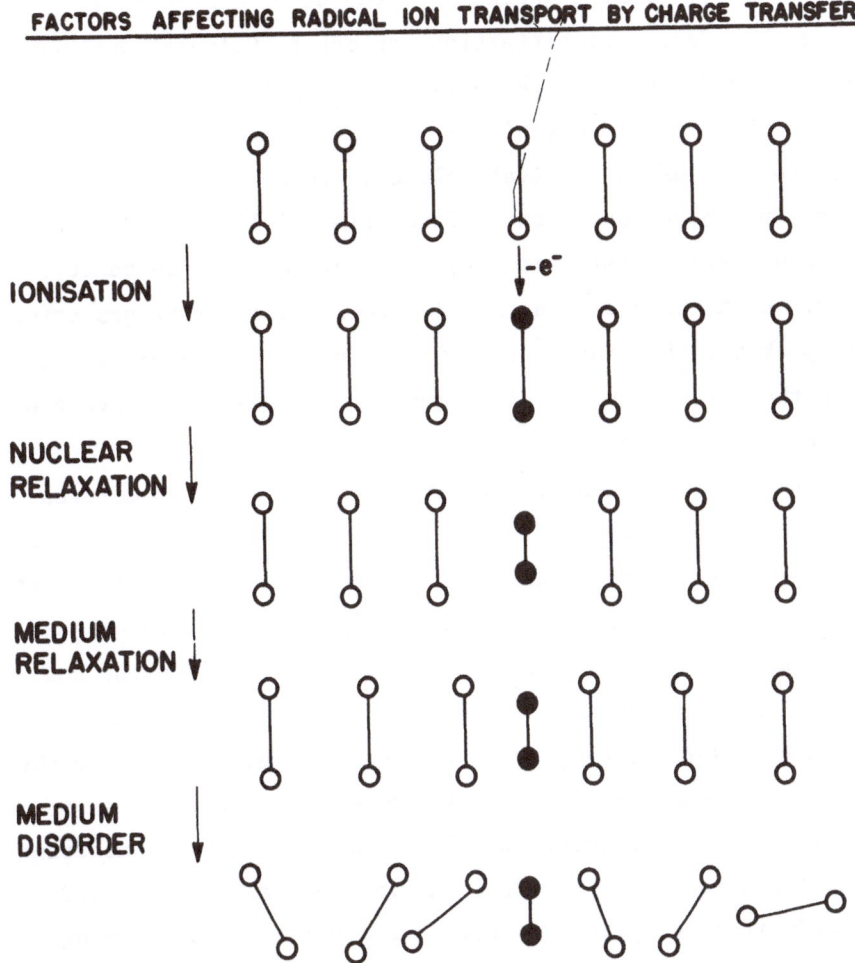

Figure 9. The dumbbells represent diatomic molecules which at the top are arranged in a perfect one-dimensional lattice. An electron is then removed from (or added to) a molecule giving initially the geometrically non-relaxed state (line 2). In line 3 nuclear relaxation has occurred to give the lowest energy equilibrium geometry of the ion. For transfer of charge a neutral neighbour must now adopt the ion geometry. In line 4 the coulombic inter-action with neighbouring molecules has produced a displacement of molecules from their equilibrium lattice positions or orientations. This distortion must now be transferred with the charge for efficient transport. In the last line free rotation has been allowed as in a liquid. Certain relative orientations of the ion and its neighbours will favour transfer.

between ion and neutral whereas for the other compounds a
considerable degree of distortion can occur. Calculations have in
fact indicated [161] the geometry of the cyclohexane radical
cation to be very close (well within thermal vibrational ampli-
tudes) to that of the neutral molecule. This is in contrast to
methane for which considerable distortion [161], resulting even
in loss of tetrahedral symmetry, is predicted for the radical
cation. In the case of benzene, and incidentally rare gas sytems,
it is probable that interaction with neutral neighbours is so
strong that stable, bound dimer ions are formed thus preventing
rapid charge transfer.

With the condition of closely similar nuclear geometries
fulfilled it then becomes a question of which other factors control
the absolute rate of charge transfer and hence the mobility. It
is in this respect of interest to obtain an impression of the
time required for charge transfer in order to explain the mobility
values of approximately 10^{-2} $cm^2V^{-1}s^{-1}$. This can be obtained from
relationship (46). Taking σ_j to be approximately 6 Å (twice the
effective spherical radius for cyclohexane) τ_j is found to be
2.4×10^{-12} s. Since molecular vibration times as well as translational and rotational relaxation times are all of this order of
magnitude it is difficult to decide which is the controlling
factor.

The temperature dependence of radical cation mobilities is
markedly different to that for either electrons or molecular ions
in the same liquid. The data for cyclohexane and trans-decalin
are shown in figure 10. The value of $\mu(+)$ is found to be almost
temperature independent and if anything decreases slightly
($\mu(+) \propto 1/T$) with increasing temperature. This is in contrast
with the thermally activated electron and molecular ion mobilities.
As an interesting consequence of this difference the radical
cation is seen to become the major charge carrier in trans-decalin
for temperatures lower than 10 °C.

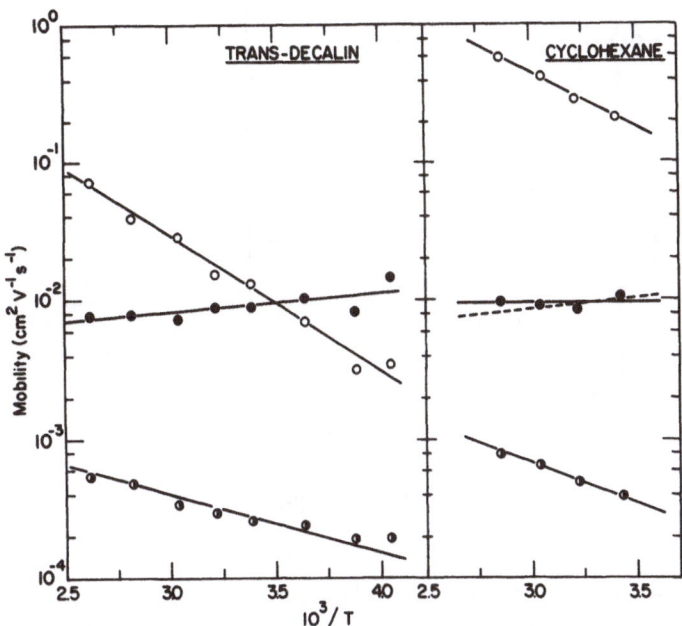

Figure 10. Arrhenius plots of the temperature dependences in trans-decalin and cyclohexane of the mobilities of: O, electrons; ●, radical cations; ◐, molecular ions. The full line drawn through the filled circles for trans-decalin and the dashed line for cyclohexane represent temperature dependences of T^{-1}.

The positive ion mobility has been measured in mixtures of cyclohexane and trans-decalin [112] and the results are shown in figure 11. The addition of a very small amount (<< 1%) of decalin to cyclohexane results in a reduction of the positive ion mobility to the magnitude expected for molecular ions. On further increasing the amount of decalin the mobility is found to rise roughly in proportion to the decalin mole fraction.

The mixture results are in accordance with resonant charge transfer from an ion to a neighbouring molecule being responsible for the high positive ion mobilities. Because of the lower ionisa-

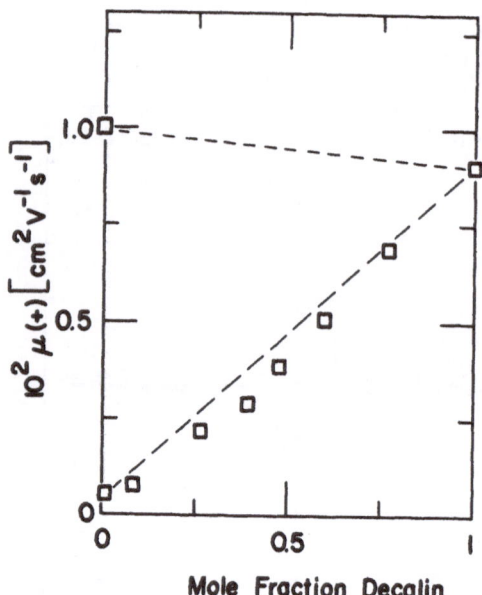

Figure 11. The change in the mobility of the positive ion with increasing mole fraction of trans-decalin in admixture with cyclohexane. The dashed lines between extreme points are "ideal" linear dependences on mole fraction. The sharp drop at very low mole fraction is due to localisation of positive charge on the lower ionisation potential decalin. Charge movement by resonant transfer can then only begin again to supercede that by molecular displacement for decalin concentrations high enough that there is a reasonable probability of a decalin$^+$ having as nearest neighbour a neutral decalin molecule.

tion potential of decalin, positive charge becomes rapidly localised on this molecule even at very low concentrations and transfer to surrounding cyclohexane molecules is energetically unfavourable. As the mole fraction of decalin increases however, the probability of a decalin radical cation having a decalin molecule as nearest neighbour to which it can charge transfer increases and the effective mobility of the positive charge rises. The almost linear increase in μ(+) with mole fraction is taken as an indication of a high degree of localisation of the charge on a single molecule. Percolation theory has recently been applied to this problem [162] and has been found to be capable of fitting the data.

The occurrence of positive charge displacement without the need for molecular motion has been accepted in low temperature glassy hydrocarbons for some time [163]. Also, it has been suggested more recently [136], on the basis of an optical pulse radiolysis study, that the radical cation or "hole" in liquid 3-methyloctane is highly mobile, at least for temperatures of 127 ^{O}K and below. Possibly in low temperature and viscous hydrocarbon media conformational changes following ionisation which would result in a lowering of the positive ion energy are sufficiently slowed down to allow resonant charge transport to occur for a substantial length of time. Obviously much more work will be required for a full understanding of this phenomenon.

Solvent Radical Anions

In the same way that removal of an electron from a solvent molecule can result in a positive ion which can move by charge transfer through the medium, the addition of an electron to a solvent molecule could result in a negative ion which could effectively move by electron transfer to neighbouring molecules. This type of charge transport is well known for the electron affinic aromatic hydrocarbons such as anthracene or naphtalene in the solid state. The only molecule for which definite evidence has been reported for rapid radical anion transport in the liquid phase is perfluorobenzene [164-166].

There is little doubt, based on gas phase electron affinity measurements [167,168] and photoemission experiments on isolated $C_6F_6^-$ ions in the liquid phase [169], that electrons are tightly bound to this molecule. The possibility that the high negative ion mobility in liquid C_6F_6 is due to thermal electron excitation into a high mobility excess electron conduction band state can therefore be excluded. The magnitude of the mobility, 1.1×10^{-2} $cm^2V^{-1}s^{-1}$ at 23 ^{o}C, is of the same order of magnitude as that found for the radical cations in the C_6 ring compounds suggesting perhaps a similar kind of electron transfer mechanism. A slight positive activation energy of the mobility, 0.11 eV [166], is however found in contrast with the lack of thermal activation for radical cations.

Additional experimental facts, such as the large red shift in the optical absorption of the negative ion in pure C_6F_6 compared with that of the isolated $C_6F_6^-$ ion and the pronounced effect on the mobility of only a few percent of a non attaching compound, suggest that the negative charge in C_6F_6 is delocalised over several molecules [165,166]. A percolation treatment of the effect of additives on the mobility [162] required the charge to be spread out over 25 surrounding molecules in order to obtained a fit to the mixture data.

A better understanding of the mechanism of negative ion transport in non-dissociative electron attaching liquids obviously requires results from a wider sample of compounds than is at present available.

Molecular Ions

The diffusion of molecular ions of dimensions similar to those of the solvent molecules is expected to be controlled by the properties of the solvent rather than of the ions themselves. This is reflected in Waldens rule which states that the product of the mobility of an ion and the viscosity of the medium is a constant. This rule is a reasonable qualitative guideline for predicting the direction of change of molecular mobility values on changing the viscosity of a medium or on changing from one medium to another. If however accurate mobility values are required for a particular medium at a given temperature then they must be separately measured. As a guide to the absolute magnitude of the sum of the molecular ion mobilities, $\mu(S^+, S^-)$, in non-polar liquids it can be said that, for a liquid of viscosity 1 cP, $\mu(S^+, S^-)$ is most likely to lie within the limits $7.5 \pm 2.5 \times 10^{-4}$ $cm^2 V^{-1} s^{-1}$.

Because of the long drift times of molecular ions in time-of-flight measurements (milliseconds) it is often difficult to be certain of the ion identity since the possibility that reactions have occurred with trace impurities during drift is large. This of course is the reason that the highly mobile primary electronic carriers in non-polar liquids remained unobserved until the advent of faster conductivity techniques combined with more rigorous purification procedures.

To date, no thorough study of the effect of ion size has been carried out. In experiments using a coaxial cell geometry, by means of which the negative and positive ion mobilities could be measured separately, no significant variation of mobility values was found for different added solutes in a given solvent [170]. An observation of some interest is that in several cases the negative ion mobility has been found to be considerably larger than that of the positive ion [170].

Recent experiments in which molecular ion mobilities have been measured from the dilute gas regime through the high density fluid to the liquid phase [171,172] should prove to be of considerable importance for the understanding of the microscopic aspects of molecular ion transport.

REACTION KINETICS

Electrons

As mentioned in the introduction, the first direct determinations of very large rate constants for electron reactions in non-polar liquids [56-59] were not particularly unexpected. Even prior to

direct measurement of mobilities, several aspects of/ steady state
scavenging results had been found which could only be explained
by a rapidly diffusing and highly reactive electron. When absolute
values of the electron mobility became known, estimates of the
order of magnitude of the rate constants to be expected could be
made by combining the Einstein relation, (31), with the expression
for diffusion controlled rate constants

$$k_D = 4\pi RD \tag{55}$$

to give

$$k_D = 4\pi R\mu k_B T/e \tag{56}$$

For R in Å, μ in $cm^2 V^{-1} s^{-1}$ and kT in electron volts (0.025 eV at
$290^\circ K$), the rate constant in the usual chemical units of litre per
mole per second ($M^{-1} s^{-1}$) is given by

$$k_D \ (M^{-1} s^{-1}) = 6.54 \times 10^{13} \ \mu R(k_B T) \tag{57}$$

For a mobility of 1 $cm^2 V^{-1} s^{-1}$ and a 5 Å reaction radius, a diffusion controlled rate constant of close to $10^{13} \ M^{-1} s^{-1}$ is predicted
at room temperature. Rate constants of the magnitude expected on
the basis of equation (57) were in fact found in the early
measurements for certain solutes.

    A solute which has received considerable attention is SF_6.
This, otherwise comparatively chemically-inert, compound is an
extremely efficient electron scavenger in the gas phase [173] with
a rate constant for attachment which is close to the maximum based
on a theoretical maximum cross section for absorption of an
electron given by [174]

$$\sigma_{max} = \pi \lambda^2 \tag{58}$$

In (58), $\lambdabar$ is the de Broglie wavelength of the electron divided by 2π. The parameter $\lambdabar$ can be seen from (58) to be the effective reaction radius for the maximum cross section. The average value of the de Broglie wavelength is

$$<\lambda> = h\left(\frac{2}{\pi mkT}\right)^{\frac{1}{2}} \tag{59}$$

which gives for $<\lambdabar>$ at room temperature a value of 13.8 Å. The value of the maximum thermal rate constant for electron capture in the gas phase, k^g_{max}, is obtained by averaging $\pi\lambdabar^2 v$, where v is the electron velocity, over the Maxwellian velocity distribution. This gives

$$k^g_{max} = <\lambdabar^2 v> \tag{60}$$

$$= h^2/(2\pi m)^{3/2}(kT)^{\frac{1}{2}} \tag{61}$$

At $296°K$ the value of k^g_{max} is 3.0×10^{14} $M^{-1}s^{-1}$. It is worthwhile noting that the attachment rate constant is independent of electron mobility in the gas phase. This is because of the long mean free paths between collisions which results in the time spent within the reaction sphere being determined simply by the thermal velocity of the electron and the size of the sphere.

In the liquid phase, at least for low electron mobility liquids, a dependence of the rate constant on mobility is observed as would be expected for reactions controlled by diffusion. In figure 12 are shown Arrhenius type plots of results obtained for electron attachment to SF_6 in four liquids which differ in $\mu(-)$ values by almost four orders of magnitude at room temperature. A corresponding large difference in the magnitudes of the rate constants in the different liquids is seen to result.

Paralleling the temperature dependences of the mobilities, the rate constants are found to increase markedly with increasing

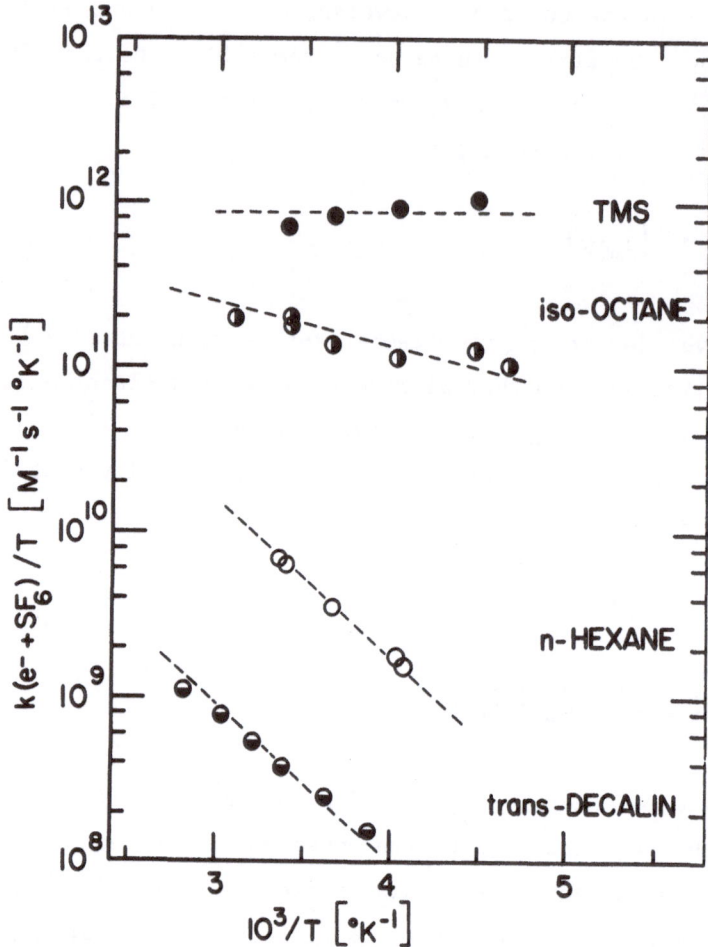

Figure 12. Arrhenius type plots of the rate constant for electron attachment to SF_6 divided by the temperature against reciprocal temperature for the liquids shown. The dashed lines are drawn with a slope which corresponds to the activation energy of the mobility in the particular liquid. The trans-decalin data is from reference (111) and the remainder from reference (65).

temperature in the lowest mobility liquids, to be slightly thermally activated in isooctane and to be almost temperature independent in TMS. The data have been plotted in figure 12 as k/T rather than k since according to (56) the former parameter should be directly proportional to the electron mobility for a diffusion controlled reaction of constant reaction radius. The dashed lines in figure 12 are in fact drawn to give the same activation energies as have been found for the mobilities for a given liquid. No extra energy of activation for the reaction over and above that of the mobility is apparently necessary. In fact, for all of the liquids the temperature dependence of k seems slightly weaker than expected on the basis of the activation energy of the mobility.

Knowing the absolute values of the electron mobilities, the reaction radii required to explain the measured rate constants can be determined. The radii found are 4.6, 13.4, 5.5 and 1.4 $\overset{o}{A}$ for trans-decalin, n-hexane, isooctane and TMS respectively at $296^{o}K$. The reaction radius in n-hexane is much larger than molecular radii and in fact comes close to the mean value of λ of 13.8 $\overset{o}{A}$ at room temperature which as pointed out above is the theoretical maximum radius for electron attachment in the gas phase.

In figure 13, the rate constant data of figure 12 have been plotted against mobility together with data available from the literature for other liquids. The solid straight line in figure 13 is the magnitude of the rate constant expected at $296^{o}K$ if it were given by the expression

$$k = 4\pi<\lambda>\mu kT/e \qquad (62)$$

As can be seen, many points do fall close to this line and those which do not are lower. This might be taken to suggest that λ also represents the maximum reaction radius attainable in the liquid phase. It has been suggested [175] that for reaction of an electron with a molecule of radius R_s, the parameter λ should be replaced by $(\lambda + R_s)$. The $1/\sqrt{T}$ dependence of $<\lambda>$ is possibly responsible for the slightly smaller activation energy associated with k/T compared to that for $\mu(-)$. A very extensive set of data for attachment to SF_6 in liquid ethane and liquid propane has recently been reported [176] in which the rate constant was measured almost continuously over two orders of magnitude change

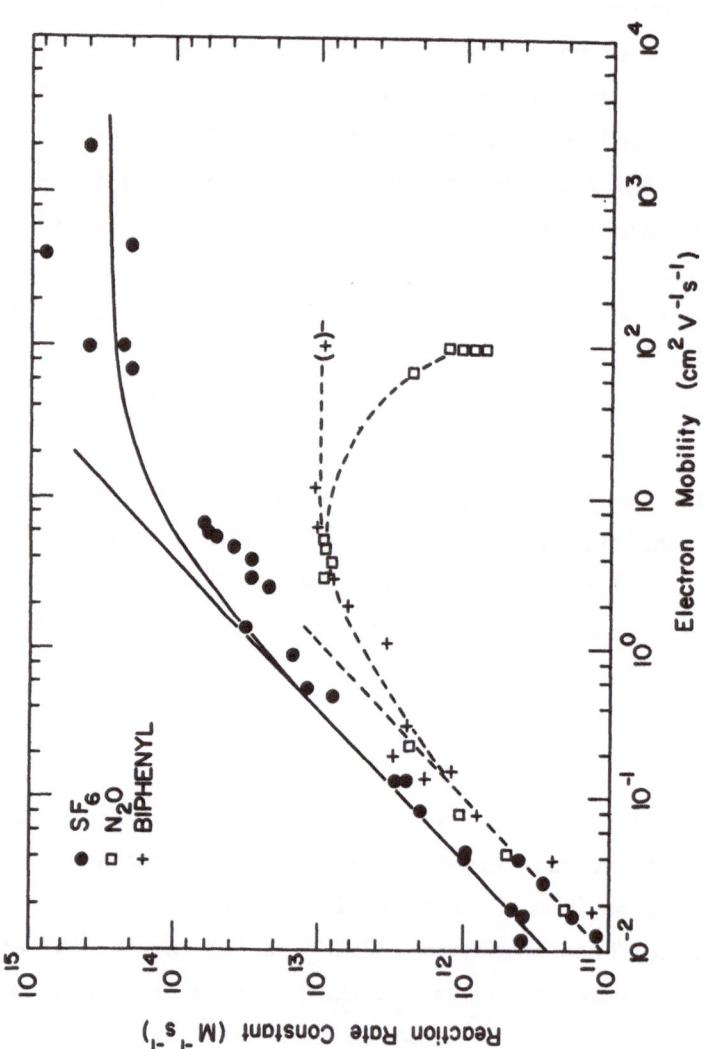

Figure 13. Illustration of the relationship(s) between the rate constants for electron attachment to SF$_6$, ●, biphenyl, +, and N$_2$O, □, and the electron mobility in dielectric liquids. The data is for different liquids and different temperatures. That for N$_2$O is from reference (65), for biphenyl from references (59) and (181) (bracketed point), and for SF$_6$ from references (65), (112), (194), (195) and (210). The full and dashed straight lines have slopes of unity with the former being calculated using $k = 4\pi R \mu kT/e$ with $R = 13.8$ Å and $T = 296°$K. The full curved line is the estimated maximum rate constant from equation (68) in the text. The dashed curved lines have no special significance.

in the electron diffusion coefficient. A good linear correlation between k and D was found in both media with the slope giving a reaction radius of 14.5 Å. This value is slightly lower than the average value of ca 18 Å for $\tilde{\lambda}$ for the temperature range covered.

For low electron mobility liquids, $\mu(-) \leq 1 \text{ cm}^2 \text{V}^{-1} \text{s}^{-1}$, it has been found, where several solutes have been studied, that if a solute reacts with electrons at all then the rate constant is close to the diffusion controlled value given by equation (60) and rarely more than an order of magnitude lower. This is in contrast to the many orders of magnitude spread in attachment rate constants found in the gas phase [177] and, as will be discussed below, in high electron mobility liquids.

This was explained by Henglein [178,179] as being due to the fact that for low mobility liquids the average time of residence, τ_D, of an electron within the reaction radius R, given by

$$\tau_D = R^2/2D \tag{63}$$

becomes much longer than the attachment time τ_A even for cases where attachment involves unfavourable Franck-Condon transitions between the neutral and negative ion states. For a mobility of $1 \text{ cm}^2 \text{V}^{-1} \text{s}^{-1}$ and R = 14 Å, τ_D is 0.4 ps at room temperature which is on the order of relaxation times for nuclear motion.

Following Henglein [178], the rate constant under conditions where the residence time becomes comparable to or shorter than the attachment time will be given by

$$k = 4\pi RD/(1 + \tau_A/\tau_D) \tag{64}$$

$$= 4\pi RD/(1 + 2D\tau_A/R^2) \tag{65}$$

The dependence of the rate constant on diffusion coefficient is therefore expected to become sublinear at high enough D and

eventually to become independent of diffusion coefficient and
given by

$$k_{D \to \infty} = 2\pi R^3 / \tau_A \tag{66}$$

This fall-off from a linear dependence on mobility and eventual
levelling off is apparent for the results on SF_6 shown in figure
13.

It is to be expected that a lower limit to τ_A will exist and
will be on the order of the timescale for free to bound electronic
transitions. An estimate of this lower limit can be obtained from
the transit time of an electron with velocity v over the reaction
diameter, $2\lambdabar$.

$$\tau_A^{min} \simeq 2\lambdabar / v \tag{67}$$

At room temperature τ_A^{min} is therefore approximately 3×10^{-14} s.
Substitution of this minimum value in (65) together with $\langle\lambdabar\rangle$
for R should then give an approximate estimate of the maximum rate
constant to be expected over the full mobility range

$$k_{max} \simeq \frac{4\pi\langle\lambdabar\rangle\mu(kT/e)}{1 + 4\pi\langle\lambdabar\rangle\mu(kT/e)/\langle\pi\lambdabar\rangle} \tag{68}$$

The full curved line in figure 13 was calculated using equation
(68) for a temperature of $296^\circ K$.

For low mobility liquids the rate constant should increase
as $\sqrt{T}$ for a given mobility whereas for very high mobility
liquids the maximum rate constant is expected to be inversely
proportional to the square root of the temperature. Only the rate
constant for liquid methane ($\mu = 400$) lies substantially higher
than the maximum rate constant line. The difference is such that
it cannot be explained away simply by the lower temperature of
the methane experiments. No explanation is apparent.

The deviation from a linear dependence of rate constant on mobility was first pointed out by Beck and Thomas [59,180] on the basis of their results on attachment to biphenyl in different liquids and at different temperatures. The crosses shown in figure 13 are from their data together with a value in parenthesis for attachment to biphenyl in TMS [181]. The data appear to level off at a value of approximately 1×10^{13} $M^{-1}s^{-1}$ which is more than an order of magnitude lower than found for SF_6. Values of 1 to $2 \times 10^{13} M^{-1}s^{-1}$ have been found for attachment to several aromatic hydrocarbons in TMS as is shown in table V. From data in low

TABLE V

Kinetic and thermodynamic parameters for equilibrium attachment reactions, $e^- + A \rightleftarrows A^-$, in liquid tetramethylsilane.

Solute	k_A(at T $^{\circ}K$)^d $[M^{-1}s^{-1}]$	$k_{-A}=B \exp(-C/RT)$		ΔH_A^o [kcal/mol]	$298\Delta S_A^o$ [kcal/mol]
		B $[s^{-1}]$	C [kcal/mol]		
triphenylene[a]	1.7×10^{13}(295)	1.1×10^{18}	17.9	−17.7	− 6.2
phenanthrene[a]	1.4×10^{13}(308)	5.3×10^{17}	16.5	−16.8	− 6.7
naphthalene[a]	1.0×10^{13}(293)	1.1×10^{19}	15.8	−16.1	− 8.5
biphenyl[b]	1.3×10^{13}(261)	9.1×10^{18}	15.0	−17.5	−10.6
styrene[a]	1.1×10^{13}(250)	1.5×10^{18}	12.9	−15.4	− 9.9
α methylstyrene[a]	1.1×10^{13}(237)	1.5×10^{20}	14.1	−16.1	−12.2
CO_2[c]	1.3×10^{11}(293)	1.1×10^{21}	21.0	−21.1	−13.7

a) ref. (182); b) ref. (181,183); c) ref. (184); d) only slightly temperature dependent

mobility liquids it would seem that this could, at least in part be explained by a reaction radius a factor of two or more lower than λ. Why the radius should be low for these compounds is not known. Photodetachment studies of aromatic anions [185] in the liquid phase may shed light on this problem [186].

An effect in the study of attachment kinetics which was

initially surprising was the quite dramatic decrease in rate
constant found for certain solutes in the higher mobility liquids
[65]. This effect is shown in figure 13 for attachment to N_2O.
Rather complex temperature dependences of the rate constant are
also a characteristic of these compounds with even, in certain
cases, maxima in Arrhenius plots for a given solvent being found
[65].

It was soon realised that the effects were related to a
marked sensitivity of the attachment rate constant to changes in
the energy level, V_o, of the electron [65]. The dramatic effects
on the forward rate constant were found to be associated particularly with compounds which have a maximum in their attachment
cross section in the gas phase at an energy much higher than
thermal. For example C_2H_5Br with a maximum at 0.7 eV in the gas
phase [177].

In 1975 Funabashi and Magee [187] and Henglein [178] presented
detailed theoretical treatments of the problem and other papers
followed [179,188,189]. While using quite different approaches
both of the original works were concerned with the effect on the
barrier to attachment of relative displacement of the ground state
energy levels of the neutral and negative ion states. The lower
series of diagrams in figure 14 is a similar representation to
that given by Funabashi and Magee for the dissociative attachment
problem. It can clearly be seen that a molecule which has a large
vertical transition energy to the negative ion state in the gas
phase will have a considerably smaller transition energy in the
liquid due to anion solvation. This effect can however be reversed
if the energy level of the electron state in the liquid is low as
is shown in the last diagram in figure 14. It is apparent that,
in liquids for which the residence time of the electron becomes
of the same order of magnitude as or longer than nuclear relaxation
times, the parameter controlling attachment will become the
height of the cross-over point in the potential energy curves
rather than the vertical transition energy.

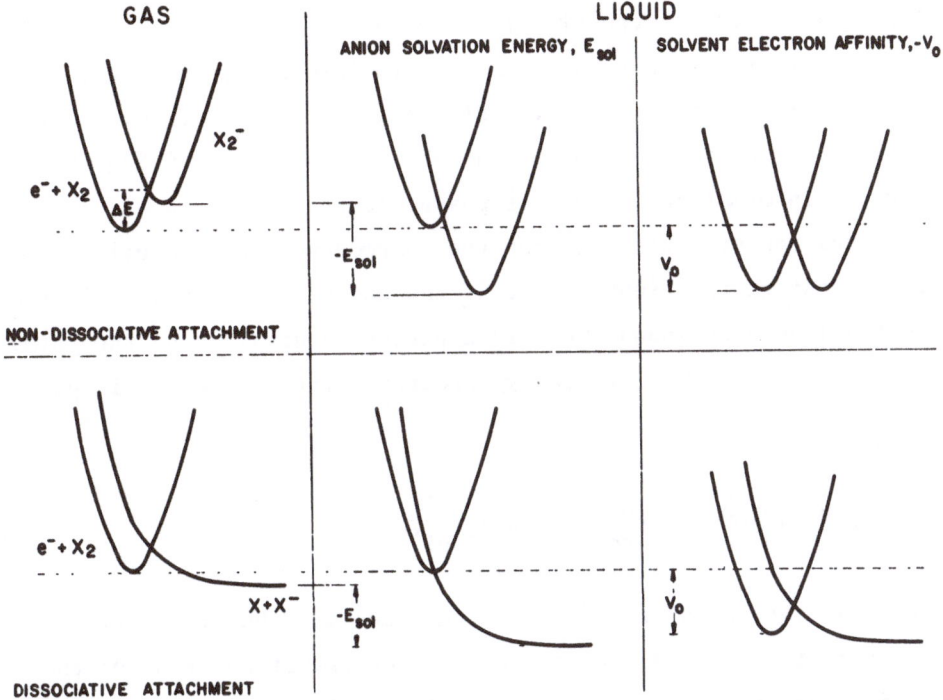

Figure 14. Illustration of the changes occurring in the relative positions of the electron plus neutral and negative ion potential energy curves of a molecule X_2 on dissolution. The gas phase situation on the left is adjusted for the solvation energy of the negative ion in the middle and further for the electron affinity of the solvent on the right.

A particularly interesting and important aspect of electron attachment to aromatic hydrocarbons has been the capability of finding conditions under which thermal emission of the electron can occur sufficiently rapidly to result in the observation of the equilibrium process [181-183,190]

$$e^- + A \underset{k_{-A}}{\overset{k_A}{\rightleftarrows}} A^- \tag{69}$$

The only example to date of a compound other than an aromatic hydrocarbon for which equilibrium kinetics have been observed is carbon dioxide [191]. Some kinetic and thermodynamic parameters for equilibrium attachment are given in table V.

The thermodynamic data can lead to information on the entropy and enthalpy of the electron state in the liquid if information on electron affinities in the gas phase is available [182]. Conversely information on the gas phase electron affinity of a compound can be extracted from the thermodanymics of equilibrium attachment in the liquid phase if the energy level of the electron in the liquid is known. This is apparent on considering the terms contributing to the free energy of attachment in the liquid phase, ΔG^o_{AL}

$$\Delta G^o_{AL} = \Delta G^o_{AG} + \Delta G^o_{sol}(A^-) - \Delta G^o_{sol}(e^-) \tag{70}$$

In (70), the terms on the right hand side are the free energy of attachment in the gas phase, the free energy of solution of the radical anion and the free energy of solution of the electron, in that order.

Of particular interest is the finding that the entropy of dissolution of an electron in TMS is close to zero whereas in n-hexane it is large and positive ($T\Delta S > 0.15$ eV at 296^oK) [182]. The, in general, extremely high preexponential factors (positive activation entropies) for thermal detachment as shown in table V are the tremendous driving forces which allow the equilibria to be observed despite the stability, in enthalpy terms, of the negative ions.

As mentioned above, information on the thermodynamics of electron attachment in the gas phase can be derived from the equilibrium kinetic data obtained in the liquid phase since $\Delta G_{sol}(A^-)$ can be estimated with reasonable accuracy using the Born expression and $\Delta G_{sol}(e^-)$ is usually known from V_o measurements or via the reverse process for compounds with well known attachment thermodynamics in the gas phase. In this way the gas phase electron affinity of molecules can be estimated even when this is quite highly negative. Using this approach a negative electron affinity of 0.4 eV has been determined for CO_2 [182,191]. The reason that this can be done is illustrated by the relative positions of the potential curves for the electron-neutral molecule and the negative ion states shown in the upper part of figure 14. Thus the solvation energy of the negative ion can reduce the level of the negative ion state sufficiently to result in a positive electron affinity. This can then be compensated to a greater or lesser degree by the "solvation" energy of the electron $-V_o$. Because of the significant variation in V_o (see table III) quite different equilibrium parameters for a given reaction can be found in different liquids.

More recently the kinetics of electron attachment and detachment to benzoquinone has been studied and found to be extremely complex in both the dependence on temperature and solvent [192]. It has been suggested that this is due to the initial formation of a relatively stable intermediate excited state of the anion

$$e^- + A \rightleftarrows A^{-*} \rightleftarrows A^- \tag{71}$$

The detachment of electrons can also result via ion molecule reactions, as was found by Bakale et al. [193] who based their experiments on the following reactions which were proposed to explain the reduction in the attachment coefficient of nitrous exide an addition of H_2 or CO.

$$e^- + N_2O \rightarrow N_2 + O^- \tag{72}$$

$$O^- + H_2 \rightarrow H_2O + e^- \tag{73}$$

$$O^- + CO \rightarrow CO_2 + e^- \tag{74}$$

The fast pulse studies of Bakale et al. on N_2O solutions in liquid

rare gases showed that these reactions could result effectively
in equilibrium kinetics.

While it is the intention to exclude as much as possible the
effects of electric fields in order to decrease confusion, it
would be unfair to finish a section on electron reaction kinetics
without having mentioned the very important work of Bakale et al.
[176,194,195] on electric field dependences of rate constants
for attachment to SF_6, N_2O and O_2 in non-polar liquids. On
applying a field, the electrons become heated up and in rare gas
liquids are found to react less rapidly with SF_6 and more rapidly
with N_2O. This can be explained at least partially by a decrease
in λ in the case of the close to maximum rate constant applicable
to the former compound and a decrease in the transition energy
in the case of the rate constant for N_2O which is much lower than
k_{max} in the rare gas media and obviously requires energy of
activation. In low mobility liquids the rate constant for attachment to SF_6 is found to increase at high electric fields [176].

Solvent Radical Cations

As pointed out previously, the large mobility and hence potentially
high reactivity of the radical cation formed on ionisation of a
non-polar liquid was first indicated indirectly for cyclohexane
on the basis of combined steady-state scavenging data (34) and
the absolute value of the electron mobility (49). In contrast
with the case history of electrons, direct evidence for a highly
mobile ion came first from kinetic measurements. Thus, from the
growth kinetics of the pyrene radical cation in a subnanosecond
pulse irradiated pyrene solution in cyclohexane, Beck and Thomas
[52] estimated a rate constant of 4×10^{11} $M^{-1}s^{-1}$ for the reaction

$$CH^+ + Py \rightarrow CH + Py^+ \tag{75}$$

This would normally have been considered sufficient evidence for
a mobile positive ion were it not for the fact that three years
prior to the Beck and Thomas measurements Capellos and Allen [196]
had reported a study of the formation of the trityl cation in
pulse irradiated solutions of triphenylmethyl chloride, Ph_3CCl,
in hydrocarbon liquids. It was suggested that cation formation
probably proceeded by the reaction

$$RH^+ + Ph_3CCl \rightarrow Ph_3C^+ + R^{\cdot} + HCl \tag{76}$$

with as intermediate a short lived protonated form of the chloride.
A rate constant of 5.3×10^9 $M^{-1}s^{-1}$, close to that expected for a
normal diffusion controlled reaction, was found in cylohexane.
The two sets of measurements were therefore contradictary and the
situation was somewhat uncertain.

Supporting evidence for the higher rate constant was forthcoming in later measurements of the growth kinetics in cyclohexane of the solute radical cations of biphenyl [53], TMPD and
again pyrene [54]. The slow rate of formation of the Ph_3C^+ ion
was later shown to be due to a slow decomposition step [197]
(still not precisely identified but possibly involving a second
molecule of Ph_3CCl) following the initial rapid formation of
Ph_3CCl^+

$$RH^+ + Ph_3 \xrightarrow{\text{FAST}} RH + Ph_3CCl^+ \tag{77}$$

$$Ph_3CCl^+ (+ Ph_3CCl) \xrightarrow{\text{SLOW}} Ph_3C^+ \tag{78}$$

The rate constants for reaction of solutes with the cylohexane radical cation determined from the increase in decay rate
of the conductivity transient in a 10^{-2} M SF_6 solution were found
to be of the same order as had been estimated from the growth
kinetics in the optical studies [66]. The rate constants deter-

TABLE VI

Rate constants for the reaction of the radical cation of cyclohexane in liquid
cyclohexane with the solutes shown.

Solute	$10^{-11}k(RH^++S)$ $[M^{-1}s^{-1}]$	solute	$10^{-11}k(RH^++S)$ $[M^{-1}s^{-1}]$
cyclopropane	1.0^a	benzene	1.9^a
diethylether	1.3^b	benzene	2.4^b
biphenyl	1.3^a	ammonia	2.6^a
triethylamine	1.3^b	triphenylmethyl chloride	2.5^c
ethanol	1.4^a	dimethylaniline	2.9^a
methanol	1.65^b	diphenylmercury	3.0^d
decalin (mixture)	1.85^b	tetraphenylsilane	3.3^d

a) ref. (66); b) J.H. Baxendale and P.H.G. Sharpe, unpublished
data; c) ref. (197); d) ref. (112)

mined to date from solvent ion decay kinetics are listed in
table VI. Both microwave and d.c. conductivity methods have been
used in the determinations and are in good agreement for the
value for benzene, the only solute for which both methods have
been applied.

Only a factor of 3 is seen to separate the least and most
reactive solutes studied in cyclohexane. As might have been expected from the almost temperature independent mobility, the
rate constants have been found to be independent of temperature
within the error limits of the measurements [66]. From the five
compounds which have been studied over the range from $23^\circ C$ to
approximately $70^\circ C$ a slight general tendency to increase with
temperature by approximately 15% can be discerned. This is no
more than would be expected on the basis of a temperature independent mobility and the first power temperature dependence given
by equation (56)

$$k = 4\pi R\mu k_B T/e \qquad (56)$$

No extra activation energy appears therefore to be associated with the reactions of the radical cation of cyclohexane with the solutes so far studied. This is perhaps particularly surprising for the reaction with cyclopropane which involves a considerable amount of molecular rearrangement.

$$c\text{-}C_6H_{12}^+ + c\text{-}C_3H_6 \rightarrow c\text{-}C_6H_{10}^+ + n\text{-}C_3H_8 \tag{79}$$

From the range of rate constants found, $2 \pm 1 \times 10^{11} \text{ M}^{-1}\text{s}^{-1}$, and the mobility of $1 \times 10^{-2} \text{ cm}^2\text{V}^{-1}\text{s}^{-1}$, the reaction radii involved are found to lie in the range 11 ± 6 Å. The residence time using (63) is therefore 24 picoseconds which is sufficiently long to possibly explain the lack of thermal activation even when considerable nuclear rearrangement is necessary. Apart from cyclopropane, the reactions with the solutes in table VI are expected to involve either charge (electron) transfer to give the solute radical cation as discussed above or proton transfer to the more polar compounds such as ammonia and the alcohols e.g.

$$c\text{-}C_6H_{12}^+ + NH_3 \rightarrow c\text{-}C_6H_{11}^\bullet + NH_4^+ \tag{80}$$

In the optical study of TMPD$^+$ and Py$^+$ formation in the six hydrocarbon solvents n-hexane, cyclopentane, cyclooctane, iso-octane, methylcyclohexane and cyclohexane [54] evidence for rate constants for reaction with solvent cations significantly greater than that expected on the basis of molecular diffusion was only found for the last two compounds. These two solvents are in fact the only two of the six given above which were later shown to display a high conductivity positive ion transient, table IV, together with the other C_6 ring compounds cis- and trans-decalin. Apart from cyclohexane, rate constants based on the effect of solutes on the decay of the conductivity transient in SF$_6$ solutions have only been determined for trans-decalin [112].

Despite the similarity of the positive ion mobility to that in cyclohexane, the kinetics of reaction with solutes have been found to be somewhat different. For example for the three solutes studied, ammonia, benzene and cyclopropane, rather than reacting with rate constants within a factor of two of each other as in cyclohexane the values found are 5.5×10^{10}, 5.0×10^{9} and 2×10^{8} $M^{-1}s^{-1}$ respectively. In addition, a definite increase in the rate constant with increasing temperature even for the reaction with NH_3 has been found. This would indicate a certain degree of thermal activation even for this proton transfer reaction.

These differences between the reaction kinetics in cyclohexane and in trans-decalin are not understood. They would appear to indicate a difference in the nature of the radical cations which does not make itself apparent in the value of the charge mobility. Further work is obviously needed in this area.

As mentioned previously, evidence has been presented from optical studies for fast positive charge migration in hydrocarbon glasses [163] and also in 3-methyloctane liquid at very low temperatures (ca 130°K) [136]. No rate constants for the reaction of the positive ion with solutes were determined in the latter study however.

Solvent Radical Anions

As pointed out in the mobility section, the only liquid to date for which electrons are known to be tightly bound to the molecules of the medium, when they are in isolation, but which nevertheless displays a high negative charge carrier mobility is perfluorobenzene [164-166]. Accordingly, rate constants for the reaction of the radical ion with certain solutes have been found which are much larger than would be expected on the basis of the viscosity alone [166].

In table VII are listed rate constants determined from the

TABLE VII

Relationship between the exothermicity of electron attachment in the liquid phase and the rate constant for reaction with the solvent radical anion of C_6F_6 [a].

Compound, S	Negative Ion Formed	Exothermicity of Attachment (ε=2.03) [eV]	$k([C_6F_6]_n^- + S)$ $[M^{-1}s^{-1}]$
SF_6	SF_6^-	1.8	$<1 \times 10^8$
CH_2Br_2	Br^-	2.5	$<1 \times 10^8$
C_6F_6	$C_6F_6^-$	2.1^b, 2.3^c, 2.9^d	-
CCl_4	Cl^-	2.6	1.6×10^9
O_2	O_2^-	2.7	7×10^9
$CHBr_3$	Br^-	2.8	9×10^9
CBr_4	Br^-	3.0	1.5×10^{11}
TCNE	$TCNE^-$	4.0	$>1.4 \times 10^{11}$

a) from ref. (166); b) electron affinity 1.2 eV [198]; c) from ref. (200); d) electron affinity 1.8 eV [199]

increase in decay rate of the conductivity transient in C_6F_6 on addition of the solutes shown. The values found for CBr_4 and TCNE (tetracyanoethylene) are of the same order of magnitude as determined for radical cation reactions in cyclohexane. For other solutes however much lower rate constants are found with only an upper limit of 1×10^8 $M^{-1}s^{-1}$ being able to be placed on the "well known" electron scavenger SF_6.

The gradation in rate constant values has been explained in terms of the energetics of the electron attachment process in the liquid phase for the different solutes [166]. In column three of table VII are given the values of the sum of the gas phase electron

affinity and solvation energy of the negative ion with the
relevant bond energy being subtracted in those cases where
dissociative attachment is considered to occur. The three values
given for C_6F_6 are due to different literature values for the
electron affinity.

The order of rate constants for reaction with the solvent
radical anion in pure C_6F_6 is seen to parallel the exothermicity
of attachment and the lack of reaction with SF_6

$$C_6F_6)_n^- + SF_6 \xrightarrow{\quad X \quad} nC_6F_6 + SF_6^- \qquad\qquad (81)$$

is seen to be explained by the endothermicity of the transfer
reaction. Because of the apparent delocalisation of the electron
over several molecules in pure C_6F_6, calculation of the exothermicity of electron attachment to C_6F_6 on the basis of a
molecular ion in a medium of dielectric constant 2.03 is obviously
only a first approximation which however appears to give
reasonable agreement with the kinetic observations.

To data no temperature dependences of rate constants for
solvent anion scavenging have been reported. A comparison of the
temperature dependence of reaction with CBr_4 and one of the close
to thermoneutral compounds for which a barrier to electron transfer apparently exists would be interesting.

Molecular Ions

The study of the reaction kinetics of molecular ions, in particular
electron transfer reactions, has been the subject of intense
experimental and theoretical effort for many years. Attention
has focussed quite naturally on reactions occurring in polar
liquids. To the authors knowledge apart from the work of Dorfman
on certain cation reactions in dichloroethane [201], no extensive,
deliberate study of the reaction kinetics of molecular ions in

very low dielectric constant media have been made. Rate constants, if they are measured, are often isolated and incidental to a broader study hence difficult to dig out of the literature. Examples are the rate constants in cyclohexane of $3.9 \times 10^9 \ M^{-1} s^{-1}$ for electron transfer from the trans-stilbene anion to SF_6 [202] and $8.7 \times 10^9 \ M^{-1} s^{-1}$ for charge transfer from the trityl cation, Ph_3C^+, to TMPD [197]. These are to be found in general pulse radiolysis studies of cyclohexane solutions of stilbene and triphenylmethylchloride. The more extensive measurements of electron transfer rates from the biphenyl radical anion to six solutes in cyclohexane by Richards and Thomas [57] are presented in a paper on solvated electrons in liquid hydrocarbons. The lack of favour given to molecular ions is no doubt a reflection of the predisposition of radiation chemists in recent times for the more exotic and fashionable species discussed in previous sections.

The study of Richards and Thomas was particularly interesting as it showed a similar gradation in rate constants for electron transfer as that in table VII for the solvent radical anion of C_6F_6. Thus no reaction was found with the electron scavengers CO_2 and N_2O and the rate constants for the remaining four solutes ranged from 0.75×10^{10} for SF_6 to $3.2 \times 10^{10} \ M^{-1} s^{-1}$ for pyrene. The latter value is extremely high for a reaction controlled by molecular diffusion in a 1 cP viscosity liquid and requires a very large reaction radius of approximately 20 Å. If this should prove to be more than an isolated or possibly misleading case then further, concentrated studies of molecular electron transfer in low dielectric constant liquids should prove to be of considerable interest and relevance to an understanding of electron transfer processes in general.

SCAVENGING YIELDS

There is no doubt that the major area of activity, in studies of electrons and ions in non-polar liquids, has been and remains the

measurement of chemical effects which result from the interaction
of usually small amounts of added compounds with the ionic
intermediates. These effects can involve product formation from
the solute as is the case for N_2O, reaction (5), ND_3, reactions
(7-9), or the alkyl halides e.g.

$$e^- + CH_3Br \rightarrow CH_3^{\cdot} + Br^- \qquad (82)$$

$$CH_3^{\cdot} + RH \rightarrow CH_4 + R^{\cdot} \qquad (83)$$

Or, the yield of a product derived from the solvent via ionic
pathways can be diminished e.g.

$$RH^+ + e^- \rightarrow RH^* \qquad (84)$$

$$RH^* \rightarrow R^{\cdot} + H^{\cdot} \qquad (85.)$$

$$H + RH \rightarrow R^{\cdot} + H_2 \qquad (86)$$

$$RH^+ + X^- \xrightarrow{\quad X \quad} RH^* \qquad (87)$$

In addition to processes leading to chemical change, an area of
considerable interest has been the formation, via ionic processes,
of relatively short lived excited state or radical species
derived from either solvent or solute molecules. These have been
investigated using optical absorption or emission pulse radiolysis
techniques.

It is not the intention in this final section to review this
elephantine mass of information but rather to show how the data
and conclusions of preceding sections are necessary in order to
arrive at a quantitative description of the eventual chemical
effects in a scavenged system. The process of understanding is by
no means complete.

Because of the proliferation of scavenging data it has proven extremely useful to have a relatively simple empirical equation which describes the yield of ions scavenged by a specific ion scavenger, $G_i(S)$, as a function of the scavenger concentration, [S], over a wide range. The extended "square root" equation of Warman, Asmus and Schuler [26,28] has fulfilled this role.

$$G_i(S) = G_{fi} + G_{gi} \sqrt{\alpha_s [S]}/(1+ \sqrt{\alpha_s [S]}) \qquad (88)$$

The WAS equation contains two empirical parameters G_{gi} and α_s (G_{fi} being separately measurable). Since according to (88) the yield of ions scavenged tends to $G_{fi} + G_{gi}$ at infinite concentration, G_{gi} has been taken to give a measure of the yield of ions which undergo geminate recombination. Because of the extrapolation procedure required to obtain G_{gi}, the latter interpretation should be treated with some caution although it must be admitted that the values obtained, 4 to 5 $(100 \text{ eV})^{-1}$, are in fact of the magnitude which might be expected for the initial yield of ionisation based on gas phase W values. The parameter α_s is called the reactivity of the solute and is the reciprocal of the concentration at which half of the effective yield of geminately recombining ions is scavenged.

While in early studies a value of approximately 4 was given for the sum $G_{fi} + G_{gi}$, later work would suggest a value of 5.0 to be more appropriate. It has been pointed out by several workers that rather than using the separate parameter G_{gi} in fitting procedures, a better choice would be to use the combined parameter $G_{gi}\sqrt{\alpha_s}$ which sould be obtainable quite accurately from low concentration data. This parameter is in fact the slope of the limiting linear dependence of the yield on square root of scavenger concentration predicted by (88) i.e.

$$G_i(S)_{s \to o} = G_{fi} + G_{gi} \sqrt{\alpha_s [S]} \qquad (89)$$

The fact that equations (88) and (89) predict a product
yield of G_{fi} even in the absence of solute is sometimes cause
for confusion for those unfamiliar with scavenging in non-polar
liquids and is strictly of course incorrect. Because the lifetime
of escaped ions is expected to be orders of magnitude longer than
for ions which return to their sibling counterion, only extremely
small concentrations of solute should be required to scavenge
the free ion component. The concentration of solute required to
scavenge at least 90% of the free ions under steady state
radiolysis conditions can be obtained from the condition

$$k[S] > 10(\alpha R)^{0.5} \tag{90}$$

where k is the scavenging rate constant, α is the homogeneous
ion recombination rate constant, which can be replaced by $\mu e/\varepsilon_o \varepsilon_r$,
and R is the rate of formation of free ions. For a dose rate of
M megarads per hour and taking the rate constant for reaction to
be the diffusion controlled value given by (56) one can readily
derive the concentration condition

$$[S] > 1.3 \times 10^{-9} \left(\frac{G_{fi}}{\mu} M\right)^{0.5} \tag{91}$$

for a reaction radius of 5 Å. In (91), for μ in $cm^2 v^{-1} s^{-1}$, [S] is
obtained in moles per litre. Even for a mobility of the scavenged
species as low as 10^{-3} $cm^2 v^{-1} s^{-1}$ and a dose rate of 1 Megarad per
hour, the concentration of [S] required for 90% scavenging of free
ions is found to be only about 2×10^{-8} M. The lowest concentrations used in scavenging studies are usually, for practical
reasons, at least two orders of magnitude larger than this.

The original theoretical treatment of Hummel [33] and subsequent work [77,79,82] indicated that, in the limit of [S] tends
to zero, geminately recombining ions should in fact be scavenged
with a yield proportional to the square root of the scavenger

concentration. As pointed out in the introduction, this was immediately (and prematurely as discussed below) taken to be a theoretical basis for the square root scavenging dependences found experimentally at low concentrations. The exact form predicted for the limiting yield-concentration dependence, which is directly related to the long time limiting reciprocal square root dependence of the survival probability on time, equation (18), is as pointed out recently by Tachiya [203] and by Crumb and Baird [204]

$$G_i(S)_{s \to o} = G_{fi} \left[1 + \left(\frac{r_c^2 k_s}{D} \right)^{\frac{1}{2}} [S]^{\frac{1}{2}} \right] \tag{92}$$

Comparison of (92) with (89) suggests that the scavenging parameter should be calculable via the relationship

$$(\alpha_s \, G_{gi}^2)_{calc.} = (r_c^2 k_s / D) G_{fi}^2 \tag{93}$$

if values of the rate constant for scavenging, the diffusion coefficients or mobilities of the primary species, and the free ion yield are known.

Values of both the parameter $\alpha_s G_{gi}^2$ from scavenging studies and the rate constant for scavenging k_s are available for certain electron and positive ion scavengers. Values for cyclohexane are listed in table VIII for some solutes for which a direct product of the scavenging reaction can be measured. Also listed in table VIII are values of $\alpha_s G_{gi}^2$ calculated on the basis of (93), using $r_c = 280$ Å and the value of D corresponding the sum of the electron and hole mobilities in cyclohexane, 0.24 $cm^2 V^{-1} s^{-1}$. The value calculated is in all cases a factor of 2 to 3 smaller than the experimentally determined value of $\alpha_s G_{gi}^2$ as was pointed out previously [203,204].

This apparent discrepancy between theory and experiment is further illustrated in figure 15 where experimental yields of

TABLE VIII

A comparison of measured values of the reactivity parameter $\alpha_s G_{gi}^2$ obtained from electron and positive ion scavenging product yields with values calculated using equation (93) together with measured values of G_{fi}, $\mu(\pm)$ and the rate constant for scavenging, k_s.

Solute[a]	$10^{-12} k_s$ [$M^{-1}s^{-1}$]	$\alpha_s G_{gi}^2$ (meas.) [M^{-1}]	$\alpha_s G_{gi}^2$ (calc.) [M^{-1}]	geminate lifetime $\tau_g = \alpha_s / k_s$ (ps)
CH_3Cl	1.4^b	78^e	32	2.32
CH_3Br	4.1^c	234^e	94	2.38
C_2H_5Br	2.0^b	113^e	46	2.35
NH_3	0.26^d	$14.3^f, 12.3^g$	5.9	2.29, 1.97
$c\text{-}C_3H_6$	0.10^d	$4.5^h, 5.8^i$	2.3	1.87, 2.42

a) only solutes have been taken for which a product results directly, and it is thought exclusively, from ion scavenging; b) ref. (65); c) ref. (112); d) ref. (66); e) ref. (26); f) ref. (205); g) ref. (206); h) ref. (207); i) ref. (34); j) based on a total ion yield of 5.0

methyl radicals from methyl bromide solutions in n-hexane, cyclo-hexane and iso-octane are plotted against the square root of the methyl bromide concentration. The scavenging yield expected on the basis of the theoretical limiting square root equation is shown as the dashed straight line in figure 15 with the parameters used being given in table IX. The experimental data lie clearly higher than the dashed lines for all three liquids. In light of the discussion of the decay kinetics of ion pairs given previously, the discrepancy is almost certainly due to the fact that the time dependence given by equation (18) on the basis of which (92) is

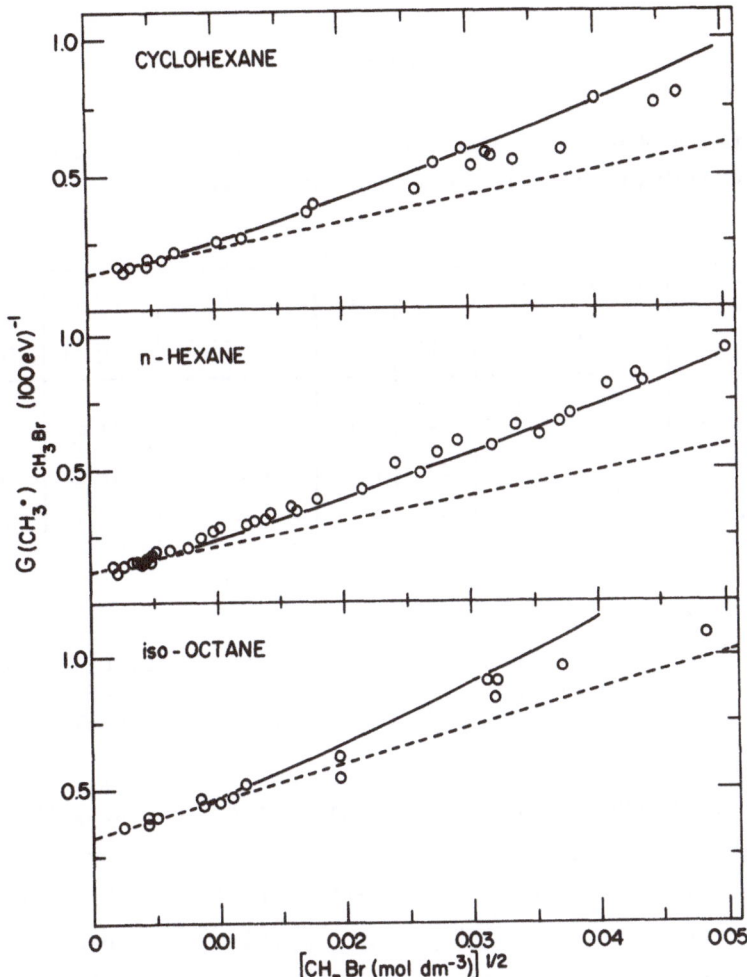

Figure 15. The circles are experimental measurements of the yield of methyl radicals formed due to electron scavenging by methyl bromide, up to a concentration of 2.5×10^{-3} M, in the hydrocarbon liquids shown as a function of the square root of the solute concentration. The dashed lines are the concentration dependences expected on the basis of the theoretical, limiting square root dependence. The full lines are predicted by the full calculations using Laplace transformation of the long time $t^{-0.6}$ behaviour shown in figure 4B. The experimental points are taken from references 25, 27, 211 and 212.

derived, is in fact only an extreme limiting case of little

practical relevance.

TABLE IX

Parameters used for calculation of the scavenging curves shown in figures (15) and (16) taking the total ion yield to be 5.0 (100 eV)$^{-1}$.

Solvent	$10^{12}k(e+CH_3Br)$ [a] $[M^{-1}s^{-1}]$	r_c [Å]	G_{fi} [b]	$\mu(-,+)$ [e] $[cm^2 v^{-1}s^{-1}]$	α_{CH_3Br} [d] $[M^{-1}]$	$\tau_g = \alpha/k$ [ps]
cyclohexane	4.1	285	0.13	0.24 [c]	11.6	2.8
n-hexane	1.2	306	0.12	0.076	18.3	15
iso-octane	28	297	0.32	5.3	19.9	0.7

a) ref. (112); b) from best fit to scavenging data; c) electron mobility plus 0.01 for hole mobility; d) reciprocal of concentration at which $G(CH_3) = G_{fi} + (5.0 - G_{fi})/2$; e) see table III.

It is probably worthwhile at this point discussing the Laplace transform method by which interchanges of time and concentration variables are carried out. The operation relies on the statistical condition that the probability that a species undergoes reaction with a solute S during the time t is $[1 - \exp(-k_s[S]t)]$. If the probability of a species having a lifetime between t + dt is F(t)dt then the fraction of all reactive species scavenged at a concentration [S], f([S]), is

$$f([S]) = \int_0^\infty F(t) \, (1 - \exp(-k_s[S]t)dt$$

$$= 1 - \int_0^\infty F(t) \, \exp(-k_s[S]t)dt \tag{93b}$$

Since the scavenging is invariably a function of the product of the rate constant and the scavenger concentration $k_s[S]$, rather than [S] alone, one can write

$$g(k_s[S]) = 1 - f([S]) \tag{94}$$

$$= \int_0^\infty F(t) \exp(-k_s[S]t)dt \tag{95}$$

The functional dependence of the variable $k_s[S]$, $g(k_s[S])$, is
called the Laplace transform of the functional dependence of
t, F(t). If one of the functions is known explicitly then, subject
to certain conditions, the other is defined and can frequently be
found by consulting tables of Laplace transforms.

To take a trivial case as illustration; if the reaction is
with a species which decays exponentially in the absence of
scavenger with a decay rate k_o, then F(t) is $\exp(-k_o t)/k_o$ and the
function $g(k_s[S])$ is found from tables   to be
$k_o/(k_o + k_s[S])$ leading to the common Stern Volmer expression
$f[S] = k_s[S]/(k_o + k_s[S])$ which of course can be readily derived
in other ways.

It was by using this approach that Rzad et al. [42] derived
the functional dependence of geminate ion decay kinetics from the
empirical scavenging equation, (88),

$$F(t) = \lambda \left[\left(\frac{1}{\pi\lambda t} \right)^{\frac{1}{2}} - e^{\lambda t} erfc(\lambda t)^{\frac{1}{2}} \right] \tag{96}$$

Equation (96) on integration yields the survival probability at
time t given by

$$W(t)/W(\infty) = 1 + \frac{G_{gi}}{G_{fi}} \int_0^\infty F(t)dt$$

$$= 1 + \frac{G_{gi}}{G_{fi}} \exp(\lambda t) erfc(\lambda t)^{\frac{1}{2}} \tag{97}$$

where λ is the ratio of the rate constant to the reactivity, k_s/α_s, or the reciprocal of the geminate ion lifetime τ_{gi} as defined in a previous section. Equation (97) is approximated to within 2% for times longer than 16 α_s/k_s by

$$W(t)/W(\infty) = 1 + G_{gi}/G_{fi}(\pi k_s t/\alpha_s)^{\frac{1}{2}} \qquad (98)$$

A comparison of the time dependence predicted for the primary ions in cyclohexane using (97) with full calculations using the single ion pair model for different separation distributions is shown in figure 5. The decay curve derived from (97) is seen to resemble most closely that calculated using an exponential distribution of initial separation distances if the initial yield is taken to be 5.0 $(100 \text{ eV})^{-1}$. Or, conversely, as shown by Abell and Funanbashi [74] and by Friauf et al. [76], the exponential distribution is found to give the best agreement with the WAS equation for the concentration dependence.

The procedure used by Rzad et al. was recently inverted by van den Ende et al. [208,209] who, having found the theoretically calculated survival probability at long times to be well described by

$$W(t)/W(\infty) = 1 + 0.6(r_c^2/tD)^{0.6} \qquad (22)$$

used the Laplace transform approach to derive the expected theoretical concentration dependence of scavenging for low concentrations viz

$$G_i(S)_{s\to o} = G_{fi} (1 + 1.33(r_c^2 k_s [S]/D)^{0.6}) \qquad (99)$$

The solid lines in figure 15 were calculated using equation (99) and the parameters listed in table IX. The fits to the experimental data are seen to be quite good and are within the combined error limits associated with the experimentally determined parameters.

It is worth emphasising that this agreement at low concentrations between experiment and the predictions of the application of the Smoluchowski equation to the diffusion of correlated ion pairs is independent of the form of the initial separation distribution function and involves no adjustable parameters. Because of this, equation (22) for the long-time survival probability and (99) for the low concentration scavenging probability should almost certainly be equally well applicable to ion pairs formed on photoionisation as to those formed using high energy radiation.

While the theoretical and experimental square root time and concentration expressions appear now to have been to a large extent red herrings, the empirical expression for scavenging given by equation (88) remains as a useful means of describing data quite accurately in terms of only two parameters over several orders of magnitude variation in concentration. Its common usage to test theoretical predictions for scavenging, particularly at elevated scavenger concentrations, remains therefore perfectly justified. The good description of experimental results given by (88) is illustrated in figure 16 where results for scavenging by methyl bromide in n-hexane, cyclohexane and iso-octane up to a concentration of 0.1 M are presented together with calculations based on equations (88), (92) and (99).

The parameters used to construct the solid lines in figure 16 are listed in table IX. Also given in the table are the values of the geminate ion lifetime, $\tau_{gi} = \alpha_s/k_s$. A large fraction of ion pairs are seen to undergo recombination on a timescale of picoseconds or in even less than a picosecond in iso-octane.

It is interesting to note in conclusion that the timescales for geminate recombination are in fact not too far removed from the original estimates of Samuel and Magee [6] on the basis of which ionic processes in non-polar liquids were dismissed as being of unlikely importance. Who would have dared propose at that time

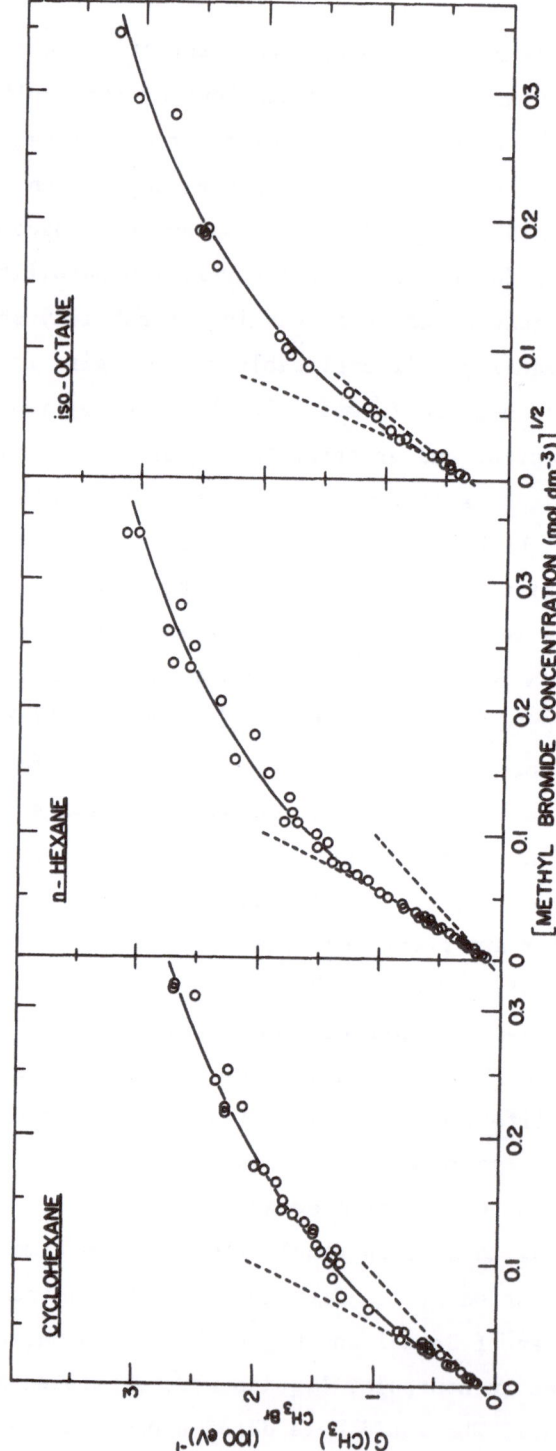

Figure 16. The circles are experimental measurements of the yield of methyl radicals formed due to electron scavenging by methyl bromide, up to a concentration of 0.1 M, in the liquids shown. The lower dashed line is the limiting theoretical square root dependence shown also as a dashed line in figure 15. The upper dashed line is an extension of the 0.6 power dependence shown as a full line in figure 15. The full lines were calculated using the empirical, extended square root equation. The experimental points are from references 25, 27, 211 and 212.

the explanation that reactions with solutes could occur with rate constants in excess of 10^{14} $M^{-1}s^{-1}$ and that 10^{12} $M^{-1}s^{-1}$ would be commonplace?

REFERENCES

(1) M. Born, Z. Physik, $\underline{1}$ (1920) 45
(2) L. Onsager, Phys. Rev., $\underline{54}$ (1938) 554
(3) I. Adamczewski, *Ionization, Conductivity and Breakdown in Dielectric Liquids, Taylor and Fracis, London, 1969*
(4) G.R. Freeman, J. Chem. Phys., $\underline{38}$ (1963) 1022; $\underline{39}$ (1963) 988
(5) A.O. Allen and A. Hummel, Disc. Farad. Soc., $\underline{36}$ (1963) 95
(6) A.H. Samuel and J.L. Magee, J. Chem. Phys., $\underline{21}$ (1953) 1080
(7) R.R. Williams Jr. and W.H. Hamill, Radiat. Res., $\underline{1}$ (1954) 158
(8) L.J. Forrestal and W.H. Hamill, J. Amer. Chem. Soc., $\underline{83}$ (1961) 1535
(9) P.R. Geissler and J.E. Willard, J. Amer. Chem. Soc., $\underline{84}$ (1962) 4627
(10) R.H. Schuler, J. Phys. Chem., $\underline{61}$ (1957) 1472
(11) G. Scholes and M. Simic, Nature, $\underline{202}$ (1964) 895
(12) G. Scholes, M. Simic, G.E. Adams, J.W. Boag and B.D. Michael, Nature, $\underline{204}$ (1964) 1187
(13) G.R. Freeman and J.M. Fayadh, J. Chem. Phys., $\underline{43}$ (1965) 86
(14) A. Hummel and A.O. Allen, J. Chem. Phys., $\underline{44}$ (1966) 3426
(15) A. Hummel, A.O. Allen and F.H. Watson Jr., J. Chem. Phys., $\underline{44}$ (1966) 3431
(16) J.P. Keene, E.J. Land and A.J. Swallow, J. Amer. Chem. Soc., $\underline{87}$ (1965) 5284
(17) R.J. Hagemann and H.A. Schwarz, J. Phys. Chem., $\underline{71}$ (1967) 2694
(18) W.R. Busler, D.H. Martin and F. Williams, Dis. Farad. Soc., $\underline{36}$ (1963) 102
(19) F. Williams, J. Amer. Chem. Soc., $\underline{86}$ (1964) 3954
(20) J.W. Buchanan and F. Williams, J. Chem. Phys., $\underline{44}$ (1966) 4377
(21) P. Ausloos, A.A. Scala and S.G. Lias, J. Amer. Chem. Soc., $\underline{88}$ (1966) 1583
(22) A.A. Scala, S.G. Lias and P. Ausloos, J. Amer. Chem. Soc., $\underline{88}$ (1966) 5701
(23) W.V. Sherman, J. Chem. Soc. A, (1966) 599
(24) S.J. Rzad and R.H. Schuler, J. Phys. Chem., $\underline{72}$ (1968) 228
(25) S.J. Rzad and J.M. Warman, J. Chem. Phys., $\underline{49}$ (1968) 2861
(26) J.M. Warman, K-D. Asmus and R.H. Schuler, J. Phys. Chem., $\underline{73}$ (1969) 931
(27) J.M. Warman and S.J. Rzad, J. Chem. Phys., $\underline{52}$ (1970) 485
(28) J.M. Warman, K-D. Asmus and R.H. Schuler, Advan. Chem. Ser. Nr. $\underline{82(II)}$, (1968) 25

(29) N.H. Sagert and A.S. Blair, Can. J. Chem., $\underline{45}$ (1967) 1351

(30) N.H. Sagert, Can. J. Chem., $\underline{46}$ (1968) 95; $3\overline{36}$

(31) L.A. Rajbenbach, J. Amer. Chem. Soc., $\underline{88}$ (1966) 4275

(32) L.A. Rajbenbach and U. Kaldor, J. Chem. Phys., $\underline{47}$ (1967) 242

(33) A. Hummel, J. Chem. Phys., $\underline{48}$ (1968) 3268

(34) S.J. Rzad, R.H. Schuler and A. Hummel, J. Chem. Phys., $\underline{51}$ (1969) 1369

(35) H.A. Schwarz, Ann. Rev. Phys. Chem., 16 (1965) 347

(36) G.R. Freeman, J. Chem. Phys., $\underline{43}$ (1965) 93

(37) G.R. Freeman, J. Chem. Phys., $\overline{46}$ (1967) 2822

(38) F. Williams and K. Hayashi, Nature, $\underline{212}$ (1966) 281

(39) F. Williams, J. Chem. Phys., $\underline{48}$ (1968) 4077

(40) A. Mozumder and J.L. Magee, J. Chem. Phys., $\underline{47}$ (1967) 939

(41) J.K. Thomas, K. Johnson, T. Klippert and R. Lowers, J. Chem. Phys., $\underline{48}$ (1968) 1608

(42) S.J. Rzad, P.P. Infelta, J.M. Warman and R.H. Schuler, J. Chem. Phys., $\underline{52}$ (1970) 3971

(43) A. Hummel, J. Chem. Phys., $\underline{49}$ (1968) 4840

(44) P.H. Tewari and G.R. Freeman, J. Chem. Phys., $\underline{49}$ (1968) 954; 4394

(45) W.F. Schmidt and A.O. Allen, Science, $\underline{160}$ (1968) 301; J. Phys. Chem., $\underline{72}$ (1968) 3730

(46) C. Capellos and A.O. Allen, Science, $\underline{160}$ (1968) 302; J. Phys. Chem., $\underline{72}$ (1968) 4265

(47) S.J. Rzad and J.M. Warman, J. Chem. Phys., $\underline{49}$ (1968) 2861

(48) R.M. Minday, L.D. Schmidt and H.T. Davis, J. Chem. Phys., $\underline{50}$ (1969) 1473; $\underline{54}$ (1971) 3112

(49) W.F. Schmidt and A.O. Allen, J. Chem. Phys., $\underline{50}$ (1969) 5037; $\underline{52}$ (1970) 4788

(50) B. Halpern, J. Lekner, S.A. Rice and R. Gonner, Phys. Rev., 156 (1967) 351

(51) L.S. Miller, S. Howe and W.E. Spear, Phys. Rev., $\underline{166}$ (1968) 871

(52) G. Beck and J.K. Thomas, J. Phys. Chem., $\underline{76}$ (1972) 3856

(53) A. Hummel and L.H. Luthjens, J. Chem. Phys., $\underline{59}$ (1973) 654

(54) E. Zador, J.M. Warman and A. Hummel, Chem. Phys. Letters, $\underline{22}$ (1973) 480

(55) M.P. de Haas, J.M. Warman, P.P. Infelta and A. Hummel, Chem. Phys. Letters, $\underline{31}$ (1975) 382

(56) J.H. Baxendale, C. Bell and P. Wardman, Chem. Phys. Letters, $\underline{12}$ (1971) 347

(57) J.T. Richards and J.K. Thomas, Chem. Phys. Letters, $\underline{10}$ (1971) 347

(58) G. Bakale, E.C. Gregg and R.D. McCreary, J. Chem. Phys., $\underline{57}$ (1972) 4246

(59) G. Beck and J.K. Thomas, J. Chem. Phys., $\underline{57}$ (1972) 3649

(60) A. Hummel, *Advances in Radiation. Chemistry*, $\underline{4}$ *(1974), John Wiley, M. Burton and J.L. Magee eds.*

(61) A. Mozumder and J.L. Magee, Radiat. Res., $\underline{28}$ (1966) 203

(62) P.H. Tewari and G.R. Freeman, 51 (1969) 1276
(63) W.F. Schmidt and A.O. Allen, J. Chem. Phys., 52 (1970) 2345
(64) W.F. Schmidt, Radiat. Res., 42 (1970) 73
(65) A.O. Allen, T.E. Gangwer and R.A. Holroyd, J. Phys. Chem., 79 (1975) 25
(66) J.M. Warman, P.P. Infelta, M.P. de Haas and A. Hummel, Chem. Phys. Letters, 43 (1972) 321
(67) R.A. Holroyd and M. Allen, J. Chem. Phys., 54 (1971) 5014
(68) R.A. Holroyd, B.K. Dietrich and H.A. Schwarz, J. Phys. Chem., 76 (1972) 3791
(69) R.A. Holroyd, S. Tames and A. Kennedy, J. Phys. Chem., 79 (1976) 2857
(70) R.A. Holroyd and L.L. Russel, J. Phys. Chem., 78 (1974) 2128
(71) W. Tauchert. H. Jungblut and W.F. Schmidt, Can. J. Chem., 55 (1977) 1860
(72) W. Tauchert and W.F. Schmidt, Z. Naturforsch., 309 (1975) 1985
(73) W.F. Schmidt, W. Döldissen, U. Hahn and E.E. Koch, Z. Naturforsch., 339 (1978) 1393
(74) G.C. Abell and K. Funabashi, J. Chem. Phys., 58 (1973) 1079
(75) A. Mozumder, J. Chem. Phys., 60 (1974) 4305
(76) R.J. Friauf, J. Noolandi and K.M. Hong, J. Chem. Phys., 71 (1979) 143
(77) K.M. Hong and J. Noolandi, J. Chem. Phys., 68 (1978) 5163, 5172
(78) A. Hummel and P.P. Infelta, Chem. Phys. Letters, 24 (1974) 559
(79) A. Mozumder, J. Chem. Phys., 48 (1968) 1659; 55 (1971) 3020
(80) G.C. Abell, A. Mozumder and J.L. Magee, J. Chem. Phys., 56 (1972) 5422
(81) P.P. Infelta, J. Chim. Phys., 69 (1972) 1526
(82) J.L. Magee and A.B. Tayler, J. Chem. Phys., 56 (1972) 3061
(83) W. de Zeeuw and A. Hummel, *Interuniversity Reactor Institute Report Nr. 133-75-09 (1975)*
(84) M. Tachiya, J. Chem. Phys., 70 (1979) 238
(85) A. Mozumder, J. Chem. Phys., 65 (1976) 3798
(86) C.A.M. van den Ende, L. Nyikos, J.M. Warman and A. Hummel, Radiat. Phys. Chem., 15 (1980) 273
(87) J.P. Smith and A.D. Trifunac, J. Phys. Chem., 85 (1981) 1645
(88) C.A.M. van den Ende, L.H. Luthjens, J.M. Warman and A. Hummel, *accepted for publication*
(89) A.O. Allen, *Nat. Bur. Standards report Nr. NSRDS-NBS57 (1976)*
(90) J-P. Dodelet, Can. J. Chem., 55 (1977) 2050
(91) W.F. Schmidt and A.O. Allen, J. Chem. Phys., 52 (1970) 2345
(92) J-P. Dodelet, K. Shinsaka and G.R. Freeman, J. Chem. Phys., 59 (1973) 1293
(93) A. Mozumder, J. Chem. Phys., 64 (1976) 912
(94) P. Debye, Trans. Electrochem. Soc., 82 (1942) 265
(95) P. Langevin, Ann. Chem., 28 (1903) 289,433
(96) W.R. Harper, Proc. Camb. Phil. Soc., 28 (1932) 219

(97) P.G. Fuochi and G.R. Freeman, J. Chem. Phys., 56 (1972) 2333

(98) A.O. Allen and R.A. Holroyd, J. Phys. Chem., 78 (1974) 796

(99) P.P. Infelta, M.P. de Haas and J.M. Warman, Radiat. Phys. Chem., 10 (1977) 353

(100) J.H. Baxendale, J.P. Keene and E.J. Rasburn, J. Chem. Soc., Farad. Trans. I, 70 (1974) 718

(101) J.H. Baxendale and E.J. Rasburn, J. Chem. Soc., Farad. Trans. I, 70 (1974) 705

(102) T. Wada, K. Shinsaka, H. Namba and Y. Hatano, Can. J. Chem., 55 (1977) 2144

(103) J.M. Warman, E.S. Sennhauser and D.A. Armstrong, J. Chem. Phys., 70 (1979) 195

(104) E.S. Sennhauser, D.A. Armstrong and J.M. Warman, Radiat. Phys. Chem., 15 (1980) 479

(105) D.R. Bates, J. Phys. B: Atom. Molec. Phys., 13 (1980) 2587

(106) D.R. Bates, J. Phys. B: Atom. Molec. Phys., 14 (1981) 3525

(107) S. Kubota, M. Hishida, M. Suzuki and J. Ruan, Phys. Rev. B: Atom. Molec. Phys., 20 (1979) 3486

(108) A. Mozumder and J.L. Magee, J. Chem. Phys., 47 (1967) 1859

(109) L.A. de Graaf, Physica, 40 (1969) 497

(110) A.O. Allen, *Nat. Bur. Standards Report Nr. NSRDS-NBS58 (1976)*

(111) J.M. Warman, P.P. Infelta, M.P. de Haas and A. Hummel, Can. J. Chem., 55 (1977) 2249

(112) *Radiation Chemistry Group, Interuniversity Reactor Institute, Delft, The Netherlands, unpublished results*

(113) W.F. Schmidt, Can. J. Chem., 55 (1977) 2197

(114) J-P. Dodelet and G.R. Freeman, Can. J. Chem., 55 (1977) 2264

(115) J.M. Warman, Radiat. Phys. Chem., 17 (1981) 21

(116) U. Sowada, J.M. Warman and M.P. de Haas, submitted for publication

(117) G. Bakale and W.F. Schmidt, Zeit. Naturforsch., 28a (1973) 511

(118) M.G. Robinson and G.R. Freeman, Can. J. Chem., 52 (1974) 440

(119) J-P. Dodelet, K. Shinsaka and G.R. Freeman, Can. J. Chem., 54 (1976) 744

(120) K. Shinsaka and G.R. Freeman, Can. J. Chem., 52 (1974) 3495

(121) R.M. Minday, L.D. Schmidt and H.T. Davis, J. Phys. Chem., 76 (1972) 442

(122) K. Shinsaka, J-P. Dodelet and G.R. Freeman, Can. J. Chem., 53 (1975) 2714

(123) H. Schnyders, S.A. Rice and L. Meyer, Phys. Rev., 150 (1966) 127

(124) J.A. Jahnke, L. Meyer and S.A. Rice, Phys. Rev., A3 (1971) 734

(125) S.S. Huang and G.R. Freeman, J. Chem. Phys., 68 (1978) 1353

(126) N. Geenand and G.R. Freeman, Chem. Phys. Letters, 60 (1979) 439

(127) J.M.L. Engels and A.J.M. van Kimmenade, Chem. Phys. Letters, 48 (1977) 451

(128) J.M.L. Engels, PhD Thesis (Technische Hogeschool, Delft, The Netherlands, 1979)

(129) N.E. Cipollini, R.A. Holroyd and M. Nishikawa, J. Chem. Phys., 67 (1977) 4636

(130) S.S. Huang and G.R. Freeman, J. Chem. Phys., 69 (1978) 1585

(131) N.E. Cipollini and A.O. Allen, J. Chem. Phys., 67 (1977) 131

(132) N. Gee and G.R. Freeman, Phys. Rev. A, 20 (1979) 1152; Chem. Phys. Letters, 60 (1979) 439

(133) U. Sowada, Dissertation Frein Universität Berlin (1976)

(134) K. Yoshino, U. Sowada and W.F. Schmidt, Technological Reports of the Osaka University, 27 (1977) No 1349

(135) G. Bakale and W.F. Schmidt, Chem. Phys. Letters, 22 (1973) 164

(136) H.A. Gillis, N.V. Klassen and R.J. Woods, Can. J. Chem., 55 (1977) 2022

(137) A.C. Prior, J. Phys. Chem. Solids, 12 (1959) 175

(138) E.J. Ryder, Phys. Rev., 90 (1953) 766

(139) J.B. Gunn, J. Electronics, 2 (1956) 87

(140) A.F. Gibson and J.W. Granville, J. Electronics, 2 (1956) 259

(141) M.B. Prince, Phys. Rev., 92 (1953) 681

(142) R.A. Holroyd and N.E. Cipollini, J. Chem. Phys., 69 (1978) 502

(143) R.A. Holroyd, Abstracts 6th Int. Cong. Rad. Res. (Tokyo, May 13, 1979, p. 7)

(144) J. Bardeen and W. Shockley, Phys. Rev., 80 (1950) 72

(145) W. Shockley, "Electrons and Holes in Semiconductors", Van Nostrand New Jersey, 1950

(146) E.M. Conwell, Solid State Physics, supplement Nr. 9, Academic press, 1967

(147) Y.A. Berlin, L. Nyikos and R. Schiller, J. Chem. Phys., 69 (1978) 2401

(148) S. Basak and M.H. Cohen, Phys. Rev. B, 20 (1979) 3404

(149) J. Lekner, Phys. Rev., 158 (1967) 130

(150) B.E. Springett, J. Jortner and M.H. Cohen, J. Chem. Phys., 48 (1968) 2720

(151) K. Shinsaka and G.R. Freeman, Can. J. Chem., 52 (1974) 3556

(152) Y. Nakamura, H. Namba, K. Shinsaka and Y. Hatano, Chem. Phys. Letters, 76 (1980) 311

(153) H.T. Davis, L.D. Schmidt and R.G. Brown, "Electrons in Fluids", eds. J. Jortner and N.R. Kestner (Springer-Verlag, 1973) p. 393

(154) R.M. Minday, L.D. Schmidt and H.T. Davis, J. Phys. Chem., 76 (1972) 442

(155) L. Nyikos and R. Schiller, Chem. Phys. Letters, 34 (1975) 128

(156) R.A. Holroyd and W. Tauchert, J. Chem. Phys., 60 (1974) 3715

(157) G. Bakale, W. Tauchert and W.F. Schmidt, J. Chem. Phys., 63 (1975) 4470

(158) T.A. Gangwer, A.O. Allen and R.A. Holroyd, J. Phys. Chem., 81 (1977) 1469

(159) J.H. Baxendale, Can. J. Chem., 55 (1977) 1996

(160) M.P. de Haas, J.M. Warman and A. Hummel, *Proc. 5th Int. Conf. on Conduction and Breakdown in Dielectric Liquids, July 1975, Noordwijkerhout, The Netherlands. Edited by J.M. Goldschvartz, Delft University Press, Delft, The Netherlands*

(161) L. Nyikos, A. Vertes and R. Schiller, Submitted for publication, Hungarian Academy of Sciences, Central Research Institute for Physics, Report Nr. KFKI-1980-56

(162) R. Schiller and L. Nyikos, J. Chem. Phys., 72 (1980) 2245

(163) W.H. Hamill, *"Radical Ions", chapt. 9 (John Wiley Interscience, 1968, eds. E.T. Kaiser and L. Kevan)*

(164) L. Nyikos, C.A.M. van den Ende, J.M. Warman and A. Hummel, J. Phys. Chem., 84 (1980) 1154

(165) C.A.M. van den Ende, L. Nyikos, U. Sowada, J.M. Warman and A. Hummel, J. Electrostatics, in press

(166) C.A.M. van den Ende, L. Nyikos, J.M. Warman and A. Hummel, Radiat. Phys. Chem., in press

(167) C. Lifshitz, T.D. Tiernan and M.B. Hughes, J. Chem. Phys., 59 (1973) 3182

(168) F.M. Page and G.C. Goode, *"Negative Ions and the Magentron", (Wiley-Interscience, 1969)*

(169) U. Sowada and R.A. Holroyd, J. Phys. Chem., 84 (1980) 1150

(170) A.O. Allen, M.P. de Haas and A. Hummel, J. Chem. Phys., 64 (1976) 2587

(171) S.S. Huang and G.R. Freeman, J. Chem. Phys., 70 (1979) 1538

(172) S.S. Huang and G.R. Freeman, J. Chem. Phys., 72 (1980) 1989

(173) F.C. Fehsenfeld, J. Chem. Phys., 53 (1970) 2000

(174) N.F. Mott and H.S.W. Massey, *"The theory of atomic collisions", (Oxford Clarendon Press, 1965) p. 185*

(175) F.J. Davis, R.N. Compton and D.R. Nelson, J. Chem. Phys., 59 (1973) 2324

(176) G. Bakale and W.F. Schmidt, Zeit. Naturforsch., 36a (1981) 802

(177) L.G. Christophorou, *"Atomic and Molecular Radiation Physics", (Wiley Interscience, 1971) chapt. 6*

(178) A. Henglein, Ber. Bunsen.-Gesellschaft Phys. Chem., 79 (1975) 129

(179) A. Henglein, Can. J. Chem., 55 (1977) 2112

(180) G. Beck and J.K. Thomas, Chem. Phys. Letters, 13 (1972) 295

(181) J.M. Warman, M.P. de Haas, E. Zador and A. Hummel, Chem. Phys. Letters, 35 (1975) 383

(182) R.A. Holroyd, Ber. Bunsen.-Gesellschaft Phys. Chem., 81 (1977) 298

(183) J.M. Warman, M.P. de Haas and A. Hummel, *"Conduction and Breakdown in Dielectric Liquids", Proc. 5th Int. Conf., Noordwijkerhout, The Netherlands, July 1975 (Delft University Press, Ed. J.M. Goldschvartz)*

(184) R.A. Holroyd, T.E. Gangwer and A.O. Allen, Chem. Phys. Letters, 31 (1975) 520

(185) U. Sowada and R.A. Holroyd, J. Phys. Chem., 85 (1981) 541

(186) U. Sowada, J. Electrostatics, in press

(187) K. Funabashi and J.L. Magee, J. Chem. Phys., 62 (1975) 4428
(188) J.K. Baird, Can. J. Chem., 55 (1977) 2133
(189) R. Schiller and L. Nyikos, J. Phys. Chem., 81 (1977) 267
(190) R.A. Holroyd, R.M. McCreary and G. Bakale, J. Phys. Chem., 83 (1979) 435
(191) R.A. Holroyd, T.E. Gangwer and A.O. Allen, Chem. Phys. Letters, 31 (1975) 520
(192) R.A. Holroyd, private communication
(193) G. Bakale, U. Sowada and W.F. Schmidt, Can. J. Chem., 55 (1977) 2220
(194) G. Bakale, U. Sowada and W.F. Schmidt, J. Phys. Chem., 80 (1976) 2556
(195) U. Sowada, G. Bakale, K. Yoshino and W.F. Schmidt, Chem. Phys. Letters, 34 (1975) 466
(196) C. Capellos and A.O. Allen, J. Phys. Chem., 73 (1969) 3264
(197) E. Zador, J.M. Warman and A. Hummel, J. Chem. Soc. Farad. Trans. I, 75 (1979) 914
(198) F.M. Page and G.C. Goode, *"Negative Ions and the Magnetron", (Wiley Interscience, New York, 1969)*
(199) C. Lifshitz, T.O. Tiernan and B.M. Hughes, J. Chem. Phys., 59 (1973) 3182
(200) U. Sowada and R.A. Holroyd, J. Phys. Chem., 84 (1980) 1150
(201) L.M. Dorfman, Y. Wang, H-Y. Wang and R.J. Sujdak, Farad. Disc. Chem. Soc., Nr. 63 (1978) 149
(202) E.A. Robinson and G.A. Salmon, J. Phys. Chem., 82 (1978) 382
(203) M. Tachiya, J. Chem. Phys., 70 (1979) 4701
(204) J.A. Crumb and J.K. Baird, J. Phys. Chem., 83 (1979) 1130
(205) S.J. Rzad, *Data presented at the 158th meeting of the American Chemical Society, New York, Sept. 1969*
(206) K-D. Asmus, Int. J. Radiat. Phys. Chem., 3 (1971) 419
(207) E.L. Davids, J.M. Warman and A. Hummel, J. Chem. Soc. Farad. Trans. I, 71 (1975) 1252
(208) C.A.M. van den Ende, PhD Thesis, University of Leiden, The Netherlands, Sept. 1981
(209) C.A.M. van den Ende, J.M. Warman and A. Hummel, to be published
(210) G. Bakale, U. Sowada and W.F. Schmidt, J. Phys. Chem., 79 (1975) 3041
(211) S.J. Rzad, private communication
(212) S.J. Rzad and K.M. Bansal, J. Phys. Chem., 76 (1972) 2374

MOLECULAR EXCITED STATES IN LIQUID SYSTEMS

Michael A. J. Rodgers

Center for Fast Kinetics Research
University of Texas at Austin
Austin, Texas 78712 USA

The primary physico-chemical events induced in liquid systems upon bombardment by high energy electrons are surveyed in the light of experimental evidence from pulse radiolysis studies. The various mechanisms for production of lower manifold states of aromatic hydrocarbon solutes in liquids such as cyclohexane and benzene are considered. Some contributions of electron pulse radiolysis to our understanding of photophysical and photochemical process in non-aqueous liquids are examined.

INTRODUCTION

A beam of high energy particles or electromagnetic photons when incident upon a molecular material can induce several effects. Occasionally such effects are observed as photo-emissions in the visible region of the spectrum.
For example:

(1) The blue glow of Cerenkov light which emanates from reactor fuel rods immersed in water is due to the energy losses from β-particles emitted from fission products into the high refractive index water medium.

(2) Solutions of fluorescent polycyclic aromatic hydrocarbons in some solvents (e.g., toluene, dioxan) show scintillation phenomena when low concentrations of radioisotopes are added to the system.

(3) High energy electrons when incident on certain inorganic solids (phosphors) cause bursts of luminescence to occur.

J. H. Baxendale and F. Busi (eds.),
The Study of Fast Processes and Transient Species by Electron Pulse Radiolysis, 535–550.
Copyright © 1982 by D. Reidel Publishing Company.

Radiation chemists and physicists study phenomena such as these and attempt to explain them by way of molecular mechanisms which stand up to quantitive examination. One aspect of these studies concerns the ways in which the energy of ionizing radiations is converted into visible light in liquid scintillator systems. This involves the formation of excited electronic levels of the molecular species which comprise the system. It is these processes and their consequences and implications which are the concern of this presentation.

PRIMARY EVENTS DURING FAST ELECTRON BOMBARDMENT

In scintillation experiments photons of visible light (ca 3 eV) are detected as a result of bombardment by high energy particles such as 1 MeV β-rays; in electron pulse radiolysis experiments molecular triplet states of 1-4 eV above the ground state are detected after irradiating with a beam of MeV range electrons. These basic observations prompt several important questions:

(1) Are these low energy molecular states the same in nature and properties as similar entities produced by low energy excitation?

(2) Is the formation of the low energy states a significant process or do other, less obvious changes dominate?

(3) How are low energy states produced from high energy bombardments?

The first question is answered unequivocally in the affirmative. Measurements of high energy electron-induced fluorescence spectra, triplet absorption spectra and energy transfer and quenching rate parameters show that the properties of molecular excited states produced by high energy electron irradiation are identical to those observed when low energy photons are used to populate these states (1).

The question concerning the magnitude of the yields cannot be given such a straightforward answer. In Table 1 are presented G-values of singlet and triplet excited states of some typical aromatic solvents in a series of common liquids. The data refer to solutes such as anthracene, biphenyl and naphthalene and have been collected from the pulse radiolysis literature. Clearly there is a large difference in excited state yields from one liquid to another. Thus, whatever mechanism is governing the excitation process, it must include processes which depend strongly on the nature of the liquid medium. Further insight into the variation in yield problem arises out of a consideration of the third question.

TABLE 1

Yields of solvents excited states as determined by measurements
of solute species.

Liquid	$G(S_1)$	$G(T_1)$	$G(S_1+T_1)$	ε
Benzene	1.6 (2)	4.2 (3)	5.8	2.3
Toluene	1.35(4)	2.8 (4)	4.2	2.4
Benzonitrile	0.7 (5)	1.4 (5)	2.1	26.0
Benzyl alcohol	0.7 (6)	1.1 (6)	1.8	13.1
Dioxane	1.4 (7)	zero (7)	1.4	2.2
Cyclohexane	0.4 (8)	0.8 (8)	1.2	2.0
Acetone	0.34(9)	1.0 (9,10)	1.34	21.0
Tetrahydrofuran	0.1 (11)	<0.1 (11)	0.14	7.3
Ethanol	<0.1 (11)	<0.1 (11)	<0.2	23.0
Acetonitrile	-	<0.1 (12)	-	39.0
Water	-probably zero-		-	78.0

For information on the mechanism of excited state production
it is necessary to consider the physical processes by which fast
moving electrons lose energy in molecular media. Fast electrons
may be introduced into a molecular liquid by several means; as
β-particles from a dissolved radioactive substance, as electrons
ejected from molecules by Compton scattering from high energy
electromagnetic photons (X- or γ-rays) or as monochromatic electrons from particle accelerating machines. Regardless of their
origin, fast electrons undergo the same energy-loss processes in
a molecular medium, the dominant one being inelastic interactions
with electrons bound in the molecular orbitals. The Coulombic
forces set up as a fast electron approaches a region of molecular
electron density have an amplitude which depends both on the rate
of approach and the nearest distance as shown in Fig.1.

With the passage of the incident particle, the bound electrons
experience time-dependent repulsive electrostatic forces which can
be resolved into components parallel (V_p) and perpendicular (V_t)
to the particle trajectory. The displacements of a bound electron
with respect to the molecular center arising out of the two components are shown in Figure 1. The net effect is that kinetic .

energy is transferred from the incident particle to molecular electrons. Up to one half of the energy of the incident electron can be transferred to a bound electron owing to the mass equality.

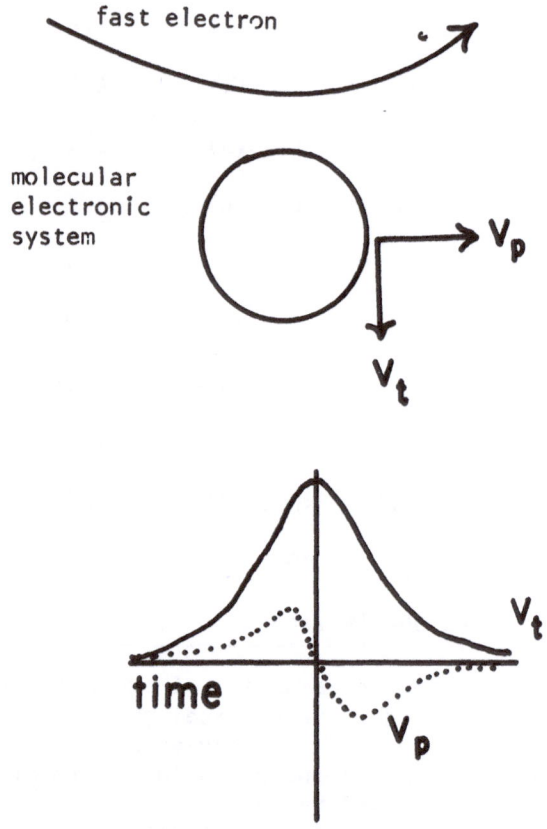

Fig. 1: electrical forces induced in molecular electronic system by fast electron.

Because electron-electron forces are responsible for the energy loss, the probability of a molecule being directly affected will depend upon the proportional contribution its electronic system makes to the total (13). Thus, in a solution the major component (solvent) undergoes most direct events. For example, in a 10^{-3} molar solution of anthracene in benzene, only about 1 in 10^4 energy exchanges with fast electrons will involve anthracene; the overwhelming majority of the energy loss is with benzene molecules. Clearly, if molecular excitation of solutes is to be significant, efficient modes of excitation transfer must be

available. A special exception occurs in two component systems in
which the minor component has a much lower excitation (ionization)
potential than the major one. There direct excitation of solute
by subexcitation electrons, those having insufficient energy to
promote solvent transitions (14), becomes a possibility.

Deposition of electronic energy into a molecular system results
in the population of electronic levels above the ground state. The
upper states may be either bound or ionized depending on the magnitude of the interaction and the ionization potential of the
molecule in the liquid phase. Bound states may be defined as those
in which the excited electron and the residual hole retain a common
molecular identity. Ionized states are regarded as having the
ejected electron in a translational continuum unrelated and distinct
from the hole which exists in one of the states in the manifold of
the molecular cation. The demarcation between bound and ionized
is somewhat vague in that molecular Rydberg states also exist (15)
in which the promoted electron is removed to a relatively large
distance from the hole while still remaining bound to the core
system. Rydberg states in liquids may well be comparable to
Wannier excitons in crystal (16) where a correlated electron–hole
pair can undergo translational migratory motions. Furthermore,
superexcited states have been proposed (17) in which the system
occupies bound levels above the ionization limit. Such states are
thought to lose energy by auto–ionization and/or molecular dissociation. There processes are summarized in Scheme 1.

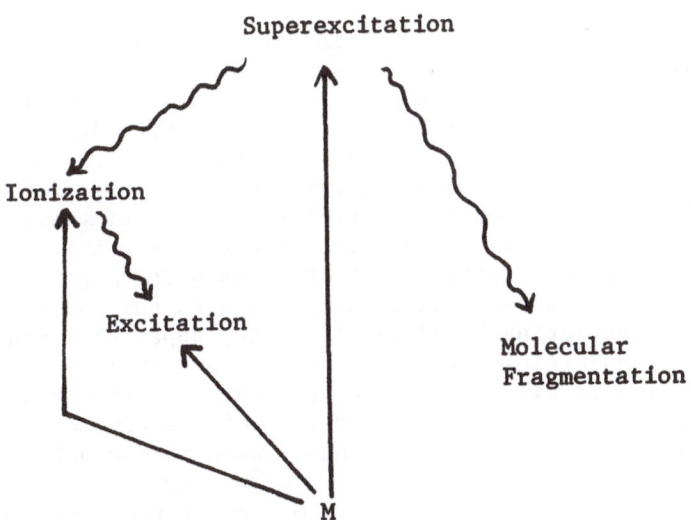

SCHEME 1. Straight arrows are immediate uphill processes induced
by fast electrons; wavy arrows are delayed, downhill processes.

Theoretical approaches have shown that the effects of the time dependent variations of the electron-electron interaction field depicted in Fig 1 are analogous to the perturbation of an electronic system by the electric vector of a photon of light (18). This leads to the conclusion that for incident energies upward of 500 eV, electron excitation closely resembles photo excitation in that the same selection rules are obeyed. Thus molecules in singlet ground states are excited within the singlet manifold predominantly. Below 500 eV or so, this optical approximation ceases to be valid and electron exchange processes can result in optically forbidden transitions being induced (19). In this way direct excitation can produce molecules in both singlet and triplet states.

Population of bound electronic states can also result from those events which in the first place populated ionization levels (either directly or through auto-ionization). Excess kinetic energy of the ejected electrons is dissipated through Coulombic encounters with bound electrons in the immediate neighborhood. Eventually the electron will attain the average thermal energy of the medium at a cation-electron distance which will reflect the distribution of secondary electron energies. The fate of the electron subsequent to thermalization will largely depend on whether it remains correlated with its parent cation by a Coulombic inter-action. The critical distance at which Coulombic and thermal randomizing forces are equal is given by

$$r_c = e^2/\varepsilon kT$$

where e is the electronic charge, ε is the local dielectric constant of the medium, k is Boltzmann's constant and T the absolute tem-perature. Liquids with large ε have small r_c values and the cor-relation volume around the cation is small. For example, water (ε = 78) has r_c = 0.7 nm, ethanol (ε = 23) has r_c = 2.1 nm, benzene (ε = 2.3) has r_c = 21 nm. Newly-thermalized electrons in benzene would therefore have a much higher probability of being within the correlation volume of their parent ion than the same species in, say, water. Correlated cation-electron pairs undergo geminate recombination in sub-nanosecond time scales with the concomitant possibility of populating bound levels in the upper electronic manifold which can efficiently cascade to lower levels. In water and other liquids of high dielectric constant, electron escape has highest probability with the formation of a cation and an electron of independent existence. Subsequent recombination is homogeneous and will occur in the 10^{-6}-10^{-3}s range when other processes may have had an opportunity to convert the parent cation to lower energy species.

These considerations notwithstanding, reference to Table 1 shows clearly that excited state yields do not correlate well with

.the dielectric constant of the liquid. Even in cases where ε is
favorable for the predominance of electron-hole combination
(e.g.,cyclohexane) other factors must be effective in preventing
formation of a high yield of molecular excited states. One major
factor concerns the reactivity of the molecular cation (RH^{+}), especially with respect to ion-molecule reactions (20)

$$RH^{+} + RH \longrightarrow RH_{2}^{+} + R\cdot$$

Such reactions are frequently encounter-controlled and in the
liquid phase will occur very rapidly (10^{-13}s). The return of an
electron to a protonated parent (RH_{2}^{+}) ion will most likely result
in hydrogen atom formation.

$$RH_{2}^{+} + e^{-} \longrightarrow RH + H\cdot$$

Rapid hole fragmentation reactions can also contibute energy degradation modes (21). In addition to the intrinsic properties of the
cation, the propensity of either upper manifold states or superexcited states to undergo bond dissociation will also be a factor
in governing what fraction of primary events lead eventually to
lower manifold states which can be observed and measured. Further,
some molecular liquids can form molecular anion states via electron
capture (e.g., acetone (9,22), benzonitrile (5)). Owing to the
lower mobilities of molecular anions compared to quasi-free electrons, even geminate recombinations will be less rapid ($10^{-9} - 10^{-7}$s)
allowing extra time for energy-degrading processes to take effect .

To summarize, molecular excitation by fast electrons is a very
significant process in some liquids (e.g., benzene and its homologues) where yields of separated ions and molecular fragmentation
products are very small, indicating that a majority of the electron
energy appears in lower manifold states exactly as if these has
been prepared directly by low energy quantum absorption. In other
liquids where excitation yields are negligible (water, alcohols)
significant yields of separated ions and molecular decomposition
products are noted. Some liquids (acetone, benzyl alcohol) show
not insignificant yields of both excited states and separated ions,
indicating intermediate cation stability. In any given liquid, the
interplay of the forces governing auto-ionization, bond rupture,
electron capture, cation reactivity and recombination dynamics is
complex and can lead to unexpected results.

CONTRIBUTIONS OF PULSE RADIOLYSIS

The advent of electron pulse radiolysis some twenty years ago
provided a very versatile technique which has been exploited by
many individuals and groups throughout the world for studying
excited state formation and decay in liquid media. The experimental
approaches fall into two general categories:

(1) Experiments designed to inform on the early events in
    energy deposition and molecular excitation in liquids.

(2) Experiments which result in increasing our knowledge of
    the nature and properties of the excited states themselves.

The first category may be labelled radiation physics/chemistry;
the second, photophysics/chemistry. Since electron pulse radiolysis
combined with some fast diagnostic techniques enables the researcher
to both induce and observe molecular excitations in very short
times (as short as 10^{-11}s), the method has proven to be of great
use in both areas. A comprehensive survey of published work in
these areas would be far beyond the scope of this article. The
approach adopted has been to select a few examples which serve to
highlight the field.

In the category of radiation physics/chemistry, the basic
approach is to reach as far back in time as possible directly and
then, if needed, use indirect methods such as addition of specific
reactive solutes to obtain information about even earlier processes.
The detection of excited states in liquids has depended on their
ability to either emit UV/visible radiation (fluorescence/phospho-
rescence spectroscopy) or to absorb low energy radiations (absorption spectroscopy). The method preferred has depended sometimes
on the molecular properties, sometimes on the equipment available
to the experimenter. Typically, fluorescent molecular singlet
states with lifetime in the 10^{-9} -10^{-7}s range have been studied
by time-resolved spectrofluorimetry, whereas triplet states having
natural lifetimes in the 10^{-7} -10^{-2}s range have been studied by
time-resolved absorption spectrophotometry. These are broad
generalizations only.

A liquid which has been the subject of many investigations
since electron pulse radiolysis became available is cyclohexane.
In the mid 1960's when electron pulses widths in the µs range were
used, it was shown that triplet states of aromatic solutes such as
naphthalene, anthracene and biphenyl were produced in high yield
(23,24). These species, studied by absorption spectrophotometry,
were found to be formed in times shorter than the pulse duration.
With the development of machines capable of delivering pulse in
the 10^{-8}s range it became possible to make measurements on the
lifetime and yields of fluorescent singlet states of aromatic
solutes (1,8). Furthermore, the aromatic triplet states were
shown (25) to be formed over the same time scale as the radical
anion of the same solute disappeared, i.e., some solute triplets
at least were coming from ion recombination processes e.g.,

$$C \rightsquigarrow C^+, e^-$$

$$A + e^- \longrightarrow A^-$$

$$C^+ + A^- \longrightarrow {}^{1,3}A^* + C$$

$$^1A^* \longrightarrow {}^3A^*$$

and possibly

$$C^+ + A \longrightarrow C + A^+$$

$$A^+ + A^- \longrightarrow {}^{1,3}A^* + A$$

In this scheme C represents solvent and A solute species.

The decay kinetics of solute ions (e.g., biphenyl) were complex but have been fitted with analytical expressions based on a geminate process (26). Under these conditions, ions are decaying very rapidly (on nanosecond time scales) where information is difficult to obtain. The addition of ion scavengers such as N_2O or SF_6, which convert electrons or solute anions to lower energy species, served to reduce the yields of solute S_1 and T_1 states in electron-irradiated cyclohexane (25).

Detailed nanosecond studies of the G-values of single and triplet states of anthracene, naphthalene and biphenyl in cyclohexane showed (8) that the total yield of S_1 and T_1 was G(Total) = 1.2 for all three solutes. Based on a similar G-values for solute ions, all solute excitation was presumed to arise out of ion-recombinations involving A^-, A^+ and C^+. However, the ratio of T_1 to S_1 did not equal the spin-statistical value of 3:1. For anthracene and biphenyl a ratio of 2 was found while naphthalene gave 1. The questions were raised as to whether the decay of a few geminate pairs within localized regions of energy deposition (spurs) properly constitute population conditions such that statistical expectations would be met. Another possibility considered was that the energy of ion-neutralization (ca 8 eV) would be sufficient to populate upper manifold levels, leading to unusual inter-system crossing properties, thereby introducing correction procedures not accounted for (8).

That an excitation channel involving solvent excited states occurs was shown when a UV emission from an upper state of cyclohexane (and other hydrocarbon liquids) was detected in nanosecond electron pulse radiolysis (27). The lifetime of this emitting state is only 300 ps but it can undergo energy transfer to aromatic solutes with a very high rate constant of 3×10^{11} 1 mol^{-1} s^{-1}. Recently, the G-value of this state has been found to be near 0.5

molecules/100 eV (28).

Controversy surrounding the contribution of excitation by ion recombination should be settled by performing experiments with higher time resolution. Observations in the picosecond regime on cyclohexane systems were first published in 1972 (29). In this work the rate of formation of fluorescence of diphenyl anthracene (DPA) was shown to occur on a sub-nanosecond timescale and be dependent on the DPA concentration. The kinetic parameters resulting from these data were consistent with those from the observations of cyclohexane luminescence (27). Very recent measurements using a single 17 ps pulse and a streak camera for fluorescence detection with less than 30 ps time-resolution have shown (30) that the fluorescence from the scintillator diphenyl oxazole (DPO) in liquid cyclohexane grows in with a time profile which is resolvable into two processes. The early component which is complete within the pulse was not clearly identified but may have resulted from an extremely rapid electron capture-recombination process. The slower process has rate parameters which conform to those of the excited state of cyclohexane (27). In other experiments with single 30 ps pulses, the rate of scintillator (e.g., DPA) fluorescence was also observed to have two components (31). The concentration dependence of S_1 yield and the results of ion scavenging studies lead to the conclusion that energy transfer from an excited state of cyclohexane to solute was not a dominant process but that ion-recombination was the major source of solute S_1 states. However, theoretical analysis of the geminate ion recombination situation has so far failed to give good fits to the experimental data (32).

In summary, exhaustive experimentation using state-of-the-art apparatus has revealed a wealth of information about early processes in liquid cyclohexane which serve to support the basic ideas developed from gas phase and theoretical considerations. In spite of this, controversy about the details still exists and critical experiments remain to be carried out.

Except for water, cyclohexane is probably the most thoroughly studied liquid in radiation chemistry. Less exhaustive work in other systems shows interesting differences, some of which have been outlined in the previous section. For example the aromatic liquids benzene and toluene are clearly different from cyclohexane. Here, pulse radiolysis studies have shown (3,4) that the total yields of excited states of the solvent, as monitored by aromatic solute probes, are near 5 molecules/100 eV (Table 1). Only very low yields (G $\sim$ 0.05) of escaped electrons have been observed in such liquids by conductivity studies (33,34). G-values near 5 mean that an average of 20 eV is expended per primary species. Since the S_1 and T_1 states of benzene (e.g.) are in the 3-5 eV range, it is concluded that ca 25% of the enrgy deposited results in an identifiable new species and therefore some 75 % is lost, presumably as heat. This appears to be about the normal level of efficiency for transferring the energy from MeV particles into the lower quantum states of molecules. The maximum total observed yield of primary chemical entities (ions, excited states, free radicals) is usually near 5 per 100 eV. In liquid toluene pulse radiolysis experiments with time-resolution of less than 1 ns have been interpreted (35) in terms of fast energy transfer to solutes from toluene states in the upper manifold, although a later report (36) suggests that the faster process, which is predominant at lowered temperature, may in fact be due to an ion formation-ion recombination cycle.

Turning to the category of pulse radiolysis work which is more concerned with the nature of the excited states themselves, a legitimate question which can be posed is: why use MeV electron beams to populate states which can be populated directly (and presumably more precisely) by ultra-violet or visible photons? Sometimes an answer to the question is related to the availability of apparatus and access to it. In some instances valid scientific reasons are apparent. For example, some states which are of interest lie at high energies and time-resolved flash experiments would necessitate the use of high energy pulsed lasers with output in the untra-violet or even vacuum ultra-violet. Only recently have such devices become widely available. On the other hand, electron irradiation of benzene produces high yields of S_1 (4.8 eV) and T_1 (3.7 eV) which are capable of sensitizing the formation of lower energy solute states by collisional energy transfer.

TABLE 2

Rate constants for singlet energy transfer from solvent to solute molecules

Liquid		Diffusion-controlled rate constant (a)	Energy transfer rate constant (a)	Acceptor
Benzene	(38)	32	51	aromatic hydrocarbons
Toluene	(38)	31	55	" "
Cyclohexane	(27)	7	300	benzene
Acetone	(41)	30	320	anthracene

(a) Units are 1. mol^{-1} ns^{-1}

The use of energy tranfer from solvent states to generate solute states has revealed interesting facts about the migration of excitation around an ensemble identical molecules in a liquid. Energy transfer by electron exchange is an encounter-dependent

process and encounter rates in liquids can be calculated from
diffusion theory or measured directly by observation of solute
excited state quenching. Table 2 shows some rate constants determined for excitation transfer from a single excited solvent
species to a guest molecule present at low concentration compared
to the diffusion-controlled values.

In benzene and its homologues continuous formation and dissociation of excimer species has been proposed (37) as being
responsible for the enhanced rate constant. Excimers of benzene
(complex of S_1 and ground states) and homologous aromatics are well
known and have been observed by both emission (38) and absorption
(39) studies. An alternative description of this process is simple
molecular diffusion with an enhanced interaction radius of the
donor state (40). The excimer-diffusion may be pictorially represented as

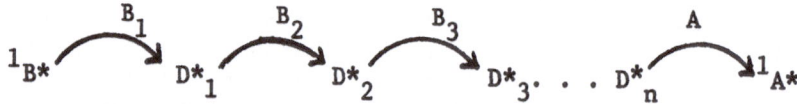

Where B_1, B_2 are neighboring benzene molecules D^*_1, D^*_2 are
excimer states and A is an energy-accepting solute. The excitation
is seen to translocate without large displacements of the molecular
sites.

In acetone (41) and cyclohexane (27), however, energy translocation via excimer states is hardly tenable since (i) excimer
states of cyclohexane and acetone are unknown and (ii) the rate
parameter for excitation transfer between acetone and anthracene
was independent of temperature over a range 196-296 K (41). An
explanation which seems to be more appropriate to these liquids is
based on a model in which a resonance interaction between an excited
solvent molecule and an adjacent unexcited identical molecule results
in transfer of the energy (42). Thus, overall energy translocation
is due to the sum of molecular translation and resonance migration.
In the time-independent Smoluchowski relationship, the overall
rate constant (k_t) is given by

$$k_t = \frac{4\pi N}{1000} R(D_A + Z_D)$$

where N is Avagadro's number, R the sum of the molecular radii, D_a
is the diffusion coefficient of the acceptor and $Z_D = D_D + \Lambda_D$, D_D
being the diffusion coefficient of the donor molecule and
Λ_D the energy migration coefficient. On substituting typical
values relevant to acetone, it is found that $Z_D = 63.5 \times 10^{-5} cm^2 s^{-1}$
and, using a Stokes' law value of $D_D = 2.4 \times 10^{-5} cm^2 s^{-1}$, $\Lambda_D = 61.1 \times 10^{-5} cm^2 s^{-1}$ was obtained (41). Thus, energy migration was the

dominant factor in k_t, explaining the lack of temperature effects.

An area in which the use of electron pulse radiolysis has
clearly made an original contribution to photophysics of molecular
states concerns the detection of the T_1 states of simple alkenes
(mono- and di-olefins) and measurements of their lifetimes. These
entities are of extreme importance in organic photochemistry but
their basic photophysical properties are relatively unknown owing
to their lack of characterizable features such as phosphorescence
or T-T absorption spectra. Mono-olefins are expected to decay from
the T_1 state extremely rapidly because of the closeness of the T_1
and S_0 energy surfaces at the equilibrium geometry of the T_1 state.
This can be modified by encompassing a single π-bond system within
a rigid σ-bonded framework to restrict facile torsional motions.
The di-olefins are different from mono-olefins and electronic
conjugation should serve to preserve the T_1 state. The S_0 -T_1
energy surfaces for 1,3-dienes are not known to approach closely.
Pulse radiolysis studies of norbornene (43), cyclopentadiene (44)
and tetramethylbutadiene (44) in benzene solution have provided

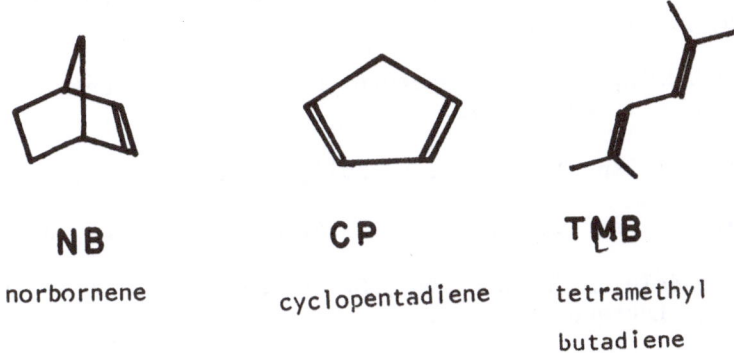

NB   **CP**   **TMB**

norbornene   cyclopentadiene   tetramethyl

                                                          butadiene

conclusive evidence for the T_1 states of these systems having
relatively long lifetimes (250 ns for NB, 1.7 μs for CP 73 ns
for TMB). These studies are made possible by the fact that the
T_1 state of benzene, produced in high yield during electron
irradiation of the liquid (Table 1), has an energy of 3.7 eV which
is capable of sensitizing even high energy olefin states. In the
case of CP, energy transfer equilibria could be established with
various aromatic substances which allowed the conclusion that the
^{3}CP* state produced in the energy transfer event had an identical
energy to that derived from oxygen-enhanced S_0-T_1 absorption spectra.
For NB and TMB triplets, however, it was shown that the relaxed
alkane T_1 state which participates in energy transfer is significantly lower(0.6 eV for NB) in energy than the spectroscopic state.

To summarize, the two decades or so since the first application
of pulse radiolysis techniques to liquid systems has produced out-
standing evidence for the occurrence of both excitation and ionization processes as primary events in energy deposition. A quantitive understanding of the processes which follow and lead to production of low energy species is not yet at hand although many
details are known. Sufficient knowledge is available for a variety
of liquid systems such that high energy electron irradiation can
be confidently used as an investigative technique, particularly
for studying the dynamic properties of molecular excited states
not readily populated direct photon absorption.

REFERENCES

1. Thomas, J. K. 1976, Int. J. Radiat. Phys. Chem., 8, p. 1 and
    references therein.

2. Cooper, R. and Thomas, J. K. 1968, J. Chem. Phys. 48, p. 5097.

3. Baxendale, J. H. and Fiti, M. 1972, J. Chem. Soc. Faraday II,
    68, p. 218.

4. Baxendale, J. H. and Rasburn, E. J. 1973, J. Chem. Soc.
    Faraday 1, 69, p. 771.

5. Kira, A. and Thomas, J. K. 1974, J. Phys. Chem., 78, p. 2094.

6. Kira, A. and Thomas, J. K. 1974, J. Chem. Phys., 60, p. 776.

7. Baxendale, J. H. Beaumond, D. and Rodgers, M. A. J. 1969, Chem.
    Phys. Letters, 4, p. 3.

8. Baxendale, J. H. and Wardman P. 1971, Trans. Faraday Soc.,
    67, p. 2997.

9. Rodgers, M. A. J. 1971, Trans. Faraday Soc., 67, p. 1029.

10. Arai, S. and Dorfman, L. M. 1965, J. Phys. Chem., 69, p. 2239.

11. Baxendale, J. H., Beaumond, D. and Rodgers, M. A. J. 1970,
    Trans. Faraday Soc.,66, p. 1996.

12. Bell, I. P., Rodgers, M. A. J. and Burrows, H. D. 1977, J.
    Chem. Soc. Faraday I, 73, p. 315.

13. Bethe, H. A. 1932, Z. Phys., 76, p. 293.

14. Platzman, R. L. 1955, Radiat. Res. 2, p. 1.

15. Robin, M. B. 1974, "Higher Excited States of Polyatomic Molecules, Vol I", Academic Press.

16. Lipsky, S. 1981, J. Chem. Educ., 58, p. 93.

17. Platzman, R. L. 1962, The Vortex 28, p. 372.

18. Inokuti, M. 1971, Rev. Mod. Phys., 43, p. 297.

19. Kupperman, A. and Ruff, L. 1962, J. Chem. Phys., 37, p. 2497

20. Ausloos, P. 1970, Progr. Reaction Kinetics, 5, p. 113.

21. Williams, F., 1963, Nature, 194, p. 348.

22. Robinson, A. J. and Rodgers, M. A. J. 1973, J. Chem. Soc. Faraday I, 69, p. 2036.

23. McCollum, J. D. and Nevitt, T. D. 1963, U. S. Tech. Report., ASD-TDR-63-616.

24. Kemp, T. J., Salmon, G. A. and Wardman, P. 1965, in "Pulse Radiolysis" (M. Ebert, J. P. Keene, A. J. Swallow and J. H. Baxendale, editors) Academic Press, p. 247.

25. Thomas, J. K., Johnson, K., Klippert, T. and Lowers, R. 1968, J. Chem. Phys., 48, p. 1608.

26. Rzad, S. J., Infelta, P. P., Warman, J. M. and Schuler, R. S. 1970, J. Chem. Phys., 52, p. 3971

27. Baxendale, J. H. and Mayer, J. 1972, Chem. Phys. Letters 17, p. 458.

28. Busi, F., Flamigni, L. and Orlandi, G. 1979, Radiat. Phys. Chem. 13, p. 165.

29. Beck, G. and Thomas, J. K. 1972, J. Phys. Chem., 76, p. 3856.

30. Katsumura, Y., Tagawa, S. and Tabata, Y. 1980, J. Phys. Chem., 84, p. 833.

31. Jonah, C. D., Sauer, M. C., Cooper, R. and Trifunac, A. C. 1979, Chem. Phys. Letters, 63, p. 535.

32. Sauer, M. C. and Jonah, C. D. 1980, J. Phys. Chem., 84, p. 2539

33. Schmidt, W. F. and Allen, A. O 1970, J. Chem. Phys., 52, p. 2345.

34. Robinson, A. J. and Rodgers, M. A. J. 1975, J. Chem. Soc. Faraday I, 71, p. 378.

35. Beck, G., Richards, J. T. and Thomas, J. K. 1976, Chem. Phys. Letters, 40, p. 300.

36. Beck, G., Ding, A. and Thomas, J. K. 1979, J. Chem. Phys., 71, p. 2611.

37. Birks, J. B. and Conte, J. C. 1960, Proc. Roy. Soc. A308, p. 85.

38. Birks, J. B. 1970, "Photophysics of Aromatic Molecules", Wiley.

39. Cooper, R. and Thomas, J. K. 1968, J. Chem. Phys., 48, p. 5097.

40. Birks, J. B., Najjar, R. Y. and Lumb, M. D., 1971, J. Phys. (B), 4, p. 1516.

41. Robinson, A. J., Rodgers, M. A. J., Keene, J. P. and Gilbert, C. W. 1972, J. Photochem,, 1, p. 379.

42. Voltz, R., Laustriat, G. and Cocke, A. 1966, J. Chem. Phys., 63, p. 1253.

43. Barwise, A. J. G., Gorman, A. A. and Rodgers, M. A. J. 1976, Chem. Phys. Letters, 38, p.313.

44. Gorman, A. A., Gould, I. R. and Hamblett, I. 1981, J. Amer. Chem. Soc., 103, p. 4553.

RADIOLYTIC STUDIES OF MICELLES AND OTHER AGGREGATED SYSTEMS

M. A. J. Rodgers and B. A. Lindig

Center for Fast Kinetics Research, University of Texas at Austin, Austin, Texas 78712 USA

ABSTRACT

Aqueous micelles, reverse micelles and vesicles are aggregates which incorporate some of the key features of cell membranes. The separation of a lipid-like region from aqueous regions by a polar or charged interface makes dispersions of these aggregates microscopically heterogeneous. Pulse radiolytic studies of such systems can 1) provide information about the organization and dynamics of the aggregates, 2) simulate radiation biological reactions in media which resemble cellular composition, and 3) initiate novel reactions in the interfacial regions.

INTRODUCTION

The aggregation of amphiphilic molecules to form organized assemblies is a structural expediency in natural systems: cell membranes provide cellular systems with their structural integrity while at the same time functioning as an active participant in chemical and biophysical processes of the cell. The cell membrane has provided the inspiration for a rapidly expanding volume of research performed on simpler, more easily fabricated assemblies which incorporate some of the characteristics of the biological prototypes. Pulse radiolytic investigations of these systems involve both studies of biochemical reactions in the aggregates, as well as exploitation of the aggregate properties for design of new chemical systems.

J. H. Baxendale and F. Busi (eds.),
The Study of Fast Processes and Transient Species by Electron Pulse Radiolysis, 551–571.
Copyright © 1982 by D. Reidel Publishing Company.

Cell Membranes (1)

A brief description of the structure and properties of the
cell membrane will serve to demonstrate some of the desirable
features of model systems such as micelles and vesicles. While
membranes are composed of comparable dry weights of proteins
(20-60%) and lipids (30-80%), there are about 30 phospholipid
molecules for every protein molecule, and these amphiphilic
lipids provide the organizing influence for the structures. The
hydrophobic effect (2) results in spontaneous aggregation of
phospholipids into a bilayer when the crystal lattice of the
lipids is exposed to an excess of water (more than 15 water molecules per lipid molecule). The driving force for this organization arises from strong attractive forces between water molecules,
which must be disrupted if individual lipid molecules are to be
water solvated. Since a water molecule has a much greater affinity for another water molecule than for a hydrocarbon species,
the lipid molecules orient themselves in such a way that a hydrocarbon region is formed in isolation from the aqueous phase, with
the charged head groups of the phospholipids forming the interface.
While at first sight it might appear that the entropy of the system has decreased because of the formation of the organized phospholipid aggregates, in fact the increase in water entropy (which
arises from a decrease in orienting influence of the hydrocarbon
chains in contact with the water) is the greater factor.

The architecture of the membrane consists in a two-dimensional matrix with the polar or charged phospatidyl groups coating
both surfaces. (Figure 1) The hydrocarbon chains are parallel to
each other and perpendicular to the plane of the bilayer. Proteins

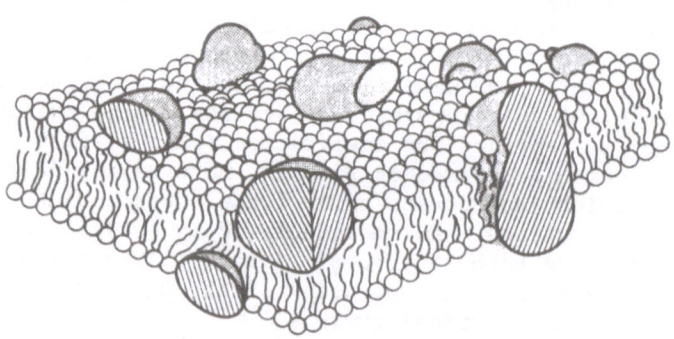

Figure 1. Schematic of a membrane (3), showing the bilayer arrangement of phospholipid molecules and embedded or surface proteins. (Reproduced by permission
of McMillan, London and Basingstoke)

Aggregation numbers generally fall within the range of 10 to 100 monomers per micelle.

The precise structure of an aqueous micelle still remains somewhat uncertain. It is generally accepted that a spherical or ellipsoidal geometry is most consistent with the accumulated body of experimental evidence. According to the classical Hartley model (Figure 2) alkyl chains radiate out from the center of the micelle and the bulky head groups occupy the surface of the structure in contact with the bulk aqueous phase (5). While the classical model suggests a hydrocarbonlike interior for the micelle, such a picture is inconsistent with results indicating a polar environment for solubilized probes, as well as a considerable degree of "wetting" of the hydrocarbon chains by the surrounding aqueous phase. Several models have been proposed to account for these observations. One model suggests a "porous cluster" structure with extensive penetration of water between the amphiphile chains (7). Other models invoke a more compacted structure in which the alkyl chains are less radially oriented; instead they can assume configurations in which segments of the chains lie along the surface of the aggregate (8,9). Distinction between these models remains of a somewhat conjectural nature since all of the experimental data require some subjective interpretation.

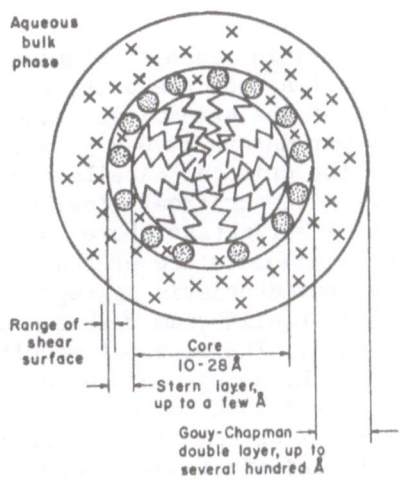

Figure 2. Idealized cross section of an aqueous micelle (6). The shaded circles represent head groups of the surfactants, while the crosses are the counterions. (Reproduced by permission of Academic Press.)

can be localized at either face of the membrane, in either layer, or can extend all the way through the membrane (4). Membranes from different parts of the cell or from different organisms can be distinguished by their chemical composition, and there may even be significant differences in the lipid composition of the inner and outer layers (asymmetry).

Several other properties of membranes deserve mention in relation to the membrane models to be discussed.

1. Membranes are selectively permeable. They provide the means of communication between compartments, but also separate the solutions on either side to maintain certain concentration balances.

2. The microviscosity of a membrane decreases with increasing unsaturation of the lipid chains, temperature, and the presence of lipid soluble agents (organic solvents, detergents, etc.). The opposite effect is rendered by increasing cholesterol content, or with greater chain length or multivalent ion concentration.

3. Localization of chemical species in the membrane or at its surface has important consequences for the dynamics of reactions. Catalytic effects can result from the proximity of reactants and the limitations on the freedom of isotropic diffusion.

Micelles

Aggregates formed in aqueous solutions of surfactants are referred to as micelles, and these systems are by far the most thoroughly studied of the assemblies being treated in this discussion. Two key parameters give a physical definition of aqueous micelles: the critical micelle concentration and the aggregation number. The critical micelle concentration (CMC) is the concentration of surfactant above which micelles are formed. Below the CMC, only monomers are present; above the CMC micelles are formed with a characteristic aggregation number (independent of concentration) in equilibrium with a monomer concentration equivalent to the CMC. The onset of micelle formation can be readily detected by the abrupt change in slope of various physical properties measured as a function of surfactant concentration. Various types of micelles can be classified according to the head group, which can be anionic ($-SO_3^-$, $-OSO_3^-$, $-CO_2^-$), cationic ($-N^+(CH_3)_3$, $-N^+C_6H_5$), zwitterionic (two alkyl chains, one with a cationic and one with an anionic head group), and nonionic (polyoxyethylene or polyoxypropylene derivatives with terminal hydroxy groups). The aggregation properties of micelles are sensitive to a number of factors: alkyl chain lengths, type of head group, ionic strength, pH, temperature and presence of alcohol additives.

Typically, 70–80% of the surfactant head groups are associated with their counterions in the Stern layer (Figure 2); the remainder of their counterions are located in the diffuse double layer (Goüy–Chapman layer) in equilibrium with the bulk aqueous phase. As a consequence of the ion concentration gradient, an exponentially decreasing (with distance) electric potential is created across the double layer. This potential field has great significance in that it can impose a catalytic or inhibitory effect on charge transfer or electron transfer reactions occurring in the interfacial region, as well as on any reactions involving charged or polar intermediates. Several reviews on the catalytic and inhibitory effects of micelles on various reactions have appeared (6, 10, 11).

Another significant feature of aqueous micelles is that they are dynamic structures. There is a continuous exchange of monomers between the micelles and the aqueous phase.

Reverse Micelles

The aggregation behavior of surfactants in apolar solvents differs considerably from that in water. Rather than the characteristic concentration and abrupt aggregation of aqueous systems, surfactants in apolar solvents often exhibit sequential self-association. That is, the solution consists of monomers in equilibrium with m–mer aggregates (m = 2,3 ... n) rather than only monomers and micelles. However, a CMC and characteristic aggregation number have been observed for some systems.

Addition of water to solutions of surfactants in apolar solvents affects the aggregation dramatically by increasing the average aggregation number. Large amounts of water can be solubilized by concentrating it in pools at the core of each micelle.(Figure 3) The resulting solutions are also called water-in-oil micro-emulsions.

The water pools are the region of interest in studies of reverse micelles – the interface between the solubilized water and the organic solvent phase is analogous to that of the aqueous micelle. However, considerable geometric constraints can be imposed by limiting the size of the water pool. In this respect, the entrapped water pool has been compared to polar pockets of folded enzymes (12). Thus, reverse micelles may provide a better model for some biological systems than the aqueous micelle. To date reverse micelle systems have not been as thoroughly studied as their aqueous counterparts.

Liposomes and Vesicles

These bilayer systems represent another step up in complexity from the aqueous and reverse micelles. The terms liposome and

bulk hydrocarbon

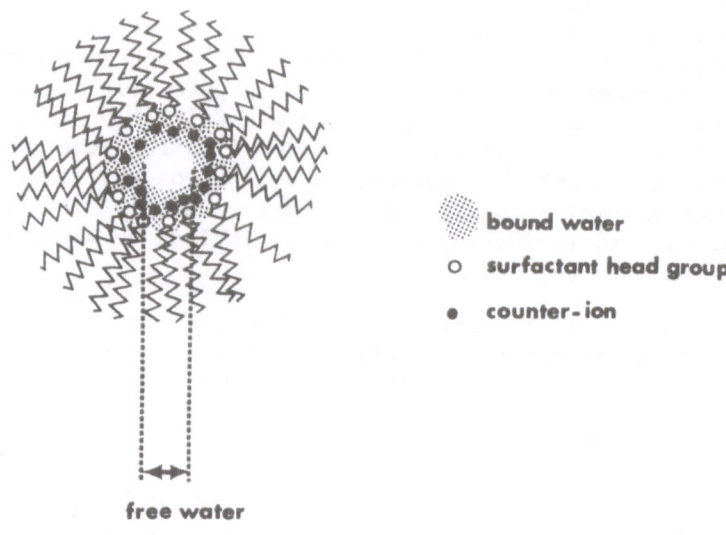

- bound water
- o surfactant head group
- • counter-ion

free water

Figure 3. Schematic representation of a water-containing reverse micelle.

vesicle are used almost interchangeably since they are essentially indentical in structure, but "vesicle" is more commonly used in relation to aggregates of synthetic surfactants, whereas "liposome' carries the connotation of a naturally occuring amphiphile, e.g. phospholipids. In the simplest terms, these aggregates are closed bilayer structures. The shape, size and number of bilayers depend on the components and the method of preparation. Thus, liposomes may be single or multi-compartment assemblies and vesicles are described as unilamellar, multilamellar, etc. In terms of creating an analogy to biological membranes, the bilayer systems may be more appropriate than micelles, but the complexity of the preparations and the relative instability of the systems have limited their practicality for use in physicochemical investigations. However, interest has recently been generated for the use of vesicles in photochemical energy storage and conversion as well as in drug encapsulation and delivery. In particular, polymerized vesicles have recently been synthesized and shown to be more stable and less leaky than the unpolymerized variety (13, 14).

While other types of aggregates could be mentioned (oil-in-water microemulsions, monolayers, black-lipid-membranes, liquid crystal, polymers, polyanions) these systems either have not been investigated by pulse radiolysis techniques or deviate from the main themes of this discussion.

The application of the pulse radiolysis technique to the aggregated assemblies described herein has a number of fundamental goals.

Characterization of the structure and dynamics of the aggregates. The pulsed production of transient species and subsequent time-resolved detection of their reactions provides a valuable tool for probing the systems. Information can be derived about the monomer-micelle equilibrium as well as about the solubilization sites of probe molecules.

Examination of biochemical reactions in a membrane-mimetic environment. Molecular mechanisms for radiation effects can be suggested by measurements of early radiolytic reactions with cellular components. It is of special importance to measure these reactions in a more complex milieu than dilute aqueous solution, which is quite remote from the realities of a living cell. The last consideration also applies to the use of pulse radiolysis in elucidating the mechanisms for reactions which are a part of normal cellular function - for example, electron transport.

Design of novel environments embodying catalytic effects for reactive systems. Just as macroscopic catalyst surfaces serve to organize and direct reactions in an efficient way, the heterogeneous character of micellar and vesicular interfaces can be exploited to accelerate reactions. This effect can result from the localization of reactants as well as from the imposition of the electric field at the charged interface on reaction dynamics. It should be realized that these same factors contribute to the remarkable efficiency of enzymatic systems.

The Hydrated Electron

The intrinsic reactivity of micellar and vesicular amphiphiles with the hydrated electron is generally negligible ($k < 10^7$ $M^{-1}s^{-1}$) (15) in comparison with the reactivity of most substrates which may be studied as additives. The only exceptions occur when the amphiphile has a functional group (e.g. aromatic) which possesses intrinsic reactivity toward e_{aq}^-. Bile acid salts exhibit moderate reactivity ($k = 10^8$ $M^{-1}s^{-1}$) toward e_{aq}^- due to the peptide linkage in the taurocholate group (16). The reactivity decreases on micellization probably due to the protective effect of the negatively charged head groups. The opposite effect is observed for cetyl pyridinium chloride micelles where micellization greatly enhances

electron addition to the pyridinium group (17). In this case,
the large positively charged aggregates act as traps for the
electrons, due to acceleration of the electron in the potential
field (~110 mV) surrounding the micelle. Observation of a rate
constant in excess of 10^{12} $M^{-1}s^{-1}$ can be rationalized by using a
scavenging radius equivalent to the Onsager radius of the micelle.
This rate constant can be lowered by increasing the ionic strength,
resulting in a decrease in the surface potential of the aggregates.

    The efficacy of electron scavenging by solubilized aromatic
substrates has been shown to be influenced analogously to the
examples described above. In the earliest investigation (15)
it was shown that the reaction of benzene with e_{aq}^{-} was affected
by the presence of micelles. The rate constant for electron
attachment was observed to increase in the various micellar media
SDS<Igepal CO-730<water (no micelles)<CTAB. (SDS: sodium dodecyl
sulfate; Igepal CO-730: polyoxyethylene(15)nonylphenol; CTAB:
cetyltrimethylammonium bromide.) The anionic micelles (SDS) significantly inhibited the reaction (relative to water) while the
cationic micelles (CTAB) produced a notable enhancement. Thus
the reactivity of micellized benzene was directly influenced by
the electron-repulsive and -attractive charged surfaces of the
micelles, respectively. Reduced reactivity in the nonionic Igepal
solutions relative to that in water seems to indicate that solubilization can reduce the reactivity of benzene in the absence of
the charge effects.

    A detailed study of a series of electron acceptors varying in
electron affinity helped to clarify the effects of the charged
interfaces on reactivity of aromatic substrates toward the electron
(18). It was shown that the rate constant for electron attachment
is very sensitive to the electron affinity of the acceptor in micelles of negative interfacial potential (SDS), but relatively
insensitive to this factor in the CTAB micelles, which possess a
positive potential. Further, rate constants for reduction of the
same substrates by the neutral radical $CH_3\dot{C}HOH$ were independent of
the charge of the micelle or the gas phase electron affinity of
the acceptor over the range 1.6 to 2.4 eV. These results emphasize the effect of the interfacial potential on the diffusion of
the electron, as opposed to a mechanism suggesting an environmental effect on the intrinsic electron affinity of the acceptor.
The "uphill" process of electron diffusion into an anionic micelle
must be countered by a high affinity of the acceptor for the
electron in order for the reaction to proceed. An electron tunnel-
ing mechanism has been invoked to account for diffusion of the
electron across the "barrier" of the micellar interface (18). (It
remains unclear however, in light of the uncertainty of micellar
structure and probe location, as to whether such a barrier truly
exists.) In some cases, the effect of the anionic micelle can be
to completely prevent the electron attachment. In the instance

of naphthalene in SDS micelles, this principle has been used to estimate the distribution coefficient of the aromatic between the micellar and aqueous phase by assuming that all reaction was due to naphthalene in the aqueous phase (19).

Micellized surfactants have also been demonstrated to affect the rates of reduction of several ionic (i.e. water soluble) species by the hydrated electron. For example, the lifetime of e_{aq}^- in the presence of Cu^{++} is markedly increased by addition of CTAB micelles (20). Anionic micelles such as SDS decrease the reactivity of metal ions toward e_{aq}^- by binding them to the micellar surface. Cu^{++} and Eu^{3+} are much more affected than Ag^+ and Tl^+ because of the higher extent of binding of the multivalent cations (21). Binding constants for Cd^{++} and Cu^{++} have been estimated by taking advantage of this effect (22). Reduction of organic cations (e.g. $Ru(bpy)_3^{++}$ and substituted viologens) (23-25) by e_{aq}^- is also less efficient in the presence of SDS micelles. However, these systems differ from the metal ion case; for $Ru(bpy)_3^{++}$ and the viologens (dimethyl-, dibenzyl- and methyl-dodecyl-) the e_{aq}^- rate constant increases with increasing SDS concentration above the CMC due to a kinetic salt effect. For metal ions, increased surfactant concentration results in increased binding of the ions, thus decreasing the reactivity toward e_{aq}^- (22).

The fates of radical anions produced by electron attachment have been examined in a study of pyrene derivatives in nonionic polyethyleneoxide micelles (26). Pyrene anion was found to decay by second order kinetics, suggesting that the anions are rapidly detached from the micelles and diffuse into the aqueous phase where they deactivate each other. In contrast to this behavior, the anions of 10-(1-pyrenyl)-dodecanoic acid and 1-pyrenylbutyric acid decay in acidic or neutral solution via a first order process, which may be an intramolecular proton transfer from the carboxyl group to the aromatic system.

Electron scavenging by pyrene in vesicular bilayers has been studied (27). While the electron reaction is relatively slow for pyrene in lecithin bilayers (in which the head groups are the zwitterions phosphatidylcholine) it is enhanced by the addition of cationic surfactant (which is evidently incorporated into the bilayer). The very low concentration of vesicles (which is lower than the pyrene concentration) is a limiting factor in the reaction. Measurements of the lifetime of e_{aq}^- as a function of pyrene concentration enabled the separate determination of the rate constants for e_{aq}^--vesicle encounter and for e_{aq}^- reaction with pyrene. The encounter rate constant was 2.7×10^{11} $M^{-1}s^{-1}$ for the lecithin vesicles and was increased 2-3 fold by the added surfactant. The reactivity of a pyrene-containing vesicle was shown to be directly proportional to the number of pyrenes residing in the aggregate. Pyrene reacted with e_{aq}^- with $k = 1.5 \times 10^{11}$ $M^{-1}s^{-1}$ in vesicles composed entirely of cationic surfactant; the same rate constant was observed for pyrene in cationic micelles (28).

The relatively small proportion of water in reverse micelle systems (<10 volume %) considerably reduces the importance of the hydrated electron. It has been demonstrated that the yield of e_{aq}^- is dependent on the water content of the solution, and that the total yield of e_{aq}^- has two contributions: direct radiolysis of the water pool, and capture by the water pool of electrons initially produced in the hydrocarbon phase (29). Subnanosecond conductivity detection has enabled direct measurement of the capture of the electron by the water pool (30). A distinction between bound and free water could be inferred from the dependence of the capture efficiency on the water content. A diffusion controlled electron attachment rate constant of about 10^{15} $M^{-1}s^{-1}$ was achieved only for water:surfactant ratios in excess of 12. Evidently, up to twelve water molecules are devoted to hydrating the sulfosuccinate head groups of the surfactant, and this bound water has a lowered ability to hydrate electrons.

It should also be noted that the reaction of e_{aq}^- with a cell membrane has also been measured (31). Unfortunately the complexity of the erythrocyte ghost system precluded any determination of the reactive site.

Hydroxyl Radical

The very high reactivity of ·OH with hydrocarbons contributes to make its chemistry in micelles and vesicles considerably more complex and less selective than that of the hydrated electron. Since surfactants need not possess a specific functional group to react with ·OH the reactions with probes located in surfactant aggregates are likely to suffer from this competition.

The thiocyanate competition method has been used to obtain values for the reaction of ·OH with surfactants above and below the CMC, as well as with benzene solubilized in the surfactant micelles (32). Table I summarizes these results. (In the case of the cationic surfactant CTAB, reactivity with Br^- was measured by optical detection of Br_2^- in the presence of methanol as a competing scavenger.)

It can be seen that the reactivity of the three surfactants toward ·OH is dramatically affected by micelle formation. Benzene is also only one-third as reactive in micelles as in water. Contrary to the case of micellar effects on the reactions of e_{aq}^-, reactions of ·OH are not likely to be influenced by charge effects on diffusion. The only remaining explanations are 1) that there

Table I (32)

Rate Constants for Hydroxyl Radicals with Monomeric and
Micellized Surfactants

Surfactants(S)	$k(\cdot OH + S)$ below CMC $(M^{-1}s^{-1})$	$k(\cdot OH + S)$ above CMC $(M^{-1}s^{-1})$	$k(\cdot OH + C_6H_6)$ above CMC $(M^{-1}s^{-1})$
SDS	7.6×10^9	0.5×10^9	3.0×10^9
Igepal(CO–730)	1.1×10^{10}	1.7×10^9	2.6×10^9
CTAB	1.0×10^{10}	2.1×10^9	2.6×10^9
None	---	---	8.2×10^9

is some change in the intrinsic reactivity of the substrates, or
2) that the geometry of the aggregates imposes some limitations
on the effectiveness of the individual molecules as ·OH scavengers.
Although this problem has not been directly addressed, it is
evident that the methylene groups of the surfactant chains which
are embedded within the lipoid region will be less effective
reactive sites simply because the ·OH radical is not able to
penetrate to these regions because of reaction with groups in the
outer layers. Thus the reactivity of each monomer in the micelle
would appear to be reduced.

The formation and decay of alkyl radicals produced by reactions of ·OH in micellar solutions can be a useful probe of the
monomer-micelle dynamics of the system. Optical absorption due
to hydroxycyclohexadienyl radicals (formed from ·OH attack on
alkyl benzene sulfonates) has been monitored as a function of the
concentration of the aromatic surfactant (33). Initial equilibration of the radical distribution between the monomers and micelles was assumed, because the time for this process is estimated to be short ($10^{-4} - 10^{-5}$s) for ionic surfactant. (The radical decay occurred over milliseconds.) The data suggest that
reaction of two radical-containing micelles or of two monomer
radicals is less important above the CMC than the reaction of a
monomer radical with a radical in a micelle. Extraction of
the rate constant for entry of a monomer into a micelle. By substituting the derived value of k_+ (4×10^8 $M^{-1}s^{-1}$) into the expression CMC = k_-/k_+, an exit rate constant k_- of 8×10^5 s^{-1} results. The method was extended to a study of polyethyleneoxide
surfactants (both nonionic and anionic) (34). In this instance
ether type radicals $-O-\dot{C}H-CH_2-O-$ produced an optical absorption.
Radicals in the nonionic system were found to undergo both intra-
micellar reaction (when more than one radical was initially formed
in the micelle) as well as intermicellar reaction of two radical-
containing micelles. At lower micelle concentrations, monomer-

micelle reactions became more important. Radical reactions in the
analogous anionic micelles differed in pathway due to inhibition
of the intermicellar reaction by Coulombic repulsion of the negatively charged micelles.

An alternative to the measurement of the intrinsic probes
(i.e. surfactant radicals) is the use of an aqueous probe:
ferricyanide has been used as a scavenger for surfactant radicals
(35). When ferricyanide is included in a solution of anionic
micelles, the $Fe(CN)_6^{3-}$ is assumed to be repelled from the micelle
surfaces and thus reacts primarily with surfactant monomer radicals. A biphasic evolution of ferricyanide bleaching could be
interpreted in terms of a fast reaction during the equilibration
time for the radical monomers and micelles, followed by a slower
process of reaction in the steady state configuration of radical
distribution, Rate constants k_+ and k_- of 1.5×10^9 $M^{-1}s^{-1}$ and
1.8×10^5 s^{-1}, respectively, could be derived from this model.

A transient UV absorption of radicals produced by ·OH attack
on linoleate micelles was observed and assigned to a conjugated
radical ($R_{conj}^{\cdot}$) produced by abstraction of a doubly allylic
hydrogen (36). The absorption evolved in two steps: the rapid
one was contributed by ·OH attack at the doubly allylic position,
while the slower one was due to $R_{conj}^{\cdot}$ formation from other radical precursors. (Since only 2 of the 31 available hydrogens are
doubly allylic, and ·OH is relatively unselective, direct formation of $R_{conj}^{\cdot}$ would be expected to be small on purely statistical grounds.) The mechanism of conversion for precursor radicals
to $R_{conj}^{\cdot}$ is likely to be intramolecular abstraction for monomers
and intermolecular abstraction from a neighboring linoleate
molecule for micellar radicals.

Another example of a micellar effect on radical reactions is
given in a CIDEP (Chemically Induced Dynamic Electron Polarization)
study (37). When surfactant radicals were produced by pulse radiolysis of micellar solutions, emissive signals assigned to $S-T_{-1}$
polarization of radical-radical interactions were observed.
This observation is an indication of restricted diffusion of the
radicals in the aggregate phase.

The reaction of ·OH with lecithin vesicles has also been
examined (38). Although the site of attack could not be pinpointed
from the transient absorptions or by the use of spin traps, the
total rate of reaction was measured. The measured rate constant
of 5×10^8 $M^{-1}s^{-1}$ could be rationalized by first calculating the
diffusion-controlled rate constant for reaction with a vesicular
particle (3.9×10^{11} $M^{-1}s^{-1}$) and then dividing by the aggregation
number of the vesicle (1000).

Pulse radiolysis of erythrocyte ghost membranes under N_2O
produced spectral transients which might be assigned as products

of the ·OH reaction with N-acetyl neuraminic acid and N-acetyl-
glucosamine, both sugars thought to be present at the membrane
surface (39). In the presence of Br^-, Br_2^- is produced by the
reactions

$$·OH + Br^- \rightarrow {}^-OH + Br· \underset{Br^-}{\rightleftharpoons} Br_2^- \tag{1}$$

By monitoring the absorption due to Br_2^-, reactivity of the membrane toward this oxidizing radical could be measured under a
variety of conditions. No reaction of Br_2^- with tyrosine or
tryptophan residues of the membrane proteins could be detected
(i.e. no optical absorption due to the oxidized amino acids) unless the membrane was partially solubilized by addition of the
anionic surfactant SDS, or treated with the enzymes α-chymotrypsin
or papain (39). Since all of these treatments result in increased
exposure of membrane proteins to the aqueous phase, it is likely
that the amino acids in the intact, non-hydrolyzed native membrane are relatively inaccessible to attack by radiolytic agents
produced in the aqueous phase. This finding emphasizes an important point to be made about the application of radiation
chemical data to radiation biology: rate constants measured for
isolated cellular components (e.g. amino acids) in dilute aqueous
solution have limited usefulness in the analysis of the molecular
mechanisms of radiation effects on living cells. The very complex
structure of cells may impose considerable geometrical restrict-
ions on the types and rates of certain reactions. For example,
highly reactive species such as ·OH are likely to react with substrates which have a high local concentration at the site at
which the radical is produced. The diffusion distance of such a
radical before reaction will not be sufficiently long for the
competitive effects of all possible substrates to come into play.

Reduction of Dimensionality

In the following paragraphs it will be demonstrated that a
micellar interface can impose order on certain ionic intermediates
in a way which acts to accelerate the course of their subsequent
reactions. The term "reduction of dimensionality" is appropriate
to describe this phenomenon, since the organizing influence of
the interface results in a restriction of the diffusion of certain
ions to the spherical surface of the micelle, which constitutes
a two-dimensional reaction space. The catalytic effect of such
oriented systems was recognized a number of years ago in a theoretical treatment by Adam and Delbrück (40):

> "We wish to propose and develop the idea that organisms
> handle some of the problems of timing and efficiency, in
> which small numbers of molecules and their diffusion are
> involved, by reducing the dimensionality in which diffus-
> ion takes place from three-dimensional space to two-
> dimensional surface diffusion."

The necessary conditions for such reactions occurring at micellar surfaces have been summarized (41): the reacting species, in this case radical ions, must be specifically adsorbed at the micellar surface, and they must be more strongly adsorbed than other counterions present in the solution. Ions which are likely to meet these conditions are those which are physically large with a low charge density.

Two reactions which are mechanistically identical have been shown to exhibit the effects predicted by the reduction of dimensionality. These are the dismutations of Ag_2^+ and Br_2^- at anionic and cationic micelles, respectively. Since the observed effects of micelles on these reactions differ slightly, they will be treated to separate discussions.

Dismutation of Br_2^- in homogeneous solution (42, 43) follows the sequence

$$2Br_2^- \rightarrow Br_2 + 2Br^- \qquad\qquad k = 1.8 \times 10^9 \ M^{-1}s^{-1} \qquad\qquad (2)$$

$$Br_2 + Br^- \rightleftharpoons Br_3^- \qquad\qquad K = 0.062 \ M^{-1} \qquad\qquad (3)$$

In a solution of the cationic surfactant cetyltrimethylammonium bromide (CTAB), two well-separated components of Br_2^- decay were observed: a fast first order decay followed by a slower second order process. Formation of Br_3^- followed the same kinetics. Examination of the statistics of Br_2^- population in micelles showed that a two-dimensional reaction between two of the radicals situated at a single micelle could account for the rapid first order process (42). The slower process was measured as a function of surfactant concentration and a rate constant for k_-, the exit rate of Br_2^- from the micelle was estimated as $6.4 \times 10^4 \ s^{-1}$ by assuming that the second order process involved diffusion of Br_2^- between micelles (43). A further notable aspect of this system is the pronounced effect on the Br_2/Br_3^- equilibrium, which is shifted toward Br_3^-, presumeably due to the high local concentration of Br^- at the surface of the micelle where the Br_2 is formed in the dismutation step.

The di-silver cation Ag_2^+ is formed by reduction of Ag^+ in solution by e_{aq}^- (41).

$$e_{aq}^- + Ag^+ \rightarrow Ag^o \qquad\qquad k = 3.1 \times 10^{10} \ M^{-1}s^{-1} \qquad\qquad (4)$$

$$Ag^o + Ag^+ \rightarrow Ag_2^+ \qquad\qquad k = 5.9 \times 10^9 \ M^{-1}s^{-1} \qquad\qquad (5)$$

The dismutation of Ag_2^+

$$2Ag_2^+ \rightarrow Ag_2 + 2Ag^+ \qquad\qquad\qquad\qquad\qquad\qquad (6)$$

occurs with a rate constant $2k = 3 \times 10^8$ $M^{-1}s^{-1}$ in homogeneous solution.

In the presence of anionic micelles, the decay of Ag_2^+ proceeds via two temporally separated steps. The decay rate for the faster process is much greater than the rate for a comparable concentration of Ag_2^+ in homogeneous solution, and this kinetic effect can be explained by invoking the reaction of two Ag_2^+, both adsorbed at the surface of the same micelle. The slower decay is the three-dimensional process, but still involves micelles. Observation of the effect of surfactant concentration showed that reaction of two free Ag_2^+ in solution is unimportant compared to the reaction of two Ag_2^+ adsorbed at different micelles, which occurs according to the scheme

$$Ag_2^+{}_{(mic\ 1)} + Ag_2^+{}_{(mic\ 2)} \rightarrow 2Ag_2^+{}_{(mic\ 1)} + mic\ 2 \qquad (7)$$

$$2Ag_2^+{}_{(mic\ 1)} \rightarrow Ag_2 + 2Ag^+ + mic\ 1 \qquad (8)$$

The exchange of Ag_2^+ between the anionic micelles appears to be more efficient for the sodium dodecyl sulfate system than for a polyethyleneoxide surfactant with a sulfate head group. This implies stronger binding of Ag_2^+ to the latter micelles.

The considerable acceleration of these dismutation reactions demonstrates that micelles can have catalytic effects on certain reactions occurring in the interfacial region. In addition to the introduction of a two-dimensional reaction which is completed in times of the order of a microsecond, the three-dimensional reaction may also be affected. For both Ag_2^+ and Br_2^-, the three-dimensional (slow second order) reaction has a lower rate constant in the presence of micelles. This is a reflection of the repulsion of the like-charged micelles which prevents reaction of two radicals in different micelles without prior transfer of one radical to the other radical-containing micelle.

A theoretical treatment of two-dimensional reactions at micellar surfaces has appeared (44).

Electron Transfer Reactions

The effects of charged micelles on the reaction of micellized and water-soluble substrates with e^-_{aq} have already been discussed. The most immediate effect of charged micelles on other electron transfer reactions between molecular species is likely to be due to the Coulombic attraction or repulsion of charged species diffusing in the aqueous phase. Further, the effect of high electric potential close to the charged interface may affect the thermodynamic parameters of redox equilibria. In addition, pulse radiolysis investigations can be motivated by several other considerations: electron transfer reactions can be measured between species which could not be jointly solubilized in water; and the possibility exists to use micellar or vesicular interfaces to inhibit back reactions (45). In addition, the compartmentalization resulting from such aggregates can be exploited for electrical energy storage (46, 47).

Analogous to the capture of e_{aq}^- by water pools in reverse micelles (30), it has also been shown that the radical anion of biphenyl diffuses to the water pools where it can subsequently reduce ions such as Cu^{++} and pyrenesulfonic acid (44). The rate constants for such reactions can be influenced by changing the water content of the micro-emulsions.

Electron transfer reactions involving pyrene excited states have been measured by exciting micellized pyrene to its triplet state (by photoexcitation) simultaneously with the application of an electron pulse to produce radicals in the aqueous phase (49). Since the triplet state has both a lower ionization potential and a higher electron affinity (by the amount of the triplet state energy) than those of the ground state, the triplet state can be both a better reducing and oxidizing agent than the ground state molecule. Electron transfer from triplet pyrene in CTAB (cationic micelles to Br_2^- in the aqueous phase has been measured (49). It has also been shown that triplet pyrene in CTAB can be reduced by CO_2^- and $\dot{C}H_2O^-$ radicals; ground state pyrene is not reduced by these radicals under the same conditions (50). In fact, the pyrene anion reduces both CO_2 and CO_2^-. While CH_3CHO^- reduces both triplet and ground state pyrene, the rate constant is fifty times greater for the triplet state.

Metal cations can be intermediaries in the reduction of $Ru(bpy)_3^{++}$ in SDS micelles by the hydrated electron (23). Under the proper conditions of concentration the divalent ions Zn^{++}, Cd^{++}, and Co^{++} are reduced to their monovalent forms by e_{aq}^-; the reduced metal ions are then electrostatically attracted to the anionic micelles where the ions reduce the ruthenium complex. The rate constant for the latter reduction is enhanced relative to its value in the absence of micelles.

Aromatic radical anions contained in vesicle bilayers can reduce ionic substrates in the aqueous phase (51). When the anion of diphenylhexatriene was produced in vesicles with positively charged head groups, Fe(III)EDTA$^-$ in the aqueous phase was very efficiently reduced by the anion, whereas no reduction of the Eu(III) cation could be detected. The radical anion of carotene produced in lecithin vesicles was able to reduce Eu^{3+}.

The electron transfer equilibrium

$$AQS^{\overline{\cdot}} + DQ \;\; \underset{k_r}{\overset{k_f}{\rightleftharpoons}} \;\; AQS + DQ^{\overline{\cdot}} \qquad\qquad (9)$$

(AQS: 9,10-anthraquinone-2-sulfonate, DQ: duroquinone)

was investigated in solutions of SDS micelles subsequent to reduction of the quinones by e_{aq}^- and $CH_3\dot{C}(CH_3)OH$ (52). The measured
equilibrium constant was an order of magnitude lower than that
reported for aqueous solution (without micelles) (53). By assuming that the micellized DQ did not participate in the electron
transfer equilibrium, this effect can be accounted for. However,
this explanation is inconsistent with the observation that k_r was
increased substantially, while k_f was only slightly reduced. The
conclusion was drawn that $DQ^{\overline{\cdot}}$ is also associated with micelles,
and that this solubilized species is a stronger reducing agent
than its counterpart in the aqueous phase.

Effects of anionic micelles were also observed on the position of the equilibrium

$$BV^{\overline{\cdot}+} + DQ \;\; \underset{k_r}{\overset{k_f}{\rightleftharpoons}} \;\; BV^{++} + DQ^{\overline{\cdot}} \qquad\qquad (10)$$

(BV: benzylviologen, DQ: duroquinone)

The equilibrium constant was lower by a factor of 3-4 for SDS
micelles than for water (54), and a decrease in k_f accounted for
most of this effect (55). In this case, BV^{++} is preferentially
adsorbed at the micellar surfaces. Association of DQ with the
micelles could account for the decreased k_f if only DQ in the
aqueous phase is reduced, leading to the conclusion that kinetic,
not energetic parameters are affected in this case.

Electron transfer to molecular oxygen from reduced dimethyl-,
dibenzyl- and methyldodecyl-viologens has been investigated for
solutions containing SDS and CTAB micelles (56, 57). The rate
constant was significantly lower in SDS micelles than in water in
each case. In addition, the rate constant for methyldodecyl
viologen was decreased in CTAB micelles relative to that in water.
These results are consistent with the reaction proceeding via an
intermediate encounter complex, the breakup of which controls the
overall rate.

$$V^{\overline{\cdot}+} + O_2 \rightleftharpoons (V^+O_2)^{\cdot} \rightarrow V^{++} + O_2^{\overline{\cdot}} \qquad\qquad (11)$$

This ion-pair forming reaction is strongly dependent on the dielectric constant of the medium as has already been shown in work

involving polar organic solvents (58). The measured rate constant provides a means of probing the dielectric nature of micellar surface regions. The effects of reduced dielectric constant and high electrical potential at sites where electron transfer reactions occur in biological systems may be far more important in determining equilibrium properties than has yet been shown.

Binding of Proteins to Vesicle Bilayers

Pulse radiolysis has been used for a number of years to obtain information about binding between biological macromolecules (proteins, DNA) and various substrates (e.g. drugs) (59-61). The basic technique was first developed to study temperature-induced configurational changes of proteins (62) and relies on the reactivity of specific sites toward radiolytically produced species such as e_{aq}^-. When these sites are obscured by a bound substrate or by being buried in the protein away from aqueous regions, their reactivity is sharply depressed. This principle was applied in the studies of erythrocyte membranes mentioned above (31, 39) and has also been more recently utilized as a tool for studying interactions of proteins with vesicle bilayers (63, 64). Oxidation of tryptophan, tyrosine, and methionine residues in the membrane binding segment of cytochrome b_5 by Br_2^- and $(SCN)_2^-$ was completely suppressed when the protein was incorporated into dipalmitoyl-phosphatidylcholine vesicles (64). It was concluded that the residue containing these proteins is buried in the lipid interior of the membrane. Cytochrome c reacts rapidly with the hydrated electron by reduction of the haeme prosthetic group. This reactivity is decreased when the protein is associated with vesicles composed of phospholipids with acidic head groups (phosphatidyl-serine, phosphatidylinositol) (63). By using the reduced reactivity toward e_{aq}^- as a test of protein binding to the bilayer, it was possible to identify ε-amino groups of lysine as being involved in the binding. These examples demonstrate the utility of pulse radiolysis in discovering protein-membrane interactions.

Epilogue

The foregoing discussion has illustrated some of the ways in which pulse radiolytically generated species can be used to probe the organization and dynamics of aggregated assemblies, as well as to initiate novel reactions in interfacial regions. Aggregated assemblies such as micelles and vesicles have been investigated much more extensively by laser photolysis techniques than by pulse radiolysis, but the latter technique has a considerable unexploited potential in this area.

References

1. Jain, M. K., and Wagner, R.C.: 1980, "Introduction to Biological Membranes", Wiley-Interscience, New York.
2. Tanford, C.: 1980, "The Hydrophobic Effect: Formation of Micelles and Biological Membranes", Wiley-Interscience, New York.
3. Quinn, P. J.: 1976, "The Molecular Biology of Cell Membranes", University Park Press, Baltimore.
4. Singer, S. J., and Nicolson, G. L.: 1972, Science 175, p. 720.
5. Hartley, G.S.: 1935, Trans. Faraday Soc. 31, p. 31; Hartley, G.S.: 1948, Quart. Rev. Chem. Soc. 2, p. 152.
6. Fendler, J. H., and Fendler, E. J.: 1975, "Catalysis in Micellar and Macromolecular Systems", Academic Press, New York.
7. Menger, F. M.: 1979, Acc. Chem. Res. 12, p. 111.
8. Dill, K. A., and Flory, P. J.: 1981, Proc. Natl. Acad. Sci. USA 78, p. 676.
9. Fromherz, P.: 1981, Chem. Phys. Lett. 77, p. 460.
10. Cordes, E. H., and Dunlap, R. B.: 1969, Acc. Chem. Res. 2, p. 329.
11. Bunton, C. A.: 1979, Catal. Rev. Sci. Eng. 20, p. 1.
12. Escabi-Perez, J. R., and Fendler, J. H.: 1978, J. Am. Chem. Soc. 100, p. 2234.
13. Regen, S. L., Singh, A., Oehme, G., and Singh, M.: 1981, Biochem. Biophys. Res. Commun. 101, p. 131.
14. Gros, L., Ringsdorf, H., and Schupp, H.: 1981, Angew. Chem. Int. Ed. Engl. 20, p. 305.
15. Fendler, J. H., and Patterson, L. K.: 1970, J. Phys. Chem. 74, p. 4608.
16. Chen, M., Grätzel, M., and Thomas, J. K.: 1975, J. Am. Chem. Soc. 97, p. 2052.
17. Grätzel, M., Thomas, J. K., and Patterson, L. K.: 1974, Chem. Phys. Lett. 29, p. 393.
18. Frank, A. J., Grätzel, M., Henglein, A. and Janata, E.: 1976, Ber. Bunsenges. Phys. Chem. 80, p. 547.
19. Evers, E. L., Jayson, G. G., Robb, I. D., and Swallow, A. J.: 1980, J. Chem. Soc., Faraday Trans. I 76, p. 528.
20. Wallace, S. C., and Thomas, J. K.: 1973, Radiat. Res. 54, p. 49.
21. Grieser, F., and Tausch-Treml, R.: 1980, J. Am. Chem. Soc. 102, p. 7258.
22. Grätzel, M., and Thomas, J. K.; 1974, J. Phys. Chem. 78, p. 2248.
23. Meisel, D., Matheson, M. S., and Rabani, J.: 1978, J. Am. Chem. Soc. 100, p. 117.
24. Rodgers, M. A. J., Foyt, D. C., and Zimek, Z. A.: 1978, Radiat. Res. 75, p. 296.
25. Rodgers, M. A. J.: unpublished results.
26. Proske, Th., Fischer, Ch.-H., Grätzel, M., and Henglein, A.: 1977, Ber. Bunsenges. Phys. Chem. 81, p. 816.

27. Henglein, A., Proske, Th., and Schnecke, W.: 1978, Ber. Bunsenges. Phys. Chem. 82, p. 956.
28. Grätzel, M., Henglein, A., and Janata, E.: 1975, Ber. Bunsenges. Phys. Chem. 79, p. 475.
29. Wong, M., Grätzel, M., and Thomas, J. K.: 1975, Chem. Phys. Lett. 30, p. 329.
30. Bakale, G., Beck, G., and Thomas, J. K.: 1981, J. Phys. Chem. 85, p. 1062.
31. Bisby, R. H., Cundall, R. B., and Wardman, P.: 1975, Biochim. Biophys. Acta 389, p. 137.
32. Patterson, L. K., Bansal, K. M., and Fendler, J. H.: 1971, Chem. Commun., p. 152.
33. Henglein, A., and Proske, Th.: 1978, J. Am. Chem. Soc. 100, p. 3706.
34. Henglein, A., and Proske, Th.: 1978, Makromol. Chem. 179, p. 2279.
35. Almgren, M., Grieser, F., and Thomas, J. K.: 1979, J. Chem. Soc., Faraday Trans. I 75, p. 1674.
36. Patterson, L. K., and Hasegawa, K.: 1978, Ber. Bunsenges. Phys. Chem. 82, p. 951.
37. Trifunac, A. D., Nelson, D. J. and Mottley, C.: 1978, J. Magn. Reson. 30, p. 263.
38. Barber, D. J. W., and Thomas, J. K.: 1978, Radiat. Res. 74, p. 51.
39. Bisby, R. H., Price, M. R., Cundall, R. B., and Wardman, P.: 1975, Int. J. Radiat Biol. 28, p. 267.
40. Adam, G., and Delbrück, M.: 1968, in "Structural Chemistry and Molecular Biology", A. Rich, and N Davidson, eds., W. H. Freeman, & Co., San Francisco, p. 198.
41. Henglein, A., and Proske, Th.: 1978, Ber. Bunsenges. Phys. Chem. 82, p. 471.
42. Frank, A. J., Grätzel, M., and Kozak, J. J.: 1976, J. Am. Chem. Soc. 98, p. 3317.
43. Proske, Th., and Henglein, A.: 1978, Ber. Bunsenges. Phys. Chem. 82, p. 711.
44. Hatlee, M. D., Kozak, J. J., Rothenburger, G., Infelta, P. P., and Grätzel, M.: 1980, J. Phys. Chem. 84, p. 1508.
45. Infelta, P. P., Grätzel, M., and Fendler, J. H.: 1980, J. Am. Chem. Soc. 102, p. 1479.
46. Willner, J., Ford, W. E., Otvos, J. W., and Calvin, M.: 1979, Nature 280, p. 823.
47. Gould, J. M., Patterson, L. K., Ling, E., and Winget, G. D.: 1979, Nature 280, p. 607.
48. Wong, M., and Thomas, J. K.: 1977, in "Micellization, Solubilization and Microemulsions", K. L. Mittal, ed., Plenum Press, New York, p. 647.
49. Frank, A. J., Grätzel, M., Henglein, A., and Janata, E.: 1976, Int. J. Chem. Kinetics 8, p. 817.
50. Frank, A. J., Grätzel, M., Henglein, A., and Janata, E.: 1976, Ber. Bunsenges. Phys. Chem. 80, p. 294.

51. Almgren, M., and Thomas, J. K.: 1980, Photochem. Photobiol. 31, p. 329.
52. Almgren, M., Grieser, F., and Thomas, J. K.: 1979, J. Phys. Chem. 83, p. 3232.
53. Meisel, D., and Neta, P.: 1975, J. Am. Chem. Soc. 97, p. 5198.
54. Wardman, P., and Clarke, E. D.: 1976, J. Chem. Soc., Faraday Trans. I 72, p. 1377.
55. Lynch, J., and Rodgers, M. A. J.: unpublished results.
56. Rodgers, M. A. J.: in press, in "Oxygen and Oxy-Radicals in Chemistry and Biology", M. A. J. Rodgers, and E. L. Powers, eds., Academic Press, New York.
57. Rodgers, M. A. J.: unpublished results.
58. Patterson, L. K., Small, R. D., and Scaiano, J. C.: 1977, Radiat. Res. 72, p. 218.
59. Phillips, G. O., Power, D. M., Robinson, C., and Davies, J. V.: 1973, Biochim. Biophys. Acta 295, p. 8.
60. Greenstock, C. L., and Ruddock, G. W.: 1975, Biochim. Biophys. Acta 383, p. 464.
61. Beaumont, P. C., Land, E. J., Navaratnam, S., Parsons, B. J., and Phillips, G. O.: 1980, Biochim. Biophys. Acta 608, p. 182.
62. Braams, B., and Ebert, M.: 1967, Int. J. Radiat. Biol. 13, p. 195.
63. Wainwright, P., Power, D. M., Thomas, E. W., and Davies, J. V.: 1978, Int. J. Radiat. Biol. 33, p. 151.
64. Pochon, F., Favaudon, V., Ferradini, C., and Pucheault, J.: 1981, Int. J. Radiat. Biol. 39, p. 207.

TRANSIENTS IN LOW TEMPERATURE ORGANIC GLASSES

Fernand Kieffer

Laboratoire de Physicochimie des Rayonnements,
associe au C.N.R.S.,
Universite de Paris-Sud, 91405 Orsay, France.

Abstract. Presolvated and solvated electrons in different organic glasses, and possible mechanisms for trap deepening are discussed. Their optical absorption spectra correspond to transitions from a bound ground state to both a bound excited state and to a conduction state. Reaction of e_t with scavenger molecules and with correlated cations seems to occur by long-distance tunnelling. Recent work on short-lived alkane cations holds a promise for future successful approaches to the initial positive species.

1. DIFFERENT TYPES OF ORGANIC GLASSES.

Observations on trapped transients in organic glasses are analogous to those made in aqueous glasses, but they are more diversified because the main matrix constituent, responsible for the absorption of the major part of radiation energy, is itself varied. Whereas in all aqueous glasses, basically the same primary products are formed, converting to the same observable species, only the initially ejected "dry" electron is common to the different organic glasses. The primary positive ions are those formed by the ejection of these electrons and are hence different from one glass to another. Similarly, trapped radicals are different in different glasses.

While aqueous glasses are all polar or very polar, the range of polarities in organic glasses covers a broader scope, from the polar methanol through less polar alcohols and ethers, to the practically non-polar hydrocarbons. These polar characteristics have a considerable influence on the properties of trapped electrons.

J. H. Baxendale and F. Busi (eds.),
The Study of Fast Processes and Transient Species by Electron Pulse Radiolysis, 573–600.
Copyright © 1982 by D. Reidel Publishing Company.

A further difference resides in the fact that, at 77 K, a very convenient working temperature, all aqueous glasses are very hard, highly viscous glasses, whereas organic glasses cover a viscosity range from similarly high values (e.g. methanol + 5% H_2O : 1.1 x 10^{25} poise) down to very low values (e.g. the very widely studied 3-methylpentane - "3 MP" - which at 77 K has a viscosity of 2.2 x 10^{12} poise, i.e. slightly less than corresponds to the usual definition of a glass : 10^{13} P). No wonder that some differences in behaviour of trapped species appear between different glasses, and that contradictory results are sometimes obtained with a glass such as 3 MP.

Furthermore , whereas most organic matrices can trap electrons, in some (e.g. alkyl halides) the electrons become attached to matrix molecules. In the case of alkyl halides, electron attachment is dissociative :

$$RX + e^- \longrightarrow R^\cdot + X^- \tag{1}$$

This process is so efficient that it has been used to obtain the absorption spectra of aromatic positive ions. In alkane glasses containing an aromatic solute, both electrons and positive charges are scavenged, producing radical-anions and -cations of the solute. In alkyl halides, all electrons undergo reaction (1) which is favoured both by the high electron affinity of the halogen atom and by the predominance of RX molecules, so that aromatic cations only are formed.

Radicals are formed in all organic glasses, generally with a yield of around 3 per 100 eV. With the exception of $H^\cdot$ and $CH_3^\cdot$ they are very stable at 77 K and their concentration goes on increasing linearly with dose up to tens of Mrad while the concentration of e_t^- passes through a maximum and then falls off to near zero. The dose corresponding to maximum e_t^- concentration varies for different glasses. It is about 1.5 Mrad for 3 MP and 6 Mrad for ethanol(1). The subsequent fall-off as dose increases must be attributed to an increasingly important capture by radicals and reaction with positive species (both e_{mob}^- + $hole_t^+$ and e_t^- + $hole_{mob}^+$, and on a longer time scale e_t^- + $hole_t^+$).

When the temperature of the glass is raised, both e_t^- and the radicals $R^\cdot$ disappear close to T_g, e_t^- disappearing at a somewhat lower temperature than $R^\cdot$, as shown in Fig.1 for 2-methyltetrahydrofuran (MTHF).

Whereas trapping and solvation of the electron are now fairly well known, the fate of the positive charge is still somewhat obscure, especially in alkane glasses. In alcohols and ethers, the positive charge is stabilised as the protonated matrix molecule, and the ionisation event can be summarised as follows :

$$CH_3OH \longrightarrow CH_3OH^+ + e^- \longrightarrow e_t^-$$
$$CH_3OH \searrow CH_3OH_2^+ + \cdot CH_2OH$$

This is probably the reason for higher e_t^- yields observed in al-

cohols and ethers ($G \approx 1.5$ to 2.5) as compared with the less polar alkanes, alkenes and amines ($G \approx 0.5$ to 1.0). Also in the former, as Kevan[3] pointed out, G (e_t^-) values measured by EPR and by the scavenger method are generally similar, in the latter G (e_t^-)$_{scav.}$ is about two or more times higher than (e_t^-)$_{EPR}$. This suggests that in the less polar glasses more electrons are lost through recombination with cations before these are stabilised. The effect of hole scavengers confirms this, for the presence of a few mole % of triethylamine or 2-methylpentene-1 in 3 MP doubles the value of G (e_t^-)[4].

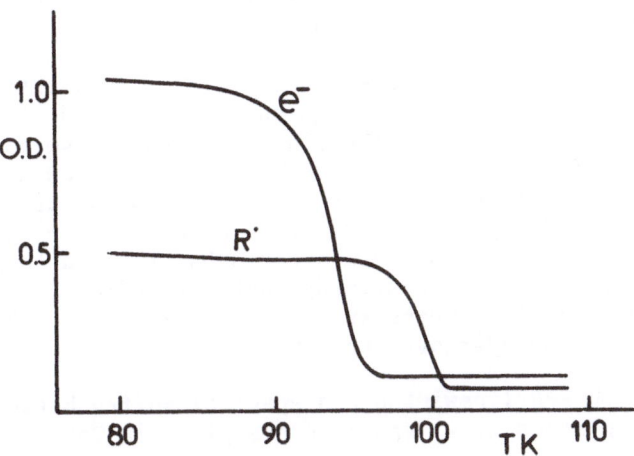

Figure 1. Decay of the absorption bands of trapped electrons and radicals during warm-up of irradiated MTHF. The e_t^- band was monitored at 11000 cm^{-1}, the R$^\cdot$ band at 40000 cm^{-1}. (Dainton and Salmon[2]).

2. PROPERTIES OF TRAPPED ELECTRONS IN ORGANIC GLASSES.

2.1 The trapping process.

Two methods are currently employed for producing trapped electrons : γ -irradiation and photoionisation of a solute with a low ionisation potential, such as tetramethyl paraphenylene diamine (TMPD), triphenylamine, indole or tryptophan. The choice between these photoionisable solutes is mainly imposed by solubility requirements. TMPD and tryptophan have been used most commonly. The mechanism of photoionisation is biphotonic, the second photon being absorbed by the molecule in its triplet state.

Since it was shown that γ -irradiation of glasses at 4 K or the application of pulse techniques at 77 K leads to the formation of presolvated electrons, these two methods have been employed to study the transformation of EPR and optical absorption spectra. After irradiation at 4 K, either a slow transformation over minutes or hours, or the transformation upon warming to 77 K, are observed. With the pulse technique, the transformation occurs spontaneously at 77 K on a time scale of nanoseconds up to seconds.

Fig. 2 and 3 due to Higashimura et al.(5,6) show extreme examples of differences in absorption spectra obtained at 4 and 77 K respectively : in ethanol the blue shift due to solvation is from 1500 to 540 nm, in 3 MP from above 2200 to 1700 nm. This means an energy difference of about 1.5 eV in ethanol and 0.2 eV in 3 MP. Upon pulse radiolysis of 3 MP at 76 K, Klassen et al.(7) observed a shift that was complete after 380s. Their data are shown in Fig. 4. It is not surprising that Willard(8) found no change between 5 and 45 minutes after γ -irradiation at 77 K, although 50% of e_t^- decayed over that period.

Some data on EPR linewidths and spectral shifts between γ -irradiated glasses at 4 and 77 K are given in Table I.

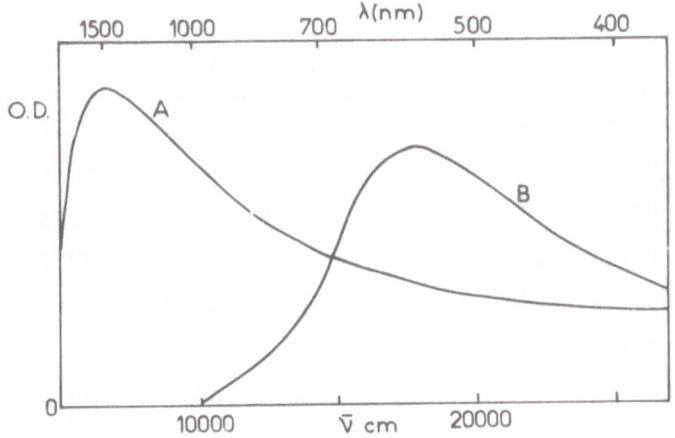

Figure 2. Absorption spectra of e_t^- in ethanol glass after irradiation at 4 K. Spectrum A was obtained at 4 K, spectrum B after rapid warming to 77 K (Hase et al. (5)).

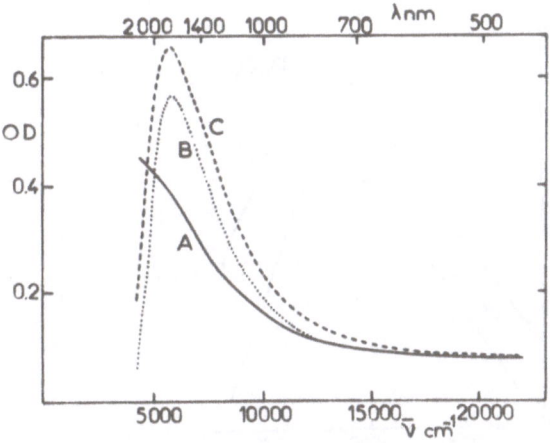

Figure 3. Absorption spectra of e_t^- in 3 MP glass. A : irradiation and measurement at 4 K; B : irradiation at 4 K, measurement after rapid warming to 73 K; C : irradiation and measurement at 73 K. (Hase et al. (6)).

Table I : EPR linewidths and optical absorption maxima of e_t^- in some glasses at 4 and 77 K (mainly from(3)).

Matrix	ΔH (G)		λ_{max} (nm)	
	4 K	77 K	4 K	77 k
Methanol (5% H_2O)			605	530
C_2H_5OH	6.7	14	1500	540
C_2H_5OD	5.6	5.6	1500[++]	540
sec-Butylamine			1220	1130
Diisopropylamine			1600	1480
Triethylamine			1680	1680
MTHF	5.6	4.4[+]	1305	1150
3 MP	4.9	3.7	>2200	1700

+ The inverse change in ΔH with respect to alcohols merely reflects the opposite signs of C-H and O-H bond dipoles.
++ This value found by Perkey et al.(9) after very careful checking of surprising data by Hase et al.(10).

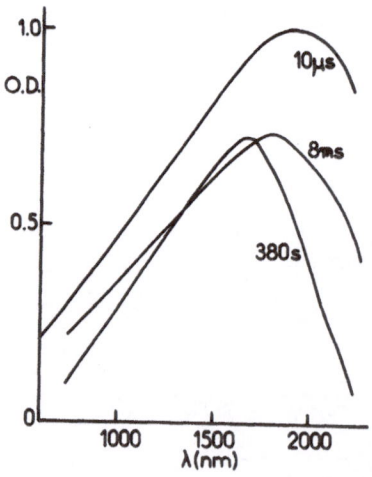

Figure 4. Shift with time of the e_t^- absorption spectrum in 3 MP after a 1 µs pulse at 76 K (Klassen et al.(7)).

Kevan(3) has summarised some pulse radiolysis data obtained at 77 K and compared them with viscosities at 77 K and static dielectric constants at 298 K. The latter may not be very significant since all D_s values are considerably decreased and not very different from each other at 77 K. Still the time to reach stable spectra tends to increase with decreasing D_s and not with increasing viscosity. Hence matrix polarity seems to favour solvation. This speaks in favour of a molecular reorientation mechanism, resulting from charge dipole interaction and polarisation interaction with surrounding molecular dipoles (or local C-H bond dipoles in alkanes). Some relevant data are summarised in Table II.

Table II : Pulse radiolysis data on solvation of presolvated electrons at 77 K (mainly from (3)).

Matrix	λ_{max}(nm) <0.2 µs > 7 µs		$t_{st.sp}$ (µs)	η(P)	D_s(298K)
Methanol (5% H_2O)	590	560	<3	1.1×10^{25}	36
Ethanol	>1300	590	7	2.1×10^{19}	24.3
2-Propanol	>850	800	100	10^{22}	18.3
MTHF	1280	1200	10	3.7×10^{20}	6.4
3 MP	1900	1700	10^6	2.2×10^{12}	2.0

The molecular reorientation mechanism for trap deepening also seems to be favoured by the effect of added scavengers such as biphenyl or benzyl chloride. If these solutes are added in concentrations sufficient to scavenge about half the electrons, absorption bands of the biphenyl anion or the benzyl radical are found in the spectrum in addition to the residual e_t^- absorption. In 2-propanol, Kevan(3) found that the intensities of these additional bands were constant through the time range of the e_t^- blue shift. Klassen et al (7) obtained similar though less clear-cut results with hydrocarbon glasses. If the mechanism underlying the blue shift were a displacement of the electron from its shallow trap to a preexisting deeper one - by thermally activated jumps or by tunnelling - one would expect some of these electrons to be captured by scavenger molecules, and the decay of the IR band should be accompanied not only by an increase of the visible band, but also by an increase of the biphenyl anion or benzyl radical bands.

Since the latter was not observed, the problem seemed to be solved in favour of the molecular reorientation mechanism.

But Baxendale and Sharpe[11] re-examined the effect of added benzyl chloride in alcohols and found that, whereas the IR decay was about the same as in the absence of benzyl chloride, there was very little growth in the visible, and only very slight growth of benzyl radical absorption. They concluded that capture by the solute must occur, but does not lead to the formation of benzyl radicals. In this case molecular reorientation could not be responsible for the blue shift, and a mobile electron mechanism seemed to impose itself. A temperature dependence appeared in the time scale of the change : at 93 K the IR absorption in 1-propanol persists for only 0.2 s, compared with 10 s at 77 K. On these grounds the authors excluded tunnelling to deeper traps and concluded that thermal excitation from shallow traps and recapture in deeper ones, i.e. "trap-hopping" is the mechanism responsible. However, it must be said that a temperature effect between 77 and 93 K is not in the least surprising : T_g of 1-propanol is 94 K and it is well known[12] that, in the neighbourhood of phase transitions, including glass transitions, molecular mobility increases considerably (with an activation energy !) so that trapping structures can be dismantled, resulting in liberation of the trapped electron. Hence B and S's argument in favour of hopping should be rejected.

Miller[13] rejects their argument altogether and replots their data - without explaining how - to show that the presence of benzyl chloride does not interfere with the normal spectral shift. He concludes in favour of the electrons "digging" deeper traps by reorienting solvent molecules, their only possibility of moving, after localisation, being direct tunnelling to an electron acceptor.

Yet another interpretation of the blue shift in alcohols has been proposed recently by Ogasawara et al.[14,15]. Through its analogy with aqueous glasses, this interpretation seems rather attractive. The authors irradiated alcohols at 4 K and then studied the evolution of the absorption spectra at 77 K. They found that the IR part of the spectrum decayed faster than the visible band grew in on top of quite an important tail. They concluded that both types of trapped electrons, e_{IR}^- and e_{vis}^-, were present at 4 K, that e_{IR}^- decayed independently without transforming into e_{vis}^-, and that the e_{vis}^- band underwent an increase in intensity due to narrowing, and a slight blue shift ($\approx 1000 cm^{-1}$). Experiments with an inefficient scavenger, toluene, added, seem to confirm this interpretation : with increasing concentrations of toluene, the 1500 nm maximum of ethanol observed at 4 K decreases and λ_{max} is displaced progressively to 600 nm ; warming to 77 K then shifts it to 520 nm and increases its intensity. Toluene scavenges e_{IR}^- about ten times

more efficiently than e_{vis}^-. The observations on ethanol + toluene
are quite comparable to what happens in methanol glass without
scavenger. It would seem that in methanol no e_{IR}^- is formed originally at 4 K, perhaps because e_{IR}^- is associated with the alkyl part
of the molecule and therefore appears only in the higher alcohols
where this part is more important relative to the hydroxyl group
with which e_{vis}^- could be associated.

2.2. Structure of trapped electrons.

Electronic absorption spectra of electrons trapped at 77 K
in many glassy matrices are known. Shida et al.(16) recorded such
spectra for over 40 glass-forming substances belonging to different chemical families and compared them with theoretical spectra.
Extinction coefficients are available in a few glasses only ;
they are about 10^4 cm^2/mole.

The spectra are generally broad and structureless, and tail
off very gradually on the short wavelength side. The asymmetry
of the spectra is most evident when they are plotted on a linear
energy scale. What are the transitions involved in these spectra ? Obviously they are not exclusively transitions from a bound
ground state to a bound excited state. To explain photobleaching,
photostimulated luminescence, and photoconductivity, it is necessary to consider also transitions to a conduction state.

Wiseall and Willard(17) found that in 3MP glass photoconductivity could be induced by infra-red illumination, with a maximum
efficiency at 1375 nm. If the sample contained enough biphenyl to
prevent the formation of e_t^- (biphenyl scavenges electrons and
forms a radical-anion absorbing at 410 and 650 nm), illumination
at 1375 nm had no effect, but at 410 nm conductivity was observed.

Baverstock and Dyne(18) obtained a photoconductivity excitation spectrum which was considerably displaced to the high energy
side of the e_t absorption spectrum in MTHF glass. As Fig.5 shows,
the threshold lies in the high energy tail (800 nm) of the spectrum which has its maximum at 1150 nm. Their data on photobleaching efficiency of e_t are situated on the same curve. The maximum
is about 600 nm.

In these two extreme cases, the difference between the absorption maxima and the quantum efficiency maxima for promoting trapped electrons into a conduction state is about 0.2 eV for the hydrocarbon and 1.1 eV for the ether.

Kevan(3) has tabulated available data on photoionisation
thresholds and efficiency maxima, determined by three methods :
photoconductivity , photobleaching and photostimulated luminescence. Table III summarises some of these data.

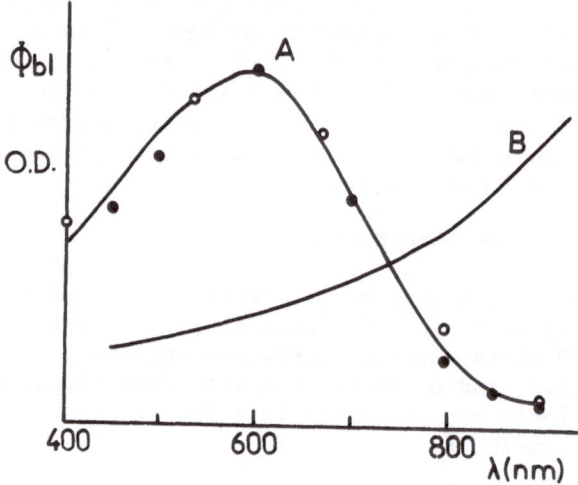

Figure 5. Quantum efficiency of bleaching (o) and photoconductivity (•) in the high energy tail of the e_t^-
absorption spectrum B in irradiated MTHF at 77 K
(Baverstock and Dyne(18)).

Table III. Comparison of optical absorption maxima(Abs.),
photoionisation thresholds(P.T.) and photoionisation efficiency
maxima(E.M.) for trapped electrons at 77 K (all data are in eV).

Matrix	Abs.	P.T.	E.M.	E.M. -P.T.	E.M. -Abs.
Methanol(5% H_2O)	2.35	2.3	>3.16	>0.9	>0.8
Ethanol	2.32	2.3	>3.10	>0.8	>0.8
MTHF	1.03	1.6	2.2	0.6	1.2
Sec-Butylamine	1.10	1.2	>1.54	>0.4	>0.5
Diiosopropylamine	0.84	1.0	1.7	0.7	0.9
Triethylamine	0.74	0.9	1.5	0.6	0.8
3 MP	0.73	0.75	1.0	0.25	0.3

It must be stressed that the data for photoionisation efficiency maxima, and even more so those for thresholds, are only approximate. Therefore the data for subtracted values need to be taken with some caution. Nevertheless a few observations can be made. In alcohol, amine and alkane glasses the photoionisation threshold seems to correspond roughly to the absorption maximum. In MTHF, the ionisation threshold lies at a higher energy. In alkanes the energy difference between ionisation threshold and maximum efficiency is particularly small(0.25 eV), for amines and MTHF it is about 0.5 eV, and it is larger still for alcohols.

In all cases, the absorption spectrum corresponds to transitions from a bound ground state not only to a bound excited state, but also, to a greater or lesser extent according to the matrix, to the conduction state. It has been shown that this is a single-photon process, and at present it seems more probable that the transition is direct to the conduction state rather than to an autoionising state which then ionises. But this question is not definitely settled.

In order to understand how the electrons move across the matrix, drift mobilities have been measured. This consists in measuring the transit time of electrons, liberated by a light flash near the cathode, from the cathode to the anode, under an applied field. Because a majority of electrons disappear on the way by retrapping or reaction, the current pulse thus measured is very small (of the order of 10^{-10} A) although the electrodes are spaced only 0.25 - 0.5 mm apart. Drift mobility is calculated from the relationship

$$\mu = d^2 / Vt$$

where d is the spacing between electrodes, V the applied voltage and t the transit time.

Such measurements have been made by Maruyama and Funabashi[19] on 3MP and by Huang and Kevan[20] on MTHF between 4 and 77 K. Fig.6 shows the reciprocal temperature plot of μ for 3 MP[19]. MTHF results were strictly comparable . From both series of experiments it appears that the process has a very small activation energy (≈ 0.01 eV) between 35 and 77 K, and no activation energy between 35 and 4 K. In the low temperature region the mobility is about 0.02 cm^2/Vs, and above 35 K it increases to reach 0.1 at 77 K. These results are compatible with a hopping model[3] : charge transport by electronic exchange interactions is hindered by competing electron-phonon interactions. When these dominate, charge transport occurs by uncorrelated phonon-assisted lattice jumps, the transport is incoherent. At lower temperature the electron-phonon interaction becomes less important, and finally the transport is coherent, which is the same as band motion. The transition temperature between hopping and band transport is in the neighbourhood of the Debye temperature for these glasses, 35 K.

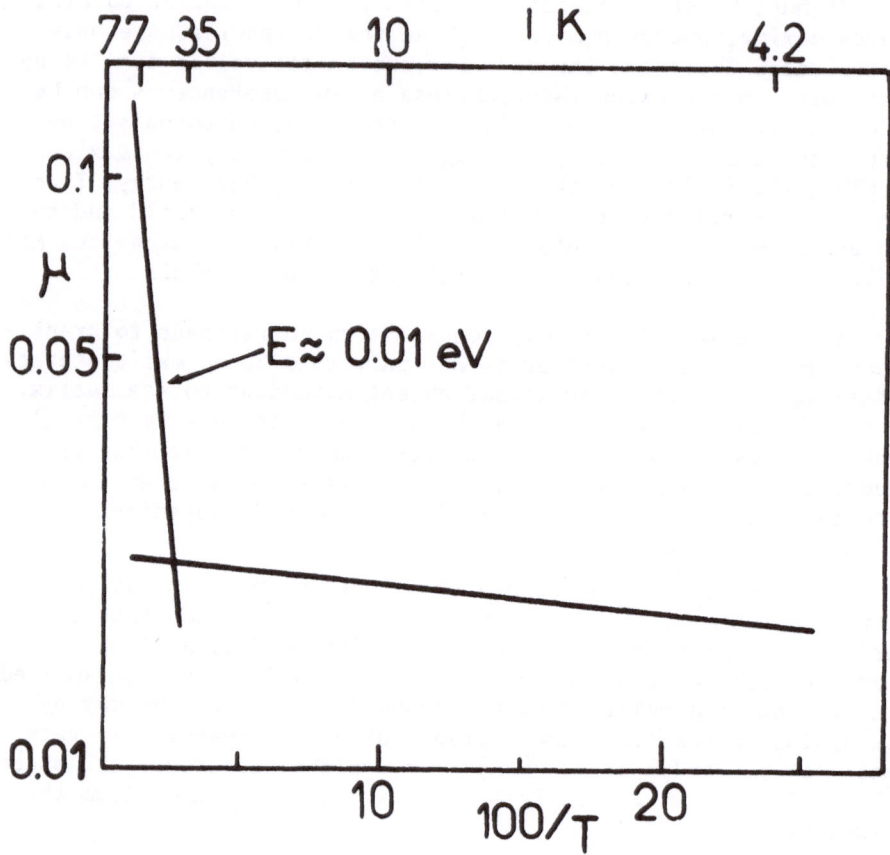

Figure 6. Temperature dependence of electron drift mobility in 3 MP glass. (Maruyama and Funabashi(19)).

The geometrical structures of electrons trapped in several organic matrices have been unravelled with a fair degree of accuracy by the use of very refined electron magnetic resonance techniques(21). Fig. 7 and 8 show the structures proposed for methanol and MTHF(21), with four, respectively three solvent molecules

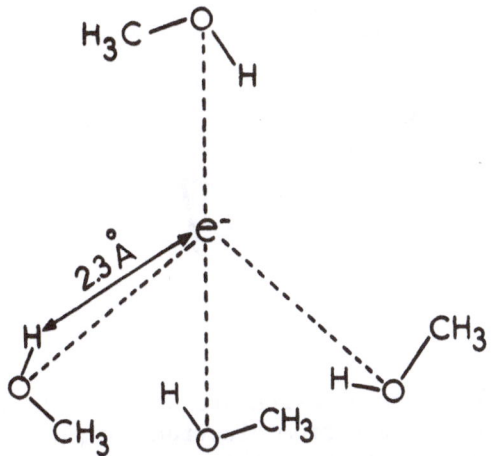

Figure 7. Geometrical structure of the solvated electron in methanol glass deduced from electron magnetic resonance data (Kevan(21)).

in the first solvation shell. For 3 MP, two to four solvent molecules seem to make up the first solvation shell. Assuming an average of three equivalent molecules, electron-proton group distances from 3.5 to 4.3 A are deduced. The electron interacts most strongly with the protons of a terminal methyl group, the second nearest being those of carbon atom 2 or of the methyl group on C_3, according to the orientation of the molecule.

3. REACTIONS OF TRAPPED ELECTRONS.

3.1. Reaction with scavengers.

In his pioneering work, Hamill observed how a number of different solutes, notably aromatics and halides, react with electrons formed by irradiation. Thus biphenyl in 3 MP which he studied in detail(22)(3) was found to compete with matrix trapping of electrons during irradiation, forming biphenyl radical-anions.

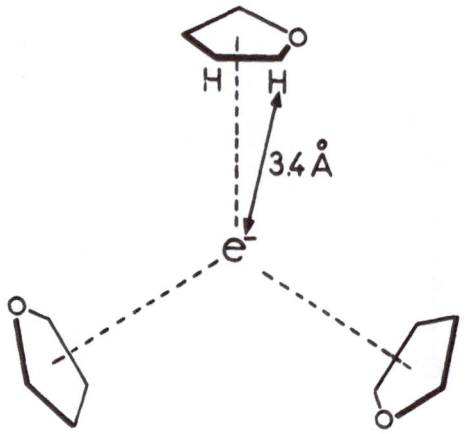

Figure 8. Geometrical structure of the solvated electron in MTHF deduced from electron spin-echo modulation
and other ESR data(Kevan(21)).

Fig. 9 shows how the initial optical density at 1600 nm (e_t^-) and
at 408 nm (Ph_2^-) varies with biphenyl concentration. Curve C was
obtained after photobleaching all solvent-trapped electrons and
shows that some of these at least are scavenged to form additional
biphenyl anion. The authors also found that this process occurs
spontaneously after irradiation, and that matrix-trapped electrons
decay faster in the presence of biphenyl than in pure 3MP, while
simultaneous growth of biphenyl anion occurs. Fig.10 reproduces
their data for pure 3 MP and a dilute solution of biphenyl, on a
reciprocal optical density versus time plot which they had shown
to be linear.

The same authors(22) found that it is possible to shift the
electrons back and forth between matrix traps and biphenyl molecules by appropriate optical bleaching, alternatively in one or the
other absorption band, showing that biphenyl can be considered as
a sort of chemical trap from which the electron can be released at
will by illumination in the anion absorption band.

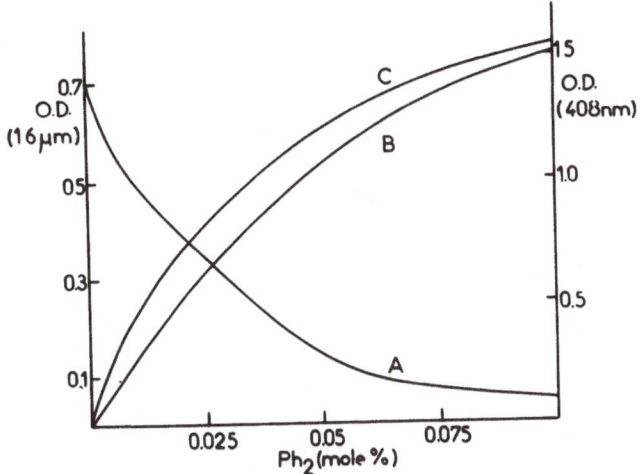

Figure 9. Yields of solvent-trapped electrons (A) and of biphenyl anions (B) with increasing biphenyl concentration in 3 MP at 77 K. (C) : biphenyl anions after full optical bleaching of e_t^- (Gallivan and Hamill(22)).

To account for the fact that 0.1 mole per cent ($\approx$ 0.01 M) biphenyl in 3 MP reduces the yield of e_t^- practically to zero (Fig.9), it is necessary to assume that mobile electrons encounter of the order of 1000 molecules and travel thousands of A before being trapped. According to theory, electrons only travel some 20 to 150 A from their parent molecules in water, and there is experimental evidence that in 3 MP most trapped electrons are close to positive ions. An elegant demonstration of this, based on polarised recombination luminescence, will be mentioned in (3.2.). Miller(23) proposed an alternative model assuming that most electrons which escape immediate geminate recombination are trapped in the matrix before encountering a scavenger molecule. Later they may tunnel to this molecule.

Homogeneous competition kinetics should apply for the capture of mobile electrons. The reciprocal of the yield of matrix-trapped electrons should then vary linearly with scavenger concentration. So should the reciprocal of the anion yield of a scavenger A, present at constant concentration, with increasing concentration of a second scavenger B. Miller found that this is not so. Instead, he observed an exponential decrease of the yield of

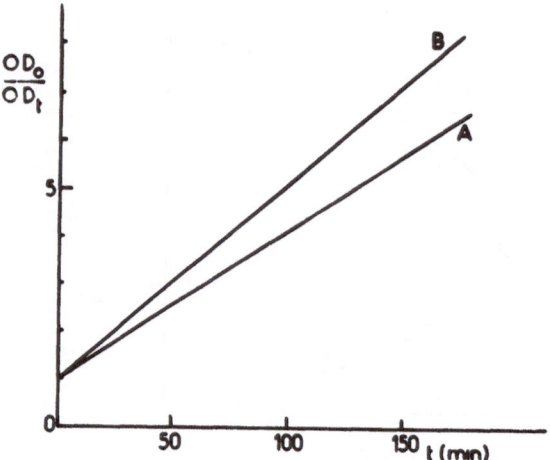

Figure 10. Post-irradiation decay of e_t^- in pure 3 MP
(A) and in 10^{-3} M biphenyl in 3 MP (B) at 77 K (data from
Hamill(22) shown as 1/OD vs. time).

e_t^- with increasing concentration of scavenger (Fig. 11). Similarly he observed an exponential decrease of A^- with increasing concentration of B.

    As early as 1928 Gamow(24) had applied a wave-mechanical treat·
ment to the problem of α -particle emission from radioactive nuclei
This had been a baffling problem, because the energy of α-particles
is much smaller than the potential barrier which they must cross.
Introducing the wave-mechanical concept of a small but non-zero
probability for a particle to pass a potential barrier of finite
height between two domains of equal energy, he succeeded in calculating radioactive decay constants which were in fair agreement
with experimental values, in spite of a number of approximations
he had to make.

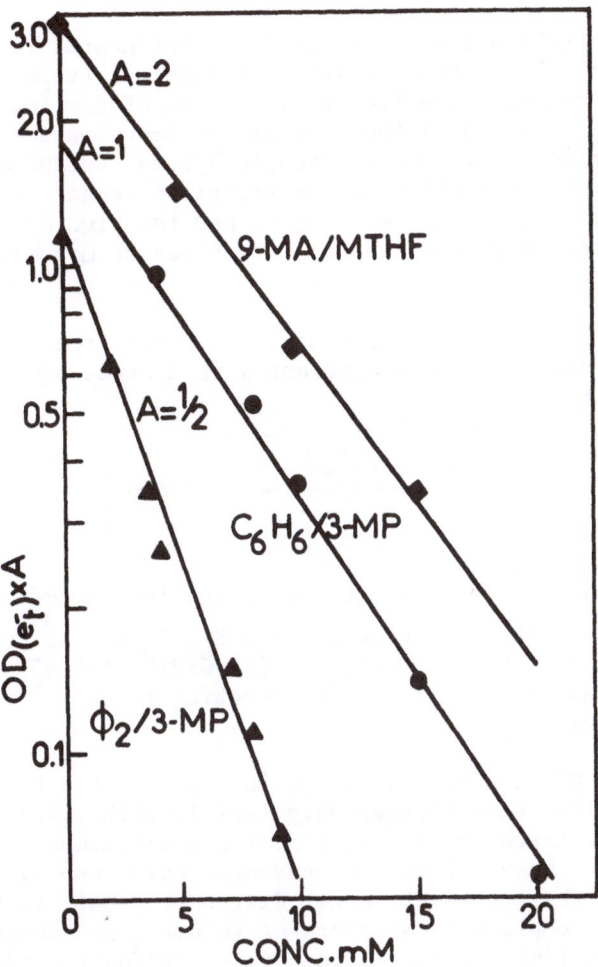

Figure 11. Exponential relationship between trapped electron yields and concentrations of scavengers : 9-methyl anthracene in MTHF, and benzene and biphenyl in 3 MP (Miller(23)).

This barrier-crossing has since been called tunnelling and has become a very familiar notion for solid state physicists. It seems to have been first introduced in radiation chemistry by Tsujikawa et al.(25) in 1965, and from about 1970 it became quite fashionable with radiation chemists and physical chemists generally (e.g. Miller, Goldanskii, Lapersonne, Kroh, Brocklehurst, Pilling and Rice).

Miller(23) noted the following experimental results in support of tunnelling from physical or chemical traps to more efficient scavengers : the linear dependence of the log of e_t^- yields on scavenger concentration, the linear dependence of the log of biphenyl anion yields on triphenylethylene concentration, the enhancement of the yield of an efficient scavenger with increasing concentration of a poor scavenger, and the time dependence of electron transfer to scavengers. To these we may add temperature-independent reaction rate.

Miller introduces plausible values into the relationship giving the tunnelling rate constant k in a crude square barrier approximation.

$$k = \left(\frac{E_o}{2md^2}\right)^{1/2} \frac{16 E_o(V_o-E_o)}{V_o^2} \times \exp\left\{\frac{-2a}{h}\left[2m(V_o-E_o)\right]^{1/2}\right\} s^-$$

where $(E_o/2md^2)^{1/2}$ is a frequency factor in a one-dimensional square well of width d and ground state E_o, V_o is the trap depth, m the electron mass, and a the barrier width. For e_t^- in 3 MP he takes V_o-E_o as equal to the photobleaching threshold, 0.53 eV, d = 6 Å , V_o = 1 eV.

For 3 MP, the rate constant is $\log k = 15 - 0.443\, a(V_o-E_o)^{1/2}$. The tunnelling rate changes very rapidly with distance (exponential term). Assuming that at time t all electrons trapped at a shorter distance than a from the scavenger have tunnelled to the scavenger and that those at larger distances remain in their traps, capture is complete in a sphere of volume $\frac{4}{3}\pi a^3$ where a is a function of the time t spent in the traps. Taking t = 1/k,

$$a(t) = \left[15+ \log t\right] / \left[0.443\,(V_o-E_o)^{1/2}\right] \quad \overset{o}{A}$$

For t = 5 min, tunnelling distances of 30-55 Å are thus found for the matrices commonly used.

With a random distribution of scavenger molecules, the probability that a trapped electron at an arbitrary position is not within the tunnelling sphere of any scavenger molecule is

$$P = \left[1 - (4\pi r^3/3V) \right]^N \quad \text{or} \quad P^{1/N} = 1 - (4\pi r^3/3V)$$

where V is the volume containing N scavenger molecules. Because N is very large ($\approx 10^{17}$), $P^{1/N} = 1 + \frac{1}{N} \ln P$.
Hence $\ln P = - 4\pi r^3 N/3V$ or $\log P = - 1.09 \times 10^{-3} r^3 M$, where M is scavenger molarity.

So here we have the semilogarithmic relationship between e_t^- yield and scavenger concentration; the slope of such a semilog plot is $1.09 \times 10^{-3} r^3$ where r is the "experimental capture radius".

3.2. Reaction with cations.

Trapped electrons could be thought to react randomly with cations. In fact, most electrons are correlated with their parent cation or another more stable cation which was formed from the parent cation by charge transfer or proton transfer.

The correlation of electrons with their parent cations was demonstrated very elegantly by McClain and Albrecht[26] who applied polarised photoselection techniques to stimulated recombination luminescence. They prepared trapped electrons by photoionising TMPD in 3 MP glass with ultra-violet light. In this case, the ejected electron is the only mobile species, since the positive charge left on TMPD cannot pass on to the neighbouring solvent molecules which have a much higher ionisation potential. The authors first confirmed this immobility of the positive charge by photoionising with polarised UV light : the recombination luminescence stimulated by infra-red light was polarised, thus allowing them to distinguish the stimulated emission from an oriented population. This also shows, by the way, that even in such a soft glass as 3 MP, TMPD$^+$ does not undergo rotational motion.

The next step consisted in carrying out two successive photoionisations with UV light polarised in different planes. First they prepared a large "vertical" population of TMPD$^+$ by irradiating for 10 min with vertically polarised light.Then they bleached exhaustively with IR, producing the "vertical " stimulated emission pattern, but this leaves quite a large fraction of "vertical" TMPD$^+$, because it requires more energetic light to detrap all e_t^-. Next they irradiated with horizontally polarised UV, for only 1 second. This produces enough fresh electrons to give an IR-stimulated emission , but adds only a negligible "horizontal" component to

the remaining "vertical" TMPD$^+$ population. Finally they stimulated the fresh electrons with IR and found that the emission was
purely "horizontal". If the electrons had recombined randomly
with the old and the new TMPD$^+$, the emission should have been
vertical.

The same result was obtained when they proceeded in reverse
order. It can be concluded that the trapped electrons remain
correlated with their parent cations and do not recombine with
any other randomly distributed cation.

Another conclusion could be drawn from these results if it
did not seem obvious anyway : there must be a distribution of
trapping distances between the correlated pairs, or else e$_t^-$
decay would follow exponential kinetics, the pairs disappearing
by a first order reaction. A distance distribution is equivalent
to a rate constant distribution, and this is precisely consistent
with the hyperbolic decay law observed for isothermal luminescence (ITL).

When the reciprocal of ITL intensity I is plotted versus
time, a linear relationship is observed in hard glasses :
$1/I = \alpha t (27)$. When $I(t)$ values are normalised with respect to
the initial value I_o, superposable plots are obtained for ITL
decays at 77 and at 4 K, $I_o/I(t)$ varying according to :

$$I_o/I(t) = 1 + \alpha (t-t_o)$$

Fig. 12 shows such superposed results (curve A) for 10^{-3} M
solutions of biphenyl in MCH at 77 and 4 K. At 77 K, data for the
3 MP glass (curve B) do not obey the same law. We interpret these
data by electron tunnelling in the case of curve A, and by at
least a strong contribution of thermal detrapping in B, since
77 K is precisely the glass transition temperature of 3 MP. This
fact, by the way, is undoubtedly responsible for many weird discrepancies in the abundant literature on 3 MP glasses.

Tunnelling calculations with first approximations led to
trapping distances of some 60-100 Å, for lifetimes of the order
of 25 minutes which were observed experimentally.

The decay of ITL was shown to be temperature-independent
below about 65 K (28) by irradiating at 77 K, recording ITL at
this temperature for some time, and then cooling to 4 K. As can
be seen in Fig. 13, decrease of light intensity was observed only down to about 65 K, whilst below 65 K the decay was again
hyperbolic. The same demonstration cannot be given by a reverse
experiment, i.e. irradiation at 4 K and subsequent warming to 77 K,

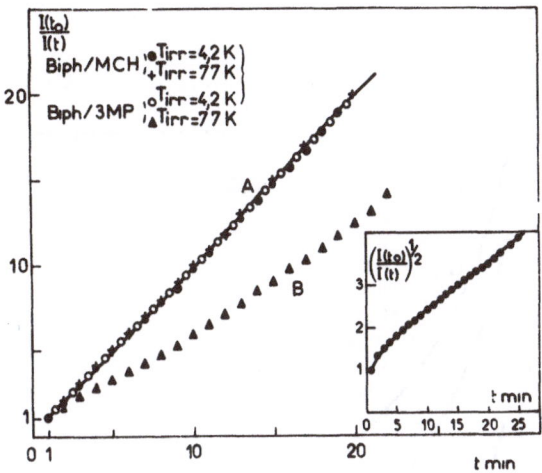

Figure 12. ITL decay kinetics expressed as $I_0/I(t)=f(t)$
Curve A contains data for 10^{-3} M biphenyl in methylcyclohexane at 77 and 4K, and in 3 MP at 4 K. Curve B is
obtained with the 3 MP solution at 77 K. Inset : curve
B replotted as $\left[I_0/I(t)\right]^{1/2} = f(t)$.(Kieffer et al.(27)).

because the trapped electron population at 4 K is different from
that at 77 K, as we saw when we dealt with presolvated and solvated electrons. And in fact, warming after irradiation at 4 K does
increase luminescence intensity, showing that at least part of
the presolvated electrons disappear by the recombination reaction.

In our endeavour to observe ITL kinetics at the shortest possible times, we applied nanosecond pulse radiolysis to MCH glass
and found that the hyperbolic law was valid at least from 1 μs
onwards (28). Using the more general Debye and Edwards formula
$I(t) = kt^{-m}$, we found that all our results fitted a linear plot
of log I vs. log t with a slope m = 1, over 10 orders of magnitude,
from 10^{-6} to 10^4 s. The fact that the same kinetics are observed
over such a vast time range is a further argument in favour of
the tunnelling mechanism, since the tunnelling rate constant
is a negative exponential function of ion separation distance.
Our results are consistent with an electron to cation (and anion
to cation) tunnelling mechanism, assuming a distance distribution between reaction partners.

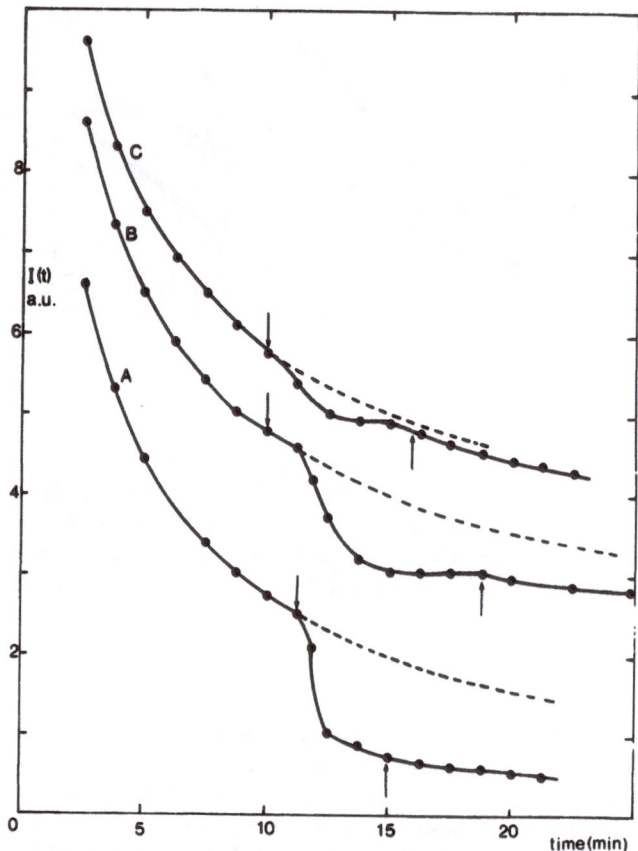

Figure 13. Effect of cooling on the decay of luminescence intensity of MCH glass containing different concentrations of biphenyl (A : 2 x 10^{-4} M, B : 10^{-3} M, C : 4 x 10^{-3}M), after γ -irradiation at 77 K. $\downarrow$ indicates beginning of temperature drop, $\uparrow$ the attainment of liquid helium temperature (Cordier et al.(28)).

We succeeded in observing ITL on the nanosecond time scale (29) only with the coincidence photon counting technique which I briefly described in my first talk. Here at last we found a deviation from the usual decay law. With α -particles, the log I vs. log t plot only shows a slope m = 1.3 between the end of prompt fluorescence ($\approx$ 20 ns) and 600 ns, the time beyond which the light

intensity could not be determined with a fair precision. With
β-particles, an initial slope of 1.3 gives way to the normal
slope 1 after a time which increases from some 50 to 500 ns with
the concentration of the scintillator p-terphenyl used as scavenger.
So in this time range the system keeps some trace of the initial
inhomogeneous deposition of energy. The larger initial value of
m can be interpreted as resulting from a contribution of "multi-
pair" recombinations in the zone of denser ionisation, whereas
the normal slope corresponds to the recombination of isolated
pairs : in the case of β particles, dense ionisations are much less
important than for α-particles, and the appearance of a "normal"
slope m = 1 permits a decomposition of the early decay with slope
1.3 into a sum of two decays, with slopes 2 and 1.(30) Computer
simulation of charge distribution and ITL kinetics has been carried
out and agreed well with experimental data(31).

Some sort of jump of the electron from an aromatic anion to
the corresponding cation has been admitted more or less tacitly
for a long time to explain the absence of excimer luminescence
in thermoluminescence experiments. In MCH glass, the anion-cation
recombination occurs at a distinctly higher temperature than the
electron-cation recombination. The reaction

$$M^+ + M^- \rightarrow M_2^*$$

could have been expected to occur, but excimer emission was
observed only when dimer cations M_2^+ were known to exist already
at 77 K. This electron "jump" occurs most probably by tunnelling,
for tunnelling was shown to occur at 77 K (32) between biphenyl
anions and cations which are totally unable to diffuse at this
temperature in MCH. On the time scale of minutes, tunnelling distances were calculated to be 24-33 Å from the anion (cf.
38-53 Å from traps).

The important role of tunnelling in reactions of trapped
electrons both with cations and with scavengers seems to be well
established. But it can be irritating to see a fashion coming
into existence, and electron tunnelling has been so frequently
invoked in radiation chemistry to explain - perhaps rightly -
all sorts of reactions that it has become something like a
fashion. It seems therefore a very healthy reaction that some
workers try to develop alternative reaction models, such as
Hamill and Funabashi's continuous time random walk(33,34). Since
diffusion of e_t^- as such, i.e. with its solvation shell, is excluded
by the rigid structure which precludes molecular mobility, this
model seems to amount to trap-to-trap hopping. In its formalism
this seems to be equivalent to trap-to-trap tunnelling as proposed
by Buxton and Kemsley(35), although physically quite different.
The major objection to trap-to-trap tunnelling is that it requires
the availability of gradually deeper traps, and since the presence
of the electron relaxes, i.e. deepens, each trap, the number of
intermediate jumps involved must be fairly small, and this again

brings in rather large transfer distances which are one of the objections to direct tunnelling.

4. CATION RADICALS.

The fate of the positive radical-ion formed initially has barely been mentioned so far. In polar glasses, such as alcohols, a stabilised ion results from hydrogen abstraction from a neighbouring molecule $\longrightarrow$ ROH_2^+. In saturated hydrocarbons, there has been much uncertainty, except in the presence of alkenes or aromatics. The positive charge is stabilised on these solute molecules, probably by a mechanism of resonant transfer. Positive aromatic radical-ions are well-known, and they are certainly responsible for the luminescence resulting from recombination. This appears very clearly in stimulated luminescence where the phosphorescence of the neutralised cation is observed. Nanosecond work on ITL caused by α-particles suggested that some other cation originating from the solvent must also participate in the recombination(30). If, as happens in alcohols, a more stable ion is produced by hydrogen abstraction from neighbouring alkane molecules, such an ion (e.g. $C_6H_{15}^+$) is not paramagnetic, and hence cannot be detected by EPR spectroscopy. Louwrier and Hamill(36) produced evidence for the presence of radical cations of higher alkanes ($\geqslant$ 7 C atoms) in γ-irradiated 3 MP and in CCl_4, characterised by broad absorption spectra in the 600-900 nm region. They also observed hole transfer to CCl_4 or to solute TMPD.

Identification by EPR is very difficult because of overlapping spectra from other radicals, but the EPR spectrum of the $C_8H_{18}^+$ radical from octane has recently been obtained in pentane by subtracting the spectrum obtained after selective photobleaching from the original spectrum(37).

Gillis et al.(38) observed a transient absorption band in pulse-irradiated viscous liquid 3-methyloctane (3 MO) at low temperatures and attributed it to the $C_8H_{18}^+$ ion because it does not appear in the presence of the positive charge scavenger triethylamine, whereas it is reinforced in the presence of electron scavengers (SF_6, N_2O, CCl_4). In pure 3 MO it disappears more rapidly than the IR-absorbing electron : less than 2/3 remain after 7 μs. Teather and Klassen(39) recently observed a broad transient absorption band peaking around 1650 nm in squalane glass at 72 K. Similarly, they observed a transient band in 3 MO glass. The 1650 nm band has been attributed to a superposition of SQ^+ and e_{IR}^- spectra, and λ_{max} for SQ^+ is deduced to be around 1400 nm. With a 1% solution of squalane in 3 MO, these authors observed a positive charge transfer from 3 MO$^+$ to SQ, i.e. in actual fact an electron transfer from the solute to a solvent cation. The growth of SQ^+ absorption has also been observed in 3 MP glass, where it

appears more clearly in the absence of any simultaneous decay of
an overlapping absorption of 3 MP$^+$ which must absorb at shorter
wavelengths than 3 MO$^+$. This growth and the beginning of the subsequent decay of SQ$^+$ are shown in Fig. 14 for a 2% solution in
3 MP. This figure also shows the decay of 3-MO$^+$ in a 1% solution
of SQ in 3-MO.

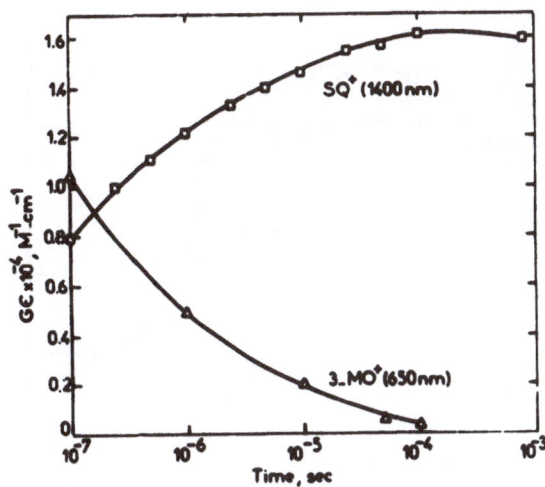

Figure 14. Decay of 3-methyloctane cation absorption and
growth of squalane cation absorption at 72 K. SQ$^+$ was
monitored at 1400 nm in 3 MP glass containing 2% squalane,
3 MO$^+$ at 650 nm in 3 MO containing 1 % squalane (Teather
and Klassen(39).

The simultaneous decay of 3 MO$^+$ and growth of SQ$^+$ are also
apparent in Fig. 15 which shows the evolution of absorption spectra of a 1% solution of squalane in 3 MO at 6 K. The process is
very rapid even at 6 K, since it is about half completed within
10 μs.

The recent developments in work on solvent cation radicals
hold a promise that further work in this direction will lead to
a much better understanding of the fate of the positive charge
in radiolysis which has been rather a weak point so far.

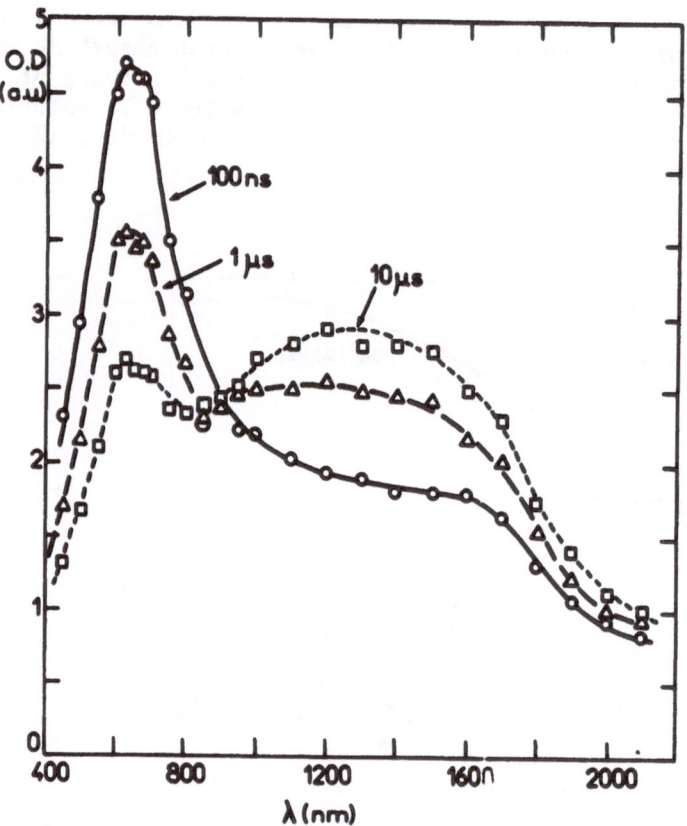

Figure 15. Spectral change (simultaneous growth of
SQ$^+$ and decay of 3 MO$^+$) observed in pulse radiolysis
of a 1% solution of squalane in 3-methyloctane at 6 K.
(Klassen, private communication).

REFERENCES

1. Willard, J.E. : 1974, Int. J. Radiat. Phys. Chem. 6,
   pp. 325-343.
2. Dainton, F.S., and Salmon, G.A. : 1965, Proc. Roy. Soc., A,
   285, pp. 319-338.
3. Kevan, L. : 1974, Adv. Radiat. Chem., 4, pp. 181-305.
4. Hamill, W.H. : 1968, in Radical Ions (Interscience),
   pp. 321-416.
5. Hase, H., Warashina, T., Noda, M., Namiki, A., and Higashimura, T. : 1972, J. Chem. Phys. 57, pp. 1039-1045.
6. Hase, H., Higashimura, T. , and Ogasawara, M. : 1972, Chem.
   Phys. Letters 16, pp. 214-216.

7. Klassen, N. V. , Gillis, H. A., and Teather, G. G. : 1972,
    J. Phys. Chem. 76, pp. 3847-3850.
8. Willard, J. E. : 1973, Science 180, pp. 553-561.
9. Perkey, L. M., Farhataziz, and Hentz, R. R. : 1974, J. Chem.
    Phys. 61, pp. 2979-2980.
10. Hase, H., and Warashina, T. : 1973, J. Chem. Phys. 59,
    pp. 2152-2153.
11. Baxendale, J. H., and Sharpe, P.H.G. : 1976, Chem. Phys.
    Letters 39, pp. 401-404.
12. Magat, M. : 1966, J. Chim. Phys. 63, pp. 142-149.
13. Miller, J. R. : 1978, J. Phys. Chem. 82, pp. 767-774.
14. Ogasawara, M., Shimizu, K., Yoshida, K. Kroh, J., and Yoshida, H.
    1979, Chem. Phys. Letters 64, pp. 43-45.
15. Noda, S., Yoshida, K., Ogasawara, M., and Yoshida, H. : 1980,
    J. Phys. Chem. 84, pp. 57-59
16. Shida, T., Iwata, S., and Watanabe, T. : 1972, J. Phys. Chem.
    76, pp. 2683-2691.
17. Wiseall, B., and Willard, J. E. : 1967, J. Chem. Phys. 46,
    pp. 4387-4399.
18. Baverstock, K. F., and Dyne, P.J. : 1970, Can. J. Chem. 48,
    pp. 2182-2191.
19. Maruyama, Y., and Funabashi, K. : 1972, J. Chem. Phys. 56,
    pp. 2342-2345.
20. Huang, T., and Kevan, L., quoted in (3).
21. Kevan, L. : 1981, Acc. Chem. Res. 14, pp. 138-145.
22. Gallivan, J.B., and Hamill, W. H. : 1966, J; Chem. Phys. 44,
    pp. 1279-1287.
23. Miller, J. R. : 1972, J. Chem. Phys. 56, pp. 5173-5183.
24. Gamow, G. :1928, Z. Physik 51, pp. 204-212.
25. Tsujikawa, H., Fueki, K., and Kuri, Z. : 1965, Bull. Chem.
    Soc. Jap. 38, p. 2210.
26. Mc Clain, W. M., and Albrecht, A. C. : 1966, J. Chem. Phys. 44,
    pp. 1594-1599.
27. Kieffer, F., Lapersonne-Meyer, C., and Rigaut, J. : 1974, Int.
    J. Rad. Phys. Chem. 6, pp. 79-84.
28. Cordier, P., Kieffer, F., Lapersonne-Meyer, C., Rigaut,J. :
    1975, Rad. Res. (Proc. 5th Int. Congr. Rad. Res.)
    pp. 426-435.
29. Kieffer, F., Klein, J., Lapersonne-Meyer,C., and Magat, M. :
    1978, Chem. Soc. Faraday Disc. 63, pp. 55-58.
30. Lapersonne-Meyer, C. : 1980, Rad. Phys. Chem. 15,
    pp. 371-376.
31. Lapersonne-Meyer, C., and Schott, M. : 1979, Chem. Phys.,
    pp. 287-296.
32. Kieffer, F., Klassen, N. V., and Lapersonne-Meyer, C. : 1979, J.
    Luminescence 20, pp.17-27
33. Hamill, W. H., and Funabashi, K. : 1977, Phys. Rev. B 16,
    pp. 5523-5527.
34. Funabashi, K., and Hamill, W. H. : 1979, Can. J. Chem. 57,
    pp, 197-206.

35. Buxton, G.V., and Kemsley, K. G. :1976, J. Chem.Soc. Faraday Trans. 1, pp. 466-480 ; 1333-1341.

36. Louwrier, P.W. F., and Hamill, W. H. : 1968, J. Phys. Chem. 72, 3878-3883.

37. Nauwelaerts, F., Lemahieu, M., and Ceulemans, J. : 1976, J. Chem Phys. 66, pp. 140-142.

38. Gillis, H. A., Klassen, N. V., and Woods, R.J. : 1977, Can. J.Chem. 55, pp. 2022-2029.

39. Teather, G. G., and Klassen, N. V. : communication of a pre-print is gratefully acknowledged.

THE USE OF PULSE RADIOLYSIS TO STUDY TRANSIENT SPECIES IN THE GAS PHASE[1]

Myran C. Sauer, Jr.

Chemistry Division
Argonne National Laboratory
Argonne, Illinois 60439, USA

ABSTRACT

Examples are chosen from the literature to show the types of species which have been investigated in the gas phase using the pulse radiolysis technique. The types of processes occurring in pure rare gases and rare gases containing additives are examined. Techniques used to study ion-recombination processes, ion-molecule reactions, and electron capture reactions are discussed and examples of typical results are given. The use of pulse radiolysis to determine rate constants for reactions of atoms and free radicals is also examined.

1. INTRODUCTION

Since the introduction of the technique of pulse radiolysis in 1960 (1) and its subsequent application to gases, a wide variety of species and reactions have been investigated using this technique. The area of gas-phase pulse-radiolysis has been the subject of two reviews (2,3). In some instances a pulse of ionizing radiation is used to create (via ion-recombination and other mechanisms) neutral reactive species in a manner which is often more convenient than other techniques (photochemical, thermal). In other instances, the charged species are studied (ion-recombination, ion-molecule reactions, electron capture).

An understanding of the ways in which pulses of ionizing radiation can be used in studying gas phase processes depends to some extent on an understanding of the time scale of various processes. This time scale is depicted in Figure 1, which shows the primary

J. H. Baxendale and F. Busi (eds.),
The Study of Fast Processes and Transient Species by Electron Pulse Radiolysis, 601–626.

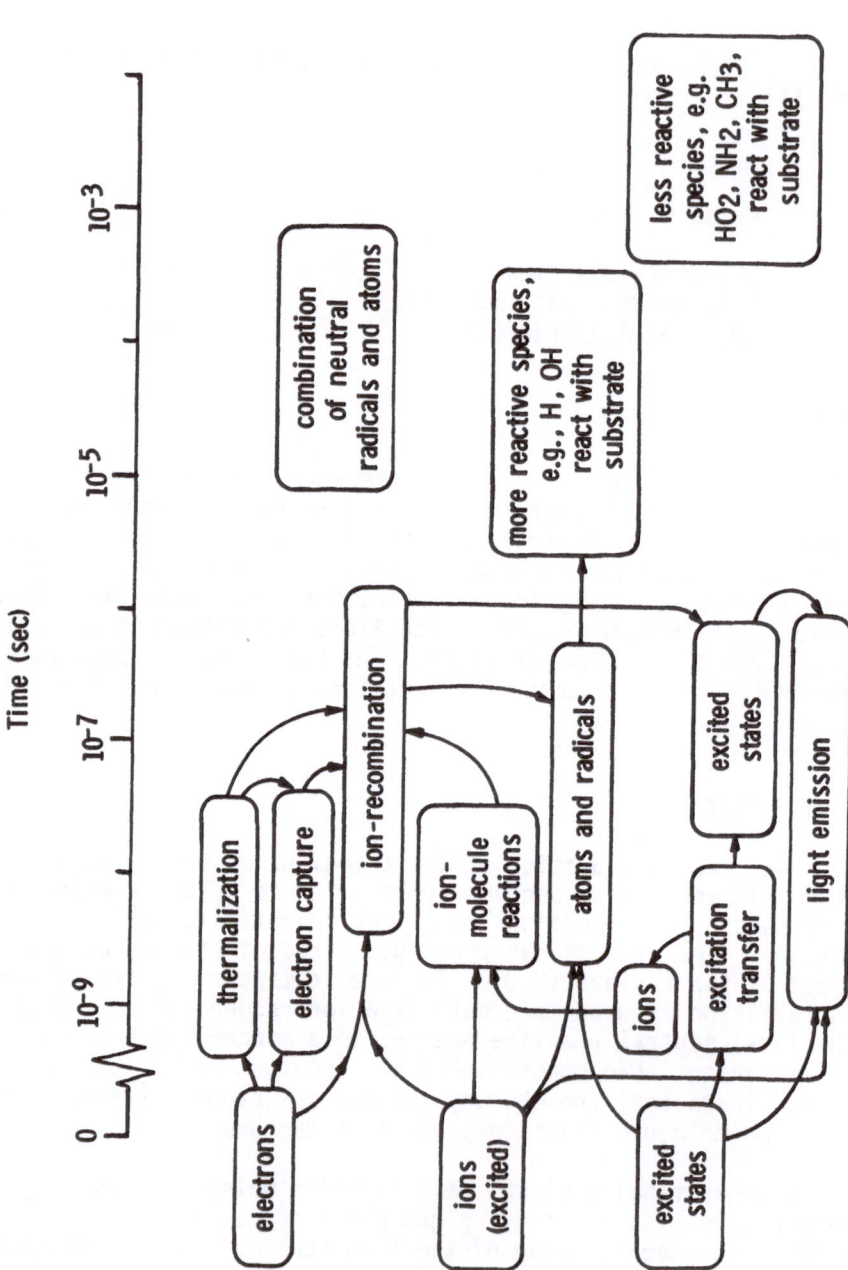

Figure 1. Approximate time scale of events in gas phase pulse radiolysis.

species produced by a pulse (time = 0) as being electrons, ions (which are, in general, excited), and neutral excited states. Light emission from excited species is, in general, observed in the range 10^{-9} to 10^{-7} sec and is in competition with other processes, e.g., fragmentation, ion-molecule reactions, and energy transfer. Kinetic information on these processes can be obtained via observation of the time dependence of the light emission.

Electron thermalization takes place in the 10^{-9} to 10^{-7} sec regime, as do ion-molecule reactions. "Recombination" reactions among oppositely charged species takes place over 10^{-9} to 10^{-6} sec, dependent, of course, on the concentration of ions produced by the pulse. These ion-recombination reactions produce atoms, radicals, and excited states. "Delayed" light emission as a result of the excited states can be observed and used to obtain information on ion-recombination rates.

Transfer of excitation energy can result in delayed production of ions, as well as light emission due to excitation of minor constituents.

In most cases ionic species have undergone recombination by 10^{-6} to 10^{-5} sec and neutral atoms and free radicals are the predominant transient species. These combine with one another to form stable molecules on a 10^{-5} to 10^{-2} sec time scale; note, however, that transformations by reaction with substrate molecules are possible on the same time scale in the case of more reactive species such as H or OH. Less reactive radicals such as CH_3, NH_2, or HO_2 generally do not react with substrate molecules (at room temperature) in less than 10^{-1} sec, and can therefore be assumed to react only by radical combination reactions.

In this presentation, examples will be given of gas-phase processes which have been investigated by pulse-radiolysis.

No attempt will be made to cover all of the work done on gas-phase pulse-radiolysis, and the work discussed will be chosen to some extent on the basis of its familiarity to the author.

Obviously, suitable detection methods are quite important; however, the detection methods will not be discussed in detail, because other chapters in this book deal with that subject.

2. EXCITED STATES

Excited state formation and decay processes have been investigated by following both light emission and optical absorption. Examples will be chosen from the large number of papers published to show the types of phenomena encountered.

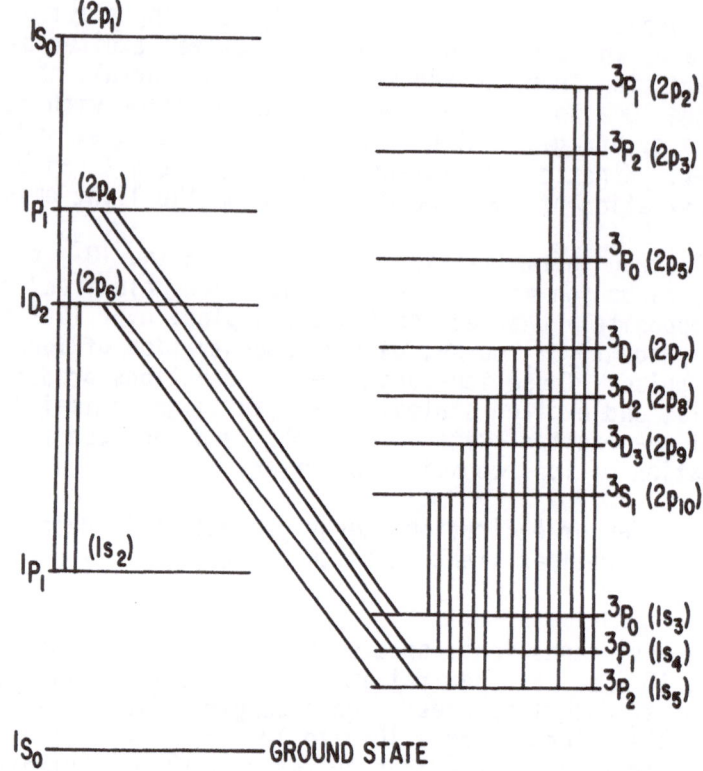

Figure 2. Grotrian diagram (2) of the 1s and 2p levels
of Ar (not to scale).

2.1. Pure Rare Gases

In pure rare gases, emissions and absorptions have been studied
(4-8) which originate from the Paschen 1s (the four lowest energy
excited states) and 2p (the next 10 excited states) atomic levels,
shown schematically in the Grotrian diagram in Figure 2. Absorptions and emissions due to rare gas dimer species have also been
studied (4,9,10).

Of the four 1s excited states, the $^1P_1$ and $^3P_1$ are resonance
states, and the corresponding transitions to the ground state can
be observed in the vacuum u.v. The $^3P_0$ and $^3P_2$ are metastable
states. All four states can be observed in absorption, corresponding to transitions to the 2p states. The 2p levels can be
followed by observing their transitions to 1s levels via the resulting light emission in the visible and near i.r. regions.
Rare gas dimer species have been directly observed by their
absorptions in the i.r. region.

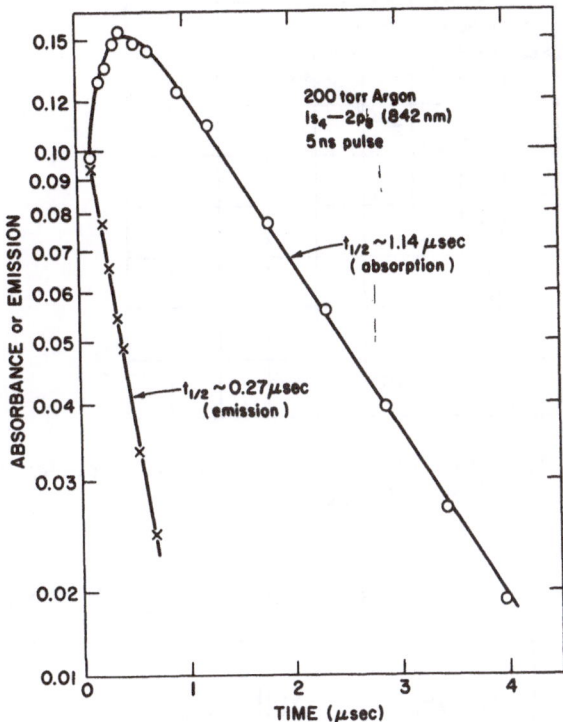

Figure 3.   Emission and absorption in Ar.

The general picture which has emerged from the experimental results is that the 1s levels are mainly populated by processes occurring after energy has been imparted to the gas by the pulse. For example, Figure 3 shows the emission and absorption at 842 nm ($1s_4$-$2p_8$) in about 200 torr argon (4). The results indicate that most of the $1s_4$ state is formed during the decay of the $2p_8$ level. The 1s levels are found to be relatively long-lived; the resonance levels have their effective lifetimes extended due to radiation imprisonment.

To continue with the general picture, the situation with respect to the 2p levels has been examined with time resolution of several nanoseconds (5). Two mechanisms of production of the 2p states are evident, as is depicted in Figure 4. The second peak in Figure 4, which is common to all of the 2p levels observed, can be shown to be due to production of the 2p level by ion-recombination. This conclusion is based on the observations (5) that decreasing the dose per pulse shifts the peak to longer times (Figure 5), and that it can be entirely eliminated by addition of a small amount of SF_6, which is an efficient electron scavenger.

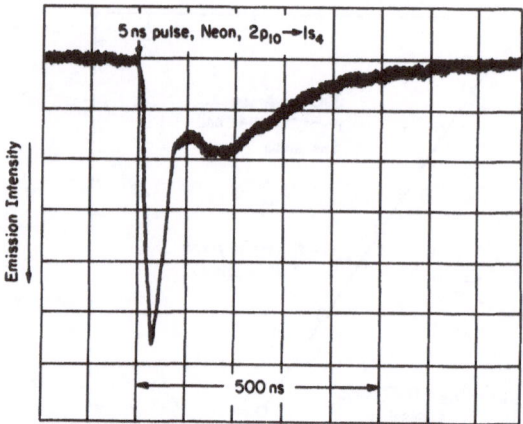

Figure 4. Emission at 725 nm in 840 torr Ne using a 5 ns pulse from a Febetron 706.

5 ns pulse, Neon, $2p_1 \rightarrow 1s_2$

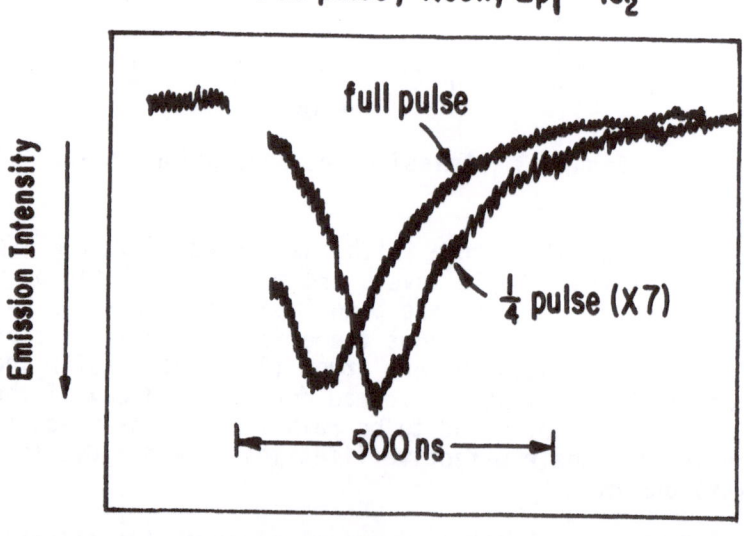

Figure 5. Emission from 840 torr Ne; $2p_1 \rightarrow 1s_2$ (585 nm). The fast component of the emission is not shown.

The first peak in Figure 4 is due to direct excitation of the rare gas to the 2p level, or to direct excitation of the rare gas to some energy level higher than the 2p level in question followed by a rapid transition. By using low dose per pulse, the two peaks in Figure 4 can be well separated, and results have been obtained on the pressure dependences of the formation and decay kinetics of the first peak. Examples of these results, which were obtained using a 30 ps linac pulse and a crossed-field photomultiplier (ca 150 ps rise-time), are shown in Figures 6 and 7 for neon (11).

Observation of all of the 2p levels was not possible, but Ar, Kr, and Xe do not show the same pattern as seen in Figures 6 and 7 for Ne. A general trend for the four rare gases is that decay rates become greater with increasing rare gas pressure. These results have not been completely interpreted; however, the variation of integrated emission with pressure suggests that the faster decay rates at higher pressure are not a result of removal of the 2p levels by dimer formation. That is, higher pressure seems simply to accelerate the emission to the 1s levels. This is in agreement with the results of Firestone and co-workers, who report that the observed initial (i.e., at times where all 2p has gone to 1s) concentrations of the $1s_2$ and $1s_4$ states are directly proportional to pressure, which indicates that the faster decay of the 2p levels does not prevent the 1s levels from being populated in proportion to the dose absorbed per unit volume, which is proportional to the pressure.

To complete the general picture on the pure rare gas systems, the work of Firestone and co-workers on the decay of the 1s levels and the formation of excited dimers will be summarized. The mechanism proposed (9) in the case of the 1s resonance levels of argon is shown below.

$$Ar(1s) + Ar \underset{2}{\overset{1}{\rightleftarrows}} Ar_2^*(v) \qquad\qquad (1,2)$$

$$Ar_2^*(v) + Ar \rightarrow Ar_2^* \qquad\qquad (3)$$

$$Ar_2^*(v) \rightarrow h\nu + Ar_2^* \qquad\qquad (4)$$

The species $Ar_2^*(v)$ is vibrationally excited. The lowest lying metastable rare gas eximer states, e.g., Ar_2^* in reactions (3) and (4), are proposed to decay by radiative processes, e.g.,

$$Ar_2(^3\Sigma_u^+) \rightarrow h\nu + 2Ar , \qquad\qquad (5)$$

with emission in the vacuum u.v.

Reactions (1)-(4) provide an explanation for the pressure dependences of the observed (7) first-order decays of the $1s_2$ and $1s_4$ levels of Ar shown in Figure 8, where the observed rate constant, k_{obs}, is defined by equation I.

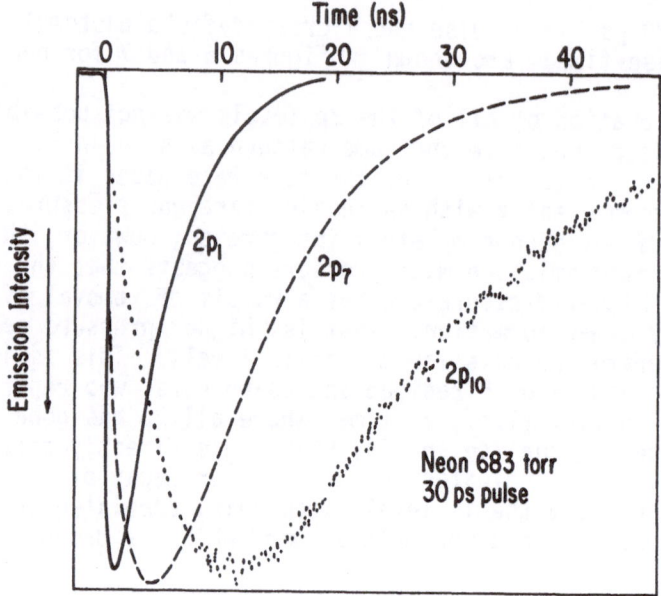

Figure 6.   Emission from 2p levels of Ne following a 30 ps pulse.

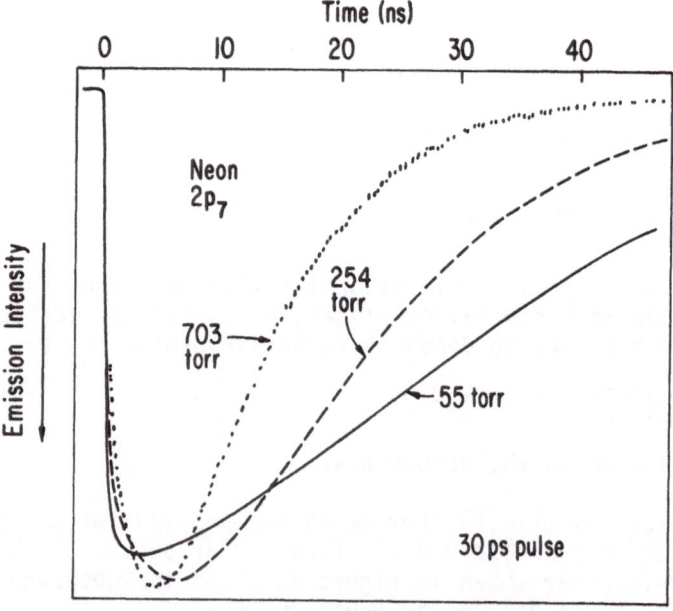

Figure 7.   Pressure dependence of emission from the $2p_7$ level of Ne following a 30 ps pulse.

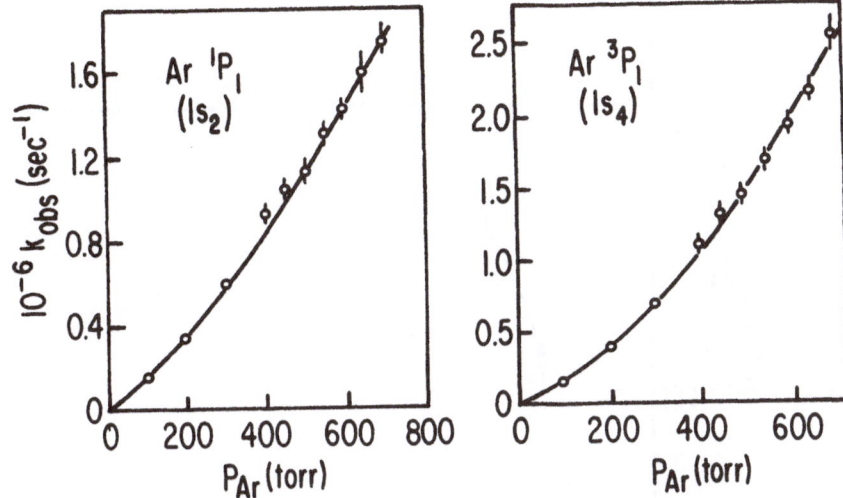

Figure 8. Pressure dependences of the observed pseudo-first-order disappearance rates of the $1s_2$ and $1s_4$ states of Ar. The curves result from equation I.

$$k_{obs} = \frac{\dfrac{k_1 k_3}{k_2 + k_4}[Ar]^2 + \dfrac{k_1 k_4}{k_2 + k_4}[Ar]}{1 + \dfrac{k_1 + k_3}{k_2 + k_4}[Ar]} \qquad (I)$$

Because of the large value of k_2 the denominator can be set equal to unity.

The decay of the relatively long lived metastable dimer species, e.g., reaction (5), which results from reaction (1) has been thoroughly investigated and found to proceed only by the radiative process indicated (9). The radiative decay constants for $Ne_2(^3\Sigma_u{}^+)$, $Ar_2(^3\Sigma_u{}^+)$, and $Kr_2(^3\Sigma_u{}^+)$ are $1.51 \pm (0.03) \times 10^5$, $3.10 \pm (0.04) \times 10^5$, and $2.83 \pm (0.08) \times 10^6$ sec^{-1}, respectively (9). It is important to note that at low rare gas pressure the radiative decay of these dimers becomes faster than their formation rate by reactions (1)-(4) and the observed decay rate of the absorption due to the dimers therefore becomes slower than the radiative decay rate (9). As shown in Figure 9, the observed decay constant reaches a plateau value equal to the radiative constant, at sufficiently high pressure.

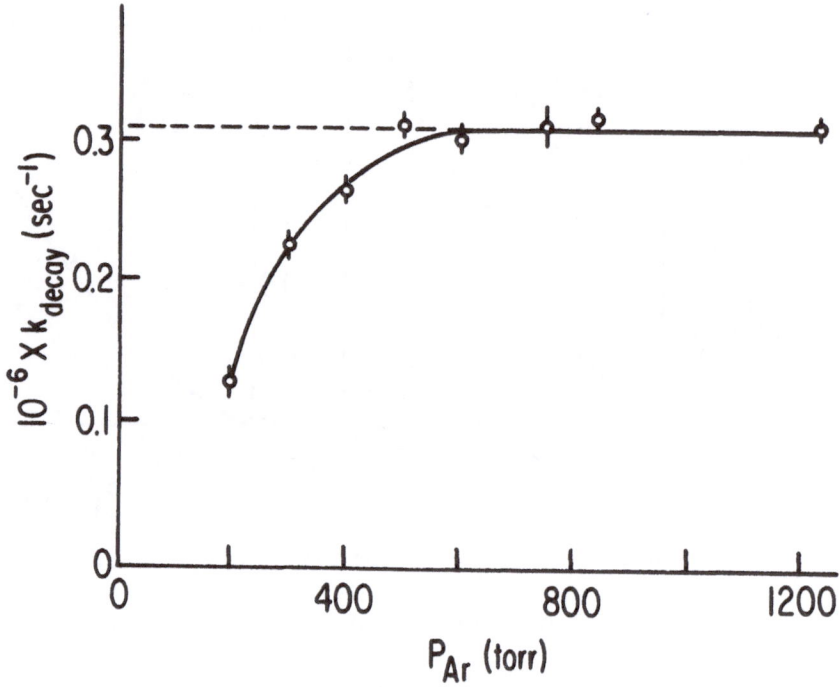

Figure 9. Pressure dependence of the decay constant of
the rate limiting slow step in the decay of $Ar_2(^3\Sigma_u^+)$.

2.2. Rare Gases Containing Additives

Only a few of the many papers on this subject will be discussed. Particular attention will be paid to illustrating the
variety of mechanisms by which excited states are produced by
reactions involving the additives. Numerous measurements of
quenching rate constants for various additives reacting with excited states have been made but these will not be discussed here.

A system which, despite its simplicity, has revealed a wealth
of interesting (but complex) data on light emission is $Ar-I_2$. Detailed observations (12,13), from less than a nanosecond to several
microseconds, in the u.v. and visible regions, have shown that the
observed spectral peaks correspond to known states of molecular
iodine. The spectrum is shown in Figure 10.

Only the strongest emission (*ca* 342 nm) could be examined
with sub-nanosecond time resolution (13). The emission at 342 nm
exhibits a rather complex behavior, as shown in Figures 11 and 12.
The "fast" peak shown in Figure 11 is also present in the cases
of the 4 and 40 ns pulses (Figure 12), but is offscale and is not
shown. The important experimental fact is that there appear to be

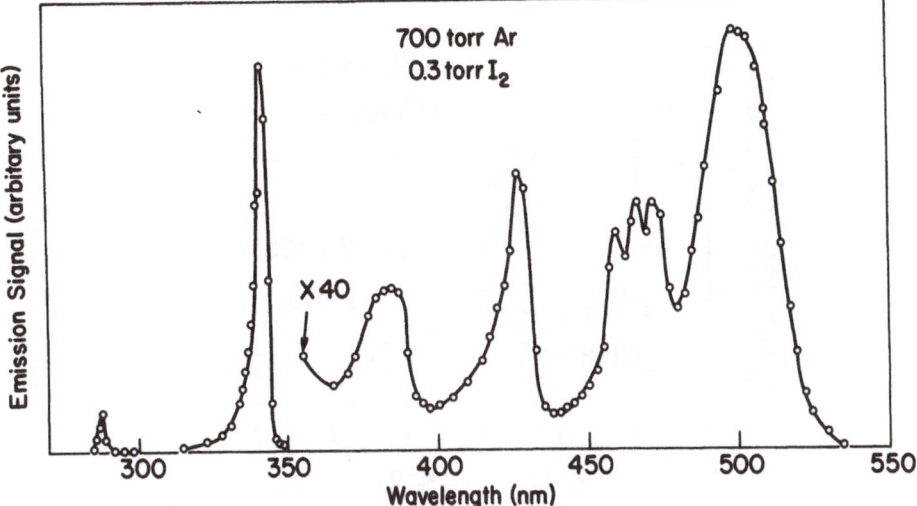

Figure 10. Emission spectrum (uncorrected) in the Ar-I_2 system observed about 10 ns after a 40 ns pulse.

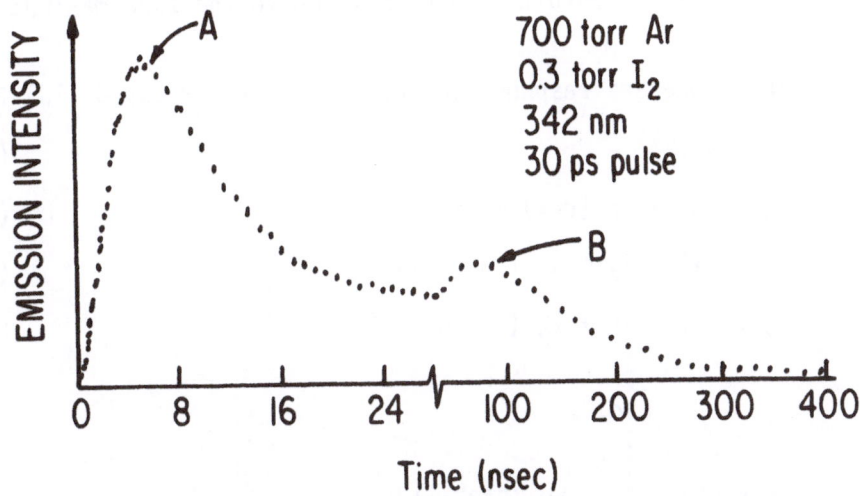

Figure 11. Emission at 342 nm following a 30 ps pulse in the Ar-I_2 system.

three processes leading to the emission at 342 nm (corresponding to the peaks A, B, and C in Figures 11 and 12).

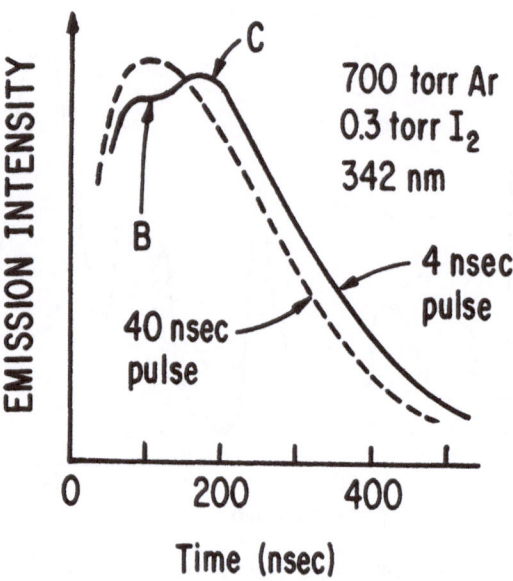

Figure 12. Emission at 342 nm following 4 ns and 40 ns pulses. Zero time corresponds to the pulse mid-point, and the results are normalized to the same maximum.

The processes responsible for peak A are proposed (13) to be:

$$I_2 \rightsquigarrow I_2^{**} \tag{6}$$

$$I_2^{**} + Ar \rightarrow I_2^* \ (v=x) + Ar \tag{7}$$

$$I_2^* \ (v=x) \rightarrow I_2 + h\nu \ (<342 \ nm) \tag{8}$$

$$I_2^* \ (v=x) + Ar \rightarrow I_2^* \ (v=x-n) + Ar \tag{9}$$

$$\vdots$$

$$etc.$$

$$\vdots$$

$$I_2^* \ (v=0) \rightarrow I_2 + h\nu \ (342 \ nm) \ . \tag{10}$$

This mechanism is based on variations in the emission kinetics with change in wavelength (320-342 nm region) and pressure, the results of which show that the emission at the low wavelength end of the band forms and decays while the 342 emission is forming. This suggests a vibrational deactivation sequence as shown in re- actions (7)-(10); the effects of pressure also support the idea of a vibrational deactivation sequence.

Figures 11 and 12 show that the "fast" 342 nm emission decays, but that the light intensity subsequently reaches a second maximum. Results with a 4 ns pulse of somewhat higher dose than the 30 ps pulse used in Figure 11 suggest an additional third emission maximum (C), as shown in Figure 12. Using a 40 ns pulse (about 10 times the dose of the 4 ns pulse) apparently causes B and C to merge. When a small amount of SF_6, an efficient electron scavenger, is added, peak C is eliminated, but A and B are still observed. Thus, B is thought to arise from reactions initiated by excited Ar produced by direct excitation. A detailed analysis of the results (12) suggests $Ar^* + I_2 \rightarrow I^*$ followed by $I^* + I_2 \rightarrow I_2^*$. Peak C is thought to arise from several ionic reactions which eventually result in I_2^*.

An interesting observation concerning this system is that the emission decays shown in Figure 12 follow pseudo-first-order kinetics from 300 to 900 ns, i.e., over several half-lives. This is unusual in view of the ionic processes, which would be expected to result in a second-order component. This finding requires that one of the ionic species must react by a first order reaction to produce an ion which doesn't result in eventual production of I_2^*. A kinetic scheme incorporating such a reaction has been found to give reasonable agreement with the experimental results (12).

A similar situation exists with the emission at 342 nm in Xe, I_2 mixtures (14). However, in this case stronger emissions in the region of 253 and 319 nm are observed, and have been ascribed to $XeI^*[B(1/2)]$ (15). Analysis of the variation of emission kinetics with wavelength and pressure indicates that, somewhat surprisingly, the XeI^* is formed by reactions of excited I_2 with Xe, the excited I_2 being formed by direct excitation. As in the case of the Ar, I_2 system, there is strong evidence of a vibrational deactivation sequence. The observations are summarized in Figures 13 and 14.

Using longer pulses (higher dose per pulse) shows that a slower sequence of reactions occurs, involving ionic precursors, which produces the same excited states (14).

The formation of XeI^* [B(1/2)] is observed also in pulse radiolysis (using 15 ns pulses) of Xe, CF_3I mixtures (15), but in contrast to the Xe, I_2 system XeI^* is formed by the reaction of Xe^* (or Xe_2^*) with CF_3I, e.g.,

$$Xe^* + CF_3I \rightarrow XeI^* + CF_3 \tag{11}$$

An ionic component to XeI^* formation is also observed in this system when the CF_3I pressure is small enough that Xe_2^+ reacts with I^- (to give XeI^*) rather than with CF_3I.

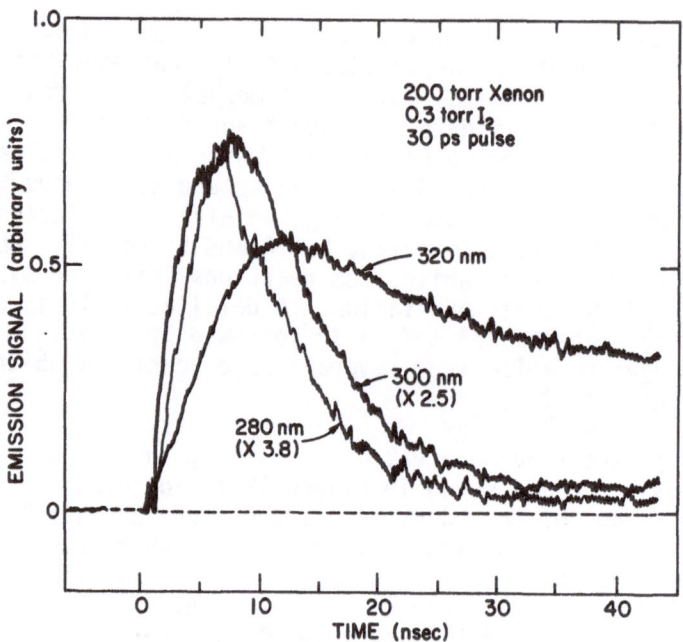

Figure 13. Dependence of the emission vs time profile on wavelength in the Xe-I$_2$ system (14).

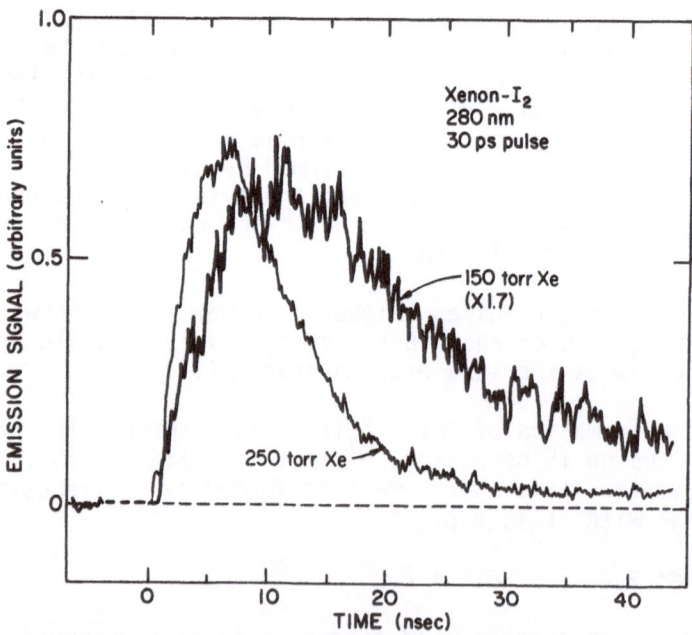

Figure 14. Effect of Xe pressure on the time dependence of emission at 280 nm in the Xe-I$_2$ system (14).

3. CHARGED SPECIES

3.1. Recombination

An early example of the use of pulse-radiolysis to study ion-recombination involved the measurement of light emission from N_2 irradiated with pulses of about 10-40 ns duration from a linear accelerator (16). Emission from the $C^3\pi_u$ state of N_2 was observed to follow kinetics indicating that this state was produced in the reaction

$$e^- + N_4^+ \rightarrow N_2 + N_2(C^3\pi_u) \tag{12}$$

The emission decayed over a time period much longer than the lifetime of the $C^3\pi_u'$ state. Therefore, assuming that $[e^-] = [N_4^+]$, or that other positive ions react with e^- with the same rate constant as k_{12}, one can easily see that (Emission Intensity)$^{\frac{1}{2}}$ should be proportional to $[e^-]$. Plots of (Emission Intensity)$^{-\frac{1}{2}}$ vs time were linear, verifying the second-order nature of the reaction. Using appropriate chemical dosimetry, (Emission Intensity)$^{\frac{1}{2}}$ was converted to electron concentration; second order plots of $[e^-]^{-1}$

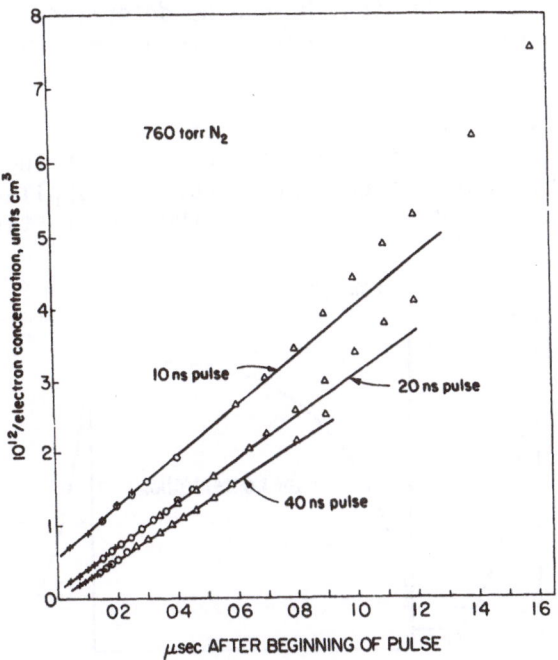

Figure 15. Second order plots of electron-ion-recombination in N_2, determined by observation of emission from $N_2(C^3\pi_u)$.

616   M. C. SAUER, Jr.

vs time are shown in Figure 15 for three pulse sizes. The value
of k_{12} obtained was 3.0×10^{-6} cm^3 s^{-1}, with no effect of pressure
between 280 and 980 torr, and $T^{-\frac{1}{2}}$ dependence from 25 to 205°C.

Ion-recombination reactions have also been studied by following light emission from naphthalene or anthracene excited singlet
states in the vapor phase (17), but methods in which the disappearance of charge is measured directly are more general. Armstrong and co-workers have used both steady-state and pulsed irradiation methods to measure recombination coefficients in a
variety of gases, i.e., NH_3, H_2O, CO_2, and others. Their charge
collection methods use relatively weak intensities of x-rays,
either pulsed or continuous, producing ion concentrations in the
range of 10^8 cm^{-3} (18). Gases are irradiated in ionization chambers, and absolute measurements of charge collected are made with
(1) continuous radiation and pulsed high voltage for charge collection, and (2) pulsed irradiation and pulsed high voltage for
charge collection. In (1), the time between collection pulses is
great enough for the steady-state ion concentration to be reached.
The high voltage pulse is made long enough that all steady-state
ions are collected. This situation is depicted in Figure 16.
From the rate of ion collection as a function of the length of
the high voltage pulse, the steady-state ion concentration, n_{ss},
and the rate of ion production, R, can be determined. These value
allow the recombination coefficient, α, to be determined, because
$\alpha = R/n_{ss}^2$.

In (2) an x-ray pulse in the tens of msec range (long enough
for the ion concentration to reach steady state during the pulse)
is used, and a high voltage collecting pulse is applied for a
variable time in a manner which enables the steady-state ion

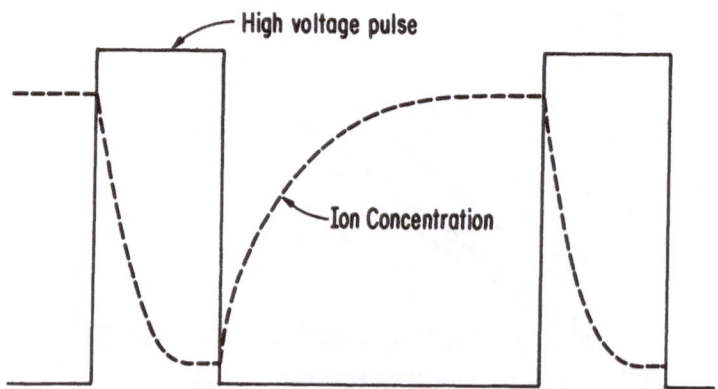

Figure 16. Schematic representation of the pulsed collection, continuous irradiation method used in measuring
ion-recombination rates (18).

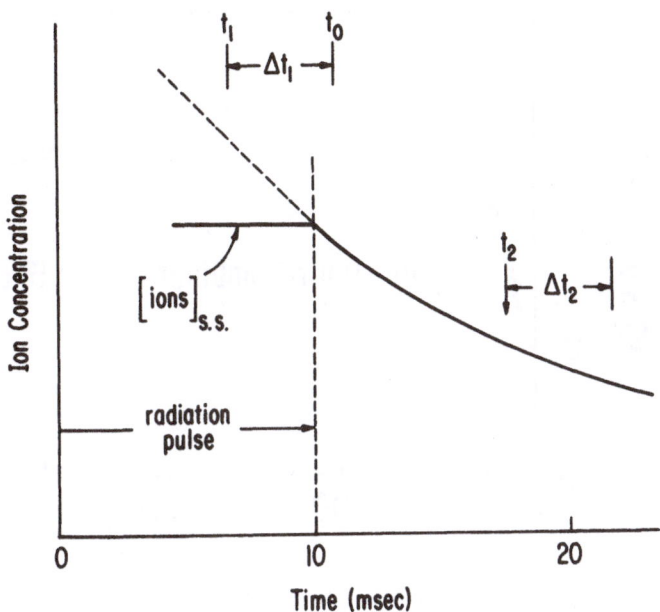

Figure 17. Depiction of the pulsed-irradiation, pulsed charge collection method used in measuring ion-recombination rates (18).

concentration during the pulse and the concentration as a function of time after the pulse to be measured. This technique is illustrated in Figure 17. Variation of the high-voltage pulse length, Δt_1, keeping t_0 fixed, allows determination of the steady state ion concentration reached during the pulse as well as R, the rate of ion formation by the pulse. Applying the high-voltage pulse Δt_2 at t_2 allows determination of the ion concentration at t_2 to be determined. The recombination coefficient can be determined from the steady state ion-concentration and R or from the curve formed by varying t_2.

Another method of determining recombination coefficients uses a microwave absorption technique to study electron decay following a pulse (19). The gas is irradiated in a cell consisting essentially of a piece of x-band waveguide. The change in microwave conductivity can be used to calculate in a straightforward manner the concentration of electrons if the mobility of the electron is known.

One of the interesting results of these recombination studies is the information obtained on the effect of pressure on recombination (18,19). Figures 18 and 19 illustrate the results for NH_3. The electron-ion recombination coefficient is found (19) to follow equation II.

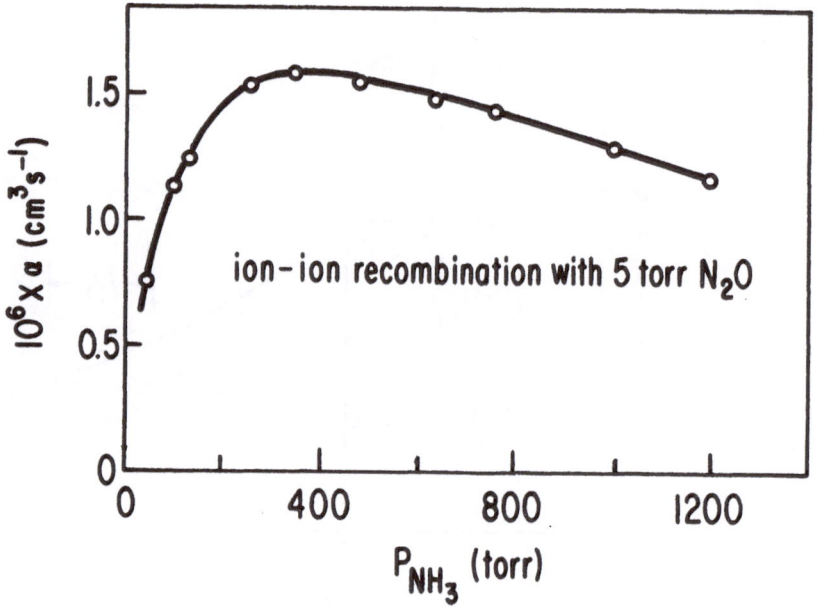

Figure 18. Ion-ion recombination coefficients in ammonia, nitrous oxide systems vs ammonia pressure (18).

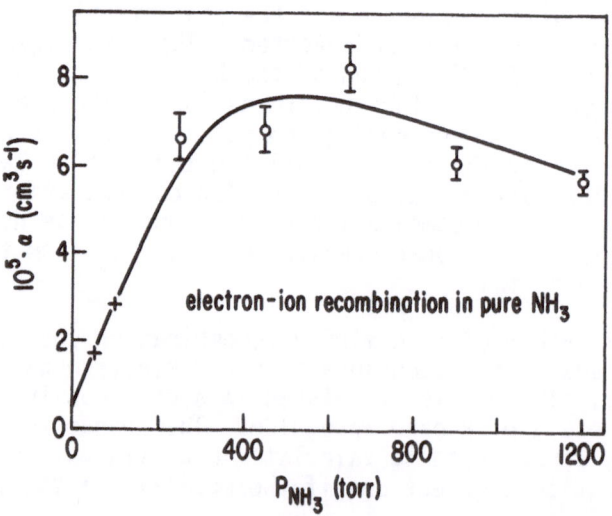

Figure 19. Electron-ion recombination coefficients in ammonia as a function of ammonia pressure. The circles are from reference 18 and the +'s are from reference 19.

$$\alpha = \alpha_2 + \alpha_3 N , \tag{II}$$

(where α_2 is the zero pressure extrapolated limit and N is the gas concentration) up to the pressures indicated in Table I. Such data are of considerable value in providing tests of theories of recombination.

Table I. Electron-Ion Recombination Coefficients (19).

Gas	$10^6 \times \alpha_2$ $(cm^3 \ s^{-1})$	$10^{25} \times \alpha_3$ $(cm^6 \ s^{-1})$	Pressure Limit (torr)
H_2O	4.1	270	20
NH_3	5.6	69	100
$(CH_3)_2CO$	6.4	55	100
CH_3Cl	7.1	26	100
CO_2	3.7	5.8	700
$(CH_3)_4C$	7.0	2.5	700

It should be noted that the identity of the recombining ions in such studies is affected by clustering. Progress has been made in understanding the effects of clustering on recombination in ammonia and water vapor (20).

3.2. Ion-molecule Reactions

An excellent example of the application of pulse-radiolysis techniques to measurement of ion-molecule reaction rates involves the reaction of N_2^+ $(X^2\Sigma_g^+, v=0)$, the ground state of N_2^+, with various compounds (21). The absorption to the $N_2^+(B^2\Sigma_u^+, v=0)$ state at 391.44 nm was monitored. A pulse from a Febetron 706 produced a sufficiently high concentration of N_2^+, and in 40 torr N_2, the reaction $N_2^+ + 2N_2 \rightarrow$ products was determined to account for the observed 19 ns half-life of the absorption. Other compounds were added in smaller concentrations, and rate constants for reaction with $N_2^+(X^2\Sigma_g^+, v=0)$ were obtained. These measurements are summarized in Table II.

Table II. Rate Constants for Reactions of N_2^+ ($X^2\Sigma_g^+$, v=0) (21).

Reactant	Rate Constant $(cm^3 s^{-1}) \times 10^{10}$	Reactant	Rate Constant $(cm^3 s^{-1}) \times 10^{10}$
O_2	< 0.4	Cyclopropane	15 ± 0.4
Xe	0.86 ± 0.12	SF_6	15 ± 1.3
CO	1.1 ± 0.3	H_2	19 ± 3.8
NO	4.1 ± 0.5	NH_3	18 ± 3.6
C_2H_2	8.8 ± 0.8	H_2O	23 ± 3.6
CH_4	10.0 ± 1.6		

3.3. Electron Capture

The microwave conductivity method coupled with pulse-radiolysis has provided great detail on electron capture rate constants. The method is very sensitive and concentrations of ~10^7 electrons per cm^3 can be used; therefore, low intensity (e.g., a Febetron 706 pulse can be converted to x-rays) can be used and interference from recombination reactions can be ignored. A "representative" sample of rate constants determined by this method taken from a review (3) is given in Table III.

Table III. Electron Capture Rate Constants at 25°C

Reactant	Rate Constant $(cm^3 s^{-1})$	Reactant	Rate Constant $(cm^3 s^{-1})$
CCl_4	4.2×10^{-7}	CH_3Br	7.0×10^{-12}
SF_6	2.2×10^{-7}	C_2H_5Br	2.8×10^{-13}
$c-C_4F_6$	1.4×10^{-7}	$c-C_3F_6$	$< 3 \times 10^{-14}$
$c-C_4F_8$	1.2×10^{-8}	CH_2F_2	1.6×10^{-14}
C_3F_6	3.0×10^{-10}	CH_3Cl	$< 2 \times 10^{-15}$
$n-C_4F_{10}$	1.0×10^{-11}	CF_4	$< 1 \times 10^{-16}$

An example of the precision and detail available using the microwave conductivity technique is the work done recently on electron attachment to O_2, and the effects of various added compounds on the process (22-26). In general, the mechanism has been shown to involve the following four reactions as the primary mechanism (Bloch-Bradbury mechanism):

$$e^- + O_2 \quad \overset{13}{\underset{14}{\rightleftarrows}} \quad O_2^-* \tag{13,14}$$

$$O_2^-* + O_2 \rightarrow O_2^- + O_2 \tag{15}$$

$$O_2^-* + M \rightarrow O_2^- + M \quad . \tag{16}$$

With M = H_2 or D_2, this mechanism was found to be satisfactory, but with M = N_2, anomalous effects of pressure were observed which were explained by the following additional reactions:

$$O_2^-* + N_2 \rightarrow [O_2^-* - N_2] \tag{17}$$

$$[O_2^-* - N_2] \rightarrow O_2 + N_2 + e^- \tag{18}$$

$$[O_2^-* - N_2] + O_2 \rightarrow O_2^- + N_2 + O_2 \tag{19}$$

$$[O_2^-* - N_2] + N_2 \rightarrow O_2^- + N_2 + N_2 \tag{20}$$

In other words, a four-body component was needed in the reaction scheme to explain the observed effects of pressure. In detail, reactions (13)-(16) yield equation III

$$\tau P_{O_2} = \frac{1}{k_{13}} + \frac{k_{14}/k_{13}}{rk_{15} + k_{16}} \times \frac{1}{P_{N_2}} , \tag{III}$$

where τ is the observed electron lifetime and r = P_{O_2}/P_{N_2}. At constant r, it was observed that plots of τP_{O_2} vs $1/P_{N_2}$ gave negative intercepts. Inclusion of reactions (17)-(20) allowed this discrepancy to be explained. The results indicated that reaction (19) is two orders of magnitude faster than reaction (20).

In addition to this "four-body" mechanism, experiments with various hydrocarbons as third bodies have been interpreted as indicating the importance of electron attachment to pre-existing van der Waals molecules ($O_2 \cdot M$). In essence, the existence of these molecules shows up as deviations from the behavior predicted by reactions (13)-(16), i.e., at high P_M anomalously high rates are observed. The four-body mechanism is not appropriate because it predicts that the observed effective (pseudo-first-order) rate constant cannot be greater than k_{13}, whereas, experimentally several cases show effective rate constants greater than k_{13}.

4. ATOMS AND RADICALS

Studies of atoms and radicals by pulse radiolysis have generally been done to determine spectra and rate constants. The work reported has been done predominantly with the high dose pulse from Febetrons. The high concentration of radicals produced is convenient in terms of the optical absorption techniques usually used, i.e., concentrations in the range 10^{14}-10^{15} radicals per cm^3 are needed to obtain reasonable signal-to-noise without use of signal averaging.

Many atomic and radical reactions have been investigated by following the absorption of either a reactant or a product. Among the species whose reactions have been studied are OH, H, O, HO_2, CH, NH_2, NH, NO_2, and CN. As an example of this type of study, we will consider some reactions of the HO_2 radical in the following

The HO_2 radical has a broad absorption spectrum which does not appear to change when H_2O or NH_3 are present (27-29), despite the fact that these substances form complexes with HO_2, as will be discussed below.

The HO_2 radical is conveniently prepared by pulse-radiolysis of 1200 torr H_2 containing about 5 torr O_2. Using a Febetron 705 ($\sim$50 nsec pulse), about 1×10^{15} H atoms per cm^3 are formed by direct excitation of H_2 and ion-recombination, which is over within about a microsecond. The reaction of H with O_2 is a three body reaction, which under the above conditions is complete within $\sim$10 μsec. The subsequent reactions of HO_2 occur on a slower time scale; the second-order decay of HO_2 proceeds with an initial $t_{\frac{1}{2}}$ of about 400 μsec under the above conditions.

A typical set of data on HO_2 is shown in Figure 20. The "white" curve through the data represents a second-order "best fit" obtained by computer analysis.

Using the known absorption coefficient of HO_2 at 230 nm, data such as these have been analyzed to determine the Arrhenius parameters for the overall reaction

$$HO_2 + HO_2 \rightarrow H_2O_2 + O_2 . \tag{21}$$

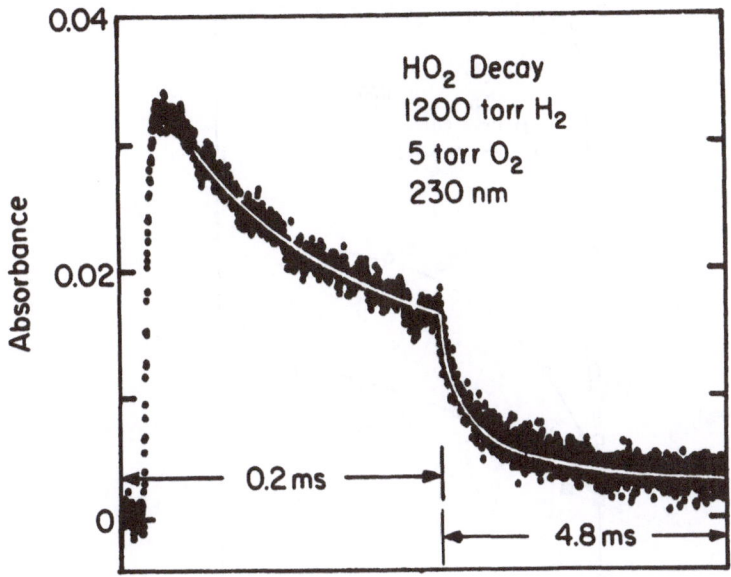

Figure 20. HO_2 formation and decay observed using a 50 ns pulse from a Febetron 705.

The result obtained is $k_{21} = (1.14 \pm 0.16) \times 10^{-13}$ exp(2100 ± 90/RT) $cm^3 s^{-1}$. The negative activation energy was explained (30) on the basis of an unstable intermediate, H_2O_4, the overall reaction being in detail

$$HO_2 + HO_2 \underset{23}{\overset{22}{\rightleftharpoons}} H_2O_4 \qquad\qquad (22,23)$$

$$H_2O_4 \rightarrow H_2O_2 + O_2 \qquad\qquad (24)$$

The rate constant for decay of HO_2 was observed to be markedly accelerated when either H_2O or NH_3 was added to the H_2,O_2 mixture. (Second-order kinetics were still followed, however.) This effect is shown in Figure 21 for the case of added NH_3, at several temperatures. The results in Figure 21 were explained on the basis of the formation of a hydrogen bonded complex between HO_2 and NH_3. It is of interest to note that the absorption spectrum in the 230 nm region does not change appreciably when the complex is formed.

The increase in k_{21} with increasing $[NH_3]$ requires that the HO_2 complex is more reactive than HO_2 itself, i.e., $k_{25} > k_{21}$.

$$H_3N \cdots HO_2 + HO_2 \rightarrow products \qquad\qquad (25)$$

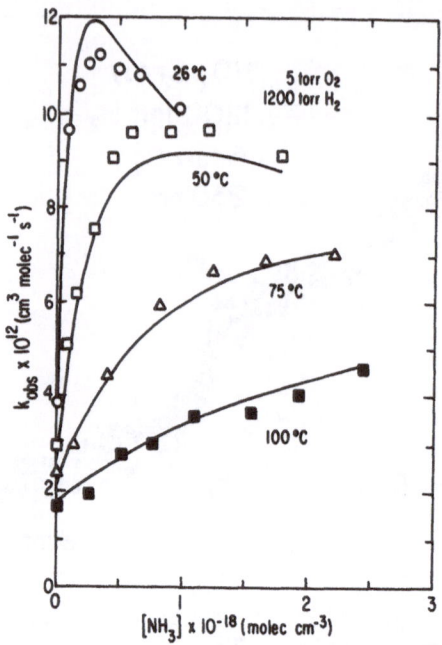

Figure 21. Rate constant for the HO_2 self-reaction as a function of NH_3 concentration and temperature (k_{obs} is the observed overall rate constant, defined by equation IV).

The results also require that $k_{26} < k_{25}$.

$$H_3N \cdots HO_2 + H_3N \cdots HO_2 \rightarrow \text{products} \tag{26}$$

The curves through the data points are the results of a detailed analysis (29) in terms of the reactions given above and the equilibrium

$$HO_2 + NH_3 \underset{28}{\overset{27}{\underset{\rightleftharpoons}{}}} H_3N \cdots HO_2 \tag{27,28}$$

The observed rate constant, k_{obs}, can be shown to be given by equation IV

$$k_{obs} = \frac{k_{21} + k_{25}K[NH_3] + k_{26}K^2[NH_3]^2}{(1 + K[NH_3])^2}, \tag{IV}$$

where K is k_{27}/k_{28}. The results of the analysis yield the following information. The value of k_{25} is about 9 times k_{21}, whereas k_{26} is only about 3 times k_{21}. The analysis also yields ΔH_{298}° = -13 ± 1.5 kcal/mol and ΔG° = -3.3 ± 0.2 kcal/mol for the formation of the complex, i.e., reaction (27).

NOTES AND REFERENCES

[1] Work supported by the Office of Basic Energy Sciences, Division of Chemical Sciences, U. S. Department of Energy, under Contract W-31-109-Eng-38.

(1) (a) Matheson, M. S. and Dorfman, L. M.: 1960, J. Chem. Phys. 32, pp. 1870-1871.
     (b) McCarthy, R. L. and MacLachlan, A.: 1960, Trans. Faraday Soc. 56, pp. 1187-1200.
     (c) Keene, J. P.: 1960, Nature 188, pp. 843-844.
(2) Firestone, R. F. and Dorfman, L. M.: 1971, in *Actions Chimiques et Biologiques des Radiations*, Ed. M. Haissinsky, Vol. 15, Masson, Paris, pp. 7-46.
(3) Sauer, M. C.,Jr.: 1976, Adv. Radiat. Chem. 5, pp. 97-184.
(4) Arai, S. and Firestone, R. F.: 1969, J. Chem. Phys. 50, pp. 4575-4589.
(5) Cooper, R., Grieser, F., Sauer, M. C.,Jr., and Sangster, D. F.: 1977, J. Phys. Chem. 81, pp. 2215-2220.
(6) Firestone, R. F. and Chen, M.: 1978, J. Chem. Phys. 69, pp. 2943-2948.
(7) Firestone, R. F., Oka, T., and Takao, S.: 1979, J. Chem. Phys. 70, pp. 123-130.
(8) Loeb, D. W., Chen, M., and Firestone, R. F.: 1981, J. Chem. Phys. 74, pp. 3270-3277.
(9) Oka, T., Rama Rao, K. V. S., Redpath, J. L., and Firestone, R. F.: 1974, J. Chem. Phys. 61, pp. 4740-4746.
(10) Thonnard, N. and Hurst, G. S.: 1972, Phys. Rev. A5, pp. 1110-1121.
(11) Cooper, R. and Sauer, M. C.,Jr.: unpublished results.
(12) Cooper, R., Grieser, F., and Sauer, M. C.,Jr.: 1976, J. Phys. Chem. 80, pp. 2138-2142.
(13) Sauer, M. C.,Jr., Mulac, W. A., Cooper, R., and Grieser, F.: 1976, J. Chem. Phys. 64, pp. 4587-4591.
(14) Cooper, R., Grieser, F., and Sauer, M. C.,Jr.: 1977, J. Phys. Chem. 81, pp. 1889-1894.
(15) Grieser, F. and Shimamori, H.: 1980, J. Phys. Chem. 84, pp. 247-250.
(16) Sauer, M. C.,Jr., and Mulac, W. A.: 1972, J. Chem. Phys. 56, pp. 4995-5004.
(17) Sauer, M. C.,Jr., and Mulac, W. A.: 1974, Int. J. Radiat. Phys. Chem. 6, pp. 55-65.

(18) Sennhauser, E. S. and Armstrong, D. A.: 1978, Radiat. Phys. Chem. 11, pp. 17-28.

(19) Warman, J. M., Sennhauser, E. S., and Armstrong, D. A.: 1979, J. Chem. Phys. 70, pp. 995-999.

(20) Sennhauser, E. S. and Armstrong, D. A.: 1978, Radiat. Phys. Chem. 12, pp. 115-123.

(21) Dreyer, J. W. and Perner, D.: 1971, Chem. Phys. Lett. 12, pp. 299-302.

(22) Shimamori, H. and Hatano, Y.: 1976, Chem. Phys. 12, pp. 439-445.

(23) Shimamori, H. and Hatano, Y.: 1976, Chem. Phys. Lett. 38, pp. 242-247.

(24) Shimamori, H. and Hatano, Y.: 1977, Chem. Phys. 21, pp. 187-201.

(25) Kokaku, Y., Hatano, Y., Shimamori, H., and Fessenden, R. W.: 1979, J. Chem. Phys. 71, pp. 4883-4887.

(26) Kokaku, Y., Toriumi, M., and Hatano, Y.: 1980, J. Chem. Phys. 73, pp. 6167-6168.

(27) Hamilton, E. J.,Jr.: 1975, J. Chem. Phys. 63, pp. 3682-3683.

(28) Hamilton, E. J.,Jr., and Lii, R. R.: 1977, Int. J. Chem. Kinet. 9, pp. 875-885.

(29) Lii, R. R., Gorse, R. A.,Jr., Sauer, M. C.,Jr., and Gordon, S.: 1980, J. Phys. Chem. 84, pp. 813-817.

(30) Lii, R. R., Gorse, R. A.,Jr., Sauer, M. C.,Jr., and Gordon, S.: 1979, J. Phys. Chem. 83, pp. 1803-1804.

INDEX OF SUBJECTS